Progress in Mathematics

Volume 173

Computational Methods for Representations of Groups and Algebras

Euroconference in Essen (Germany), April 1–5, 1997

P. Dräxler
G. O. Michler
C. M. Ringel
Editors

Birkhäuser Verlag
Basel · Boston · Berlin

Editors:

P. Dräxler
C.M. Ringel
Fakultät für Mathematik
Universität Bielefeld
Universitätsstr. 25
D-33615 Bielefeld
Germany

G.O. Michler
Institut für Experimentelle Mathematik
Universität GH Essen
Ellernstr. 29
D-45326 Essen
Germany

1991 Mathematics Subject Classification 20C40

A CIP catalogue record for this book is available from the Library of Congress, Washington D.C., USA

Deutsche Bibliothek Cataloging-in-Publication Data

Computational methods for representations of groups and algebras ; euroconference in Essen /Germany), April 1 - 5, 1997 /
P. Dräxler … ed. - Boston ; Basel ; Berlin : Birkhäuser, 1999
(Progress in mathematics ; Vol. 173)
ISBN 3-7643-6063-1 (Basel …)
ISBN 0-8176-6063-1 (Boston)

Printed on acid-free paper produced of chlorine-free pulp. TCF ∞
Printed in Germany
ISBN 3-7643-6063-1
ISBN 0-8176-6063-1

9 8 7 6 5 4 3 2 1

Preface

A joint research project of algebraists from the universities of Antwerp, Bielefeld, Essen, Leeds, Paris VI and Trondheim on "Invariants and Representations of Algebras" has been supported from 1991 to 1997 by the European Union programmes "Science" and "Human Capital and Mobility", it was coordinated by Mme M.-P. Malliavin (Paris VI). Later, algebraists from the universities of Edinburgh, Ioannina, Murcia and Toruń joined the collaboration. This network is now coordinated by C.M. Ringel (Bielefeld). It has received funds from the European Commission in order to organize four conferences as part of the programme "Training and Mobility of Researchers", to be held during the period 1997–1999 at Essen, Murcia, Bielefeld and Ioannina. The first Euroconference of this series took place at the University of Essen, April 1–4, 1997. It was devoted to "Computational Methods for Representations of Groups and Algebras". The organizers were P. Dräxler (Bielefeld) and G. Michler (Essen).

This volume collects most of the material presented at the conference. There had been an additional introductory lecture by H. Gollan; it is not included here, since its contents is available in the lecture notes: P. Fleischmann, G.O. Michler, P. Roelse, J. Rosenboom, R. Staszewski, C. Wagner, M. Weller, "Linear algebra over small finite fields on parallel machines", Vorlesungen Fachbereich Math. Univ. Essen, 23 (1995). Together with these notes, this volume will provide a survey on the present state of art. It aims at a general audience wanting to become familiar with the latest developments. The lecturers had been invited in order to get a coverage of all the central themes. We hope that our collection provides an informative account of the tremendous possibilities of computational methods in the representation theory of groups and algebras. The papers are based on these lectures, some have been expanded and revised; all have been refereed.

The aim of the Euroconference "Computational Methods for Representations of Groups and Algebras" was to give a survey on general algorithmic and computational methods. Recent advances in the representation theory of finite simple groups of Lie type and new applications of the computational methods in linear algebra to the revision of the classification of the finite simple sporadic groups were presented. It turns out that computational and probabilistic tools (including high performance computations on supercomputers) have become more and more important for the classification and investigation of exceptional structures like the sporadic simple groups or the sincere representation-directed algebras. They are also inevitable for the construction of projective resolutions of finitely generated modules over finite-dimensional algebras, the study of group cohomology and the explicit construction of generators for rings of invariants.

A major part of the meeting was devoted to give a survey on algorithms for computing special examples in the study of Grothendieck groups, quadratic forms and derived categories of finite-dimensional algebras of finite global dimension. Open questions on Bruhat orders, Coxeter groups and Kazhdan-Lusztig polynomials are investigated with the aid of computer programs. Commutative and non commutative Gröbner basis methods are applied to determine complete intersections and Hilbert series. Similarly vector enumeration is used for the algorithmic investigation of finitely presented groups. Finally, the representation theory of finite groups and finite dimensional algebras are linked by the condensation technique.

The editors thank the European community for its financial support in the name of all participants of the meeting; but it has to be mentionned that there have been problems due to political, but also burocratic regulations of the EU which are in conflict with good scientific traditions, in particular the insistence of the EU on nationality questions. Without the additional financial help of the IEM Essen and the SFB 343 Bielefeld, it would not have been possible to present at the meeting all the topics in an optimal way. As the reader will see, the volume (as the meeting) reflects the truely international character of present research in computational representation theory, not restricted by the narrow boundaries of the policies of individual countries or communities. Finally, it is our pleasure to acknowledge the help and cooperation the publisher Birkhäuser has provided in the production of this text.

P. Dräxler
G. O. Michler
C. M. Ringel

Bielefeld and Essen, May 6, 1998

Contributors

M. Barot
Institudo de Matematicas, Universidad National
Autonoma de Mexcio, Ciudad Universitario,
Mexico, 04510 D. F., Mexico
barot@@gauss.matem.unam.mx

J. F. Carlson
Department of Mathematics, The University of Georgia
Athens, GA 30602-7403, USA
jfc@@sloth.math.uga.edu

A. M. Cohen
Dept. of Mathematics and Computing Science, TU Eindhoven
PO Box 513, NL-5600 MB Eindhoven, The Netherlands
amc@@win.tue.nl

S. Cojacaru
Institute of Mathematics, Academy of Sciences of Moldava
5 Academiei str., Kishinev, 277028, Moldava
sanya @@add.moldava.su, 22svet @@math.moldava.su

G. Cooperman
College of Computer Science, Northeastern University,
M/S 215 CN, Boston, MA 02115, USA
gene@@ccs.neu.edu.

W. Decker
Fachbereich 9, Mathematik Universität des Saarlandes
Postfach 15 11 50, D-66041 Saarbrücken, Germany
decker@@math.uni-sb.de

P. Dowbor
Dept. of Mathematics, Nicholas Copernicus University
Chopina 12/18, 87-100 Torun, Poland
dowbor@@mat.uni.torun.pl

P. Dräxler
Fakultät für Mathematik, Universität Bielefeld
Postfach 100 131, D-33501 Bielefeld
draexler@@mathematik.uni-bielefeld.de

F. du Cloux
Institut de Mathematiques et Informatique, Universite Claude
Bernard F-69622 Villeurbanne Cedex, France
ducloux@@celeste.univ-lyon1.fr

P. Fleischmann
Institute of Mathematics and Statistics, University of Kent
Canterbury, CT 27 NF, England
P.Fleischmann@ukc.ac.uk

H. Gollan
Institut für Experimentelle Mathematik, Universität GH Essen
Ellernstr. 29, D-45326 Essen, Germany
holger@@exp-math.uni-essen.de

E. Green
Mathematics Department, Virginia Institute of Technology
426 McBryde Hall, Blacksburg, VA 24061-0123, USA
green@@math.vt.edu

G.-M. Greuel Fachbereich Mathematik, Universität Kaiserslautern
Erwin-Schrödinger-Straße, D-67663 Kaiserslautern, Germany
greuel@@mathematik.uni-kl.de

G. Havas Dept. of Computer Science, Universiy of Queensland
Queensland 4072, Australien
havas@@cs.uq.edu.au

H. von Höhne Institut für Mathematik II, Freie Universität Berlin
Arnimallee 3, D-14195 Berlin, Germany
hoehne@@math.fu-berlin.de

D. Holt Mathematics Institute, University of Warwick
Coventry CV4 7 AL, England
dfh@@maths.warwick.ac.uk

T. Hübner Fachbereich 17, Universität GH Paderborn
Warburger Str. 100, D-33095 Paderborn, Germany
huebner@@uni-paderborn.de

T. de Jong Fachbereich 9, Mathematik, Universität des Saarlandes
Postfach 15 11 50, D-66041 Saarbrücken, Germany
dejong@@math.uni-sb.de

G. Kemper Interdiziplinäres Zentrum für Wissenschaftliches Rechnen
Universität Heidelberg, Im Neuenheimer Feld 368, D-69120
Heidelberg, Germany
gregor.kemper@@iwr.uni-heidelberg.de

W. Lempken Institut für Experimentelle Mathematik, Universität GH Essen
Ellernstr. 29, D-45326 Essen, Germany
lempken@@exp-math.uni-essen.de

H. Lenzing Universität Paderborn, Fachbereich Mathematik
Wartburger Str. 100, D-33098 Paderborn
helmut@@uni-paderborn.de

R. de Man Dept. of Mathematics and Computing Science, TU Eindhoven
PO Box 513, NL-5600 MB Eindhoven
deman@@win.tue.nl

G. O. Michler Institut für Experimentelle Mathematik, Universität GH Essen
Ellernstr. 29, D-45326 Essen, Germany
archiv@@exp-math.uni-essen.de

J. Müller Lehrstuhl D für Mathematik, RWTH Aachen
Templergraben 64, D-52062 Aachen, Germany
Juergen.Mueller@@Math.RWTH-Aachen.De

R. Nörenberg Fakultät für Mathematik, Universität Bielefeld
Postfach 100 131, D-33501 Bielefeld, Germany
noerenberg@@mathematik.uni-bielefeld.de

S. Ovsienko Faculty of Mechanics and Mathematics, University of Kiev
Vladimirskaja No. 64, Kiev, 252617, Ukraine
ovsienko @@uni-alg.kiev.ua

J. A. de la Peña Institudo de Matematicas, Universidad National Autonoma de Mexcio, Ciudad Universitario, Mexico, 04510 D. F.
jap@@penelope.matem.unam.mx

G. Pfister Fachbereich Mathematik, Universität Kaiserslautern Erwin-Schrödinger-Straße, D-67663 Kaiserslautern, Germany
pfister@@mathematik.uni-kl.de

A. Podoplelov Institute of Mathematics, Academy of Sciences of Moldava 5 Academiei str., Kishinev, 277028, Moldava
sanya@@add.moldava.su

J. Rosenboom Institut für Experimentelle Mathematik, Universität GH Essen Ellernstr. 29, D-45326 Essen, Germany
jens@@exp-math.uni-essen.de

C. Sims Department of Mathematics, Rutgers University, 110 Frelinghuysen Road, Piscataway, NJ 08854, USA
sims@@math.rutgers.edu

A. Steel School of Mathematics and Statistics, University of Sydney NSW 2006, Australia
allan@@maths.usyd.edu.au

V. Ufnarovski Department of Mathematics, Lund Institute of Technology BX 118, S-22100 Lund, Sweden
ufn@@maths.lth.se

M. Weller Institut für Experimentelle Mathematik, Universität GH Essen Ellernstr. 29, D-45326 Essen, Germany
eowmob@@exp-math.uni-essen.de

R. A. Wilson Department of Mathematics, University of Birmingham P.O. Box 363, Birmingham B152TT, England
r.a.wilson@@bham.ac.uk

Contents

I Introductory Articles

II Keynote Articles

Part I

Introductory Articles

Progress in Mathematics, Vol. 173, © 1999 Birkhäuser Verlag Basel/Switzerland

Chapter 1

Classification Problems in the Representation Theory of Finite-Dimensional Algebras

Peter Dräxler and Rainer Nörenberg

1.1 Introduction

1.1.1 Combinatorics arising in the Representation Theory of Algebras The aim of the representation theory of algebras is to understand the category Mod$-A$ of modules over a given associative unital k-algebra A where k is a commutative ring. We will restrict ourselves to the case that k is a field and A is finite-dimensional over k. Familiar examples for algebras of this kind are the group algebras kG for a finite group G or the factor algebras $k[X]/I$ of the polynomial ring $k[X]$ by a non-zero ideal I.

Frequently, we will assume that the field k is algebraically closed. Moreover, we will mainly deal only with the full subcategory mod$-A$ of Mod$-A$ given by all right A-modules M which are finite-dimensional as vector spaces over k.

The key opening representation theory of finite-dimensional algebras to combinatorial methods is the observation that these algebras as well as their categories of finite-dimensional modules can be described to a large extent by locally finite quivers. Recall that 'quiver' is the short notation for 'oriented graph' which is used commonly in representation theory. The quiver describing the algebra is sometimes called the 'Gabriel quiver' whereas for the quiver of the category of finite-dimensional modules meanwhile 'Auslander-Reiten quiver' is the standard name.

In the Auslander-Reiten quiver an indecomposable module shrinks to a mere vertex. A compromise between this total loss of structure and the wish to work only with combinatorial data is furnished by looking at the corresponding element of the module in the Grothendieck group (with respect to short exact sequences). Note, that this Grothendieck group is always of the shape $\mathbb{Z}^n$ for some $n \in \mathbb{N}$.

For general algebras the Auslander-Reiten quiver together with the Grothendieck group does not lead to many effective combinatorial algorithms. The situation improves drastically if one considers strongly simply connected algebras.

Associated with the representation theory of these algebras another combinatorial structure namely partially ordered sets arise. Quivers, Grothendieck groups, partially ordered sets together with various actions on these structures form the combinatorial environment of the classifications and algorithms which we will present.

1.1.2 Representation Types, Classifications and CREP In the last 30 years a lot of effort was made to study and classify finite-dimensional algebras with respect to their representation type. An algebra is said to be of finite representation type if there are only finitely many isomorphism classes of indecomposable finite-dimensional modules. A generalization is given by the tame algebras which are characterized by having for any fixed dimension only finitely many 1-parameter families of isomorphism classes of indecomposable modules. By a theorem of Drozd algebras which are not tame have to be wild which means that the classification of their indecomposable modules is as difficult as the classification of pairs of square matrices under simultaneous conjugation.

In the study of finite representation type it turned out that it is better to pass to the universal covering of a suitable degeneration called the 'standard form' of a given algebra. Thus one arrives at a strongly simply connected algebra. For the study of tame algebras a similar approach seems to work although up to now there are no complete results in this case.

In order to understand finite representation type for strongly simply connected algebras, the aim was to classify the minimal representation infinite algebras and those algebras which have a faithful indecomposable module. This led to the classifications of the tame concealed algebras and the sincere representation directed algebras. At this place computers had to enter the business because it turned out that there were so many exceptional structures yielding huge amounts of combinatorial data that otherwise the explicit classifications would have been impossible. Our main aim is to review these classifications introducing the necessary material to understand the algorithms which were used.

The original classifications were performed by stand-alone programs. Meanwhile we have developed the program system CREP (Combinatorial REPresentation theory) which allows to deal with algorithms, examples and classifications in a more conceptual and systematic way (see [26]). Another aim of these notes is to illustrate this progress and to indicate new developments. In particular, we will complete a classification of certain tame incidence algebras associated with partially ordered sets using our algorithms on thin start modules and rather recent results of Skowroński on tame strongly simply connected algebras of polynomial growth.

1.1.3 Outline of Contents Presently there exist three books offering an introduction to the representation theory of finite-dimensional algebras, namely [3], [35] and [49]. Although, [49] is the oldest one, its main topic are advanced examples of algebras of tame representation type, and therefore the introduction to the general theory is short and several proofs (e.g. existence of almost split sequences) are skipped. In our opinion [3] gives the most comprehensive introduction to the theory. Concerning computer applications the disadvantage of this book is that it deals with Artin algebras and their module categories whereas currently the most powerful computer algorithms work for finite-dimensional algebras over an algebraically closed field. However, the representation theory of algebras over an algebraically closed field is the main topic in [35]. Using the language of functors and categories even for the basic concepts, this book seems to be meant for more advanced readers. Nevertheless, we feel that this approach is really also the best-suited for the computer applications which we want to present since it leads immediately to concepts like quivers and partially ordered sets.

Therefore, in Sections 1.2 to 1.4 we introduce the basic material about representation theory of finite-dimensional algebras following [35] in notation and referring to this monograph for proofs. In Section 1.5 we give the precise definitions of the representation types and provide a brief account about covering theory and degenerations. Section 1.6 is concentrated on the classifications of the tame concealed, sincere representation directed and minimal wild concealed algebras. We explain how the completely separating algebras i.e. the schurian strongly simply connected algebras are made accessible for a computer. Moreover, we present some few results from tilting theory which we need. The final Section 1.7 is devoted to our algorithms using partially ordered sets of thin start modules. We relate the module category of an algebra to the subspace category of a vector space category using fiber sum functors and explain that linear vector space categories are completely described by partially ordered sets. After recalling the Theorems of Kleiner, Nazarova and Zavadskij on the representation type of partially ordered sets, we present the classification result about tame incidence algebras of polynomial growth mentioned above as a current application.

For all definitions and results which are used explicitely but not introduced or stated completely, we will give precise references. On the other hand, we will use several arguments between the lines and refer to the books mentioned above for proofs of these.

1.2 Algebras, Aggregates and Modules

1.2.1 Finite-Dimensional Algebras and their Basic Algebras Algebras appearing in 'nature' tend not to be basic. In the representation theory of algebras we usually pass to a Morita equivalent basic algebra which can be described by a finite quiver with relations provided that the field k is algebraically closed. Later we will explain how this passage is performed on the categorical level, but for the convenience of the reader we start out by recalling the classical approach using idempotents.

Let A be a unital associative algebra over a field k. The *Jacobson radical* $\operatorname{rad} A$ of A is by definition the intersection of all maximal right ideals of A. One obtains the same set by intersecting all maximal left ideals. Consequently, $\operatorname{rad} A$ is a two-sided ideal of A. If A is finite-dimensional, this ideal is nilpotent and the factor algebra $A/\operatorname{rad} A$ is semisimple, hence isomorphic to a finite product of matrix algebras of skew fields which are finite-dimensional k-algebras.

The set of non-invertible elements of an algebra A always contains the Jacobson radical. If this inclusion is an equality and the unit element of A is non-zero, then the algebra is said to be *local*. For a local algebra the factor algebra $A/\operatorname{rad} A$ is a skew field.

An element e of an algebra A is called *idempotent* if $e = e^2$. Two idempotents e, f are said to be *orthogonal* if $0 = ef = fe$. An idempotent e is called *primitive* if $e \neq 0$ and e cannot be written as $e = f + g$ where $f \neq 0 \neq g$ and f, g are orthogonal idempotents.

If A is finite-dimensional over the field k, then the unit element 1_A of A can be written as a sum $1_A = \sum_{i=1}^{n} e_i$ of primitive idempotents e_i which are mutually orthogonal. For each such idempotent e_i the subset e_iAe_i together with the induced addition and multiplication becomes a finite-dimensional algebra with unit element e_i. The primitivity of e_i implies that e_iAe_i is a local algebra.

On the set $\{e_1, \dots, e_n\}$ we obtain an equivalence relation by defining $e_i \sim e_j$ if the ideal $e_iAe_jAe_i$ of the algebra e_iAe_i coincides with e_iAe_i. We choose a minimal set $\{f_1, \dots, f_m\}$ of representatives of the equivalence classes. The sum $f := \sum_{j=1}^{m} f_j$ is again an idempotent of A and $B := fAf$ is a finite-dimensional k-algebra with unit element f. Any algebra isomorphic to B is called a *basic algebra* of A. The subset fA of A becomes a B-A-bimodule in the obvious way. Tensoring with this bimodule over B furnishes equivalences $\text{Mod}{-}B \to \text{Mod}{-}A$ and $\text{mod}{-}B \to \text{mod}{-}A$ which are called *Morita equivalences*. These equivalences allow us to consider only basic algebras A i.e. algebras which coincide with their basic algebra B for the conceptual study of $\text{Mod}{-}A$ and $\text{mod}{-}A$. Nevertheless, it has to be emphasized that it is usually non-trivial to construct B from a practically given A. One may consider group algebras to get a feeling for the difficulties.

1.2.2 Aggregates and Spectroids Let us now reformulate and generalize the construction of the basic algebra in categorical terms following [35]. Remember that k is a field. A category $\mathcal{A}$ is called *k-linear* if all morphism spaces $\mathcal{A}(x, y)$ are vector spaces over k and the composition is k-bilinear. In analogy to the algebra case one defines the *Jacobson radical* $\operatorname{rad} \mathcal{A}$ of a k-linear category as the intersection of all maximal right (or maximal left) ideals of $\mathcal{A}$. The category $\mathcal{A}$ is called *k-additive* if it is k-linear and additive i.e. has finite coproducts. A skeletally small k-additive category $\mathcal{A}$ is said to be an *aggregate* if all morphism spaces $\mathcal{A}(x, y)$ are finite-dimensional over k and each object of $\mathcal{A}$ is a coproduct of finitely many objects with local endomorphism algebras. Due to the Theorem of Krull-Remak-Schmidt the objects forming such a coproduct are unique up to isomorphism. Note that in an aggregate the *indecomposable* objects, which are those non-zero objects that cannot be written as a coproduct of two non-zero objects, coincide with the objects having a local endomorphism algebra. It is well-known that the category mod$-A$ of finite-dimensional right modules over a finite-dimensional algebra A is an aggregate.

Given an arbitrary k-linear category $\mathcal{A}$ we can consider the smallest k-additive category $\operatorname{add}\mathcal{A}$ which contains $\mathcal{A}$. Usually, idempotent morphisms in a k-linear category $\mathcal{A}$ will not split i.e. will not factorize as $\iota\pi$ where π is a retraction and ι is a section. But again we can form the smallest category $\operatorname{splt}\mathcal{A}$ containing $\mathcal{A}$ in which all idempotent morphisms split. In particular, if we start out with a skeletally small k-linear category $\mathcal{A}$ where all morphism spaces are finite-dimensional over k, then $\operatorname{splt}\operatorname{add}\mathcal{A}$ will be an aggregate.

Each finite-dimensional k-algebra A furnishes a k-linear category which has only one object ω whose endomorphism algebra is just A and the composition is the multiplication. By abuse of language this category is also denoted by A. Obviously, it is skeletally small and has finite-dimensional morphism spaces. Thus $\operatorname{splt}\operatorname{add}A$ is an aggregate.

A small k-linear category $\mathcal{S}$ is called a *spectroid* if it has finite-dimensional morphism spaces, the endomorphism algebra of each object is local and two objects are isomorphic if and only if they are equal. Note that for a spectroid $\mathcal{S}$ the Jacobson radical $\operatorname{rad}\mathcal{S}(x,y)$ coincides with $\mathcal{S}(x,y)$ if $x \neq y$ and coincides with the set of non-invertible elements of the local algebra $\mathcal{S}(x,x)$ for $x = y$. If for an aggregate $\mathcal{A}$ we choose a complete unshortenable set of representatives of the indecomposable objects and denote by $\mathcal{S}(\mathcal{A})$ the full subcategory of $\mathcal{A}$ associated with this set of objects, then $\mathcal{S}(\mathcal{A})$ is a spectroid and $\operatorname{add}\mathcal{S}(\mathcal{A})$ is equivalent to $\mathcal{A}$. Conversely, if $\mathcal{S}$ is a spectroid, then $\operatorname{add}\mathcal{S}$ is an aggregate and $\mathcal{S}(\operatorname{add}\mathcal{S})$ is isomorphic to $\mathcal{S}$. Hence the equivalence classes of aggregates are in one-to-one correspondence with the isomorphism classes of spectroids.

If we start out with a finite dimensional algebra A, then $\mathcal{S}(\operatorname{add}A)$ is a finite spectroid i.e. has a finite set $\{x_1, \dots, x_n\}$ of objects. The space B of matrices $(b_{ij})_{i,j=1,\dots,n}$ where $b_{ij} \in \mathcal{S}(\operatorname{add}A)(x_j, x_i)$ becomes a finite-dimensional k-algebra via the usual matrix multiplication. This algebra B is a basic algebra of A and $\mathcal{S}(\operatorname{add}B)$ can be identified with $\mathcal{S}(\operatorname{add}A)$. Thus basic algebras and finite spectroids are just two aspects of the same concept. We think that finite spectroids are more appropriate for computational purposes as we will see that they are closely related to quivers. Therefore we will continue using this terminology.

1.2.3 Modules over Spectroids Given a skeletally small k-linear category $\mathcal{A}$, we denote by Mod$-\mathcal{A}$ the category of all contravariant k-linear functors $M : \mathcal{A} \to$ Mod$-k$ where a functor is called k-linear if it induces k-linear maps on the morphism spaces. By mod$-\mathcal{A}$ we denote the full subcategory of all locally finite-dimensional objects M i.e. $\dim_k M(x) < \infty$ for each object x of $\mathcal{A}$. Since k-linear functors preserve finite (co)products, we obtain that Mod$-\mathcal{A}$ is equivalent to Mod$-(\operatorname{add}\mathcal{A})$ and Mod$-(\operatorname{splt}\mathcal{A})$. In particular, we obtain that for an aggregate $\mathcal{A}$ the module categories Mod$-\mathcal{A}$ and Mod$-\mathcal{S}(\mathcal{A})$ are equivalent which yields an equivalence of mod$-\mathcal{A}$ and mod$-\mathcal{S}(\mathcal{A})$. This is the reformulation of Morita equivalence in the categorical context.

For a finite-dimensional algebra A now the symbol Mod$-A$ seems to have two meanings. But obviously, the two categories associated with this symbol actually coincide. Therefore we obtain that Mod$-A \cong$ Mod$-\mathcal{S}(\operatorname{add} A)$ and also mod$-A \cong$ mod$-\mathcal{S}(\operatorname{add} A)$. This justifies to replace the study of mod$-A$ for finite-dimensional algebras A by the study of mod$-\mathcal{S}$ for finite spectroids $\mathcal{S}$ which as seen above is nothing else but passing to a basic algebra. Moreover, thanks to the Krull-Remak-Schmidt Theorem most of the information about the aggregate mod$-\mathcal{S}$ will be contained in the spectroid $\mathcal{S}(\text{mod}-\mathcal{S})$ which we will denote by ind$-\mathcal{S}$.

Let $\mathcal{S}$ be a finite spectroid. The full subcategory proj$-\mathcal{S}$ of *projective* objects in mod$-\mathcal{S}$ consists of those modules P for which the functor $\operatorname{Hom}_{\mathcal{S}}(P,-)$: mod$-\mathcal{S} \to$ mod$-k$ is exact i.e. preserves short exact sequences. Since it is even a full subaggregate we can ask for a spectroid of proj$-\mathcal{S}$. For this reason let us consider the functor $-^{\wedge} : \mathcal{S} \to$ mod$-\mathcal{S}$ which sends the object x of $\mathcal{S}$ to the representable functor $x^{\wedge} := \mathcal{S}(-,x)$ and a morphism $f : x \to y$ to $f^{\wedge} := \mathcal{S}(-,f) : \mathcal{S}(-,x) \to \mathcal{S}(-,y)$. The functor is actually fully faithful. This follows from the Yoneda-Lemma which states that the functor Mod$-\mathcal{S} \to$ mod$-k$ sending a module M to $M(x)$ is isomorphic to the representable functor $\operatorname{Hom}_{\mathcal{S}}(x^{\wedge},-)$. Thus the representable functors $\operatorname{Hom}_{\mathcal{S}}(x^{\wedge},-)$: mod$-\mathcal{S} \to$ mod$-k$ are exact for all $x \in \mathcal{S}$. Moreover, since every module in mod$-\mathcal{S}$ is an epimorphic image of a finite coproduct of modules of the shape $x^{\wedge}$, the full subcategory of proj$-\mathcal{S}$ whose objects are the $x^{\wedge}$ is a spectroid of proj$-\mathcal{S}$.

Of course the categories $\mathcal{S}-$Mod resp. $\mathcal{S}-$mod of covariant k-linear functors $\mathcal{S} \to$ Mod$-k$ resp. $\mathcal{S} \to$ mod$-k$ behave completely dually to Mod$-\mathcal{S}$ and mod$-\mathcal{S}$. If we denote by $\mathrm{D} = \operatorname{Hom}_k(-,k)$ the usual vector space duality, then $\mathcal{S}-$mod $\to$ mod$-\mathcal{S}$, $M \mapsto \mathrm{D}\,M$ is also a duality functor which will be called D as well. We again obtain a fully faithful functor $\mathcal{S} \to$ mod$-\mathcal{S}$ by sending an object x of $\mathcal{S}$ to the functor $x^{\vee} := \mathrm{D}\,\mathcal{S}(x,-)$. Let us invoke the dual version of the Yoneda-Lemma saying that the functor Mod$-\mathcal{S} \to$ mod$-k$ which sends a module M to $\mathrm{D}\,M(x)$ is isomorphic to the functor $\operatorname{Hom}_{\mathcal{S}}(-,x^{\vee})$. Consequently, if we denote by inj$-\mathcal{S}$ the full subcategory of mod$-\mathcal{S}$ formed by the *injective* objects, which are those Q in mod$-\mathcal{S}$ such that the representable functor $\operatorname{Hom}_{\mathcal{S}}(-,Q)$: mod$-\mathcal{S} \to$ mod$-k$ is exact, then inj$-\mathcal{S}$ is a full subaggregate which has a spectroid whose objects are the $x^{\vee}$ for x in $\mathcal{S}$.

A module M in mod$-\mathcal{S}$ is said to be *semisimple* if $0 = \operatorname{rad} M := M(\operatorname{rad} \mathcal{S})$. Clearly $\operatorname{top} M := M/\operatorname{rad} M$ is semisimple for all M in mod$-\mathcal{S}$. The full subcategory of semisimple modules forms a subaggregate of mod$-\mathcal{S}$. If we put $x^{-} := \operatorname{top} x^{\wedge}$ for all x in $\mathcal{S}$, then the full subcategory of mod$-\mathcal{S}$ whose objects are the modules x^{-} forms a spectroid of the aggregate of semisimple modules. Modules isomorphic to some x^{-} are called simple. They are precisely the modules in mod$-\mathcal{S}$ which have exactly two submodules. For each module M in mod$-\mathcal{S}$ the largest semisimple submodule is called the *socle* of M and denoted by $\operatorname{soc} M$. As $\operatorname{rad} x^{\wedge}$ is the unique maximal submodule of $x^{\wedge}$ by duality the module $x^{\vee}$ has a simple socle which is isomorphic to x^{-} again.

1.3 Quivers and Posets

1.3.1 The Path Category of a Quiver and the k-linear Category of a Poset A *quiver* Δ consists of a set of vertices Δ_0 and a set of arrows Δ_1 which is the disjoint union of sets $\Delta(x, y)$ where the elements of $\Delta(x, y)$ are the arrows $\alpha : x \to y$ from the vertex x to the vertex y. Thus a quiver is nothing but an oriented graph. Given a quiver Δ its *path category* $\mathrm{Pth}\,\Delta$ has as objects the vertices of Δ and the morphisms $x \to y$ are the *paths* from x to y which are by definition the formal compositions $\alpha_n \cdots \alpha_1$ where α_1 starts in x, α_n ends in y and the end point of α_i coincides with the start point of α_{i+1} for all $i = 1, \dots, n-1$. The positive integer n is called the *length* of the path. We also introduce a path 1_x of length $= 0$ from each vertex x to itself. The composition in $\mathrm{Pth}\,\Delta$ of paths of positive length is just concatenation whereas the 1_x act as identities.

With any category $\mathcal{C}$ we can associate a k-linear category $k\mathcal{C}$ by taking as objects the objects of $\mathcal{C}$ and by using the elements of $\mathcal{C}(x, y)$ as k-basis of the morphism space $k\mathcal{C}(x, y)$. The composition in $k\mathcal{C}$ is of course obtained by k-linear extension of the composition in $\mathcal{C}$. In particular, if we start with a quiver Δ, then the category $k\Delta := k(\mathrm{Pth}\,\Delta)$ is called the *k-linear path category* of Δ.

Let $k^1\Delta$ be the ideal of $k\Delta$ generated by the arrows of Δ which are considered as paths of length $= 1$ and for all $n \in \mathbb{N}_0$ let $k^n\Delta$ be the n-th power of the ideal $k^1\Delta$. Usually the category $k\Delta$ will not be a spectroid. But we can obtain a spectroid from $k\Delta$ by factoring out a suitable ideal and imposing suitable finiteness conditions on Δ. Namely, let us require that Δ is *locally bounded* i.e. only finitely many arrows start and stop in each vertex. Let us call an ideal I of $k\Delta$ *weakly admissible* if $I \subseteq k^2\Delta$ and for any two objects x, y of Δ there is a positive integer $n(x, y)$ such that $k^{n(x,y)}(x, y) \subseteq I(x, y)$. Then $k\Delta/I$ is a spectroid, if Δ is a locally bounded quiver and I is a weakly admissible ideal of $k\Delta$. Note that in this case the Jacobson radical of $k\Delta/I$ coincides with $k^1\Delta/I$.

Another instructive class of spectroids can be constructed from partially ordered sets which we briefly call *posets*. We consider a poset S as category by taking as usual the underlying set as set of objects and introducing a morphism $(y \mid x)$ for all elements x, y of the poset satisfying $x \geq y$. The composition $(z \mid y) \cdot (y \mid x)$ is defined to be $(z \mid x)$. For a poset S the k-linear category kS is always a spectroid. The radical of kS is formed by all k-multiples of morphisms $(y \mid x)$ such that $x > y$. If we factor out an ideal I of kS with $I \subseteq \mathrm{rad}\, kS$, then kS/I is again a spectroid. In case all intervals in the poset S are finite, we can construct kS from the Hasse quiver Δ_S of S by forming $k\Delta_S/I$ where I is the ideal of $k\Delta_S$ formed by all differences $u - v$ where u, v are paths in Δ_S with the same start and end point.

1.3.2 Modules over Path Categories and Posets Quivers and posets are discrete structures. Therefore finite quivers and posets are easily encoded by matrices with integer coefficients which are well-adapted for computers. However, ideals of a path category kQ will usually be generated by linear combinations of paths involving scalars from the field k. Thus a typical element ρ of a generating set of an ideal I of $k\Delta$ is of the shape $\rho = \sum_{i=1}^{n} \lambda_i u_i$ where $\lambda_i \in k$ and the $u_i = \alpha^{(i)}_{n(i)} \dots \alpha^{(i)}_1$ are paths from x to y in Δ.

Modules over a path category $k\Delta$ can be studied via mere linear algebra. Namely, they are the *k-linear representations* of the quiver Δ. This means that such a module M is given by vector spaces $M(x)$ over k for each vertex x of Δ and a k-linear map $M(\alpha) : M(y) \to M(x)$ for each arrow $\alpha : x \to y$ in Δ. This is inherited by modules over $k\Delta/I$ where we assume I to be generated by a set $\{\rho\}$ of generators. The modules in Mod$-k\Delta/I$ are the k-linear representations M of the quiver Δ *satisfying the relations* $\{\rho\}$ i.e. $0 = M(\rho) = \sum_{i=1}^{n} \lambda_i M(u_i)$ for all ρ where $M(u_i) := M(\alpha^{(i)}_1) \dots M(\alpha^{(i)}_{n(i)})$.

In contrast to path categories, for a poset S an ideal of kS can always be generated by morphisms $\rho = (y \mid x)$. Modules M over kS are *k-linear representations of S* i.e. are given by vector spaces $M(x)$ over k for each element x of S and k-linear maps $M(y \mid x) : M(y) \to M(x)$ for all $x \geq y$ which satisfy $M(z \mid x) = M(y \mid x)M(z \mid y)$ for all $x \geq y \geq z$. Consequently, modules over kS/I where I is generated by a set of morphisms $\rho = (y \mid x)$ are k-linear representations of S *satisfying the relations* $\{\rho\}$ i.e. $M(y \mid x) = 0$ for all $\rho = (y \mid x)$.

1.3.3 The Quiver of a Spectroid *From now on we assume that the field k is algebraically closed!*

The quiver $\mathrm{Qv}\,\mathcal{S}$ of a spectroid $\mathcal{S}$ has as set of vertices the set of objects of $\mathcal{S}$. For two objects x, y there are precisely $\dim_k \operatorname{irr} \mathcal{S}(x,y)$ arrows from x to y where $\operatorname{irr} \mathcal{S}(x,y)$ is the factor space $\operatorname{rad} \mathcal{S}(x,y)/(\operatorname{rad} \mathcal{S})^2(x,y)$. The morphisms in $\operatorname{rad} \mathcal{S}(x,y)$ whose classes are non-zero in $\operatorname{irr} \mathcal{S}(x,y)$ are called *irreducible*. Since $\mathcal{S}$ has finite-dimensional morphism spaces the quiver $\mathrm{Qv}\,\mathcal{S}$ is *locally finite* i.e. there are only finitely many arrows between any two vertices.

If we consider the spectroid $\mathcal{S}$ as quiver in the obvious way, then a quiver morphism $\Phi : \mathrm{Qv}\,\mathcal{S} \to \mathcal{S}$ is said to be a *presentation* if it acts as identity on the vertices and sends the arrows from x to y in $\mathrm{Qv}\,\mathcal{S}$ to morphisms in $\mathcal{S}(x,y)$ whose classes form a k-basis of $\operatorname{irr} \mathcal{S}(x,y)$. A presentation $\Phi : \mathrm{Qv}\,\mathcal{S} \to \mathcal{S}$ extends uniquely to a k-linear functor $k(\mathrm{Qv}\,\mathcal{S}) \to \mathcal{S}$ which we will call a presentation of $\mathcal{S}$ as well.

Unfortunately, a presentation needs not to be full. For instance, consider the spectroid $k\mathbb{Q}$ where $\mathbb{Q}$ are the rationals totally ordered in the usual way. Since $\operatorname{rad} k\mathbb{Q} = (\operatorname{rad} k\mathbb{Q})^2$, the quiver of $k\mathbb{Q}$ does not have any arrows. Therefore no presentation of $k\mathbb{Q}$ can be full. This example suggests to consider the ideal $\operatorname{rad}^\infty \mathcal{S}$ of the spectroid $\mathcal{S}$ which is defined as the intersection of all powers $(\operatorname{rad} \mathcal{S})^n$ where $n \in \mathbb{N}$. Since $\operatorname{rad}^\infty \mathcal{S} \subseteq \operatorname{rad}^2 \mathcal{S}$, we obtain that a presentation $\Phi : k(\mathrm{Qv}\,\mathcal{S}) \to \mathcal{S}$ induces a presentation $\Phi : k(\mathrm{Qv}\,\mathcal{S}) \to \mathcal{S}/rad^\infty\mathcal{S}$ and $\mathcal{T} := \mathcal{S}/rad^\infty\mathcal{S}$ is still a spectroid. Moreover, this spectroid $\mathcal{T}$ is *generalized standard* i.e. $\operatorname{rad}^\infty \mathcal{T} = 0$. It is easy to prove that a presentation $\Phi : k(\mathrm{Qv}\,\mathcal{S}) \to \mathcal{S}$ of a spectroid $\mathcal{S}$ is full and the kernel of Φ is a weakly admissible ideal of $k(\mathrm{Qv}\,\mathcal{S})$ if and only if the spectroid $\mathcal{S}$ is generalized standard.

Since finite spectroids $\mathcal{S}$ are of particular interest for us, we remember that this finiteness forces $\operatorname{rad} \mathcal{S}$ to be nilpotent. Hence $\mathcal{S}$ is generalized standard and consequently can be identified with $k(\mathrm{Qv}\,\mathcal{S})/I$ where I is some weakly admissible ideal of $k(\mathrm{Qv}\,\mathcal{S})$.

Roughly speaking, the basic aim of Combinatorial REPresentation theory is to study the quiver of ind$-\mathcal{S}$ for a given finite spectroid $\mathcal{S}$. However, usually this quiver gives only rather poor information about the category ind$-\mathcal{S}$ because ind$-\mathcal{S}$ fails to be generalized standard in most cases. In the next two sections we will outline how to get deeper insight on one hand by looking at additional data and on the other hand by passing to special situations.

1.4 The Auslander-Reiten Quiver

1.4.1 Almost split Sequences A morphism $g : Y \to Z$ in an aggregate $\mathcal{A}$ is called *right almost split* if it lies in rad $\mathcal{A}$ and any other morphism $g' : Y' \to Z$ in rad $\mathcal{A}$ can be factorized as $g' = gh$ for some $h : Y' \to Y$ in $\mathcal{A}$. The morphism g is said to be *right minimal*, if every endomorphism s of Y satisfying $gs = g$ has to be an automorphism. Left almost split and left minimal morphism are defined in the dual fashion. Morphisms which are both right almost split and right minimal are briefly called *right minimal almost split*. Minimal almost split morphism are basically unique: if $g : Y \to Z$ is a minimal right almost split morphism, then any other minimal right almost split morphism ending in Z is of the shape gs for some isomorphism s. A right minimal almost split morphism $g : Y \to Z$ in an aggregate $\mathcal{A}$ with Z in $\mathcal{S}(\mathcal{A})$ stores all information about the predecessors of Z in $\mathrm{Qv}\,\mathcal{S}(\mathcal{A})$. Namely, for any object U in $\mathcal{S}(\mathcal{A})$ the number of arrows from $U \to Z$ in $\mathrm{Qv}\,\mathcal{S}(\mathcal{A})$ coincides with the multiplicity of U as direct summand of Y.

In a module category left minimal almost split monomorphisms and right minimal almost split epimorphisms starting resp. ending in indecomposable objects are closely related. Let $0 \to X \xrightarrow{f} Y \xrightarrow{g} Z \to 0$ be a short exact sequence in mod$-\mathcal{S}$ for a finite spectroid $\mathcal{S}$. Then the following two properties are equivalent:

(i) The module Z is indecomposable and g is right minimal almost split.

(ii) The module X is indecomposable and f is left minimal almost split.

Short exact sequences as above are called *almost split sequences* or *Auslander-Reiten sequences*. They may be interpreted as a device to construct a 'new' indecomposable module X by forming the kernel of the right minimal almost split epimorphism g ending in the 'known' indecomposable module Z. That a right minimal almost split epimorphism g exists in all possible cases follows from the theorem below which is most important for modern representation theory.

Theorem 1.1 (Auslander,Reiten[2]). *Let $\mathcal{S}$ be a finite spectroid. For each module Z in* ind$-\mathcal{S}$ *there exists a right minimal almost split morphism $g : Y \to Z$ in* mod$-\mathcal{S}$ *which is an epimorphism if Z is not projective and can be chosen as the canonical inclusion* $\mathrm{rad}\, x^\wedge \hookrightarrow x^\wedge$ *if $Z = x^\wedge$ is projective.*

By the uniqueness of almost split morphisms an almost split sequence $0 \to X \longrightarrow Y \longrightarrow Z \to 0$ is uniquely determined up to isomorphism by Z and dually also by X. Thus by mapping Z to $\tau Z := X$ we obtain a bijection from the set of non-projective modules in ind$-\mathcal{S}$ to the set of non-injective modules in ind$-\mathcal{S}$. The inverse of this bijection is usually denoted by τ^{-}.

Using the map τ the Auslander-Reiten formula $\mathrm{D}\,\mathrm{Ext}^1_{\mathcal{S}}(Z, U) \cong \overline{\mathrm{Hom}}_{\mathcal{S}}(U, \tau Z)$ allows to calculate Ext-functors via Hom-functors. This will be very useful for calculations since Hom-functors will turn out to be computable. Note, that $\overline{\mathrm{Hom}}_{\mathcal{S}}(U, \tau Z)$ is the factor of $\mathrm{Hom}_{\mathcal{S}}(U, \tau Z)$ by the subspace of all morphisms factoring through an injective module. In fact, the Auslander-Reiten formula can be used for proving the existence of almost split sequences. Of course for this purpose the map τ has to be introduced before as the 'dual of the transpose'.

1.4.2 The Auslander-Reiten Quiver The existence of almost split sequences shows that for a finite spectroid $\mathcal{S}$ the quiver Qv(ind$-\mathcal{S}$) is endowed with an additional structure, namely the translation map τ which has the property that for each non-projective vertex Z and for any vertex U the number of arrows from U to Z equals the number of arrows from τZ to U. Quivers endowed with such a translation are called *translation quivers*. The translation quiver of ind$-\mathcal{S}$ for a finite spectroid $\mathcal{S}$ is called the *Auslander-Reiten quiver* of $\mathcal{S}$. It is usually denoted by $\Gamma_{\mathcal{S}}$ rather than Qv(ind$-\mathcal{S}$) and we will follow this tradition in the following.

In general the Auslander-Reiten quiver $\Gamma_{\mathcal{S}}$ will not be connected and various kinds of components appear. We will mention some special kinds of components for later use. Note, that a quiver is said to be *directed* it it does not admit an oriented cycle i.e. a path of positive length which starts and ends in the same vertex. A component of $\Gamma_{\mathcal{S}}$ is called preprojective if it is directed and any vertex is of the shape $\tau^{-m}x^{\wedge}$ for some indecomposable projective vertex $x^{\wedge}$. Preinjective components are defined in the dual way. Preprojective components can be 'knitted' inductively starting from their projective vertices. An infinite component of $\Gamma_{\mathcal{S}}$ as in the figure below (the right and the left rim have to be identified) is said to be a *stable tube.* As usual the translation τ is indicated by dotted arrows. The number $r \in \mathbb{N}$ of vertices at the 'mouth' (which is 6 in the example in Figure 1.1) is called the *rank* of the tube. Stable tubes of rank $= 1$ are said to be *homogeneous.*

The translation structure of the Auslander-Reiten quiver imposed by the existence of almost split sequences can be used for the calculation of dimensions of Hom-functors. If Z is a non-projective vertex from $\Gamma_{\mathcal{S}}$, the modules $Y_1, \dots, Y_l$ are the predecessors of Z and there are n_i arrows from Y_i to Z, then for any U in ind$-\mathcal{S}$ the formula

$$\dim_k \mathrm{Hom}_{\mathcal{S}}(U, \tau Z) + \dim_k \mathrm{Hom}_{\mathcal{S}}(U, Z) = \sum_{i=1}^{l} n_i \dim_k \mathrm{Hom}_{\mathcal{S}}(U, Y_i) + \delta_{U,Z}$$

holds. This allows to compute the values of dimensions of Hom-functors on stable tubes from the values on the 'mouth'. Also one can calculate dimensions of Hom-

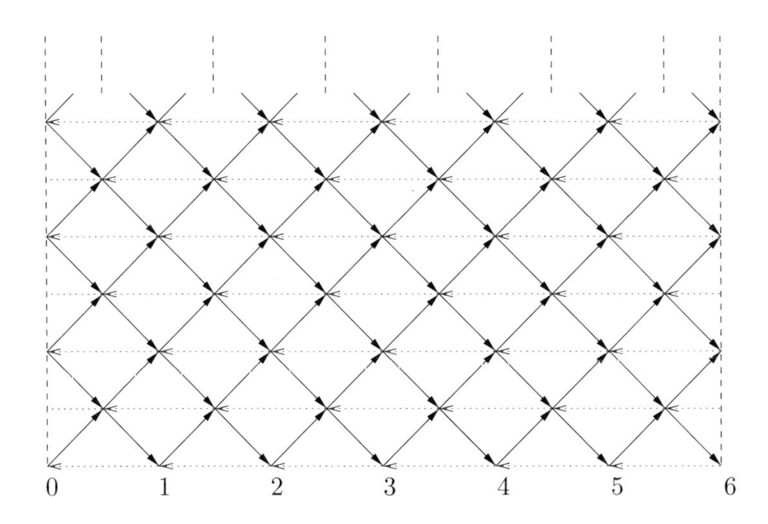

Figure 1.1: Stable tube

functors on preinjective components by induction starting with the injective modules where the value is given by the dual Yoneda-Lemma.

1.4.3 The Grothendieck Group For a finite spectroid $\mathcal{S}$ the *Grothendieck group* $K_0(\mathcal{S})$ with respect to short exact sequences is the factor of the free abelian group with basis mod$-\mathcal{S}$ by the subgroup generated by all differences $Y - X - Z$ such that there is a short exact sequence $0 \to X \longrightarrow Y \longrightarrow Z \to 0$. The structure of this group is easy to determine. For any module M in mod$-\mathcal{S}$ we define its *dimension vector* $\mathbf{dim}\, M$ as the vector in $\mathbb{Z}^{\mathcal{S}}$ whose entry at x is the dimension $\dim_k M(x)$. Then the map $\mathbf{dim} : \text{mod}-\mathcal{S} \to \mathbb{Z}^{\mathcal{S}}$ induces an isomorphism of groups $K_0(\mathcal{S}) \to \mathbb{Z}^{\mathcal{S}}$. In particular, this shows that for any short exact sequence $0 \to X \longrightarrow Y \longrightarrow Z \to 0$ in mod$-\mathcal{S}$ the equation $\mathbf{dim}\, X + \mathbf{dim}\, Z = \mathbf{dim}\, Y$ holds. Applying this to almost split sequences allows to calculate the dimension vectors of modules in stable tubes from those at the mouth and inductively the dimension vectors of modules in preprojective (resp. preinjective) components from the dimension vectors of the indecomposable projective (resp. injective) modules. Note that $(\mathbf{dim}\, x^\wedge)_y = \dim_k \mathcal{S}(y, x)$ and $(\mathbf{dim}\, x^\vee)_y = \dim_k \mathcal{S}(x, y)$ for all x, y in $\mathcal{S}$.

The dimension vector isomorphism can also be used to calculate certain dimensions of Hom- and Ext-functors by multiplication of integer matrices: The *Cartan matrix* $C_{\mathcal{S}}$ of the finite spectroid $\mathcal{S}$ is an integer matrix indexed over $\mathcal{S} \times \mathcal{S}$ whose entry at (x, y) is given by the dimension $\dim_k \mathcal{S}(x, y)$. Thus, if $\mathcal{S}$ has n objects, then $C_{\mathcal{S}}$ is an $n \times n$-matrix. In case the spectroid $\mathcal{S}$ has finite global dimension, on one hand the Cartan matrix is invertible over the integers and on the other hand we obtain a well-defined bilinear form $K_0(\mathcal{S}) \times K_0(\mathcal{S}) \to \mathbb{Z}$ by putting $\langle M, N \rangle_{\mathcal{S}} := \sum_{\nu=0}^{\infty} (-1)^\nu \dim_k \operatorname{Ext}^\nu_{\mathcal{S}}(M, N)$. These two constructions are related by the formula $\langle M, N\rangle_{\mathcal{S}} = (\mathbf{dim}\, M)C_{\mathcal{S}}^{-T}(\mathbf{dim}\, N)^T$ valid for all M, N in mod$-\mathcal{S}$. Later we will encounter the quadratic form $\chi_{\mathcal{S}}$ associated with this bilinear form which is called the *Euler form* of $\mathcal{S}$.

1.5 Representation Types

1.5.1 Finite Representation Type A finite spectroid $\mathcal{S}$ is said to be *of finite representation type* (or shorter *representation finite*) if its associated spectroid ind$-\mathcal{S}$ is finite i.e. there exist up to isomorphism only finitely many indecomposable modules in mod$-\mathcal{S}$. The Auslander-Reiten quiver gives a handy criterion to check finite representation type.

Theorem 1.2 (Auslander[1]). *If the Auslander-Reiten quiver $\Gamma_{\mathcal{S}}$ of a finite spectroid $\mathcal{S}$ with connected quiver* Qv $\mathcal{S}$ *has a component such that there is a common bound for the dimensions of the modules in this component, then the component is finite and coincides with $\Gamma_{\mathcal{S}}$.*

This leaves the problem to find at least one component of the Auslander-Reiten quiver. Later we will see a class of spectroids which always have at least one preprojective component. For spectroids with small global dimension deciding finite representation type is reduced to a combinatorial problem by the following result.

Theorem 1.3 (Ringel[49]; Bongartz[8]). *Suppose $\mathcal{S}$ is a finite spectroid whose quiver* Qv $\mathcal{S}$ *is directed and connected, the global dimension of $\mathcal{S}$ is at most* 2 *and the Auslander-Reiten quiver $\Gamma_{\mathcal{S}}$ has a preprojective component. Then $\mathcal{S}$ is representation finite if and only if the Euler form $\chi_{\mathcal{S}}$ is weakly positive. Moreover, in this case the dimension vector map is a bijection between* ind$-\mathcal{S}$ *and the set of positive* 1*-roots of $\chi_{\mathcal{S}}$.*

We have to make up for some notations used in the above theorem. The Euler form $\chi_{\mathcal{S}}$ is said to be *weakly positive* if $\chi_{\mathcal{S}}(x) > 0$ for all $x \in \mathbb{Z}$, $x > 0$ where we endow $\mathbb{Z}^{\mathcal{S}}$ with the usual product order. The *positive* 1*-roots* of $\chi_{\mathcal{S}}$ are the vectors $x \in \mathbb{Z}^{\mathcal{S}}$ satisfying $x > 0$ and $\chi_{\mathcal{S}}(x) = 1$. To find out in an efficient algorithmic way, whether $\chi_{\mathcal{S}}$ and similar quadratic forms are weakly positive, will be the topic of other contributions in this collection. Also generalizations of weak positivity will be treated. Therefore we will not go deeper into this subject. Let us mention however that the above theorem applies to hereditary spectroids because their Auslander-Reiten quivers always possess a preprojective component. We recall that a finite spectroid is hereditary if and only if its quiver Qv $\mathcal{S}$ is directed and $\mathcal{S} \cong k(\text{Qv}\,\mathcal{S})$. Thus we obtain the following classical result which is one of the starting points of the modern representation theory of algebras.

Theorem 1.4 (Gabriel[29]). *Let $\mathcal{S} = k\Delta$ be a finite hereditary spectroid with connected quiver Δ. Then $\mathcal{S}$ is of finite representation type if and only if the underlying graph of the quiver Δ is one of the Dynkin graphs $\mathbb{A}_n$ $(n \geq 1)$, $\mathbb{D}_n$ $(n \geq 4)$, $\mathbb{E}_n$ $(8 \geq n \geq 6)$.*

1.5.2 Tame and wild Representation Type To define tame and wild representation type we have to introduce some geometric concepts. Let $\mathcal{S} = k\Delta/I$ be a finite spectroid and $\{\rho\}$ be a finite set of generators of the ideal I. For fixed $d \in \mathbb{N}_0^{\mathcal{S}}$ the set $\mathrm{mod_d}{-}\mathcal{S}$ of modules M in $\mathrm{mod}{-}\mathcal{S}$ satisfying $\mathbf{dim}\, M = d$ can be considered as a Zariski closed subvariety of the linear variety $\prod_\alpha k^{d_x \times d_y}$ where α runs through all arrows $\alpha : x \to y$ in the quiver Δ. To see this, we recall first that such a module M can be identified with a representation of Δ such that $M(\rho) = 0$ for each relation ρ and $\dim_k M(x) = d_x$ for all x in $\mathcal{S}$. By fixing bases in the spaces $M(x)$ any arrow $\alpha : x \to y$ in Δ is represented by a $d_x \times d_y$-matrix $M(\alpha)$ over k and the equation $M(\rho) = 0$ amounts to a finite collection of polynomial equation on the matrix entries.

There is an analogous way to associate varieties with finite spectroids whose object set without loss of generality is chosen as $\{1, \dots, n\}$ for some $n \in \mathbb{N}$. For this purpose let us now fix a $n \times n$-matrix B over the non-negative integers and define $C := B + I$ where I is a unit matrix of the same shape as B. The set alg_C of spectroids with Cartan matrix C can be considered as a Zariski closed subvariety of the linear variety $\prod_{(x,y,z)} k^{B_{y,z} \times B_{x,y} \times B_{x,z}}$ where the triples (x, y, z) run through $\mathcal{S}^3$. To see this, one chooses bases in all spaces $\mathrm{rad}\,\mathcal{S}(x, y)$ and represents the composition $\mathrm{rad}\,\mathcal{S}(y, z) \times \mathrm{rad}\,\mathcal{S}(x, y) \to \mathrm{rad}\,\mathcal{S}(x, z)$ by an $B_{x,z}$-tuple of $B_{y,z} \times B_{x,y}$-matrices. The associativity of the composition and the nilpotency of $\mathrm{rad}\,\mathcal{S}$ impose polynomial equations on the entries of these matrices.

A finite spectroid $\mathcal{S}$ is said to be of *tame representation type* (or shorter *tame*) if for each possible dimension vector $d \in \mathbb{N}_0^{\mathcal{S}}$ there exist finitely many regular 1-parameter families $\gamma_1, \dots, \gamma_{\mu(d)}$ with the property that up to isomorphism and finitely many failures each indecomposable module in $\mathrm{mod_d}{-}\mathcal{S}$ lies in the image of one of the maps $\gamma_i : k \to \mathrm{mod_d}{-}\mathcal{S}$.

Let us choose these 1-parameter families in a way such that $\mu(d)$ is minimal for all $d \in \mathbb{N}_0^{\mathcal{S}}$ and let us define a function $\bar{\mu} : \mathbb{N}_0 \to \mathbb{N}_0$ by $\bar{\mu}(n) := \sum_d \mu(d)$ where we sum over all d satisfying $\sum_{x \in \mathcal{S}} d_x = n$. Thus $\bar{\mu}$ measures the number of 1-parameter families needed to parameterize the indecomposables M of total dimension $n = \sum_{x \in \mathcal{S}} \dim_k M(x)$. If the function $\bar{\mu}$ grows at most polynomially, then the spectroid $\mathcal{S}$ is said to be *tame of polynomial growth.* In particular, $\mathcal{S}$ is called *domestic* if μ is of constant growth. Note, that representation finite spectroids are tame. In fact, the confirmation of the second Brauer-Thrall conjecture means that the representation finite spectroids are precisely the tame spectroids such that $\bar{\mu} = 0$.

Let Ω be the double loop quiver which is shown in Figure 1.2. A finite spectroid $\mathcal{S}$ is called *of wild representation type* (or shorter *wild*) provided there is an embedding functor $\mathrm{mod}{-}k\Omega \to \mathrm{mod}{-}\mathcal{S}$ which preserves indecomposability and isomorphism classes.

Recall that the modules in $\mathrm{mod}{-}k\Omega$ can be identified with pairs of square matrices of the same size. Furthermore, isomorphism means equivalence under simultaneous conjugation.

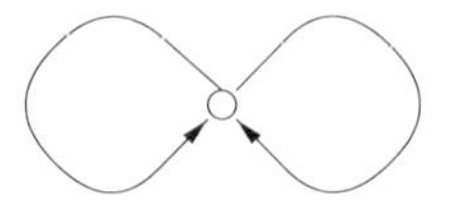

Figure 1.2: Double loop

Theorem 1.5 (Drozd[27] see also [14]). *A finite spectroid is tame or wild but not both.*

1.5.3 Degenerations and Coverings A well-tried strategy to investigate complicated structures in mathematics is to 'simplify' the structure without changing too much the property one wants to study. This applies as well to finite spectroids as to their representation type. Let us first make use of the geometry of alg_C.

Theorem 1.6 (Gabriel[30]). *The subset of representation finite spectroids in the variety* alg_C *is Zariski open.*

Recall that a spectroid $\mathcal{T}$ in alg_C is said to be a *degeneration* of another spectroid $\mathcal{S}$ in alg_C provided $\mathcal{T}$ lies in the Zariski closure of the subset of alg_C formed by all spectroids isomorphic to $\mathcal{S}$. Thus the above theorem implies that for checking finite representation type one may pass to a degeneration of the given spectroid. For the time being it is only conjectured that also the subset of tame spectroids is open, however it has already been proved by Geiß (see [36]) that degenerations of wild spectroids are wild which shows that also tame representation type can be decided via degenerations.

If a spectroid is given by a quiver with relations, then degenerations usually allow to simplify the relations. In contrast, covering theory allows to simplify the quiver by 'untwisting oriented cycles'. Let us make this a little more precise. Let $\mathcal{T}$ be a *locally bounded* spectroid i.e. for all x in $\mathcal{T}$ the sum $\sum_y(\dim_k \mathcal{T}(x,y) + \dim_k \mathcal{T}(y,x))$ is finite where y runs through all objects y of $\mathcal{T}$. Moreover, let G be a group of automorphism of $\mathcal{T}$ which acts freely on the objects of $\mathcal{T}$. Then the orbits of objects and morphisms of $\mathcal{T}$ under G are made into a spectroid $\mathcal{T}/G$ in the natural way. The canonical functor $\mathcal{T} \to \mathcal{T}/G$ is called a *covering functor*. There are various theorems asserting that $\mathcal{T}/G$ inherits the representation type of $\mathcal{T}$ under additional assumptions on $\mathcal{T}$ and G. We refer e.g. to [32], [17] and [18] for more details.

Whereas for finite representation type coverings also allow complete control over the spectroid of indecomposable representations, this is still an open problem for tame type. Moreover, the covering $\mathcal{T}$ usually fails to be finite. Although the representation type is determined by the finite full subspectroids of $\mathcal{T}$, for algorithmic purposes one of course does not want to consider all of them but looks for some finite subcollection. This can be done for finite representation type. In this case one also has a canonical degeneration the so-called 'standard form' which admits a strongly simply connected 'universal covering'. Therefore in the next section we will study strongly simply connected spectroids from the computational point of view. Examples show that there is good hope that also for tame type a similar simplification using a covering of a standard form may be possible. Nevertheless, the right concept of a standard form has to be developed and the notion of 'strong simple connectedness' needs some further generalization.

Coming back to finite representation type, the formation of the standard form and of the universal covering lead to quite clear combinatorial algorithms. Unfortunately, up to now these algorithms are only partially implemented.

1.6 Strongly simply connected Spectroids of Finite and Tame Representation Type

1.6.1 Completely separating Spectroids There are various ways to introduce strongly simply connected spectroids. We will use one which is best suited for computational purposes. Let $\mathcal{S}$ be a finite spectroid whose quiver $\mathrm{Qv}\,\mathcal{S}$ is directed. For x in $\mathcal{S}$ we decompose the radical $\mathrm{rad}\, x^\wedge = \oplus_{i=1}^r R_i$ of the indecomposable projective modules $x^\wedge$ into a direct sum of indecomposable modules R_i in ind$-\mathcal{S}$. The object x is said to be *separating* if the supports $\mathrm{supp}\, R_i := \{y \in \mathcal{S} \mid R_i(y) \neq 0\}$ lie in r different connected components of the full subquiver of $\mathrm{Qv}\,\mathcal{S}$ supported by the vertices z which are not terminus of a path in $\mathrm{Qv}\,\mathcal{S}$ starting in x. The spectroid $\mathcal{S}$ itself is called *separating* if all objects x in $\mathcal{S}$ are separating. Finally, the spectroid $\mathcal{S}$ is defined to be *strongly simply connected* provided each full convex subcategory is separating where convexity is required with respect to the partial ordering imposed on $\mathcal{S}$ by the directed quiver $\mathrm{Qv}\,\mathcal{S}$.

We have seen before that the existence of a preprojective component allows the checking of the representation type with combinatorial tools. From this follows the importance of the following result for computational purposes.

Theorem 1.7 (Bautista,Larrión[5]; Bongartz[10]). *If the spectroid $\mathcal{S}$ is separating, then the Auslander-Reiten quiver $\Gamma_{\mathcal{S}}$ has a preprojective component.*

It is easy to see that a finite spectroid $\mathcal{S}$ with directed quiver $\mathrm{Qv}\,\mathcal{S}$ cannot be representation finite if there are two objects x, y satisfying $\dim_k \mathcal{S}(x,y) > 1$. Analogously, $\mathcal{S}$ cannot be tame if there exist x, y such that $\dim_k \mathcal{S}(x,y) > 2$. Therefore concerning finite type we do not loose generality if we consider only spectroids $\mathcal{S}$ which are *schurian* i.e. $\mathcal{S}(x,y) \leq 1$ for each pair x, y of objects. Schurian strongly simply connected spectroids are called *completely separating.* As tame strongly simply connected spectroids $\mathcal{S}$ such that $\dim_k \mathcal{S}(x,y) = 2$ appear, we really loose some generality by considering only the schurian case. Nevertheless, all the known interesting phenomena for tame type already occur in the completely separating situation. On the other hand there is not yet a good data structure for strongly simply connected spectroids in general whereas a computer can easily deal with completely separating spectroids due to the following result.

Theorem 1.8 ([22]). *If $\mathcal{S}$ is a finite completely separating spectroid, then there is a finite poset S and an admissible ideal I of kS such that $\mathcal{S} \cong kS/I$.*

Thus one has a simple combinatorial data structure given by the Hasse quiver of S and a finite set of relations $\rho = (y \mid x)$ for I. The above result is the reason why many CREP procedures deal with spectroids given by a finite poset with relations. Clearly this concept will have to undergo generalizations in the future.

1.6.2 Tilting Spectroids Before coming to classifications we make a small detour to an area in representation theory called tilting theory. Let $\mathcal{S}$ be a finite spectroid with n objects. A full subspectroid $\mathcal{T}$ of ind$-\mathcal{S}$ is called *tilting* if the following three conditions are satisfied:

(i) The projective dimension of each module in $\mathcal{T}$ is at most 1.

(ii) $\mathrm{Ext}^1_{\mathcal{S}}(X, Y) = 0$ for any pair of objects X, Y in $\mathcal{T}$.

(iii) $\mathcal{T}$ has precisely n objects.

The importance of tilting theory comes from the fact that the representation theory of $\mathcal{S}$ and $\mathcal{T}$ are closely related. In fact, the derived categories of $\mathcal{S}$ and $\mathcal{T}$ even coincide. We will only need the following special aspect of these facts.

Theorem 1.9 (Brenner, Butler[4]; Happel, Ringel[38]). *Let $\mathcal{S} = k\Delta$ be a hereditary spectroid and $\mathcal{T}$ a tilting spectroid in* ind$-\mathcal{S}$. *Then there are two disjoint subsets $\mathcal{G}$ and $\mathcal{F}$ of* ind$-\mathcal{S}$ *such that* ind$-\mathcal{T}$ *is the disjoint union of the the set of modules* $\mathrm{Hom}_{\mathcal{S}}(-, X)$, *$X \in \mathcal{G}$ and the set of modules* $\mathrm{Ext}^1_{\mathcal{S}}(-, Y)$, *$Y \in \mathcal{F}$.*

In particular, this means that ind$-\mathcal{T}$ has less objects than ind$-\mathcal{S}$. More precisely, $\mathcal{T}$ is representation finite provided $\mathcal{S}$ is representation finite and $\mathcal{T}$ is tame or representation finite provided $\mathcal{S}$ is tame.

We have seen before that we can calculate dimensions of Hom-functors on certain components of the Auslander-Reiten quiver in a combinatorial way. Thus the Auslander-Reiten formula can be used to check condition (ii) by a computer. Fortunately, the same applies to condition (i) because X in ind$-\mathcal{S}$ has projective dimension at most 1 if and only if $\mathrm{Hom}_{\mathcal{S}}(y^{\vee}, \tau X) = 0$ for all y in $\mathcal{S}$. Finally, the calculation of Hom-dimensions provides the Cartan matrix $C_{\mathcal{T}}$. If $\mathcal{T} = kT/I$ is completely separating, then it is easy to restore the combinatorial data namely the poset T and the combinatorial generators for I from $C_{\mathcal{T}}$.

1.6.3 Classifications The proper substructures to consider of strongly simply connected spectroids are the convex full subspectroids because they inherit the defining property. This is the motivation for the following definitions. A completely separating spectroid $\mathcal{S} = kS/I$ is called

- *critical* if $\mathcal{S}$ is not representation finite but each proper convex full subspectroid is representation finite.
- *hypercritical* if $\mathcal{S}$ is wild but each proper convex full subspectroid is tame.

- *sincere* if there exists M in ind$-\mathcal{S}$ which is sincere i.e. $M(x) \neq 0$ for all x in S.
- *weakly sincere* if there exists M in ind$-\mathcal{S}$ which is weakly sincere i.e. $M(x) \neq 0$ for all extremal x in S.

Obviously, for showing that a given completely separating spectroid $\mathcal{S}$ is representation finite (resp. tame) it is necessary and sufficient that no critical (resp. hypercritical) completely separating spectroid appears as convex full subspectroid. Moreover, once one knows all weakly sincere completely separating spectroids and their weakly sincere indecomposables, then one knows all indecomposables over any completely separating spectroid. The reason for this is that for a convex full subspectroid $\mathcal{T}$ of $\mathcal{S}$ there is a canonical full embedding mod$-\mathcal{T} \to$ mod$-\mathcal{S}$ which extends a module M in mod$-\mathcal{T}$ to a module in mod$-\mathcal{S}$ by putting $M(x) := 0$ for all x in $\mathcal{S} \setminus \mathcal{T}$.

Theorem 1.10 (Bongartz [9]). *The critical completely separating spectroids coincide with the preprojective schurian tilting spectroids $\mathcal{T}$ in* ind$-k\Delta$ *where Δ is an extended Dynkin tree i.e. the underlying graph of Δ is one of $\widetilde{\mathbb{D}}_n$ $(n \geq 4)$ or $\widetilde{\mathbb{E}}_n$ $(8 \geq n \geq 6)$ (see Figure 1.3).*

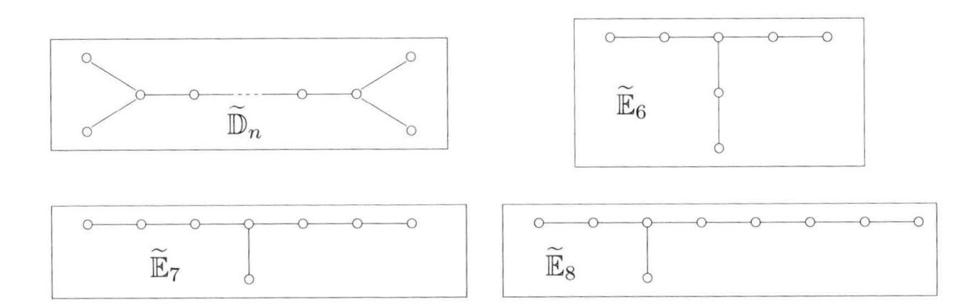

Figure 1.3: Extended Dynkin trees

Using this characterization the critical completely separating spectroids were classified by Happel and Vossieck in [39]. There are three series arising from $\widetilde{\mathbb{D}}_n$ $(n \geq 4)$ and 4302 isomorphism classes of exceptional cases coming from $\widetilde{\mathbb{E}}_n$ $(8 \geq n \geq 6)$ which were classified using a computer program based on the combinatorial techniques which we outlined before.

Theorem 1.11 (de la Peña [45]). *The hypercritical completely separating spectroids contain the preprojective schurian tilting spectroids $\mathcal{T}$ in* ind$-k\Delta$ *where Δ is a hyperextended Dynkin tree i.e. the underlying graph of Δ is one of $\widetilde{\widetilde{\mathbb{D}}}_n$ $(8 \geq n \geq 4)$, $T_{2,2,2,2,2}$ or $\widetilde{\widetilde{\mathbb{E}}}_n$ $(8 \geq n \geq 6)$ (see Figure 1.4).*

It is conjectured that the the hypercritical completely separating spectroids really coincide with the tilting spectroids from the above theorem. These tilting

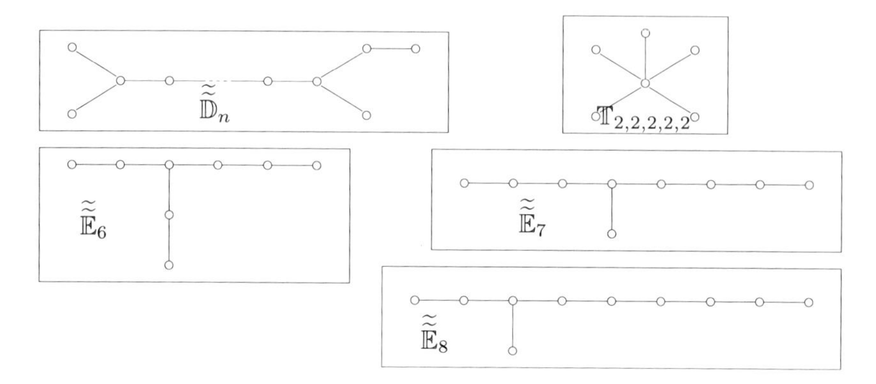

Figure 1.4: Hyperextended Dynkin trees

spectroids were classified by Unger in [55]. There are 13747 isomorphism classes coming from $\widetilde{\widetilde{\mathbb{E}}}_n$ $(8 \geq n \geq 6)$ and a small list from $\widetilde{\widetilde{\mathbb{D}}}_n$ $(8 \geq n \geq 4)$.

Only for the representation finite case there exists a complete classification for the weakly sincere spectroids. Note, that an indecomposable module over a completely separating spectroid of finite representation type is weakly sincere if and only if it is sincere and if and only if it is faithful.

Theorem 1.12 (Bongartz[7]; Ringel[49]). *The sincere completely separating representation finite spectroids are contained in the schurian tilting spectroids $\mathcal{T}$ in* ind$-k\Delta$ *having preprojective and preinjective objects where Δ is a tree with at most four endpoints.*

In [7] Bongartz gave a bound N such that the sincere completely separating representation finite spectroids with more than N objects lie in 24 infinite series. This bound was optimized to 13 by Ringel in [49]. The 16344 isomorphism classes of exceptional cases are constructed in [19] (see also [20]). Note, that the above theorem did not help too much in that construction because there are too many of these trees. An inductive construction proposed in [49] was used for [19].

It would be an exercise to repeat the above classifications using CREP. In practice one is usually interested in the lists themselves e.g. in order to test some conjectures on them. The lists are accessible in CREP by calling the procedures **critical_alge**, **hypercritical_algd**, **hypercritical_alge** and **esrdalg**. Note that in these data bases not all isomorphism classes but only representatives in certain equivalence classes called 'frames' are stored. Roughly speaking two spectroids belong to the same frame if up to duality they differ only by 'trivial branches'. CREP also offers procedures **iscriticald**, **iscriticale**, **ishypercriticald**, **ishypercriticale** and **esrdtest** to verify if a given spectroid is in one of the classification lists. We refer to [26] for details about the use of the programs mentioned. In particular, by checking the criticality of all convex full subspectroids of a given completely separating spectroid using **iscriticald** and **iscriticale** one would get an algorithm for deciding finite representation type. We will see in the next section how this can be done in a more efficient way.

1.7 Thin Modules over completely separating Spectroids

1.7.1 Vector Space Categories As seen above, modules over a finite spectroid can be identified with representations of quivers. After choosing bases, one obtains finite sequences of matrices satisfying some relations. In this language morphisms in the category mod$-\mathcal{S}$ correspond to certain admissible transformations on the rows and columns of the matrices. Clearly there are many similar 'matrix problems' and some of them actually play an important role in representation theory. Putting them into categorical language these problems deal with 'representations of bimodules'. We refer to [15] for an account on the various concepts which are used.

A special class of bimodules which is very well studied is furnished by vector space categories. A *vector space category* is a pair $(\mathcal{A}, M)$ where $\mathcal{A}$ is an aggregate and $M : \mathcal{A} \to \text{mod}-k$ is a k-linear functor. The vector space category $(\mathcal{A}, M)$ is said to be *faithful* if the functor M acts faithful. The corresponding category of representations is called the *subspace category* of $(\mathcal{A}, M)$ and denoted by $\check{\mathcal{U}}(\mathcal{A}, M)$. It is again an aggregate whose objects are the pairs $V = (V_\omega, \gamma_V, V_0)$ such that V_ω is a finite-dimensional k-space, V_0 is an object in $\mathcal{A}$ and $\gamma_V \in \text{Hom}_k(V_\omega, M(V_0))$. A morphism $V \to W$ is a pair $f = (f_\omega, f_0)$ such that $f_\omega : V_\omega \to W_\omega$ is a k-linear map, $f_0 : V_0 \to W_0$ is a morphism in $\mathcal{A}$ and $\gamma_W f_\omega = M(f_0)\gamma_V$. For finite vector space categories $(\mathcal{A}, M)$, which means that the spectroid $\mathcal{S}(\mathcal{A})$ is finite, the representation types are defined in an analogous way as for spectroids.

Vector space categories arise in representation theory because there are various 'reduction functors' relating mod$-\mathcal{S}$ to a suitable subspace category. We only present one example of such a reduction functor which we will need later. Suppose the finite spectroid $\mathcal{S}$ has an objects x such that the algebra $\mathcal{S}(x, x)$ is isomorphic to k. This happens for all vertices of spectroids whose quiver $\text{Qv}\,\mathcal{S}$ is directed thus in particular for completely separating spectroids. Let us denote by K_x the full subaggregate of mod$-\mathcal{S}$ given by all M satisfying $\tau M(x) = 0$. Then the functor $F : \check{\mathcal{U}}(K_x, \text{Hom}_{\mathcal{S}}(x^\wedge, -)) \to \text{mod}-\mathcal{S}$ which maps an object $V = (V_\omega, \gamma_V, V_0)$ to the cokernel of the adjoint homomorphism $V_\omega \otimes_k x^\wedge \to V_0$ of the map $\gamma_V : V_\omega \to \text{Hom}_{\mathcal{S}}(x^\wedge, V_0)$ is called *fiber sum functor.*

Theorem 1.13 ([23]). *The fiber sum functor* $F : \check{\mathcal{U}}(K_x, \text{Hom}_{\mathcal{S}}(x^\wedge, -)) \to \text{mod}-\mathcal{S}$ *is full, dense and its kernel is an ideal generated by finitely many indecomposable objects.*

Note, that the classical one-point extension (resp. one-point coextension) technique (see e.g. [49]) is a special case of the reduction via fiber sum functors for the case that x is a source (resp. sink).

1.7.2 Linear Vector Space Categories It follows from the above theorem that F preserves and reflects finite and tame representation type. Nevertheless, the result is only helpful provided $\check{\mathcal{U}}(K_x, \text{Hom}_{\mathcal{S}}(x^\wedge, -))$ is easier to study than mod$-\mathcal{S}$. We will outline an interesting situation where this is actually the case.

A faithful vector space category $(\mathcal{A}, M)$ is said to be *linear* if $\dim_k M(X) = 1$ for all X in $\mathcal{S}(\mathcal{A})$. Linear vector space categories are easy to understand. They can always be identified with $(\operatorname{add} k\mathcal{P}, can)$ where $\mathcal{P}$ is a poset and *can* is the canonical functor sending each point x of $\mathcal{P}$ to k and each morphism $(y \mid x)$ to the identity of k. It is trivial to construct the poset $\mathcal{P}$ from $\mathcal{A}$. Namely, the object set of $\mathcal{P}$ is $\mathcal{S}(\mathcal{A})$ and $x \geq y$ holds whenever $\mathcal{A}(x, y) \neq 0$.

The importance of linear vector space categories comes mainly from the fact that it is much easier to determine their representation type. This is due to theorems originating from the 'Kiev school' of representation theory.

Theorem 1.14 (Kleiner[40]). *Suppose $\mathcal{P}$ is a finite poset. Then the linear vector space category* $(\operatorname{add} k\mathcal{P}, can)$ *is of finite representation type if and only if $\mathcal{P}$ does not contain a full subposet from the list $\mathcal{K}$ of* 5 *posets in Figure 1.5.*

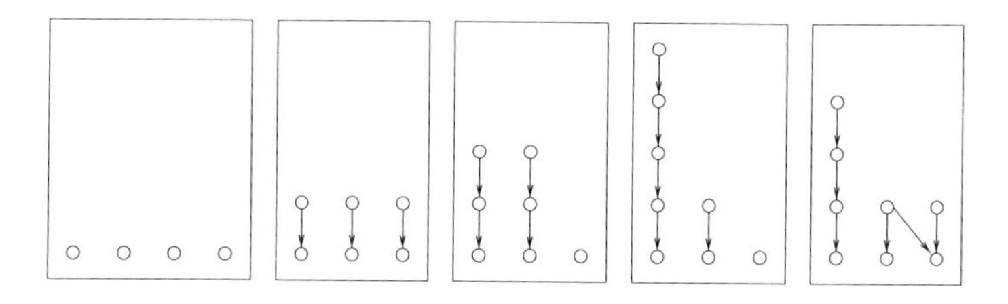

Figure 1.5: Kleiner's list $\mathcal{K}$

Theorem 1.15 (Nazarova[42]). *Suppose $\mathcal{P}$ is a finite poset. Then the linear vector space category* $(\operatorname{add} k\mathcal{P}, can)$ *is of tame representation type if and only if $\mathcal{P}$ does not contain a full subposet from the list $\mathcal{N}$ of* 6 *posets in Figure 1.6.*

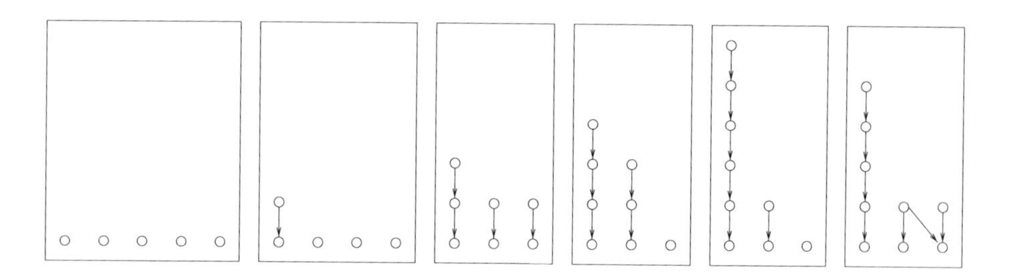

Figure 1.6: Nazarova's list $\mathcal{N}$

Theorem 1.16 (Nazarova,Zavadskij[43]). *Suppose $\mathcal{P}$ is a finite poset. Then the linear vector space category* $(\operatorname{add} k\mathcal{P}, can)$ *is tame of polynomial growth if and only if $\mathcal{P}$ does not contain as full subposet a poset from the list $\mathcal{N}$ or the poset $\mathcal{Z}$ in Figure 1.7.*

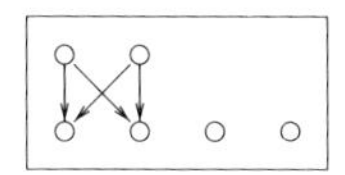

Figure 1.7: Zavadskij's poset $\mathcal{Z}$

1.7.3 Thin Modules In the representation finite situation the vector space category obtained from $(K_x, \mathrm{Hom}_{\mathcal{S}}(x^\wedge, -))$ by factoring out the kernel of $\mathrm{Hom}_{\mathcal{S}}(x^\wedge, -)$ from K_x (in order to have a faithful functor) is linear for arbitrary objects x of $\mathcal{S}$. In the tame situation this is not true any more. Moreover, the spectroid $\operatorname{ind} K_x$ may also be infinite in this case. Finally, it seems to be impossible to construct K_x from $\mathcal{S}$ without good knowledge about ind$-\mathcal{S}$. In the following we will introduce a subaggregate of K_x which can be computed directly from $\mathcal{S}$ and gives rise to a finite linear vector space category.

Call a module M in mod$-\mathcal{S}$ *thin* if it satisfies $\dim_k M(x) \leq 1$ for all x in $\mathcal{S}$. Indecomposable thin modules over a completely separating spectroid $\mathcal{S} = kS/I$ are easy to describe. We say that a convex subset T of S is *strictly convex* if $\mathcal{S}(x, y) \neq 0$ for all $x \geq y$ in T. In terms of generators of I this conditions means that $x \geq s \geq t \geq y$ does not happen for any generator $\rho = (t|s)$. With this notation the map $M \mapsto \operatorname{supp} M$ furnishes a bijection from the subset of thin modules in ind$-\mathcal{S}$ to the set of strictly convex connected subsets of S. How can one check via $\operatorname{supp} M$ if an indecomposable thin modules M belongs to K_x? We can assume $M(x) \neq 0$ since otherwise $\mathrm{Hom}_{\mathcal{S}}(x^\wedge, -)$ vanishes on M. It turns out that M with $T := \operatorname{supp} M$ lies in K_x if and only if T is *x-irreducible*. This means by definition that the sum of the number of connected components of $T \setminus \{y \in T \mid x \leq y\}$ and the number of x-extension points of T is at most 1. An *x-extension point* of T is an element y of $S \setminus T$ such that $y \geq x$ and $T \cup \{y\}$ is still strictly convex. Denoting by $\mathcal{P}_x$ the set of x-irreducible strictly convex connected subsets of S containing x, we have seen that we can embed the linear vector space category $(\operatorname{add} k\mathcal{P}_x, can)$ into $(K_x, \mathrm{Hom}_{\mathcal{S}}(x^\wedge, -))$ (at least after factoring out the kernel of the functor $\mathrm{Hom}_{\mathcal{S}}(x^\wedge, -)$ from K_x). It is still necessary to express the order on $\mathcal{P}_x$ in a easily computable way: we have to put $T_1 \geq T_2$ if and only if the connected component C of $T_1 \cap T_2$ which contains x is an ideal in T_1 (i.e. closed against smaller elements in T_1) and a coideal in T_2 (i.e. closed against bigger elements in T_2). Thus with each objects x of a completely separating spectroid $\mathcal{S} = kS/I$ we have associated a finite partially ordered set $\mathcal{P}_x$ which can be calculated from S and the generators of I in an algorithmic way.

Theorem 1.17 ([25]). *Let $\mathcal{S}$ be a finite completely separating spectroid.*

(a) $\mathcal{S}$ is representation finite if and only if for each object x of $\mathcal{S}$ the finite poset $\mathcal{P}_x$ does not contain a full subposet from the list $\mathcal{K}$.

(b) If $\mathcal{S}$ is tame, then for each object x of $\mathcal{S}$ the finite poset $\mathcal{P}_x$ does not contain a full subposet from the list $\mathcal{N}$.

The value of the result comes from the fact that the lists $\mathcal{K}$ and $\mathcal{N}$ are remarkably shorter than the list of critical and hypercritical spectroids so that one can implement efficient procedures which construct the posets $\mathcal{P}_x$ and look for full subposets from the lists $\mathcal{K}$ and $\mathcal{N}$. This is done in CREP with the procedures *mchrepfin* and *mchwild*.

Unfortunately part (b) does not give a necessary and sufficient condition. The situation improves if one considers tameness of polynomial growth.

Theorem 1.18. *A finite completely separating spectroid $\mathcal{S}$ is tame of polynomial growth if and only if for each object x of $\mathcal{S}$ the finite poset $\mathcal{P}_x$ does not contain as full subposet a poset from the list $\mathcal{N}$ or the poset $\mathcal{Z}$.*

The proof of this rests on a recent Theorem of Skowroński (see [54]) who showed that a strongly simply connected spectroid is tame of polynomial growth if it does not contain a convex full subspectroid which is hypercritical or pg-critical. We already encountered the hypercritical spectroids. The pg-critical spectroids are certain extensions of critical spectroids of type $\widetilde{\mathbb{D}}_n$ which were classified in [44]. To get the above result it was checked that each pg-critical completely separating spectroid has an object x such that $\mathcal{P}_x$ contains $\mathcal{Z}$ as a full subposet.

The algorithm suggested by the above theorem is implemented in CREP as the procedure *mchpg*. Let us finish by an exercise using this program. We wish to classify the tame completely separating connected spectroids of the shape $\mathcal{S} = kS$ containing a convex full subspectroid which is critical of type $\widetilde{\mathbb{E}}_n$ ($n = 6, 7, 8$). This classification was announced in [25]. Note, that any such spectroid has a characteristic sincere indecomposable module M where $M(x) = k$ for all objects x and all $(y \mid x)$ are mapped to id_k. Following [47] there should be only finitely many isomorphism classes of these spectroids and all the appearing ones should be tame of polynomial growth. We try to construct the corresponding posets S be induction starting with the critical ones as list L_0. For constructing L_{n+1} from L_n we write a program which calculates all possible posets which are a connected one-point extension of the ones in the list L_n obtaining a list L''. From L'' we eliminate all those which are not completely separating using the CREP-procedure *complsep* and obtain a list L'. We apply the function *mchwild* to all elements of L' and put into L_{n+1} only those where the answer is 'false'. We observe that for $n = 9$ the list L_n is empty. Hence all L_n for $n \geq 9$ are empty. Finally we apply *mchpg* to all elements of the lists $L_1, \ldots, L_8$ and always get the answer 'true'.

The classification list obtained can be seen in the latest release of CREP using the procedure *tpe_alg*. Let us present in a table the number of frames of spectroids in this list which are not themselves tame concealed.

number of objects	8	9	10	11	12	13	14	15
number of frames	28	155	482	224	47	14	5	1

Bibliography

[1] M. Auslander: *Applications of morphisms determined by objects*, Proc. Conf. Represent. Theory, Philadelphia (1976), Lect. Notes Pure App. Math. 37 (1978), 245–327.

[2] M. Auslander, I. Reiten: *Representation theory of Artin algebras III*, Commun. Algebra 3 (1975), 239–294.

[3] M. Auslander, I. Reiten, S.O. Smalø: *Representation theory of Artin algebras*, Cambridge, 1995.

[4] S. Brenner, M.C.R. Butler: *Generalizations of the Bernstein-Gelfand-Ponomarev reflection functors*, Lect. Notes Math. 832 (1980),103–169.

[5] R. Bautista, F. Larrión: *Auslander-Reiten quivers for certain algebras of finite representation type*, J. London Math. Soc. 26 (1982), 43–52.

[6] R. Bautista, P. Gabriel, A.V. Roiter, L. Salmerón: *Representation-finite algebras and multiplicative bases*, Invent. Math. 81 (1985), 217–285.

[7] K. Bongartz: *Treue einfach zusammenhängende Algebren I*, Comment Math. Helv. 57 (1982), 282–330.

[8] K. Bongartz: *Algebras and quadratic forms*, J. London Math. Soc. 28 (1983), 461–469.

[9] K. Bongartz: *Critical simply connected algebras*, Manuscr. Math. 46 (1984), 117–136.

[10] K. Bongartz: *A criterion for finite representation type*, Math Ann. 269 (1984), 1–12.

[11] K. Bongartz, P. Gabriel: *Covering spaces in representation theory*, Invent. Math. 65 (1982), 331–378.

[12] S. Brenner: *Decomposition properties of some small diagrams of modules*, Symposia Math. Inst. Naz. Alta Mat. 13 (1974), 127–141.

[13] O. Bretscher, P. Gabriel: *The standard form of a representation-finite algebra*, Bull. Soc. math. France 111 (1983), 21–40.

[14] W. Crawley-Boevey: *On tame algebras and bocses*, Proc. London Math. Soc. (3) 56 (1988), 451–483.

[15] W. Crawley-Boevey: *Matrix problems and Drozd's theorem*, in: Topics in Algebra, Banach Center Publ. 26, part I (1990), 199–222.

[16] V. Dlab, C.M. Ringel: *Indecomposable representations of graphs and algebras*, Mem. Amer. Math. Soc. 173 (1973).

[17] P. Dowbor, A. Skowroński: *On the representation type of locally bounded categories*, Tsukuba J. Math. 10 (1986), 63–72.

[18] P. Dowbor, A. Skowroński: *Galois coverings of representation infinite algebras*, Comment. Math. Helv. 62 (1987), 311–337.

[19] P. Dräxler: *Aufrichtige gerichtete Ausnahmealgebren*, Bayreuther Math. Schriften 29 (1989).

[20] P. Dräxler: *Sur les algèbres exceptionalles de Bongartz*, C.R. Acad. Sci. Paris, t. 311, Série I, (1990), 495–498.

[21] P. Dräxler: *Fasersummen über dünnen s-Startmoduln*, Arch. Math. 54 (1990), 252–257.

[22] P. Dräxler: *Completely separating algebras*, J. Algebra 165 (1994), 550–565.

[23] P. Dräxler: *On the density of fiber sum functors*, Math. Z. 216 (1994), 645–656.

[24] P. Dräxler: *Combinatorial vector space categories*, Comm. Alg. 22 (1994), 5803–5815.

[25] P. Dräxler, R. Nörenberg: *Thin start modules and tame representation type*, in: Representations of Algebras, CMS Conference Proceedings 14 (1993), 149–163.

[26] P. Dräxler, R. Nörenberg: *CREP Manual – Version 1.0 using Maple as surface*, Preprint E 96-002 of the SFB 343 Bielefeld.

[27] Ju. A. Drozd: *Tame and wild matrix problems*, Lecture Notes in Math. 832 (1980), 242–258.

[28] K. Erdmann: *Blocks of tame representation type and related topics*, Lecture Notes in Math. 1428 (1980).

[29] P. Gabriel: *Unzerlegbare Darstellungen I*, Manuscr. Math. 6 (1972), 71–103.

[30] P. Gabriel: *Finite representation type is open*, Lecture Notes Math. 488 (1975), 132–155.

[31] P. Gabriel: *Auslander-Reiten sequences and representation-finite algebras*, Lecture Notes in Math. 831 (1980), 1–71.

[32] P. Gabriel: *The universal cover of a representation-finite algebra*, Lecture Notes in Math. 903 (1981), 68–105.

[33] P. Gabriel, L.A. Nazarova, A.V. Roiter, A.V, V.V.Sergejchuk, D. Vossieck: *Tame and wild subspace problems*, Ukr. Math. J. 45 (1993), 313–352.

[34] P. Gabriel, J.A. de la Peña: *On algebras, wild and tame*, in: Duration and change, fifty years at Oberwolfach, Springer (1994), 177–210.

[35] P. Gabriel, A.V. Roiter: *Representations of finite-dimensional algebras*, Encyclopedia of the Mathematical Sciences, Vol. 73, Algebra VIII, A.I. Kostrikin and I.V. Shafarevich (Eds.), Berlin, Heidelberg, New York, 1992.

[36] Ch. Geiß: *On degenerations of tame and wild algebras*, Arch. Math. 64 (1995), 11–16.

[37] D. Happel: *Triangulated categories in the representation theory of finite dimensional algebras*, LMS lecture notes series 119 (1988).

[38] D. Happel, C.M. Ringel: *Tilted algebras*, Trans. Amer. Math. Soc. 274 (1982), 399–443.

[39] D. Happel, D. Vossieck: *Minimal algebras of infinite representation type with preprojective component*, Manuscr. Math. 42 (1983), 221–243.

[40] M. M. Kleiner: *Partially ordered sets of finite type*, J. Sov. Math. 23 (1975), 607–615.

[41] B. Mitchell: *Rings with several objects*, Advances in Math. 8 (1972), 1–161.

[42] L. A. Nazarova: *The representations of partially ordered sets of infinite type*, Izv. Akad. Nauk SSSR, Ser. Math. 39 (1975), 963–991.

[43] L.A. Nazarova, A.G. Zavadskij: *Partially ordered sets of tame type*, in: Matrix Problems, Kiev (1977), 122–143 (in Russian).

[44] R. Nörenberg, A. Skowroński: *Tame minimal non-polynomial growth strongly simply connected algebras*, Colloq. Math. 73 (1997), 301–330.

[45] J. A. de la Peña: *Algebras with hypercritical Tits form*, in: Topics in Algebra, Banach Center Publ. 26, part I (1990), 353–369.

[46] J. A. de la Peña: *The Tits form of a tame algebra*, in: Representation Theory of Algebras and related Topics, CMS Conf. Proc. 19 (1996), 159–183.

[47] J. A. de la Peña, A. Skowroński: *Forbidden subcategories of non-polynomial growth tame simply connected algebras*, Can. J. Math. 48 (1995), 1018–1043.

[48] C. M. Ringel: *Tame algebras*, in: Representation Theory I, Lecture Notes in Math. 831 (1980), 134–287.

[49] C. M. Ringel: *Tame algebras and integral quadratic forms*, Lecture Notes in Math. 1099 (1984).

[50] A. Skowroński: *Algebras of polynomial growth*, in: Topics in Algebra, Banach Center Publ. 26, Part I, (1990), 535–568.

[51] A. Skowroński: *Simply connected algebras and Hochschild cohomologies*, in: Representations of Algebras, CMS Conf. Proc. 14 (1993), 431–447.

[52] A. Skowroński: *Cycles in module categories,*, in: Finite Dimensional Algebras and Related Topics, ed.: V. Dlab, L.L Scott, Dordrecht/Boston/ London (1994), 309–345.

[53] A. Skowroński: *Module categories over tame algebras,*, in: Representation Theory of Algebras and related Topics, CMS Conf. Proc. 19 (1996), 281–313.

[54] A. Skowroński: *Simply connected algebras of polynomial growth,*, to appear in Compositio Math.

[55] L. Unger: *The concealed algebras of the minimal wild hereditary algebras*, Bayreuther Math. Schr. 31 (1990), 145–154.

Progress in Mathematics, Vol. 173, © 1999 Birkhäuser Verlag Basel/Switzerland

Chapter 2

Noncommutative Gröbner Bases, and Projective Resolutions

Edward L. Green

2.1 Overview

These notes consist of five sections. The aim of these notes is to provide a summary of the theory of noncommutative Gröbner bases and how to apply this theory in representation theory; most notably, in constructing projective resolutions.

Section 2.2 introduces both linear Gröbner bases and Gröbner bases for algebras. Section 2.3 surveys some of the basis algorithms of Gröbner basis theory. These include the Division Algorithm, and the Termination Theorem (Bergman's Diamond Lemma). Section 2.4 presents the noncommutative version of Buchberger's algorithm, universal Gröbner bases and considers the special case of finite dimensional algebras. Section 2.5 applies the theory of Gröbner bases to the study of modules. Projective presentations and resolutions are considered. A method of constructing projective resolutions for finite dimensional modules is given. Section 2.6 considers further theoretical applications. The study of Koszul algebras via Gröbner bases is presented.

2.2 Gröbner Bases

2.2.1 Linear Gröbner Bases In this section we consider only vector spaces. The ideas introduced here underlie much of what follows.

Throughout these notes, K will denote a field. Let V be a vector space. We fix a K-basis $\mathcal{B} = \{b_i\}_{i \in \mathcal{I}}$ where $\mathcal{I}$ is an index set. One of the essential features of the theory of Gröbner basis is the selection of a well-ordered basis. Recall that $>$ is a *well-order* on $\mathcal{B}$ if $>$ is a total order on $\mathcal{B}$ and every nonempty subset of $\mathcal{B}$ has a minimal element. The standard axioms of set theory imply that every set can be well-ordered.

Let $>$ be a well-order on $\mathcal{B}$. We recall basic properties of $>$.

Proposition 2.1. *[15] If $\mathcal{B}$ is a set then $>$ is a well-order on $\mathcal{B}$ if and only if for each descending chain of elements of $\mathcal{B}$, $b_1 \geq b_2 \geq b_3 \geq \cdots$, there exists some $N > 0$, such that $b_N = b_{N+1} = b_{N+2} = \cdots$.*

We will keep the following convention for the remainder of these lectures. If $\mathcal{I}$ is an index set and $v_i \in V$ for $i \in \mathcal{I}$ then $\sum_{i\in\mathcal{I}} v_i$ implies that all but a finite number of $v_i = 0$. Thus, if I write $\sum_{i\in\mathcal{I}} \alpha_i b_i$ with $\alpha_i \in K$ and $b_i \in \mathcal{B}$, then all but a finite number of $\alpha_i = 0$.

One of the main uses of a well-order is to be able to have the notion of a largest basis element in a vector. If $v = \sum_{i\in\mathcal{I}} \alpha_i b_i$ we say b_i *occurs in* v if $\alpha_i \neq 0$. This leads to the following important definition.

Definition 2.1. If $\mathcal{B} = \{b_i\}_{i\in\mathcal{I}}$ is a basis of a vector space V and $>$ is a well-order on $\mathcal{B}$, then if $v = \sum_{i\in\mathcal{I}} \alpha_i b_i$ is a nonzero element of V, we say b_i is the *tip of* v if b_i occurs in v and $b_i \geq b_j$ for all b_j occurring in v.

We denote the tip of v by $\mathrm{Tip}(v)$. If X is a subset of V we let

$$\mathrm{Tip}(X) = \{b \in \mathcal{B} \,|\, b = \mathrm{Tip}(x) \text{ for some nonzero } x \in X\}.$$

We let

$$\mathrm{NonTip}(X) = \mathcal{B} \setminus \mathrm{Tip}(X).$$

Thus, both $\mathrm{Tip}(X)$ and $\mathrm{NonTip}(X)$ are subsets of the fixed basis $\mathcal{B}$. Both sets are dependent of the choice of well order on $\mathcal{B}$.

It is not easy to see what the tip set of a subspace is from a generating set. Consider the following example where K is the field of rational numbers. Let $V = K^7$ and $\mathcal{B}$ be standard basis ordered by $e_1 > e_2 > \cdots > e_7$. Let W be the subspace spanned by $(1,2,-1,0,2,1,5), (-1,-2,0,0,1,-1,-3), (1,2,-1,0,5,1,6)$. Then the tip set of W is $\{e_1, e_3, e_5\}$. Hence $\mathrm{NonTip}(W) = \{e_2, e_4, e_6, e_7\}$.

We give a fundamental result which will be used often in what follows.

Theorem 2.1. *Let V be a vector space over the field K with basis $\mathcal{B}$. Let $>$ be a well-order on $\mathcal{B}$. Suppose that W is a subspace of V. Then*

$$V = W \oplus \mathit{Span}(\mathit{NonTip}(W)).$$

Proof. First we show that $W \cap \mathrm{Span}(\mathrm{NonTip}(W)) = (0)$.
Let $x \in \mathrm{Span}(\mathrm{NonTip}(W)) \setminus \{0\}$. If $x \in W$ then $\mathrm{Tip}(x) \in \mathrm{Tip}(W)$. But $\mathrm{Tip}(x) \in \mathrm{NonTip}(W)$ since $x \in \mathrm{Span}(\mathrm{NonTip}(W))$ and we would obtain a contradiction.

Now we show that $W + \mathrm{Span}(\mathrm{NonTip}(W)) = V$. This will use that $>$ is a well-order on $\mathcal{B}$. Let $v \in V$ be such that $\mathrm{Tip}(v)$ is minimal with respect to the property that $v \notin W + \mathrm{Span}(\mathrm{NonTip}(W))$. We wish to show this leads to a contradiction. Consider $\mathrm{Tip}(v) = b$. Let α be the coefficient of b in v. Note that

$$\mathrm{Tip}(v - \alpha \cdot b) < \mathrm{Tip}(v).$$

If $b \in \mathrm{NonTip}(W)$ then by the above remarks and the minimality condition on v, $v - \alpha \cdot b = w + z$ with $w \in W$ and $z \in \mathrm{Span}(\mathrm{NonTip}(W))$. But since $b \in \mathrm{NonTip}(W)$, we see

$$v = w + (z + \alpha \cdot b) \in W + \mathrm{Span}(\mathrm{NonTip}(W)),$$

a contradiction.

On the other hand, if $b \in \text{Tip}(W)$ let $w \in W$ such that $\text{Tip}(w) = b$. Let α be the coefficient of b in v and β be the coefficient of b in w. Then $v - (\alpha/\beta)w$ has smaller tip than v. Again, by the minimality condition on v,

$$v - (\alpha/\beta) \cdot w = w' + z,$$

where $w' \in W$ and $z \in \text{Span}(\text{NonTip}(W))$. Thus, we get a contradiction, since then

$$v = (w' + (\alpha/\beta)w) + z \in W + \text{Span}(\text{NonTip}(W)).$$

□

We now give the definition of a linear Gröbner basis. Let W be a subspace of V. We say a set of vectors $\mathcal{G} \subset W$ is a *linear Gröbner basis for W with respect to $>$* if

$$\text{Span}(\text{Tip}(\mathcal{G})) = \text{Span}(\text{Tip}(W)).$$

The reader should prove that if $\mathcal{G}$ is linear Gröbner basis of W in V then $\text{Span}(\mathcal{G}) = W$.

Considering the fundamental Theorem 2.1, we see that every nonzero vector $v \in V$ can be written UNIQUELY as $w_v + N(v)$, where $w_v \in W$ and $N(v) \in \text{Span}(\text{NonTip}(W))$.

Definition 2.2. We call $N(v)$ the *normal form* of v.

Let the coefficient of the tip of a vector v be denoted $\text{CTip}(v)$. There is a "best" linear Gröbner basis which can be defined as follows.

Definition 2.3. Let W be a subspace of V. We say a set $\mathcal{G}$ of vectors in W is a *reduced linear Gröbner basis for V (with respect to $>$)* if the following conditions hold:

1. $\mathcal{G}$ is a linear Gröbner basis of W.
2. If $g \in \mathcal{G}$ then $\text{CTip}(g) = 1$.
3. If g and g' are distinct elements of $\mathcal{G}$ then $\text{Tip}(g) \neq \text{Tip}(g')$.
4. If $g \in \mathcal{G}$ then $g - \text{Tip}(g) \in \text{Span}(\text{NonTip}(W))$.

The next result shows the existence and uniqueness of reduced linear Gröbner bases.

Proposition 2.2. *Let V be a vector space and $>$ a well-order on a basis $\mathcal{B}$ of V. Let W be a subspace of V. Then there is a unique linear reduced Gröbner basis of V.*

Proof. Let $\mathcal{T} = \text{Tip}(W)$. Then define $\mathcal{G} = \{t - N(t) \,|\, t \in \mathcal{T}\}$. By Theorem 2.1, $t - N(t) \in W$. Now, $\text{Tip}(t - N(t)) = t$ since no basis element occuring in $N(t)$ can be the tip of an element in W. Thus $\mathcal{G}$ is a linear Gröbner basis of W. The rest of the properties of a reduced linear Gröbner basis are easy to check. □

Note that by Proposition 2.3 below, a reduced linear Gröbner basis is in fact a basis. The next result gives both an algorithm to find the reduced linear Gröbner basis in the finite dimensional case and also, in conjunction with Proposition 2.2, proves the uniqueness of the reduced row echelon form of a matrix. We omit the details of row reduction techniques except to remind the reader that there are three row operations:

1. Multiply a row by a nonzero constant.
2. Interchange two rows.
3. Change a row by adding a multiple of another row to it.

Given an $m \times n$ matrix M we view the m rows as vectors in K^n. The *row space of* M is the span of the row vectors. It is easy to see that performing row operations to a matrix does not change the row space.

Since we are fixing a K-basis of our vector space V, if $\dim_K V = n$ then we will identify V with K^n by using the ordering of $\mathcal{B}$. Let $\{e_1, \ldots, e_n\}$ be the standard basis for K^n and we assume that $>$ is the order $e_1 > e_2 > \cdots > e_n$. Now if W is a subspace of V spanned by vectors $w_1, \ldots, w_m$ then we view W as the row space of the matrix that has $w_1, \ldots, w_m$ as the rows.

An $m \times n$ matrix M is reduced row-echelon form if

1. For some $1 \leq r \leq m$ the first r rows are nonzero and the last $m - r$ are zero.
2. There is an increasing sequence $1 \leq c_1 < c_2 < \cdots < c_r \leq n$ such the first nonzero column in the ith row is c_i with entry 1.
3. If $i \neq j$ then entry in the jth column of the ith is 0.

The proof of the following proposition is left as an exercise.

Proposition 2.3. *Let $e_1, \ldots, e_n$ be the standard basis of K^n with well-order $e_1 > e_2 > \cdots > e_n$. If M is an $m \times n$ matrix is reduced row echelon form with r nonzero rows, then the reduced linear Gröbner basis for the row space M is given by the first r row vectors.*

Note that as a consequence of the above discussion, using Theorem 2.1 and the uniqueness of the reduced Gröbner basis, we have an easy proof the uniqueness of the reduced row echelon form of a matrix.

2.2.2 Rings and Admissible Orders We now turn our attention to K-algebras. Let R be a K-algebra and let $\mathcal{B}$ be a K-basis. We assume that $\mathcal{B}$ is a semigroup with 0. That is, assume that under the multiplication of the ring, we have

$$b, b' \in \mathcal{B} \text{ implies } b \cdot b' \in \mathcal{B} \text{ or } b \cdot b' = 0.$$

We call such a K-basis a *multiplicative basis* of R.

Examples:

a): Let $R = K[x_1, \ldots, x_n]$, the commutative polynomial ring in n variables. Let $\mathcal{B} = \{\text{monomials}\}$.

b): Let $R = K\langle x_1, \ldots, x_n \rangle$ be the free associative algebra in n noncommuting variables. Let $\mathcal{B} = \{\text{monomials}\}$.

c): **Path algebras**: Let $\Gamma = (\Gamma_0, \Gamma_1)$ be a finite directed graph. Here $\Gamma_0 = \{v_1, \ldots, v_n\}$ will be the vertex set (which we give some arbitrary order) and $\Gamma_1 = \{a_1, \ldots, a_m\}$ is the arrow set (which we also give some arbitrary order). Technically, we need two functions from $\Gamma_1 \to \Gamma_0$ corresponding to the origin vertex of the arrow and the terminus vertex of the arrow. We will denote these by $o(a)$ and $t(a)$ respectively. Furthermore, for vertices we set $o(v_i) = v_i = t(v_i)$.

Let $\mathcal{B}$ denote the set of finite directed paths in Γ, including the vertices, which are viewed as paths of length 0. Each path p has a length, $l(p)$, which is the number of arrows in p. We will write paths as follows:

$$p = a_{i_1} a_{i_2} \cdots a_{i_r}$$

where $a_{i_j} \in \Gamma_1$, $t(a_{i_j}) = o(a_{i_{j+1}})$ for $1 \leq j \leq r-1$. We let $o(p) = o(a_{i_1})$ and $t(p) = t(a_{i_r})$ and say that p is a (directed) path from $o(a_{i_1})$ to $t(a_{i_r})$. Of course in this case $l(p) = r$.

We give $\mathcal{B}$ structure of a semigroup with 0 via concatenation. That is, if $p = a_1 \ldots a_r \in \mathcal{B}$ and $q = b_1 \ldots b_s \in \mathcal{B}$ where $a_i, b_j \in \Gamma_1$ then

$$p \cdot q = \begin{cases} a_1 \ldots a_r b_1 \ldots b_s & \text{if } t(a_r) = o(b_1) \\ 0 & \text{otherwise} \end{cases}.$$

Note that under the above definition, $v_i \cdot v_j$ is 0 if $i \neq j$ and is $v_i v_i = v_i$ if $i = j$.

The fact that $\mathcal{B}$ is a semigroup with a relatively easy structure (for computational purposes) is one of the essential features of Gröbner bases. As we will see, reducing computations to $\mathcal{B}$ is one of the underlying reasons that the techniques are so powerful.

We are now in a position to introduce the *path algebra* which we will denote by $K\Gamma$. As a K-vector space, $K\Gamma$ has K-basis $\mathcal{B}$. Thus the elements of $K\Gamma$ are K-linear combinations of paths and look like

$$x = \sum_{i=1}^{r} \alpha_i p_i$$

where $\alpha_i \in K$ and p_i are paths in $\mathcal{B}$. We call the α_i the *coefficient of* p_i and say that p_i *occurs* in x if $\alpha_i \neq 0$. Having described the additive structure of $K\Gamma$ we need to give the multiplicative structure. For this, we define the multiplication on $\mathcal{B}$ and then extend linearly to all of $K\Gamma$. If $p, q \in \mathcal{B}$ we define via the semigroup multiplication in $\mathcal{B}$. Thus, $p \cdot q$ is 0 if the terminus of p is different than the origin of q, otherwise $p \cdot q$ is the concatenated path pq. We let K act centrally in $K\Gamma$ by $\alpha \cdot \sum \alpha_i p_i = \sum (\alpha\alpha_i) p_i$. The action of K commutes with all elements of $K\Gamma$. In particular, $K\Gamma$ is the semigroup algebra for the the semigroup $\mathcal{B}$.

Theorem 2.2. *Basic Properties Let K be a field and Γ a finite directed graph. Then*

1. *$K\Gamma$ is a K-algebra with 1.*
2. *$1 = \sum_{v \in \Gamma_0} v$.*
3. *$\{v \,:\, v \in \Gamma_0\}$ is a full set of primitive orthogonal idempotents for $K\Gamma$. (A* primitive idempotent *is an idempotent that cannot be written as a sum of two orthogonal nonzero idempotents and we say two idempotents are orthogonal if their product, in either order, is 0)*
4. *$K\Gamma$ is a positively $\mathbf{Z}$-graded ring with the elements of $\mathcal{B}$ homogeneous and if $p \in \mathcal{B}$ has length l then $p \in (K\Gamma)_l$.*
5. *$K\Gamma$ is a tensor algebra.*

Now assume that R is a K-algebra with multiplicative basis $\mathcal{B}$. We do not want an arbitrary well-order on $\mathcal{B}$. We want one that works well with the multiplicative structure of $\mathcal{B}$. For this, we introduce the following definition.

We say a well-order $>$ on $\mathcal{B}$ is *admissible* if it satisfies the following two conditions where $p, q, r, s \in \mathcal{B}$:

1. if $p < q$ then $pr < qr$ if both $pr \neq 0$ and $qr \neq 0$.
2. if $p < q$ then $sp < sq$ if both $sp \neq 0$ and $sq \neq 0$.
3. if $p = qr$ then $p > q$ and $p > r$.

Note that in case R is the commutative or noncommutative polynomial ring, the requirements that $pr \neq 0$, $qr \neq 0$, $sp \neq 0$, and $sq \neq 0$ all can be dropped. If R is a path algebra, then we can obtain 0 when multiplying basis elements.

The above properties restrict the orderings under consideration to those that "work well" with the multiplicative structure of $\mathcal{B}$. In the commutative theory of Gröbner bases, admissible orders are usually called "term orders" or "monomial orders".

We now give some examples which show that there are "natural" admissible orders on $\mathcal{B}$ in the case of path algebras. For the following examples, assume we are given a finite directed graph Γ.

Example 2.1. The (left) length-lexicographic order:

Order the vertices and arrows arbitrarily and set the vertices smaller than the arrows. Thus

$$v_1 < \cdots < v_n < a_1 < \cdots < a_m.$$

If p and q are paths of length at least 1, set $p < q$ if $l(p) < l(q)$ or if $p = b_1 \dots b_r$ and $q = b'_1 \dots b'_r$ with $b_i, b'_j \in \mathcal{B}$ and for some $1 \leq i \leq r$, $b_j = b'_j$ for $j < i$ and $b_i < b'_i$.

The right length-lexicographic order is defined similarly.

Example 2.2. The (left) weight-lexicographic order:

Let $W : \Gamma_1 \to \{1, 2, 3, \dots\}$ be a set map. Define $W : \mathcal{B} \to \{1, 2, 3, \dots\}$ to be the natural extension. That is, $W(v) = 0$ if v is a vertex and if $a_i \in \Gamma_1$ define $W(a_1 \cdots a_r) = \sum_{i=1}^{r} W(a_i)$. Next, order the vertices and set the vertices smaller than the arrows. Order the arrows in such a fashion that if $W(a) < W(b)$ then $a < b$.

Finally, define $p < q$ if $W(p) < W(q)$ or if $W(p) = W(q)$ then use the left lexicographic order. Notice that the length-lexicographic order is a special case of the weight-lexicographic order. Mainly, give every arrow weight 1.

This order is sometimes called the *degree-lexicographic* order also.

It should be pointed out that the (left) lex order is NOT admissible in general. For example, if Γ has one vertex and two loops, a and b with $b > a$ then we get

$$b > ab > aab > aaab > \cdots .$$

Hence it is not a well-order. This is different from the commutative case.

There are some other, less obvious admissible orders.

Example 2.3. The (left) weight-reverse-lex order:

Take $W : \Gamma_1 \to \{1, 2, 3, \dots\}$ and $>$ define on the arrows and vertices as in the weight-lex order. Note that arrows and paths of positive length have positive weights.

Define $p < q$ if $W(p) < W(q)$ or if $W(p) = W(q)$ then $p > q$ in the right lex order. This is a well-order since there are only a finite number of paths of any weight.

One final example:

Example 2.4. The Total Lexicographic Order:

Label the arrows arbitrarily, say $a_1, \dots, a_m$. Arbitrarily order the vertices and let them be smaller than any path of positive length. If p and q are paths, then $p < q$ if there is some i, $1 \leq i \leq m$, such that the number of a_j's occuring in p and q are the same for $j < i$ and the number of a_i's occurring in p is less than the number of a_i's occurring in q. If p and q have the same number of each arrow then $p < q$ in lexicographic ordering (for some choice of ordering on the arrows).

2.2.3 Gröbner Bases Let R be a K-algebra with multiplicative basis $\mathcal{B}$ and admissible order $>$. Let I be an ideal in R.

Definition 2.4. We say that a set $\mathcal{G} \subset I$ is a *Gröbner basis for I with respect to* $>$ if

$$<\text{Tip}(\mathcal{G})>=<\text{Tip}(I)> .$$

That is, the two-sided ideal generated by the tips of $\mathcal{G}$ equals the two-sided ideal generated by the tips of I. Equivalently, $\mathcal{G} \subset I$ is a Gröbner basis for I if for every $b \in \text{Tip}(I)$ there is some $g \in \mathcal{G}$ such that $\text{Tip}(g)$ *divides* b; i.e., there are basis elements $p, q \in \mathcal{B}$ such that $p\text{Tip}(g)q = b$.

We end this section by recalling Theorem 2.1 in the setting of rings. Suppose that R is K-algebra with multiplicative basis $\mathcal{B}$ and admissible order $>$. Then if I is an ideal then

$$R = I \oplus \text{Span}(\text{NonTip}(I))$$

as vector spaces. In particular, every nonzero element r of R can be written uniquely as $r = i_r + N(r)$ where $i_r \in R$ and $N(r) \in \text{Span}(\text{NonTip}(I))$. $N(r)$ is still called the *normal form of* r. In the next section, I will address the existence of a reduced Gröbner basis and how to construct Gröbner bases.

2.3 Algorithms

2.3.1 Monomial Ideals and Reduced Gröbner Bases We begin with a proposition which shows that not all rings with a multiplicative basis admit an admissible ordering on the basis. Recall, if $\mathcal{B}$ is a subset of R we say b_1 *divides* b_2 (in $\mathcal{B}$) if $b_1, b_2 \in \mathcal{B}$ and there are elements $c, d \in \mathcal{B}$ such that $b_2 = cb_1d$.

Proposition 2.4. *If a multiplicative basis $\mathcal{B}$ of a ring R admits an admissible order then every infinite sequence of elements of $\mathcal{B}$, $b_1, b_2, b_3, \ldots$, such that b_i divides b_{i-1} for $i \geq 1$ stabilizes; that is, for some N, $b_N = b_{N+1} = b_{N+2} = \cdots$.*

Proof. The proof follows from the properties of an admissible order. Namely, suppose $>$ is an admissible order for $\mathcal{B}$. Then, if b_i divides b_{i-1}, we get $b_1 \geq b_2 \geq b_3 \geq \cdots$. Since $>$ is a well-ordering of $\mathcal{B}$, we get the desired result. □

Let R be a ring with multiplicative basis $\mathcal{B}$. We will call the elements of $\mathcal{B}$ *monomials*. We say an ideal I in R is a *monomial ideal* if it can be generated by elements of $\mathcal{B}$. The next proposition is important in the definition of a reduced Gröbner basis.

Proposition 2.5. *Let R be K-algebra with multiplicative basis $\mathcal{B}$ which admits an admissible order. If I is a monomial ideal then I has a unique minimal monomial generating set. That is, there is a unique set of generators of I, none can be omitted and still generate I.*

Proof. Consider the set A of all monomials in I. Let $M = \{p \in A |$ if $q \in A$ divides p then $q = p\}$. Note that by Proposition 2.4, M is not empty. We claim that M is the unique minimal generating set of I. To show that M generates I, let B be a set of monomials that generates I. If $b \in B$, then for some $m \in M$, m divides b. Hence, $B \subset <M>$. Thus $I \subset <M>$.

If M' is another set of monomial set of generators of I, then every $m \in M$ is divisible by some $m' \in M'$. But then, by definition of M, $m = m'$ and we have shown that $M \subset M'$. □

Note that the minimal set of generators given in the above proposition is independent of any particular admissible order. The existence of an admissible order is necessary to show that M is nonempty. Furthermore, it is possible that the unique minimal monomial generating set is not finite. This differs from the commutative case in that Dickson's Lemma [15] proves that every monomial ideal in commutative polynomial ring is can be generated by a finite number of monomials. Good references for the commutative theory are [15, 9, 1].

Let I be an ideal in R and $>$ an admissible order on a multiplicative basis $\mathcal{B}$. Then if $\mathcal{G}$ is a Gröbner basis of I, then $\mathrm{Tip}(\mathcal{G})$ must contain the minimal monomial generating set of $<\mathrm{Tip}(I)>$.

If I is an ideal in R, we let I_{MON} be the ideal generated by $\mathrm{Tip}(I)$ (given some admissible order $>$). Then there is a unique minimal monomial generating set $\mathcal{T}$ of I_{MON}.

Definition 2.5. The *reduced Gröbner basis for I with respect to* $>$ is

$$\mathcal{G} = \{t - N(t) \,|\, t \in \mathcal{T}\}.$$

The proof of the next result is left as an exercise.

Proposition 2.6. *Let $>$ be an admissible order on a multiplicative basis $\mathcal{B}$ of a K-algebra R. Let I be an ideal in R. Let $\mathcal{G}$ be the reduced Gröbner basis for I. Then the following holds.*

1. *$\mathcal{G}$ is a Gröbner basis for I.*
2. *If $g \in \mathcal{G}$ then $CTip(g) = 1$.*
3. *If $g \in \mathcal{G}$ then $g - Tip(g) \in Span(NonTip(I))$.*
4. *$Tip(\mathcal{G})$ is the minimal monomial generating set for I_{MON}.*

2.3.2 The Division Algorithm In this section, we present a "division algorithm" in the rings we are studying. Throughout this section, we fix a K-algebra R with multiplicative basis $\mathcal{B}$ and admissible order $>$.

Given an ORDERED set of elements $X = \{x_1, x_2, \ldots, x_n\}$ of R and another element $y \in R$ we show how to "divide" y by the set. We emphasize that the order of the elements affects the outcome of the division algorithm. What should division mean in this context? We mean that we find nonnegative integers $m_1, \ldots, m_n$ and elements $u_{i,j}, v_{i,j}, r \in R$ for $1 \leq i \leq n$ and $1 \leq j \leq m_i$ such that

1. $y = \sum_{i=1}^{n}(\sum_{j=1}^{m_i} u_{i,j}x_iv_{i,j}) + r$.
2. $\mathrm{Tip}(y) \geq \mathrm{Tip}(u_{i,j}x_iv_{i,j})$ for all i and j.
3. For $b \in \mathcal{B}$ occurring in r, $\mathrm{Tip}(x_i)$ does not divide b for $1 \leq i \leq n$.

Note that it follows that $\text{Tip}(r) \leq \text{Tip}(y)$. We will call r a *remainder by division by* $x_1, \dots, x_n$.

We now present the algorithm in pseudocode. Our presentation here takes the commutative division algorithm in [15] and changes it to make it noncommutative.

The Division Algorithm

INPUT: $x_1, \dots, x_n$ (ordered), y
OUTPUT: $m_1, \dots, m_n, u_{i,j}, v_{i,j}, r$
INITIALIZE: $m_1 := 0, \dots, m_n := 0, r := 0, z := y$, DIVOCCUR $:=$ *False*
WHILE ($z \neq 0$ and DIVOCCUR $==$ *False*) **DO**
  FOR $(i = 1)$ **TO** n **DO**
    IF $\text{Tip}(z) = u\text{Tip}(x_i)v$ for $u, v \in \mathcal{B}$ **THEN**
      $m_i := m_i + 1$
      $u_{i,m_i} := [\text{CTip}(z)/\text{CTip}(x_i)]u$
      $v_{i,m_i} := v$
      $z := z - [\text{CTip}(z)/\text{CTip}(x_i)]ux_iv$
      DIVOCCUR $:=$ *True*
    ELSE $i := i + 1$
  IF DIVOCCUR $==$ *False* **THEN**
    $r := r + \text{CTip}(z)\text{Tip}(z)$
    $z := z - \text{CTip}(z)\text{Tip}(z)$
    DONE
  DONE
DONE

We leave it to the reader to analyze the algorithm and show that it does what it is supposed to do. We give one small example.

Example 2.5. Let R be the noncommutative polynomial ring in three noncommuting variables x, y, z over a field K. Let $\mathcal{B}$ be the set of monomials and $>$ the length-lexicographic order with $x > y > z$. Let's divide $xy - x = f_1, xx - xz = f_2$ into $zxxyx$. Note that the tip of f_1 is xy and tip of f_2 is xx.

Beginning the algorithm, we see that $zxxyx = (zx)\text{Tip}(f_1)x$.

Thus $u_{1,1} = zx, v_{1,1} = x$ and we replace $zxxyx$ by $zxxyx - zx(f_1)x = zxxx$. Now $\text{Tip}(f_1)$ does not divide $zxxx$. Continuing, $\text{Tip}(f_2)$ does. There are two ways to divide $zxxx$ by xx and for the algorithm to be precise we must choose one. Say we choose the "left most" division. Then $zxxx = z(\text{Tip}(f_2))x$ and we let $u_{2,1} = z$, $v_{2,1} = x$ and replace $zxxx$ by $zxzx$. Neither $\text{Tip}(f_1)$ nor $\text{Tip}(f_2)$ divide $zxzx$ so we let $r = zxzx$ and $zxzx$ is replaced by 0 and the algorithm stops. We have

$$zxxyx = (zx)f_1x + zf_2x + zxzx.$$

The remainder is $zxzx$.

Note that if we interchange the order of f_1 and f_2 (so we start with f_2 first) we see the outcome of the division algorithm is

$$zxxyx = z(f_2)yx + zxzyx,$$

which gives a different remainder, namely, $zxzyx$.

Definition 2.6. If $X = \{x_1, \ldots, x_n\}$ (as an ordered set) and y is divided by X, we will denote the remainder r by $y \Rightarrow_X r$.

If one had an infinite set $X = \{x_1, x_2, \ldots\}$ with $\text{Tip}(x_1) \leq \text{Tip}(x_2) \leq \text{Tip}(x_3) \leq \cdots$ and only a finite number of x_i's with a given tip, we could perform division by X. That is, only a finite number of elements of X are needed for division and these can be determined prior to division. For, given y, there can only be a finite number of x_i's such that $\text{Tip}(x_i) \leq \text{Tip}(y)$, say $x_1, \ldots, x_n$. By the desired properties of division, $x_{n+1}, x_{n+2}, \ldots$ would never occur since we want $\text{Tip}(y) \geq \text{Tip}(u_{i,j}x_iv_{i,j})$. Thus, we may divide by an infinite set of elements of R provided that that the set has the desired properties described earlier in the paragraph.

Now we show that when we have a Gröbner basis, the order of the x_i's does not affect the remainder!

Proposition 2.7. *If $\mathcal{G}$ be a Gröbner basis for an ideal I in R. Let $y \in R$ and assume that $X = \{g_1, \ldots, g_n\} = \{g \in \mathcal{G} \mid Tip(g) \leq Tip(y)\}$. If $y \Rightarrow_X r$ then r is independent of the order of $g_1, \ldots, g_n$ in X. In fact, $r = N(y)$.*

Proof. Consider $y \Rightarrow_X r$. Then, since $\text{Tip}(r) \leq \text{Tip}(y)$, we see that for each $g \in \mathcal{G}$, $\text{Tip}(g)$ does not divide any basis element occurring in r. Hence $r \in \text{Span}(\text{NonTip}(I))$. Now $y = \sum_i \sum_j u_{i,j}g_iv_{i,j} + r$. But Theorem 2.1 implies that $y = i_y + N(y)$ with $i_y \in I$ and $N(y) \in \text{Span}(\text{NonTip}(I))$ unique. But $\sum_i \sum_j u_{i,j}g_iv_{i,j} \in I$ and $r \in \text{Span}(\text{NonTip}(I))$. Hence $r = N(y)$. $\square$

Corollary 2.1. *If $\mathcal{G}$ is a Gröbner basis of an ideal I in R such that for each $b \in \mathcal{B}$ there are only a finite number of elements g in $\mathcal{G}$ with $Tip(g) \leq b$ then there is an algorithm to find the normal form of elements of R.*

In practice, Gröbner bases are found with only a finite number of terms with a given tip and hence the division algorithm gives an algorithm to find normal forms. Moreover, this is usually a "fast algorithm". The difficulty is in constructing Gröbner bases!

Note that we also now have a method to find the reduced Gröbner basis once we have found a Gröbner basis with only a finite number of terms with a given tip. This is an algorithm if I_{MON} has a finite set of monomial generators. The method proceeds as follows:

1. Given a Gröbner basis $\mathcal{G}$ such that only a finite number terms of $\mathcal{G}$ have a given tip.

2. Output will be the reduced Gröbner basis.

3. Find the minimal monomial generating set $\mathcal{T}$ of I_{MON}.
4. For each $t \in \mathcal{T}$, using the division algorithm, calculate $t \Rightarrow_{\mathcal{G}} N(t)$.
5. The reduced Gröbner basis is $\{t - N(t) \mid t \in \mathcal{T}\}$.

2.3.3 Overlap Relations and the Termination Theorem In this section, we give the noncommutative version of the S-polynomial found in commutative theory [15, 9]. We call these overlap relations. Throughout this section, R will be a K-algebra with multiplicative basis $\mathcal{B}$ and admissible order $>$ for $\mathcal{B}$.

Definition 2.7. Let $f, g \in R$ and suppose that there are elements $b, c \in \mathcal{B}$ such that

1. $\text{Tip}(f)c = b\text{Tip}(g)$.
2. $\text{Tip}(f)$ does not divide b and $\text{Tip}(g)$ does not divide c.

Then the *overlap relation of f and g by b, c* is

$$o(f, g, b, c) = (1/\text{CTip}(f))fc - (1/\text{CTip}(g))bg.$$

Note that $\text{Tip}(o(f, g, b, c)) < \text{Tip}(f)c = b\text{Tip}(g)$. We give some examples to help clarify overlap relations.

Example 2.6. Let $R = K\Gamma$ where Γ is the graph:

$$\begin{array}{ccc} v_1\bullet & \overset{a}{\rightarrow} & \bullet v_2 \\ d\uparrow & & \downarrow b \\ v_4\bullet & \overset{c}{\leftarrow} & \bullet v_3 \\ g\uparrow & & \downarrow e \\ v_6\bullet & \overset{f}{\leftarrow} & \bullet v_5 \end{array}$$

Use the length-lexicographic order with $v_1 < \cdots < v_6 < a < b < c < \cdots < g$. Consider $p = abcdabcd - abefgd$ and $q = cdabcda - efgda$. There are a number of overlap relations between p and q. We list them below.

1. $o(p, q, a, ab) = p \cdot a - ab \cdot q = (-abefgd)a + abefgda = 0$.
2. $\begin{aligned} o(p, q, abcda, abcdab) &= p \cdot abcda - abcdab \cdot q \\ &= -abefgdabcda + abcdabefgda \neq 0. \end{aligned}$
3. $o(q, p, bcd, cd) = -efgdabcd + cdabefgd \neq 0$.
4. $o(q, p, bcdabcd, cdabcd) = -efgdabcdabcd + cdabcdabefgd \neq 0$.

As the above example shows, we need to look at all overlaps, including self overlaps; that is, overlaps of the form $o(f, f, p, q)$. In this example, we have self overlaps $o(p, p, abcd, abcd)$ and $o(q, q, cdab, bcad)$. In the commutative case, one only uses the least common multiple of the leading monomials. In the noncommutative case, we must look at all overlaps.

Next we consider "tip reduction". Suppose that $X = \{x_1, \ldots, x_n\}$ is a set of elements in R. Let I be the ideal generated by X. If $\text{Tip}(x_i) = u\text{Tip}(x_j)v$ for some $u, v \in \mathcal{B}$, then letting

$$X' = \{x_1, \ldots, x_{i-1}, x_i - [\text{CTip}(x_i)/\text{CTip}(x_j)]ux_jv, x_{i+1}, \ldots, x_n\}$$

we that X' is a generating set of I also. Continuing in this fashion, we obtain a finite set of generators of I such that no tip of a generator divides the tip of another generator. This process is finite by the well-ordering assumption on $>$. Thus it is not unreasonable to assume that we have no tip divisions on a set of generators of an ideal.

Definition 2.8. We say a set of elements X is *tip reduced* if for distinct elements $x, y \in X$, $\text{Tip}(x)$ does not divide $\text{Tip}(y)$.

Before giving the termination Theorem, we need one more concept: uniformity.

Definition 2.9. We say an element $\sum_{i=1}^{n} \alpha_i b_i$, with $\alpha_i \in K^*$ and $b_i \in \mathcal{B}$, is *(left) uniform* if for each $c \in \mathcal{B}$, either $cb_i = 0$ for all i, $1 \leq i \leq n$ or $cb_i \neq 0$ for all i, $1 \leq i \leq n$.

Note that in the case of a noncommutative polynomial ring with $\mathcal{B} = \{\text{monomials}\}$, all elements are uniform. In the case of a path algebra, with $\mathcal{B} = \{\text{finite directed paths}\}$, then an element $\sum_{i=1}^{n} \alpha_i p_i$ is uniform if and only if there are vertices v, w such that for $1 \leq i \leq n$, the origin vertex of p_i is v and the terminus vertex is w. Note that every element $x \in K\Gamma$ is a sum of uniform elements. Since $1 = v_1 + \cdots + v_n$, we have $x = \sum_{i,j=1}^{n} v_i x v_j$ and we see that $v_i x v_j$ is uniform since all the paths that occur must have origin vertex v_i and terminus vertex v_j. It follows that every ideal in a path algebra can be generated by uniform elements.

We also note that if $f \in R$ is a uniform element and $b \in \mathcal{B}$ then both bf and fb are either 0 or uniform elements. We also remark that elements of $\mathcal{B}$ are uniform.

We now give the termination Theorem which is a version of G. Bergman's Diamond Lemma [11, 16].

Theorem 2.3. *Let R be a K-algebra with multiplicative basis $\mathcal{B}$ and admissible order $<$. Suppose the $\mathcal{G}$ is a set of uniform, tip reduced elements of R. Suppose for every overlap relation*

$$o(g_1, g_2, p, q) \Rightarrow_{\mathcal{G}} 0,$$

with $g_1, g_2 \in \mathcal{G}$. Then $\mathcal{G}$ is a Gröbner basis for $<\mathcal{G}>$.

Proof. Assume that $\mathcal{G}$ has the property that all overlap relations have remainder 0 under division by $\mathcal{G}$. Let $x \in I$ and we assume that $tip(x)$ is not divisible by the tip of any element of $\mathcal{G}$. We show this leads to a contradiction. Without loss of generality, we amy assume that x is uniform. In this way, all multiplications are in effect nonzero. Since we are assuming that $\mathcal{G}$ is a generating set, we may write

$$x = \sum_{i,j} \alpha_{i,j} p_{i,j} g_i q_{i,j} \qquad (*)$$

where g_i varies over $\mathcal{G}$ and $p_{i,j}, q_{i,j} \in B$. Consider all such ways of writing x. Let p^* be the largest path occuring on the rhs of $(*)$. Since we are assuming that $tip(x)$ is not divisible by the tip of any element of $\mathcal{G}$, by uniformity it follows that p^* is larger than $tip(x)$ in the $<$ order. Thus, the p^*'s terms cancel each other out.

Considering all ways of writing x as in $(*)$, choose one such that p^* is as small as possible and has the fewest occurrences in the rhs of $(*)$.

Since p^* does not occur on the lhs, it must appear in two summand of the rhs. So there exist i, j, i', j' so that

$$p^* = p_{i,j} tip(g_i) q_{i,j} = p_{i',j'} tip(g_i') q_{i',j'}.$$

To simplify notation, write $p = p_{i,j}, g = g_i, q = q_{i,j} p' = p_{i',j'}, g' = g_i'$ and $q' = q_{i',j'}$. We proceed by a case by case study of the possible scenarios.

Case 1: length $p <$ length p'.

In this case either, length $q \geq$ length q' or not.

Case 1.1: Length $q <$ length q'.

Then $tip(g')$ contains $tip(g)$ and hence $tip(g)$ divides $tip(g')$ contradicting the hypothesis.

Case 1.2: length $q \geq$ length q'. We consider two possibilities.

Case 1.2.1: length $p' \geq$ length $ptip(g)$.

Then there is no overlap of $tip(g)$ and $tip(g')$ in p^*. By the choice of lengths, it follows that there is a path q'' such that $p^* = ptip(g)q''g'q'$. Now if $g = \sum \alpha_i p_i + \alpha tip(g)$, and $g' = \sum \beta_i p_i' + \beta tip(p')$, then

$$\begin{aligned} pgq &= pgq''(1/\beta)g'q' - pgq''(1/\beta)(g' - tip(g'))q' \\ &= (\alpha/\beta)ptip(g)q''g'q' + \sum(\alpha_i/\beta)pp_iq''g'q' - \sum(\beta_i/\beta)pgq''(p_i'). \end{aligned}$$

Thus, in writing pgq this way, can combine its tip with the tip of $p'g'q'$ and lower the number of occurrances of p^* which contradicts the minimality assumption.

Case 1.2.2: length $p' <$ length $ptip(g)$.

Then there is an overlap of $tip(g)$ and $tip(g')$ is p^*. Say $tip(g)r = s \cdot tip(g')$. Thus $p^* = ptip(g)rq' = pstip(g')q'$. Then

$$pgq = c_g(tip(g))po(g, g', r, s)q + (c_g(tip(g))/c_g'(tip(g')))p'g'q'.$$

By assumption $o(g, g', r, s)$ totally reduces over $\mathcal{G}$ so is a K-linear combination of terms of the form $\hat{p}\hat{g}\hat{q}$ for paths $\hat{p}, \hat{q} \in B$ and $\hat{g} \in \mathcal{G}$, all of whose tips are smaller that $tip(g)r = s \cdot tip(g')$. So we may combine the occurrance of p^* in pgq with its occurrance in $p'gq'$. This again contradicts the minimality assumption.

Case 2: length $p =$ length p'.

Then $tip(g)$ divides $tip(g')$ or vice versa - which contradicts the assumption on $\mathcal{G}$.

Case 3: length $p >$ length p'.

Same as Case 1.

□

Note also that if division by $\mathcal{G}$ of some overlap relation is NOT 0, then $\mathcal{G}$ is not a Gröbner basis. This is easy to see since the remainder will have tip which is not divisible by the tip of any element of $\mathcal{G}$. Thus, when the hypothesis of Theorem 2.3 are met, we have an algorithm to determine if $\mathcal{G}$ is a Gröbner basis (assuming $\mathcal{G}$ is finite).

Example 2.7. Let $R = K\Gamma$ where Γ is the graph:

$$\begin{array}{ccccc} & a & & b & \\ v_1 & \rightarrow & v_2 & \rightarrow & v_3 \\ \bullet & & \bullet & & \bullet \\ & \leftarrow & & \leftarrow & \\ & d & & c & \end{array}$$

Let $>_1$ be the length-lexicographic order with $v_1 < v_2 < v_3 < d < c < b < a$. Consider $\mathcal{G} = \{cdab - cb, bc - da\}$. Both elements are uniform and tip reduced. Then

$$o(bc - da, cdab - cb, dab, b) = -dadab + bcb \Rightarrow_{\mathcal{G}} -dadab + dab \neq 0.$$

Hence $\mathcal{G}$ is not a Gröbner basis with respect to $>_1$.

But if we change $>_1$ to $>_2$ which is the length-lexicographic order with $v_1 < v_2 < v_3 < a < b < c < d$ then $\text{Tip}(cdab - cb) = cdab$ and $\text{Tip}(bc - da) = da$. Thus the set is not tip reduced. If we tip reduce $\mathcal{G}$ we get $\mathcal{G}' = \{cbcb - cb, -da + bc\}$. Then $\text{Tip}(cbcb - cb) = cbcb$ and $\text{Tip}(-da + bc) = da$. The only overlap relation is

$$o(cbcb - cb, cbcb - cb, cb, cb) = 0.$$

Hence all overlap relations have remainder 0 and we conclude that under $>_2$, $\mathcal{G}'$ is a Gröbner basis.

2.4 Computational Aspects

2.4.1 Construction of Gröbner bases In this section we present the noncommutative analog of Buchberger's algorithm [13] for constructing Gröbner bases. Given uniform elements $f_1, \ldots, f_n$ in R, let $I = < f_1, \ldots, f_n >$. Hence, by construction, I is generated by a set of uniform elements. The algorithm produces a (possibly infinite) sequence of uniform elements $g_1, g_2, \ldots$ where $g_i = f_i$ for $1 \leq i \leq n$ and, for $i > n$, $g_i \in I$ such that

$$\mathrm{Tip}(g_i) \notin < \mathrm{Tip}(g_1), \mathrm{Tip}(g_2), \ldots, \mathrm{Tip}(g_{i-1}) > .$$

It can be shown that $\{g_1, g_2, \ldots, g_m, g_{m+1}, \ldots\}$ is in fact a Gröbner basis for I. We present the algorithm in pseudocode.

INPUT: $f_1, \ldots, f_n$
OUPUT: $g_1, g_2, g_3, \cdots$

FOR $i = 1$ TO n DO
  $g_i := f_i$
  $\mathcal{G} := \{g_1, \ldots, g_n\}$
  Count $:= n$
DO
  $\mathcal{H} := \mathcal{G}$
  FOR each pair of elements $h, k \in \mathcal{H}$ **AND** each overlap relation of h, k
    DO
    IF $o(h, k, p, q) \Rightarrow_{\mathcal{H}} r$ AND $r \neq 0$ **DO**
      Count := Count + 1
      $g_{\text{Count}} = r$
      $\mathcal{G} := \mathcal{G} \cup \{g_{\text{Count}}\}$
    DONE
  DONE
WHILE $(\mathcal{H} \neq \mathcal{G})$

Modifying the proof of the termination Theorem of the previous section, it can be shown that if $b \in \mathcal{B}$ is a minimal monomial generator of I_{MON} then for some m, $\mathrm{Tip}(g_m) = b$. From this we get the next result.

Proposition 2.8. *If I_{MON} has a finite set of monomial generators then the above algorithm terminates in a finite number of steps and yields a finite Gröbner basis.*

Proof. If I_{MON} has a finite set of mononial generators, then the unique minimal monomial generating set must be finite. Suppose $\mathcal{T} = \{t_1, \ldots, t_s\}$ is the finite generating set of minimal monomials for I_{MON}. Then, by the remarks preceeding the proposition, $\mathcal{T} \subset \{\text{Tip}(g_1), \ldots, \text{Tip}(g_N)\}$ for sufficiently large N. But then $\text{Tip}(\{g_1, \ldots, g_N\})$ generates I_{MON} and hence is a Gröbner basis of I. But then division by $\{g_1, \ldots, g_N\}$ can only have remainder 0 since all overlap relations are elements of I. Thus the algorithm terminates in a finite number of steps outputting a Gröbner basis. □

For a discussion of improvements of the Buchberger algorithm in the commutative case we refer to [9]. Most of the discussion there can be translated to the noncommutative case.

2.4.2 Basic Computational Use of Gröbner Bases We are interested in studying quotient rings R/I. Elements of R/I consist of equivalence classes of the form $f + I$ where $f \sim g$ if and only if $f - g \in I$. The K-algebra has addition and multiplication given by

$$\begin{aligned}(f + I) + (g + I) &= ((f + g) + I) \\ (f + I) \cdot (g + I) &= ((fg) + I)\end{aligned}$$

To study R/I we need a way studying the equivalence classes $f+I$. Assuming we have R represented on a computer via the given multiplicative basis $\mathcal{B}$, we would like to be able to find "good" representatives of equivalence classes. But given an admissible order $>$ on $\mathcal{B}$ we have such representatives.

Proposition 2.9. *Let R be a K-algebra with multiplicative basis $\mathcal{B}$ and admissible order $>$ on $\mathcal{B}$. Let I be an ideal in K.*

1. *$f + I = g + I$ if and only $N(f) = N(g)$.*
2. *$f + I = N(f) + I$.*
3. *The map $\sigma : R/I \to R$ with $\sigma(f + I) = N(f)$, is a vector space splitting to the canonical surjection $\pi : R \to R/I$.*
4. *σ is a K-linear isomorphism between R/I and Span(NonTip(I)).*
5. *Identifying R/I with Span(NonTip(I)), then NonTip(I) is a K-basis of R/I contained in $\mathcal{B}$.*

Proof. The first two parts are immediate consequences of Theorem 2.1 which states that $R = I \oplus \text{Span}(\text{NonTip}(I))$. The remaining parts are left as an exercise. □

Thus, normal forms solves the problem of finding representatives of the equivalence classes. Furthermore, addition of two equivalence classes $(f + I) + (g + I)$ is given simply by $N(f) + N(g) = N(f + g)$ since Span(NonTip(I)) is a linear subspace of R. By Proposition 2.9, since $N(f) \in f + I$ we see that multiplication of classes is given by

$$N(N(f) \cdot N(g)),$$

that is, to find the representative of $(f+I)\cdot(g+I)$ simply multiply $N(f)N(g)$ in R and then take the normal form of the result.

As pointed out in the last section, once a Gröbner basis has been found, the division algorithm provides an algorithm to find normal forms. In this way, one can use the computer to study quotient rings R/I.

2.4.3 Finite Dimensional Algebras Suppose that R/I is finite dimensional. What further can be said in this case?

Proposition 2.10. *Let R be a finitely generated K-algebra with multiplicative basis $\mathcal{B}$ and admissible order $>$. Suppose that I is an ideal such that $\dim_K(R/I) = N$. Then I_{MON} has a finite set of monomial generators.*

Proof. Since R/I is isomorphic to Span(NonTip(I)) as vector spaces, it follows that NonTip(I) is a finite set since it is a basis of Span(NonTip(I)). Next, since R is finitely generated as an algebra, $\mathcal{B}$ is finitely generated as a semigroup.

Let $X = \{b_1, \ldots, b_k\}$ generate $\mathcal{B}$. We show

$$\{bc \,|\, b \in X \text{ and } c \in \text{NonTip}(I)\} \cap \text{Tip}(I)$$

generates I_{MON}. Suppose that t is a element of the minimal monomial generating set of I_{MON}. Then $t = b_{i_1}b_{i_2}\cdots b_{i_l}$ with $b_{i_j} \in X$. By minimality, $b_{i_2}\cdots b_{i_l} \notin \text{Tip}(I)$. Thus, $b_{i_2}\cdots b_{i_l} \in \text{NonTip}(I)$. Hence $t = b_{i_1}c$ with $c \in \text{NonTip}(I)$. This completes the proof. □

The above result has the following immediate consequence.

Corollary 2.2. *Let R be a finitely generated K-algebra with multiplicative basis and admissible order $>$. Suppose that I is an ideal generated by uniform elements and R/I is finite dimensional. Then I has a finite uniform Gröbner basis with respect to $>$ and can be computed in a finite number of steps by the above algorithm.*

Open Question: Given R and a multiplicative basis $\mathcal{B}$. Find necessary and sufficient conditions on an ideal I in R such that there is some admissible order $>$ for which I has a finite Gröbner basis.

2.4.4 Universal Gröbner Bases If R is a K-algebra with multiplicative basis $\mathcal{B}$. Assume that R is finitely generated and hence so is B. Fix a set of generators of the semigroup $\mathcal{B}$, $b_1, \ldots, b_n$.

Definition 2.10. If $b \in \mathcal{B}$ we define the *length of* b to be k if b is a product of k generators and cannot be written as a product of fewer than k generators. We denote the length of b by length(B).

Thus length$(b_i) = 1$ for $i = 1, \ldots, n$ and if $b \in \mathcal{B}$, length$(b) \geq 1$.

Let I be an ideal in R.

Definition 2.11. We say a set of elements $\mathcal{G}$ in I is a *universal Gröbner basis for* I if for every admissible order $>$, $\mathcal{G}$ is a Gröbner basis with respect to $>$.

In some cases, finite universal Gröbner bases exist.

Example 2.8. Let R be the noncommutative polynomial ring in three noncommuting variables x, y, z over the rational numbers. Let $\mathcal{B}$ be the set of monomials. Let I be the ideal generated by $xy - 2yx, zx, zy$. This set is a universal Gröbner basis since all overlap relations reduce to 0 whether the tip of $xy - 2yx$ is xy or is yx. Hence the order doesn't affect that the set is a Gröbner basis. Note that R/I is not even noetherian.

In general, it is not known which ideals have finite universal Gröbner bases. One of the difficulties with this problem in the noncommutative case is that admissible orders are not classified.

We will show that finite universal Gröbner bases exist if R/I is finite dimensional. For this we introduce some terminology. If V is a vector space with basis $\mathcal{B}$, then the *support of a vector* v, $\mathrm{supp}_{\mathcal{B}}(v)$, is defined to be

$$\mathrm{supp}_{\mathcal{B}}(v) = \{b \in \mathcal{B} \mid b \text{ occurs in } v\}.$$

If X is a subset of V, the *support of* X is

$$\mathrm{supp}_{\mathcal{B}}(X) = \{\mathrm{supp}_{\mathcal{B}}(x) \mid x \in X\}.$$

Proposition 2.11. *Let R be a finitely generated K-algebra with multiplicative basis $\mathcal{B}$ which admits an admissible order. Let I be an ideal such that R/I is finite dimensional. Then there exists a finite universal Gröbner basis for I.*

Proof. Let $d = \dim_K(R/I)$ and let $\pi : K\Gamma \to K\Gamma/I$ be the canonical surjection. Assume that $\mathcal{B}$ is generated by $X = \{b_1, \ldots, b_n\}$. We claim that every $b \in \mathcal{B}$ with of length longer than $D = (d+2)n$ has a factor which is the tip of an element in I of smaller length. That is, if the length of b is greater than D, then there is some monomial q of smaller length such that $b = sqt$ and such that q is the tip of some element in I. Suppose b is a monomial of length longer than D. Write $b = b_{i_1} b_{i_2} \ldots b_{i_E}$ with $b_{i_j} \in X$ and $E > D$. Then some generator of $\mathcal{B}$, say b^*, must occur at least $d+2$ times. Hence we have a factorization of b into monomials, $b = p_1 c_1 \cdots c_{d+1} p_2$, where each c_i is of the form $b^* b_{i_j} \ldots b^*$ and is of length at least 1 but of length less than the length of b. Since $\dim_K(R/I) = d$, the set $\{\pi(c_1), \ldots, \pi(c_{d+1})\}$ is linearly dependent over K. Therefore there is a nontrivial linear combination $x = \sum_{j=1}^{d+1} \alpha_i c_i \in I$. Thus, some c_i is a tip in the given ordering and we have proven the claim.

Now let V be the vector subspace of R with basis Ω consisting of monomials of length bounded by D. Let Y be the subspace $I \cap V$ of V. Since V is finite dimensional, the support of Y is a finite set, say $\mathrm{supp}_{\Omega}(Y) = \{s_1, \ldots, s_m\}$. For each $s_i \in \mathrm{supp}_{\Omega}(X)$, choose an element $f_i \in I$ such that $\mathrm{supp}(f_i) = s_i$. We now show that $\mathcal{G} = \{f_1, \ldots, f_m\}$ is a universal Gröbner basis for I.

Let $>$ be an admissible order and let $h \in I$. Suppose the remainder h^* of division of h by $\mathcal{G}$ is not 0. No monomial occuring in h^* is divisible by any $\mathrm{Tip}(f_i)$, $i = 1, \ldots, m$. Let h^* have that property that the longest monomial in h^* is minimal in the set of all $h \in I$ such that no monomial in h is divisible by any $\mathrm{Tip}(f_i)$. Let C be the length of the longest monomial in h^*. If $C \leq D$

then $\text{supp}(h^*) \in \text{supp}_\Omega(Y)$ and hence there is some $f_i \in \mathcal{G}$ such that $\text{supp}(f_i) = \text{supp}(h^*)$. But then $\text{Tip}(f_i) = \text{Tip}(h^*)$, a contradiction. If $C > D$, then by an earlier argument, if $p \in \mathcal{B}$ has length greater than D and occurs in h^*, then there is some $f \in \mathcal{G}$ such that $\text{Tip}(f)$ is a subpath of p. Thus, p can be reduced by an element in $\mathcal{G}$. But this contradicts our assumption on h^* and we are done. □

2.5 Modules, Presentations and Resolutions

In this section I will address the problem of studying R/I-modules. If R/I is not finite dimensional and the module is not finite dimensional, then the only possibility of computationally handling this situation is using generators and relations; a setup ideally suited for the use of the theory of Gröbner bases. Throughout this section we fix a K-algebra R, a multiplicative basis $\mathcal{B}$, and an admissible order $>$. Also fix an ideal I in R and assume we have a Gröbner basis $\mathcal{G}$ of I with respect to $>$.

2.5.1 Modules Suppose that M is a *finitely presented* right R/I-module. This means that M is the cokernel of a map between finitely generated projective R/I-modules. This is equivalent to assuming that M is the cokernel of a map between two finitely generated free R/I-modules.

Let $f_1 : F_1 \to F_0$ be an R/I-homomorphism between finitely generated free right R/I-modules. We choose bases for F_0 and F_1, say $\{e_1, \dots, e_n\}$ for F_0 and $\{d_1, \dots, d_m\}$ for F_1. Then f_1 is given by an $m \times n$ matrix $A_{f_1} = (a_{i,j})$ where $a_{i,j} \in R/I$. Then

$$f_1(d_i) = \sum_j^n e_j a_{i,j}.$$

Let $M = \text{Coker}(f_1) = F_0/\text{Im}(f_1)$ and let $f_0 : F_0 \to M$ be the canonical map.

Note that computationally, we can represent the matrix A_{f_1} by an $m \times n$ matrix with entries in Span(NonTip(I)).

Before specializing, I present a way of viewing M as an ideal in a ring where we can use Gröbner basis theory to study M. In the representation theory of finite dimensional algebras, this is called "one point extension".

Let $\bar{S} = \begin{pmatrix} K & M \\ 0 & R/I \end{pmatrix}$. Note that multiplication is just matrix multiplication; namely,

$$\begin{pmatrix} k & m \\ 0 & r \end{pmatrix} \cdot \begin{pmatrix} k' & m' \\ 0 & r' \end{pmatrix} = \begin{pmatrix} kk' & km' + mr' \\ 0 & rr' \end{pmatrix},$$

which makes sense in that we view M as a K-R/I-bimodule (with elements of K commuting will elements of M).

We now show how to view $\bar{S}$ as quotient of a K-algebra for which we can find a multiplicative basis and an admissible order related to that of R.

Let $G_0 = \coprod_{i=1}^n R$ with basis $e_1^*, \dots, e_n^*$.

Define S to be the K-algebra by $S = \begin{pmatrix} K & G_0 \\ 0 & R \end{pmatrix}$. Set $\pi : S \to \bar{S}$ by

$$\pi\left(\begin{pmatrix} k & \sum_{i=1} e_i^* r_i \\ 0 & r \end{pmatrix}\right) = \begin{pmatrix} k & f_0(\sum_{i=1} e_i^* N(r_i)) \\ 0 & N(r) \end{pmatrix}.$$

Note that we have identified R/I with Span(NonTip(I)). Let $I^* = \mathrm{Ker}(\pi)$. We now give a multiplicative basis for S. Let $\mathcal{B}^*$ be the set

$$\begin{pmatrix} 1 & 0 \\ 0 & 0 \end{pmatrix} \cup \begin{pmatrix} 0 & 0 \\ 0 & b \end{pmatrix} \mid b \in \mathcal{B}\} \cup \begin{pmatrix} 0 & e_i^* b \\ 0 & 0 \end{pmatrix} \mid 1 \leq i \leq n, b \in \mathcal{B}\}.$$

We can define an order $<^*$ on $\mathcal{B}^*$ by $\left(\begin{smallmatrix} 1 & 0 \\ 0 & 0 \end{smallmatrix}\right) <^* \left(\begin{smallmatrix} 0 & 0 \\ 0 & b \end{smallmatrix}\right)$ for each b. $\left(\begin{smallmatrix} 0 & 0 \\ 0 & b \end{smallmatrix}\right) <^* \left(\begin{smallmatrix} 0 & 0 \\ 0 & b' \end{smallmatrix}\right)$ if $b < b'$. $\left(\begin{smallmatrix} 0 & 0 \\ 0 & b \end{smallmatrix}\right) <^* \begin{pmatrix} 0 & e_i^* b' \\ 0 & 0 \end{pmatrix}$ for all b, b' and all i. Finally, $\begin{pmatrix} 0 & e_i^* b \\ 0 & 0 \end{pmatrix} <^* \begin{pmatrix} 0 & e_j^* b' \\ 0 & 0 \end{pmatrix}$ if $i < j$ or if $i = j$, $b < b'$.

We leave it to the reader to check that $<^*$ is an admissible order on $\mathcal{B}^*$. Next we give a set of generators for I^*. Assume I is generated by $\{h_i\}_{i \in \mathcal{I}}$ in R. Then I^* is generated by

$$(*) \qquad \{\left(\begin{smallmatrix} 0 & 0 \\ 0 & h_i \end{smallmatrix}\right) \mid i \in \mathcal{I}\} \cup \{\begin{pmatrix} 0 & \sum_{j=1}^n e_j^* a_{i,j} \\ 0 & 0 \end{pmatrix} \mid 1 \leq i \leq m\}.$$

Note that we view the elements $a_{i,j} \in$ Span(NonTip(I)) in this setting as elements of R. Again the reader may check that (*) is a generating set for I^*. Under reasonable circumstances, we can find a Gröbner basis for I^* with respect to $>^*$. For example, if R is a path algebra, then S is also a path algebra and hence we can algorithmically find Gröbner bases.

Applying Theorem 2.1, we get $S = I^* \oplus$ Span(NonTip(I^*)). Remember that $S/I^* = \left(\begin{smallmatrix} K & M \\ 0 & R/I \end{smallmatrix}\right)$. In particular, M can be identified with $\left(\begin{smallmatrix} 0 & M \\ 0 & 0 \end{smallmatrix}\right)$. It follows that the elements of NonTip(I^*) of the form $\left(\begin{smallmatrix} 0 & c \\ 0 & 0 \end{smallmatrix}\right)$ form a K-basis of M. Noting that if $x \in I$ then $\left(\begin{smallmatrix} 0 & 0 \\ 0 & x \end{smallmatrix}\right) \in I^*$ and hence $\begin{pmatrix} 0 & e_i^* x \\ 0 & 0 \end{pmatrix} \in I^*$ we conclude that the c's occurring in the K-basis of M are all of the form $\sum_{i=1}^n e_i^* c_i$ with $c_i \in$ Span(NonTip(I)).

Finally, assuming that a Gröbner basis of uniform elements of I^* can be computed, we note that the normal form of an element of M, $\left(\begin{smallmatrix} 0 & c \\ 0 & 0 \end{smallmatrix}\right)$, when multiplied by an element of R/I, $\left(\begin{smallmatrix} 0 & 0 \\ 0 & b \end{smallmatrix}\right)$, yields an element of the form $\left(\begin{smallmatrix} 0 & c' \\ 0 & 0 \end{smallmatrix}\right)$ with $c' = \sum_{i=1}^n e_i^* c_i^*$ with $c_i^* \in$ Span(NonTip(I)).

We identify R/I with Span(NonTip(I)) (as usual) and M with the set of elements $c = \sum_{i=1}^n e_i^* c_i \in G_0$ where $c_i \in$ Span(NonTip(I)) with $\left(\begin{smallmatrix} 0 & c \\ 0 & 0 \end{smallmatrix}\right) \in$ Span(NonTip(I^*)). Note that such c can in fact be viewed as in F_0 and with this identification, we get a vector space splitting $\sigma : M \to F_0$ of $f_0 : F_0 \to M$ by $\sigma(m) = N(\left(\begin{smallmatrix} 0 & g \\ 0 & 0 \end{smallmatrix}\right))$ where $f_0(g) = m$.

Stepping back from the above details, what we have is the following, assuming we can compute a Gröbner basis of uniform elements of I^* algorithmically:

1. One can compute a K-basis of M inside F_0 algorithmically; namely, the Gröbner basis of I^* with respect to $>^*$.

2. One can algorithmically compute the action of R/I on this basis; namely, the computation of the normal form of the product.

In practice, we can use a Gröbner basis of I with respect to $>$ in that the elements $\left(\begin{smallmatrix} 0 & 0 \\ 0 & g \end{smallmatrix}\right)$ are part of a Gröbner basis of I^*. Once the Gröbner basis for I^* is computed, operations by elements of R/I on M can be performed using the division algorithm.

2.5.2 Projective Resolutions In this section, we use the results of the previous section and Gröbner bases to construct projective resolutions of modules. We need to require more properties of R.

Our goal is to show how to algorithmically construct a projective R/I-resolution of a right R/I-module M. This construction will use both the construction of Gröbner bases, overlap relations and the division algorithm.

In this section we assume

1. R has a multiplicative basis $\mathcal{B}$ and an admissible order $>$.
2. R is a finitely generated K-algebra and $\mathcal{B}$ is generated by $X = \{b_1, \ldots, b_n\}$.
3. Every element of $\mathcal{B}$ is a unique product of elements of X; that is, if $c_1 \cdots c_r = d_1 \cdots d_s$ with $c_i, d_j \in \mathcal{B}$ then $r = s$ and $c_i = d_i$.
4. I is an ideal with Gröbner basis $\mathcal{G}$.

As in the previous section, given a finitely presented right R/I-module M with presentation $f_1 : F_1 \to F_0$, we form the ring $\bar{S} = \left(\begin{smallmatrix} K & M \\ 0 & R/I \end{smallmatrix}\right)$. Viewing $\bar{S}$ as a quotient of

$$S = \left(\begin{smallmatrix} K & F_0 \\ 0 & R \end{smallmatrix}\right)$$

by I^* , we find a Gröbner basis $\mathcal{G}^*$ of I^* with respect to $>^*$. Recall that $\mathcal{G}^* = \{\left(\begin{smallmatrix} 0 & 0 \\ 0 & g \end{smallmatrix}\right) \mid g \in \mathcal{G}\} \cup \{\left(\begin{smallmatrix} 0 & h_i^* \\ 0 & 0 \end{smallmatrix}\right)\}_{i \in \mathcal{I}}$ where $\mathcal{G}$ is a Gröbner basis for I with respect to $>$ and h_i^* are of the form $\sum_{i=1}^n e_i^* c_i$ with $c_i \in \text{Span}(\text{NonTip}(I))$. We assume each $\sum_{i=1}^n e_i^* c_i$ and each $g \in \mathcal{G}$ are are uniform elements of R. Furthermore, assume F_0 has basis $e_1, \ldots, e_n$

Note that if $h_i^* = \sum_j e_j^* c_{i,j}$, we define F_1' to be the free R/I-module with basis $\{d_i' \mid i \in \mathcal{I}\}$. Define

$$f_1' : F_1' \to F_0$$

by $f_1'(d_i') = \sum_j e_j c_{i,j}$. We see that $f_1' : F_1' \to F_0$ is also a free presentation of M.

Let f_1' be represented by the matrix $(a_{i,j})$ as in the last section. We also let $f_0 : F_0 \to M$ be the canonical surjection. Note that the modules $(\,0 \ R/I\,)$ and $(\,K \ M\,)$ are projective $\bar{S}$-modules. This follows from $(\,0 \ R/I\,) \simeq \left(\begin{smallmatrix} 0 & 0 \\ 0 & 1 \end{smallmatrix}\right) \bar{S}$ and $(\,K \ M\,) \simeq \left(\begin{smallmatrix} 1 & 0 \\ 0 & 0 \end{smallmatrix}\right) \bar{S}$.

Let U be the right $\bar{S}$-module $(K \ \ 0)$ (with the action of $\bar{S}$ given by right matrix multiplication). We have a projective presentation of U given as follows.

$$(\,0 \ F_0\,) \xrightarrow{(\,0 \ f_0\,)} (\,K \ M\,) \to U \to 0.$$

Note that the image of $\left(\begin{smallmatrix} 0 & f_0 \end{smallmatrix}\right)$ is $\left(\begin{smallmatrix} 0 & M \end{smallmatrix}\right)$. We may continue the resolution using f_1' as follows:

$$\left(\begin{smallmatrix} 0 & F_1' \end{smallmatrix}\right) \xrightarrow{\left(\begin{smallmatrix} 0 & f_1' \end{smallmatrix}\right)} \left(\begin{smallmatrix} 0 & F_0 \end{smallmatrix}\right) \xrightarrow{\left(\begin{smallmatrix} 0 & f_0 \end{smallmatrix}\right)} \left(\begin{smallmatrix} K & M \end{smallmatrix}\right) \to U \to 0. \qquad (**)$$

We show how to algorithmically find F_2, a free R/I-module and a map $f_2 : F_2 \to F_1'$ so that

$$F_2 \xrightarrow{f_2} F_1' \xrightarrow{f_1'} F_0 \xrightarrow{f_0} M \to 0$$

is an exact sequence. Once this is done, we can continue the projective resolution by replacing $F_1' \xrightarrow{f_1'} F_0$ with $F_2 \xrightarrow{f_2} F_1'$ and repeat the construction. This will be done by continuing the sequence (**) over $\bar{S}$ and finding the next map in the resolution

$$\left(\begin{smallmatrix} 0 & F_2 \end{smallmatrix}\right) \xrightarrow{\left(\begin{smallmatrix} 0 & f_2 \end{smallmatrix}\right)} \left(\begin{smallmatrix} 0 & F_1' \end{smallmatrix}\right).$$

For this we first describe F_2. Let $\mathcal{O}$ be the set of all overlap relation of form $o(\alpha,\beta,p^*,q^*)$ where α is of the form $\left(\begin{smallmatrix} 0 & h_i^* \\ 0 & 0 \end{smallmatrix}\right)$ and β is of the form $\left(\begin{smallmatrix} 0 & 0 \\ 0 & g \end{smallmatrix}\right)$.

Let $o(\alpha,\beta,p^*,q^*)$ be an overlap relation in $\mathcal{O}$. Let $\alpha = \left(\begin{smallmatrix} 0 & h \\ 0 & 0 \end{smallmatrix}\right)$ and $\beta = \left(\begin{smallmatrix} 0 & 0 \\ 0 & g \end{smallmatrix}\right)$. It follows that from the structure of $\mathcal{B}^*$ that $p^* = \left(\begin{smallmatrix} 0 & p \\ 0 & 0 \end{smallmatrix}\right)$ for some $p \in \mathcal{B}$ and that $q^* = \left(\begin{smallmatrix} 0 & 0 \\ 0 & q \end{smallmatrix}\right)$ for some $q \in \mathcal{B}$. Furthermore, we have

$$o(\alpha,\beta,p^*,q^*) = (1/\mathrm{Tip}(\alpha))\alpha \cdot q^* - (1/\mathrm{Tip}(\beta))p^*\beta =$$

$$\begin{pmatrix} 0 & (1/\mathrm{Tip}(h))hq - (1/\mathrm{Tip}(g))pg \\ 0 & 0 \end{pmatrix}.$$

Since $\mathcal{G}^*$ is a Gröbner basis, the remainder of $o(\alpha,\beta,p^*,q^*)$ by division by $\mathcal{G}^*$ is 0. Hence

$$o(\alpha,\beta,p^*,q^*) =$$

$$\textstyle\sum_i \sum_j \begin{pmatrix} 0 & h_i \\ 0 & 0 \end{pmatrix} \begin{pmatrix} 0 & 0 \\ 0 & q_{i,j} \end{pmatrix} + \sum_{g\in\mathcal{G}} \sum_j \begin{pmatrix} 0 & u_{i,j} \\ 0 & 0 \end{pmatrix} \begin{pmatrix} 0 & 0 \\ 0 & g \end{pmatrix} \begin{pmatrix} 0 & 0 \\ 0 & v_{i,j} \end{pmatrix}$$

where $q_{i,j}, u_{i,j}, v_{i,j} \in \mathcal{B}$.

We can now describe F_2 and $f_2 : F_2 \to F_1'$. F_2 is the free R/I-module with basis $\hat{e}_i$, where i is indexed by the overlap set $\mathcal{O}$. To describe $f_2 : F_2 \to F_1'$ we need only define $f_2(\hat{e}_i)$ for $i \in \mathcal{O}$. If $i = o(\alpha,\beta,p^*,q^*)$, then, keeping the notation of the previous paragraph,

$$f_2(\hat{e}_i) = \sum_{j\in\mathcal{I}} d_j' q_{i,j}.$$

The above description of the construction of projective resolutions is a variation on the results found in [17] which use results from [5]. The proof of the next result can be obtained by vary the proof found in [17] of Theorem 4.1 to our setup.

Theorem 2.4. *Keeping the above notations,*

$$F_2 \xrightarrow{f_2} F_1' \xrightarrow{f_1'} F_0 \xrightarrow{f_1} M \to 0$$

is an exact sequence.

2.5.3 Finite Dimensional Algebras and Modules In this section, we show how to find projective presentations from "standard" representations of modules and vice versa. R will denote a finitely generated K-algebra with multiplicative basis $\mathcal{B}$, and generators $b_1, \ldots, b_n$ and admissible order $>$. Let I be an ideal such that R/I is finite dimensional. Let M be a finite dimensional right R/I-module. As usual, we identify R/I with Span(NonTip(I)) and let Span(NonTip(I)) $= \overline{\mathcal{B}}$. Without loss of generality, we may assume that the generators $b_1, \ldots, b_n$ of $\mathcal{B}$ are in $\overline{\mathcal{B}}$. (If some $b_i \notin \overline{\mathcal{B}}$ then b_i acts as 0 on M and the following can be modified accordingly.)

Let $\{m_j\}_{j\in\mathcal{J}}$ be a K-basis for M. Well-order the m_j's arbitrarily. Let $D = |\mathcal{J}|$. Viewing M as a D-dimensional vector space, each generator b can be represented by a $D \times D$-matrix $(c^b_{i,j})$ with entries in K, where

$$m_i \cdot b = \sum_{j=1}^{D} m_j c^b_{i,j}.$$

The $D \times D$ matrices $(c^{b_l}_{i,j})$, $l = 1, \ldots, n$ might be called the "standard" way of representing M.

First we show how to obtain the $(c^{b_l}_{i,j})$ from an R/I-free presentation of M, $f_1 : F_1 \to F_0$. Suppose F_0 has R/I-basis $\{e_1, \ldots, e_r\}$ and F_1 has R/I-basis $\{d_1, \ldots, d_s\}$. Then, viewing f_1 as a K-linear map M is the cokernel of f_1. Note that the K-bases of F_0 and F_1 are just $\{e_i b \,|\, 1 \le i \le r \text{ and } b \in \overline{\mathcal{B}}\}$ and $\{d_i b \,|\, 1 \le i \le s \text{ and } b \in \overline{\mathcal{B}}\}$ respectively. Hence knowledge of f_1 as an R/I-homomorphism easily yields viewing f_1 as a K-linear map. Furthermore, for each generator b_l we have a commutative diagram:

$$\begin{array}{ccc} F_1 & \xrightarrow{f_1} & F_0 \\ \downarrow b_l & & \downarrow b_l \\ F_1 & \xrightarrow{f_1} & F_0 \end{array}$$

where the vertical morphisms are right multiplication by b_l. It follows that we get a map $g_{b_l} : M \to M$ representing multiplication by b_l. g_{b_l} yields the matrix $(c^{b_l}_{i,j})$.

To go the other way, assume that we have a K-basis $\{m_j\}_{j\in\mathcal{J}}$ of M and the $D \times D$ K-matrices $(c^{b_l}_{i,j})$ representing multiplication by the n generators b_l, $1 \le l \le n$. We show how to construct an R/I-projective presentation $f_1 : F_1 \to F_0$ of M. First let F_0 be the free R/I-module with basis $\{e_j \,|\, j \in \mathcal{J}\}$ and F_1 be the free R/I-module with basis $\{d_{x,j} \,|\, x \in \overline{\mathcal{B}}, j \in \mathcal{J}\}$. Define $f_0 : F_0 \to M$ by $f_0(e_j) = m_j$. Next, define $f_1 : F_1 \to F_0$ as follows. Given $j \in \mathcal{J}$ and $x \in X$, since x is a product of b_l's, using the matrices $(c_{u,v}^{b_l})$, we may calculate $m_j \cdot x = \sum_{i \in \mathcal{J}} m_i \alpha_{i,j}^x$ where $\alpha_{i,j}^x \in K$. We now define $f_1(d_{x,j}) = e_j x - \sum_{i \in \mathcal{J}} e_i \alpha_{i,j}^x$.

Proposition 2.12. *Keeping the notations above,*

$$F_0 \xrightarrow{f_1} F_1 \xrightarrow{f_0} M \to 0$$

is exact.

Proof. We have seen that $f_0(F_0) = M$. Hence the sequence is exact at M. It is immediate that $f_0 f_1(d_{x,j}) = 0$ for each $x \in \overline{\mathcal{B}}$ and $j \in \mathcal{J}$. Thus $\mathrm{Im}(f_1) \subset \mathrm{Ker}(f_0)$. It remains to show that $\mathrm{Ker}(f_0) \subset \mathrm{Im}(f_1)$.

Consider $\{z = \sum_{j \in \mathcal{J}} e_j w_j \in \mathrm{Ker}(f_0) \mid z \notin \mathrm{Im}(f_1)\}$. We wish to show this set is empty. Let $z = \sum_{j \in \mathcal{J}} e_j w_j$ be in this set. Suppose $w_j = \sum_{x \in \overline{\mathcal{B}}} \beta_{j,x} x$. Consider $z' = z + f_1(\sum_{j,x} d_{j,x} \beta_{j,x})$. Since $z \notin \mathrm{Im}(f_1)$, $z' \notin \mathrm{Im}(f_1)$. But, from the definition of f_1, we see $f_1(\sum_{j,x} d_{j,x} \beta_{j,x}) = \sum_{j \in \mathcal{J}} e_j \gamma_j - z$ for some $\gamma_j \in K$. Thus $z' = \sum_{j \in \mathcal{J}} e_j \gamma_j$. But $f(z') = 0$ and hence $\sum_{j \in \mathcal{J}} m_j \gamma_j = 0$. It follows that $\gamma_j = 0$ for all $j \in \mathcal{J}$. This is a contradiction and we have shown no such z can exist. □

2.6 Applications of Gröbner Bases

Throughout this section, we will tacitly assume that $R = K\Gamma$, $\mathcal{B}$ is the set of paths, and $>$ is an admissible order. Hence Gröbner bases exist for ideals in R.

2.6.1 Topics not covered in Detail

1. Computation of $\mathrm{Hom}_{R/I}(M, N)$ for two right R/I-modules can easily be performed. More precisely, Gröbner bases yield K-bases of right modules which behave "nicely" with respect to the ring structure of R/I. That is, viewing R/I as Span(NonTip(I)) with "rewrite formulas" given by the Gröbner basis, the module bases constructed in the last lecture come with how the basis of R/I acts on basis elements of the module because that is precisely the information contained in the Gröbner basis of I^* in $\begin{pmatrix} K & G_0 \\ 0 & R \end{pmatrix}$ of the last lecture. Using these bases, the computation of $\mathrm{Hom}_{R/I}(M, N)$ as a vector space is straightforward.

2. The study of submodules and quotient modules can be easily handled using the bases of last lecture.

3. Computation of $M \otimes_{R/I} N$ for a left R/I-module M and a right R/I-module N can be performed. For this one must be able to find bases of both left and right modules (which Gröbner bases can do). One method of attacking the computation $M \otimes_{R/I} N$ is to find $M \otimes_K N$ using the K-bases provided by the Gröbner bases of the last lecture and then put in the relations over R/I and find quotient.

4. Using the above two parts, it should be apparent that one can computationally study module theory of noncommutative rings that are quotients of rings that have a theory of Gröbner bases. This class includes algebras that are quotients of path algebras and therefore includes quotients of free algebras.

5. One can study the growth of algebras. By Theorem 2.1, R/I can be identified with Span(NonTip(I)) and NonTip(I) is a K-basis of R/I. If R is the free algebra and $\mathcal{B}$ is the set of monomials or if R is a path algebra and $\mathcal{B}$ is the set of paths then the growth of the number of nontips of a given degree or length measures the growth of the algebra. More precisely, if $H_n = \dim_K(\mathrm{Span}(\{x \in \mathrm{NonTip}(I) \mid \mathrm{length}(x) = n\})$, then the *Hilbert series of R/I* is

$$H(R/I) = \sum_{n=o}^{\infty} H_n z^n.$$

Here z is a variable. Questions like the rationality of $H(R/I)$ are being studied [2, 3, 4, 30].

One of the important tools to study $H(R/I)$ and the numbers H_n is the *Ufnarovski graph*. Space doesn't permit me to say much about it, but it is a powerful tool to study the nontips of I. We refer to [32].

6. Noncommutative Gröbner bases in free algebras have been applied to study H^∞ control problems in the work of Helton, Stankus and Varvik.

7. As described in the last section, Gröbner bases allows the construction of projective resolutions of a right module M. Then, applying $\mathrm{Hom}_{R/I}(-, N)$ to the resolution and taking cohomology of the resulting complex, one can obtain the ext-groups $\mathrm{Ext}^n_{R/I}(M, N)$. For a left R/I-module N, applying $- \otimes_{R/I} N$ to the resolution and taking cohomology of the resulting complex, one can obtain the tor-groups $\mathrm{Tor}^{R/I}_n(M, N)$. Hence, one can study homological questions computationally.

One important invariant of an algebra is the *Poincaré series of a module.* If M is a right R/I-module and S is a simple right R/I-module (usually K), let $P_n = \dim_K \mathrm{Ext}^n_{R/I}(M, S)$. The Poincaré series is

$$P_S(M) = \sum_{n=0}^{\infty} P_n z^n.$$

Again, questions like rationality and connections with the Hilbert series are of interest. [4, 30]

8. Michael Bardzell [8] has used the theory of Gröbner basis to study the Hochschild cohomology of a monomial K-algebra.

2.6.2 Algebras and their associated Monomial Algebras One of the interesting connections one obtains from the theory of Gröbner bases is the associated monomial algebra to an algebra. As we have done, let $I_{MON} = \text{Span}(\text{Tip}(I))$. Then I_{MON} is a monomial ideal. We say R/I_{MON} is the *associated monomial algebra to* R/I. Note that R/I_{MON} is dependent on the choice of basis $\mathcal{B}$ of R and on the choice of admissible order $>$.

If R/I is a finite dimensional K-algebra, there is an important invariant of R/I called the *Cartan matrix*. Let $S_1, \ldots, S_n$ be a full set of nonisomorphic simple R/I-modules and let $P_1, \ldots, P_n$ be indecomposable projective R/I-modules such that there are surjections $P_i \to S_i$. Then we define the Cartan matrix of R/I to be the $n \times n$ matrix $(C_{i,j})$ where $C_{i,j}$ is the number of times S_i occurs as a composition factor of P_j.

Since R is a path algebra, it is sometimes natural to assume that I is contained in the ideal $<\Gamma_1>^2$ where Γ_1 is the set of arrows in Γ. In this case there are certain simple modules of R/I we distinguish, called the *vertex simple modules*. A vertex simple module is a module of the form $S_i = v_i R/ <\Gamma_1>$ where v_i is a vertex. It is easy to see that if there are n vertices in Γ, $S_1, \ldots, S_n$ are right simple R/I-modules which are 1-dimensional over K.

If M is a right R-module and

$$\cdots P_n \overset{f_n}{\to} P_{n-1} \to \cdots \to P_1 \overset{f_1}{\to} P_0 \overset{f_0}{\to} M \to 0$$

is a projective resolution of M, then the *projective dimension of* M, $\text{pd}_{R/I}(M)$, is the smallest n such that $\text{Im}(f_n)$ is a projective module. If no such n exists, we say that M has *infinite projective dimension* and write $\text{pd}_{R/I}(M) = \infty$. The *global dimension of* R/I, $\text{gl.dim}(M)$, is N if

$$N = \text{Sup}\{\text{pd}_{R/I}(M) \mid M \text{ is an } R/I\text{-module}\}$$

if such an N exists. Otherwise, $\text{gl.dim}(M) = \infty$.

We can state a result relating R/I and R/I_{MON}.

Theorem 2.5. *[23] Let $R = K\Gamma$ be a path algebra with K-basis $\mathcal{B}$ the finite directed path in Γ and with admissible order $>$. Let I be an ideal in R and I_{MON} be the associated monomial ideal. Then*

1. *$H(R/I) = H(R/I_{MON})$.*
2. *If R/I is finite dimensional, then*

$$\dim_K(R/I) = \dim_K(R/I_{MON}) = |NonTip(I)|.$$

3. *Assume that $I \subset <\Gamma_1>^2$.*
   (a) *The construction of a projective resolution given in [17] for each vertex simple module S_i is minimal for R/I_{MON}.*
   (b) *For each $i = 1, \ldots, n$, $\text{pd}_{R/I}(S_i) \leq \text{pd}_{R/I_{MON}}(S_i)$.*

(c) *If* R/I *is finite dimensional, then* $\operatorname{gl.dim}(R/I) \leq \operatorname{gl.dim}(R/I_{MON})$.

(d) *If* R/I *is finite dimensional, then the Cartan matrices of* R/I *and* R/I_{MON} *are equal.*

(e) *If* S *is a vertex simple module viewed as a module over both* R/I *and* R/I_{MON}*, and if* $\to P_1 \to P_0 \to S \to 0$ *and* $\to P_1' \to P_0' \to S \to 0$ *are minimal projective resolutions over* R/I *and* R/I_{MON} *respectively, then* $\dim_K(P_n) \leq \dim_K(P_n')$ *for all* $n \geq 0$.

2.6.3 Koszul Algebras Let $R = K\Gamma$ and I be an ideal in R. We say that I is a *quadratic ideal* if there is a set of generators $\{f_1, \ldots, f_n\}$ of I such that each f_1 is a K-linear combination of paths in Γ of exactly length 2. In this case, we say that $\{f_1, \ldots, f_n\}$ is a set of *quadratic generators* and that R/I is a *quadratic algebra.*

The path algebra R has a natural positive $\mathbf{Z}$-grading given by letting vertices be homogeneous of degree 0, and paths be homogeneous of degree equal to the length. We will view R as a graded ring in this way. Note that an element $x = \sum_{i=1}^{m} \alpha_i p_i$ with $\alpha_i \in K$ and $p_i \in \mathcal{B}$ is *homogeneous* if each p_i occurring in x has the same length. We call this the *length grading* of R. Let I be a *graded ideal in* R; that is, an ideal generated by homogeneous elements. Then R/I is a $\mathbf{Z}$-graded algebra by the induced grading.

For the remainder of this section, we assume that I is graded ideal in R and let $R/I = S_0 \oplus S_1 \oplus S_2 \oplus \cdots$ as a graded ring. We let $\operatorname{Gr}(S)$ denote the *category of* $\mathbf{Z}$*-graded* S*-modules and degree* 0 S*-module maps.* That is, the objects of $\operatorname{Gr}(S)$ are

$$M = \cdots \oplus M_{-2} \oplus M_{-1} \oplus M_0 \oplus M_1 \oplus M_2 \oplus \cdots$$

where each i, j, M_i is a right S_0-module and if $s_j \in S_j$ and $m_i \in M_i$ then $m_i s_j \in M_{i+j}$ such that, forgetting the graded structure, M is a right S-module. A *degree* 0 *map* $f : M \to N$ between graded S-modules is an S-module map such that if $m_i \in M_i$ then $f(m_i) \in N_i$.

We say a graded S-module is *generated in degree* n if $M_j = 0$ for $j < n$ and for all $i \geq 0$, the multiplication maps $M_n \otimes_{S_0} S_i \to M_{n_i}$ are surjective. We say that an S-module X is *gradable* if there is a graded S-module M, such that X is isomorphic to M when one forgets the grading on M and views M as an S-module. We have the following result whose proof is standard.

Proposition 2.13. *Let* I *be a graded ideal in a path algebra* $K\Gamma$ *where* $K\Gamma$ *has the length grading. Assume that* $I \subset <\Gamma_1>$. *Let* $K\Gamma/I = S_0 \oplus S_1 \oplus S_1 \oplus \cdots$ *be the graded quotient ring. Then*

1. S_0 *is isomorphic to* $K\Gamma/<\Gamma_1>$ *and hence is semisimple.*

2. *Each* S_i *is finite dimensional over* K.

3. *The vertex simple modules are gradeable modules.*

4. $S_1 \oplus S_2 \oplus \cdots$ *is the graded Jacobson radical of* S.

5. *The category of finitely generated graded S-modules has projective covers.*
6. *Graded S-modules have graded projective resolutions.*
7. *A graded S-projective resolution of a graded S-module M forgets to a S-projective resolution of X where X is the S-module M when we forget the graded structure.*

We say R/I is *Koszul algebra* if S_0, viewed as a graded S-module generated in degree 0 has a graded projective resolution

$$\cdots P_n \to P_{n-1} \to \cdots \to P_1 \to P_0 \to M \to 0$$

such that for each n, P_n is generated in degree n. Koszul algebras are an important class of algebras that naturally occur in algebraic geometry, topology, and the theory of quantum groups [10, 12, 14, 29, 31, 33]. Proofs of many of the basic results about Koszul algebras can be found in [21, 22, 10]. My goal here is to demonstrate that Gröbner bases can be used in studying such a class of Koszul algebras.

It is well-known that if R/I is a Koszul algebra then I must be a quadratic ideal [21, 10]. At this time, there is no classification known of which quadratic ideals I have the property that $K\Gamma/I$ is a Koszul algebra at this time, in terms of the generators of the quadratic ideals. We do have the following result though.

Theorem 2.6. *[20] Let I be a quadratic ideal in a path algebra $K\Gamma$. Let $>$ be an admissible order on the paths such that I has a quadratic Gröbner basis. Then $K\Gamma/I$ is a Koszul algebra.*

The proof of the result, although too technical for these lectures, involves an investigation of the projective resolution of the vertex simple modules given by the construction discussed in the last lecture. Analysis of the construction shows that if the algebra is graded and the module is graded, then the constructed resolution is, in fact, a graded resolution of the module. It is also shown that if the generators of I are quadratic, then the construction of a graded projective resolution of a vertex simple module has the desired degree properties.

The next result follows from the above theorem and also from [23].

Corollary 2.3. *Let I be a monomial ideal generated by some paths of length 2. Then $K\Gamma/I$ is a Koszul algebra.*

Proof. If J is a monomial ideal in a path algebra, since overlap relations are in fact 0, we see by the Termination Theorem that any generating set of monomials for J is a Gröbner basis under any admissible order. Since I can be generated by paths of length 2, it follows that I has a quadratic Gröbner basis. □

We end with another application which is more fully described in [19]. Let $R = K\langle x_1, \ldots, x_n \rangle$ be the free associative algebra in n noncommuting variables. Let $>$ be the degree-lexicographic order with $x_1 < x_2 < \cdots < x_n$. For $1 \leq i < j \leq n$, let

$$q_{i,j} = x_j x_i - c_{i,j} x_i x_j + r_{i,j}$$

where $r_{i,j}$ is a quadratic polynomial, each of whose terms is less than x_ix_j. Thus $\mathrm{Tip}(q_{i,j}) = x_jx_i$ and $\mathrm{Tip}(q_{i,j} - \mathrm{Tip}(q_{i,j}) = x_ix_i$. Let I be the ideal generated by $\{q_{i,j}\}$.

Note that I is a quadratic ideal. Consider R/I. We denote the image of x_i in R/I by $\bar{x}_i$. We say R/I has a *Poincaré-Birkhoff-Witt basis* or PBW basis if $\{\bar{x}_1^{a_1}\bar{x}_2^{a_2}\dots\bar{x}_n^{a_n}\}$ where a_i are nonnegative integers, is a K-basis of R/I.

The set $\{q_{i,j}\}_{1\le i,j\le n}$ can be viewed as rewriting rules in the sense that if a monomial $\bar{m} = \bar{x}_{i_1}\bar{x}_{i_2}\dots\bar{x}_{i_s}$ has the property that for some j, $i_j > i_{j+1}$ then in R/I, $\bar{x}_{i_j}\bar{x}_{i_{j+1}} = c_{i_j,i_{j+1}}\bar{x}_{i_{j+1}}\bar{x}_{i_j} - \bar{r}_{i_j,i_{j+1}}$,

$$\bar{m} = \bar{x}_{i_1}\bar{x}_{i_2}\dots\bar{x}_{i_{j-1}}\bar{x}_{i_{j+1}}\bar{x}_{i_j}\bar{x}_{i_{j+2}}\dots\bar{x}_{i_s} - \bar{x}_{i_1}\bar{x}_{i_2}\dots\bar{x}_{i_{j-1}}\bar{r}_{i,j}\bar{x}_{i_{j+2}}\dots\bar{x}_{i_s}.$$

Thus, by replacing $\bar{x}_j\bar{x}_i$ as above if $j > i$ and by the fact that $>$ is a well-order, we see that $\{\bar{x}_1^{a_1}\bar{x}_2^{a_2}\dots\bar{x}_n^{a_n}\}$ generate R/I as a K vector space. It is natural to ask if R/I has a PBW basis. The next result answers the question.

Proposition 2.14. *Keeping the above notations, R/I has a PBW basis if and only if $\{q_{i,j}\}$ is a Gröbner basis of I. Thus, if R/I has a PBW basis, R/I is a Koszul algebra.*

Proof. Suppose that R/I has a PBW basis. Then $\{\bar{x}_1^{a_1}\bar{x}_2^{a_2}\dots\bar{x}_n^{a_n}\}$ is a K-basis for R/I. To show that $\{q_{i,j}\}$ is a Gröbner basis for I it suffices to show that overlap relations have remainder 0 when divided by $\{q_{i,j}\}$.

Consider division by $\{q_{i,j}\}$. Any monomial m occuring in a remainder cannot have x_jx_i with $j > i$ in it since if so, $\mathrm{Tip}(q_{i,j})$ would divide m. Thus, the remainder of division by $\{q_{i,j}\}$ of any overlap relation is in the span of $\{x_1^{a_1}\dots x_n^{a_n}\}$. By the PBW basis assumption, no element of this span is in I other than 0.

If $\{q_{i,j}\}$ is a Gröbner basis for I, then

$$\mathrm{NonTip}(\{q_{i,j}\}) = \{m \in \mathcal{B} \mid \mathrm{Tip}(q_{i,j}) \text{ does not divide } m\}.$$

But it immediate that $\mathrm{NonTip}(\{q_{i,j})$ is $\{x_1^{a_1}\dots x_n^{a_n} \mid a_i \ge 0\}$. But $\mathrm{NonTip}(I)$ is a K-basis of R/I under the usual identification. Thus, R/I has a PBW basis.

The last result of the proposition is a consequence of Theorem 2.6. □

Bibliography

[1] W. Adams and P. Loustaunau: *An Introduction to Gröbner bases*, Graduate St. in Math, AMS **3**, 1994.

[2] D. Anick: *Noncommutative graded algebras and their Hilbert series*, J. of Algebra **78**, (1982), 120–140.

[3] D. Anick: *On monomial algebras of finite global dimension*, Transactions AMS **291**, (1985), 291–310.

[4] D. Anick: *Recent progress in Hilbert and Poincaré series*, LNM 1318, (1986) Springer-Verlag, 1–25.

[5] D. Anick and E.L. Green: *On the homology of path algebras*, Comm. in Algebra **15**, (1985), 641–659.

[6] M. Auslander, I. Reiten, and S. Smalø: *Representation Theory of Artin Algebras* , Cambridge Studies in Advanced Math. **36**, (1995), Cambridge Univ. Press.

[7] J. Backelin, R. Froberg: *Koszul algebras, Veronese subrings and rings with linear resolutions*, Rev. Roumaine Math. Pures Appl. **30**, (1980), 85–97.

[8] M. Bardzell: *The alternating syzygy behavior of monomial algebras*, J. of Algebra **188**, (1997), no. 1, 69–89.

[9] T. Becker and V. Weispfenning: *Gröbner bases. A Computational Approach to Commutative Algebra*, GTM **141**, Springer-Verlag, 1993.

[10] A. Beilinson, V. Ginsburg, & W. Soergel: *Koszul Duality Patterns in Representation Theory*, J. Amer. Math.Soc. **9**, (1996) 473–527.

[11] G. Bergman: *The diamond Lemma for ring theory*, Adv. Math. **29**, (1978) 178–218.

[12] A.I. Bondal: *Helices, representations of quivers and Koszul algebras*, London Math. Soc. Lecture Note Ser. **148**, (1990), 75–95.

[13] Buchberger: *An algorithm for finding a basis for the residue class ring of a zero-dimensional ideal*, Ph.D. Thesis, University of Innsbruck, (1965).

[14] E. Cline, B. Parshall, and L. Scott: *Finite dimensional algebras and highest weight categories*, J. Reine Angew. Math. **391**, (1988), 85–99.

[15] D. Cox, J. Little, and D. O'Shea: *Ideals, Varieties, and Algorithms*, UTM Series, Springer-Verlag (1992).

[16] D.R. Farkas, C. Feustel, and E.L. Green: *Synergy in the theories of Gröbner bases and path algebras*, Canad. J. of Mathematics **45**, (1993), 727–739.

[17] C. Feustel, E.L. Green, E. Kirkman, and J. Kuzmanovich: *Constructing projective resolutions*, Comm. in Alg. **21**, (1993) 1869–1887.

[18] E.L. Green: *Representation theory of tensor algebras*, J. Algebra **34**, (1975), 136–171.

[19] E.L. Green: *Poincaré-Birkhoff-Witt bases and Gröbner bases*, preprint.

[20] E.L. Green and R. Huang: *Projective resolutions of straightening closed algebras generated by minors*, Adv. in Math. **110**, (1995), 314–333.

[21] E.L. Green, and R. Martínez Villa: *Koszul and Yoneda algebras*, Canadian Math. Soc. **18**, (1994), 247–298.

[22] E.L. Green, and R. Martínez Villa: *Koszul and Yoneda algebras II*, in Yoneda algebras II, Canadian Math. Soc., Proceedings of ICRA, Ed. Reiten, Smalö, Solberg, 1998.

[23] E.L. Green and D. Zacharia: *The cohomology ring of a monomial algebra*, Manuscripta Math. **85**, (1994).

[24] R. Hartshone: *Residues and Duality*, LNM 20, Springer-Verlag, (1966).

[25] C. Löfwall: *On the subalgebra generated by the one-dimensional elements in the Yoneda ext-algebra*, LNM 1183, Springer-Verlag, (1986), 291–338.

[26] R. Martínez Villa: *Applications of Koszul algebras: the preprojective algebra*, Canadian Math. Soc. **18**, (1994), 487–504.

[27] S. McLane: *Homology*, Springer-Verlag, 1963.

[28] T. Mora: *Gröbner bases for non-commutative polynomial rings*, Proc. AAECC3 L.N.C.S. **229**, (1986).

[29] S. Priddy: *Koszul resolutions*, Trans. AMS **152**, (1970), 39–60.

[30] J.E. Roos: *Relations between the Poincaré-Betti series of loop spaces and of local rings*, LNM **740**, Springer-Verlag, 285–322.

[31] M. Rosso: *Koszul resolutions and quantum groups*, Nuclear Phys. B Proc. Suppl. **18b**, (1990), 269–276.

[32] V. Ufnarovskii: *A growth criterion for graphs and algebras defined by words*, Mat. Zemati **31**, (1980) 465–472; Math. Notes **37**, (1982) 238–241.

[33] Y. Yoshino: *Modules with linear resolutions over a polynomial ring in two variables*, Nagoya Math. J. **113**, (1989), 89–98.

Progress in Mathematics, Vol. 173, © 1999 Birkhäuser Verlag Basel/Switzerland

Chapter 3

Construction of Finite Matrix Groups

Robert A. Wilson

Abstract

We describe various methods of construction of matrix representations of finite groups. The applications are mainly, but not exclusively, to quasisimple or almost simple groups. Some of the techniques can also be generalized to permutation representations.

3.1 Introduction

It is one thing to determine the *characters* of a group, but quite another to construct the associated *representations*. For example, it is an elementary exercise to obtain the character table of the alternating group A_5 by first determining the conjugacy classes, then writing down the trivial character and the permutation characters on points and unordered pairs, and using row orthogonality to obtain the irreducibles of degree 4 and 5, and finally using column orthogonality to complete the table. The result (see Table 3.1) shows that there are two characters of degree 3, but how do we construct the corresponding 3-dimensional representations?

In general, we need some more information than just the characters, such as a presentation in terms of generators and relations, or some knowledge of subgroup structure, such as a generating amalgam, or something similar.

Table 3.1: The character table of A_5

Class name	$1A$	$2A$	$3A$	$5A$	$5B$
Class size	1	15	20	12	12
χ_1	1	1	1	1	1
χ_2	3	-1	0	τ	σ
χ_3	3	-1	0	σ	τ
χ_4	4	0	1	-1	-1
χ_5	5	1	-1	0	0

$$\tau = \frac{1}{2}(1+\sqrt{5}), \sigma = \frac{1}{2}(1-\sqrt{5})$$

If we have a presentation for our group, then in a sense it is already determined, and there are various algorithms which in principle at least will construct more or less any desired representation. The most important and well-known is Todd–Coxeter coset enumeration, which converts a presentation into a permutation representation, and is well described in many places, such as [11]. From a (faithful) permutation representation it is then (in principle) possible to obtain at least any irreducible representation over any finite field, by using 'Meataxe' techniques described by Richard Parker [16]. A generalization of these methods to characteristic zero has recently been described by Parker [17]. Basically these methods enable one to chop any given representation into its irreducible constituents, and then tensor representations together to produce new ones to chop up, and so on.

What we are concerned with here, however, is something more basic, namely how to construct a matrix representation of a group from scratch, when no representation at all is known to begin with. We also assume that no presentation is known, or at least that no presentation can be used to produce a sufficiently small permutation representation.

Note: the original title of my lectures was something like "Computer construction of matrix representations of sporadic simple groups over finite fields", but it gradually became clear that almost every word of the title was redundant, and that "Construction of groups" was the most the various ideas had in common (and even the word "groups" was a trifle restrictive). The present title is a (somewhat unhappy) compromise between the two extremes.

3.2 A Small Example

We illustrate the basic ideas by considering the 3-dimensional representations of A_5 mentioned above. Now we know that A_5 has a subgroup A_4 obtained by fixing one of the five points, and that this subgroup is maximal since it has prime index. Fixing another point we obtain a subgroup A_3, and by 2-transitivity there is an element of A_5 interchanging these two points, and therefore normalizing A_3 to S_3. By maximality, these subgroups generate A_5, and the situation is as shown in Figure 3.1. In the figure, we also give generating permutations for these subgroups, although the figure can equally well be understood at the level of abstract groups.

If we now wish to construct one of the 3-dimensional representations of A_5, we can start by constructing its restriction to A_4. For simplicity we assume that the characteristic of the underlying field is not 2 or 3. Then it is easy to see that the character restricts to A_4 as the unique 3-dimensional irreducible, and the

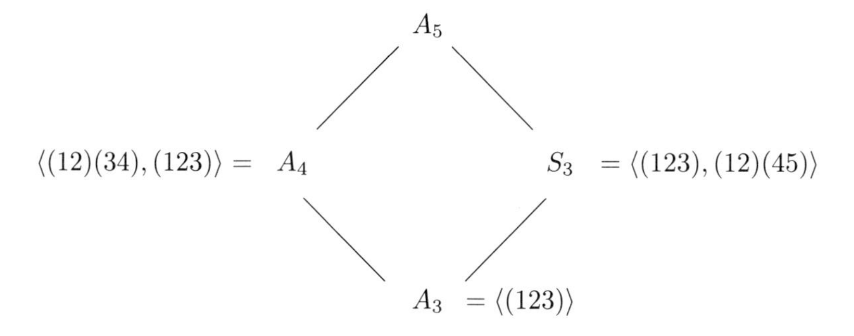

Figure 3.1: Generating A_5 from subgroups

corresponding representation may be defined by

$$(12)(34) \mapsto \begin{pmatrix} 1 & 0 & 0 \\ 0 & -1 & 0 \\ 0 & 0 & -1 \end{pmatrix}$$

$$(123) \mapsto \begin{pmatrix} 0 & 1 & 0 \\ 0 & 0 & 1 \\ 1 & 0 & 0 \end{pmatrix}$$

(In general, an inductive approach allows us to assume anything we wish about properties of the proper subgroups of our group.)

The next step is to adjoin an element of order 2 conjugating $\left(\begin{smallmatrix} 0 & 1 & 0 \\ 0 & 0 & 1 \\ 1 & 0 & 0 \end{smallmatrix}\right)$ to its inverse. One obvious possibility is $\left(\begin{smallmatrix} 1 & 0 & 0 \\ 0 & 0 & 1 \\ 0 & 1 & 0 \end{smallmatrix}\right)$, but this extends A_4 to S_4, not to A_5. We actually need to look at *all* such involutions, in order to pick the right one(s). In this particular case, it is easy to show from first principles that the elements inverting $\left(\begin{smallmatrix} 0 & 1 & 0 \\ 0 & 0 & 1 \\ 1 & 0 & 0 \end{smallmatrix}\right)$ are exactly those of the form $\left(\begin{smallmatrix} \alpha & \beta & \gamma \\ \beta & \gamma & \alpha \\ \gamma & \alpha & \beta \end{smallmatrix}\right)$. (To see this, note that if the matrix $\left(\begin{smallmatrix} a & b & c \\ d & e & f \\ g & h & i \end{smallmatrix}\right)$ conjugates the given element to its inverse, then

$$\begin{pmatrix} a & b & c \\ d & e & f \\ g & h & i \end{pmatrix}\begin{pmatrix} 0 & 1 & 0 \\ 0 & 0 & 1 \\ 1 & 0 & 0 \end{pmatrix} = \begin{pmatrix} 0 & 0 & 1 \\ 1 & 0 & 0 \\ 0 & 1 & 0 \end{pmatrix}\begin{pmatrix} a & b & c \\ d & e & f \\ g & h & i \end{pmatrix}$$

$$\Rightarrow \begin{pmatrix} c & a & b \\ f & d & e \\ i & g & h \end{pmatrix} = \begin{pmatrix} g & h & i \\ a & b & c \\ d & e & f \end{pmatrix},$$

whence $c = g = e$, $a = h = f$, $b = i = d$, as required.) The extra condition that it is an involution gives

$$\alpha^2 + \beta^2 + \gamma^2 = 1,$$

$$\text{and} \quad \alpha\beta + \beta\gamma + \gamma\alpha = 0,$$

while the fact that all involutions in A_5 have trace -1 gives

$$\alpha + \beta + \gamma = -1.$$

In fact these three equations are not independent, and we obtain a one-parameter family of solutions. To find which ones actually give rise to A_5, we need some more information. If the field is finite, we can of course try all cases until we have eliminated all but the correct answer(s). If the field is infinite, however, we need to use some more structure of the group.

In the present case, for example, we could use what we know about A_5 to say that without loss of generality our extra element represents (12)(45), and therefore has product of order 3 with our original element $\begin{pmatrix} 1 & 0 & 0 \\ 0 & -1 & 0 \\ 0 & 0 & -1 \end{pmatrix}$, representing (12)(34). This product is

$$\begin{pmatrix} 1 & 0 & 0 \\ 0 & -1 & 0 \\ 0 & 0 & -1 \end{pmatrix} \begin{pmatrix} \alpha & \beta & \gamma \\ \beta & \gamma & \alpha \\ \gamma & \alpha & \beta \end{pmatrix} = \begin{pmatrix} \alpha & \beta & \gamma \\ -\beta & -\gamma & -\alpha \\ -\gamma & -\alpha & -\beta \end{pmatrix},$$

and if this is to have order 3 then it has trace 0. Thus we obtain $\alpha - \gamma - \beta = 0$, which together with $\alpha + \beta + \gamma = 1$ gives $\alpha = \beta + \gamma = \frac{1}{2}$, and substituting into $\alpha^2 + \beta^2 + \gamma^2 = 1$ we obtain $\beta\gamma = -\frac{1}{4}$, whence β and γ are the roots of the quadratic equation $x^2 - \frac{1}{2}x - \frac{1}{4} = 0$, namely

$$\beta, \gamma = \frac{1}{4}(1 \pm \sqrt{5}).$$

These two possibilities for the extra element, namely

$$\frac{1}{2}\begin{pmatrix} 1 & \sigma & \tau \\ \sigma & \tau & 1 \\ \tau & 1 & \sigma \end{pmatrix} \quad \text{and} \quad \frac{1}{2}\begin{pmatrix} 1 & \tau & \sigma \\ \tau & \sigma & 1 \\ \sigma & 1 & \tau \end{pmatrix}$$

where $\tau = \frac{1}{2}(1 + \sqrt{5})$ is the golden ratio, and $\sigma = \frac{1}{2}(1 - \sqrt{5})$, correspond to the two different 3-dimensional representations of A_5.

3.3 Some Variants

We have already described all the important points in the method used to construct from scratch matrix representations of the sporadic groups, from Janko's construction of J_1 published in 1965 [6], to my construction of the Monster in collaboration with Linton, Parker and Walsh [12], [27]. However, to give more of the flavour of what was actually done in the finite-field cases, we describe slightly different methods for (a) obtaining the complete parametrization of all extending elements, and (b) determining which one(s) of these extending elements give rise to the desired group. We keep the same example as above.

Suppose we have two elements x and y which both invert the 3-cycle (123). Then xy^{-1} centralizes it. Thus the essential step is to find the centralizer of this 3-cycle (or of whatever group appears in its place in larger examples). Now, provided the field has characteristic not 3, and provided there is a primitive cube root ω of unity in the field, we find that the representation restricted to A_3 is the direct sum of its three 1-dimensional representations, acting on the three 1-spaces spanned by the three eigenvectors $(1,1,1)$, $(1,\omega,\omega^2)$ and $(1,\omega^2,\omega)$. By changing to a basis of eigenvectors, we immediately see that the centralizer consists of all diagonal matrices, constituting a 3-parameter family. Changing basis back again we obtain the centralizer as the set of matrices

$$\begin{pmatrix} \alpha & \beta & \gamma \\ \gamma & \alpha & \beta \\ \beta & \gamma & \alpha \end{pmatrix}$$

and can continue as before. (In practical terms it is often easiest to use the basis of eigenvectors instead, from this point onwards—or, more generally, a suitable "standard basis" for the representation restricted to the subgroup in question. In this example, it means that the inverting elements all have the form $\left(\begin{smallmatrix} \lambda & 0 & 0 \\ 0 & 0 & \mu \\ 0 & \nu & 0 \end{smallmatrix}\right)$, which makes them particularly easy to deal with.)

Our second variant concerns the final elimination of cases. To illustrate the points, we need to work over a finite field in which there are both cube roots of unity and solutions to the equation $x^2 - \frac{1}{2}x - \frac{1}{4} = 0$ that arose later in the calculation. The field $F = GF(19)$ of order 19 suits the purpose. Let us suppose that we do not know quite as much about the structure of A_5 as we used before. In particular, suppose we do not know which elements of A_5 are represented by the new matrices. We do know however that every element of A_5 has character value 3, -1, 0, τ or σ, and therefore has trace 3, 18, 0, 15 or 5 in F. (Here we note that σ and τ are the roots of $x^2 - x - 1$, which modulo 19 factorizes as $(x-15)(x-5)$.) In particular, our new element $\left(\begin{smallmatrix} \alpha & \beta & \gamma \\ \beta & \gamma & \alpha \\ \gamma & \alpha & \beta \end{smallmatrix}\right)$ can be multiplied by one of the old elements

$$\begin{pmatrix} 1 & 0 & 0 \\ 0 & -1 & 0 \\ 0 & 0 & -1 \end{pmatrix}, \begin{pmatrix} -1 & 0 & 0 \\ 0 & 1 & 0 \\ 0 & 0 & -1 \end{pmatrix} \text{ or } \begin{pmatrix} -1 & 0 & 0 \\ 0 & -1 & 0 \\ 0 & 0 & 1 \end{pmatrix},$$

and in each case the trace, namely $\alpha-\beta-\gamma$, $\beta-\alpha-\gamma$ or $\gamma-\alpha-\beta$, must be one of the allowed numbers. We had already reduced to a 1-parameter family of examples, although this is a little messy to describe as it involves solving quadratic equations. Substituting $\gamma = -1-\alpha-\beta$ into $\alpha^2+\beta^2+\gamma^2 = 1$ gives $\alpha^2+\beta^2+\alpha+\beta+\alpha\beta = 0$, whose solutions can be expressed as

$$\alpha = \frac{1}{2}\left(-1-\beta \pm \sqrt{(1+\beta)(1-3\beta)}\right).$$

Some arithmetic then leads to the 18 solutions as follows: (α,β,γ) is any permutation of $(2,7,9)$, $(3,5,10)$, $(0,0,18)$ or $(12,12,13)$. These four possibilities give the following triples of values for $\alpha-\beta-\gamma$, $\beta-\alpha-\gamma$ and $\gamma-\alpha-\beta$: $(5,15,0)$, $(7,11,2)$, $(1,1,18)$, $(6,6,8)$. It therefore follows that $\{\alpha,\beta,\gamma\}=\{2,7,9\}$. Now a cyclic permutation of α,β,γ corresponds to conjugating by an element of the subgroup A_3 we wish to normalize, so there are only two cases, which must therefore give rise to the two 3-dimensional representations of A_5 over F.

Explicitly, we have over the field F of order 19,

$$\left\langle \begin{pmatrix}1&0&0\\0&18&0\\0&0&18\end{pmatrix}, \begin{pmatrix}0&1&0\\0&0&1\\1&0&0\end{pmatrix}, \begin{pmatrix}2&7&9\\7&9&2\\9&2&7\end{pmatrix} \right\rangle \cong A_5,$$

and the other representation is given by

$$\left\langle \begin{pmatrix}1&0&0\\0&18&0\\0&0&18\end{pmatrix}, \begin{pmatrix}0&1&0\\0&0&1\\1&0&0\end{pmatrix}, \begin{pmatrix}2&9&7\\9&7&2\\7&2&9\end{pmatrix} \right\rangle.$$

The point to notice is that rather than solving equations to find the exact solution, we have some weaker conditions which we can test to eliminate some possibilities, until we are left with only the final answer. In this case we see that the trace ($\alpha-\beta-\gamma$, say) of a single element is enough to rule out all the unwanted cases, as in the six correct cases, $\alpha-\beta-\gamma=0,5$ or 15, while in the twelve incorrect cases $\alpha-\beta-\gamma=1,2,6,7,8,11$ or 18.

3.4 Towards a General Method

In the above example, in order to construct a particular representation, we first chose some generating subgroups, A_4, and S_3, intersecting in A_3, as in Figure 3.1, and then we

1. constructed the restriction of the representation to A_4,
2. restricted further to A_3,
3. extended from A_3 to S_3—here we used the centralizer of the A_3 to help us construct all possibilities,
4. looked at all possible cases, to eliminate all but the right one—this might involve finding further equations and solving them, or (in the finite case) eliminating all cases but one with a finite series of *ad hoc* tests.

This suggests a general method for constructing a representation of a group G on a vector space V. First choose some suitable subgroups $L < H < G$, with $N_G(L) \not\leq H$, and then:

1. construct the restriction of the representation to H,
2. restrict further to L,

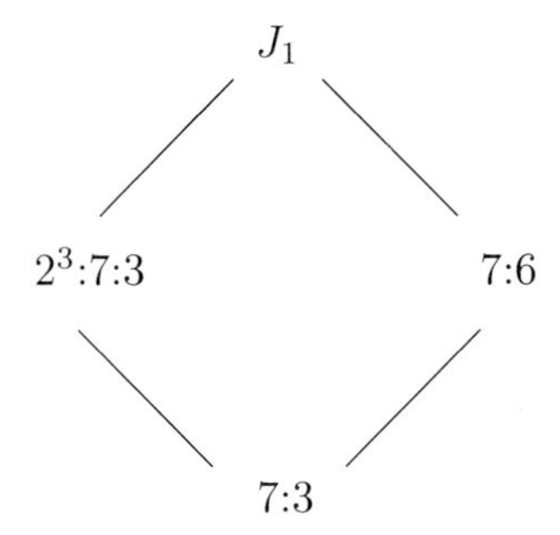

Figure 3.2: Generating J_1 from subgroups

3. extend from L to all groups isomorphic to $N_G(L)$, by first finding the full normalizer $N_{GL(V)}(L)$, or a suitable subgroup thereof,

4. look at all these cases, and eliminate all but the right one.

Before we go any further, we describe a genuine application. This is Janko's original construction (i.e. existence proof) of his first group J_1. We have to assume a few properties of the group, which Janko deduced from the basic hypothesis that it is a simple group with involution centralizer $2 \times A_5$. It is quite straightforward to deduce that the Sylow 2-normalizer is 2^3:7:3, and somewhat harder to show that the Sylow 7-normalizer is 7:6. These two subgroups, moreover, are both maximal, and therefore together they generate the group. After calculating the complete class list, and the character table, Janko calculated the Brauer trees for all primes dividing the group order exactly once, up to a few ambiguities. In particular he was able to show that there is a 7-dimensional representation over the field $F = GF(11)$ of order 11.

Armed with only this information, we embark on the construction, using the subgroups 2^3:7:3 and 7:6, intersecting in 7:3, as shown in Figure 3.2.

First, there is, up to equivalence, only one 7-dimensional faithful representation of 2^3:7:3 over F. To see this, we reproduce the character table in Table 3.2. Since F contains no cube root of 1, as the multiplicative group has order 10, we must have the character χ_6 from the table.

This representation may be easily constructed, for example as the induced representation from the 1-dimensional representation of $2 \times A_4$ which has kernel A_4, or otherwise. We take

$$A = \begin{pmatrix} -1 & 0 & 0 & 0 & 0 & 0 & 0 \\ 0 & 1 & 0 & 0 & 0 & 0 & 0 \\ 0 & 0 & 1 & 0 & 0 & 0 & 0 \\ 0 & 0 & 0 & -1 & 0 & 0 & 0 \\ 0 & 0 & 0 & 0 & 1 & 0 & 0 \\ 0 & 0 & 0 & 0 & 0 & -1 & 0 \\ 0 & 0 & 0 & 0 & 0 & 0 & -1 \end{pmatrix}$$

Table 3.2: The character table of $2^3{:}7{:}3$

Class name	$1A$	$2A$	$7A$	$7B$	$3A$	$3B$	$6A$	$6B$
Class size	1	7	24	24	28	28	28	28
χ_1	1	1	1	1	1	1	1	1
χ_2	1	1	1	1	ω	$\overline{\omega}$	ω	$\overline{\omega}$
χ_3	1	1	1	1	$\overline{\omega}$	ω	$\overline{\omega}$	ω
χ_4	3	3	η	$\overline{\eta}$	0	0	0	0
χ_5	3	3	$\overline{\eta}$	η	0	0	0	0
χ_6	7	-1	0	0	1	1	-1	-1
χ_7	7	-1	0	0	ω	$\overline{\omega}$	$-\omega$	$-\overline{\omega}$
χ_8	7	-1	0	0	$\overline{\omega}$	ω	$-\overline{\omega}$	$-\omega$

$$\omega = \frac{1}{2}(-1+\sqrt{-3}), \eta = \frac{1}{2}(-1+\sqrt{-7}).$$

$$B = \begin{pmatrix} 0&1&0&0&0&0&0\\ 0&0&1&0&0&0&0\\ 0&0&0&1&0&0&0\\ 0&0&0&0&1&0&0\\ 0&0&0&0&0&1&0\\ 0&0&0&0&0&0&1\\ 1&0&0&0&0&0&0 \end{pmatrix}, C = \begin{pmatrix} 1&0&0&0&0&0&0\\ 0&0&1&0&0&0&0\\ 0&0&0&0&1&0&0\\ 0&0&0&0&0&0&1\\ 0&1&0&0&0&0&0\\ 0&0&0&1&0&0&0\\ 0&0&0&0&0&1&0 \end{pmatrix}$$

Now B and C generate the subgroup 7:3, and the representation restricted to this subgroup has three inequivalent constituents, 1, 3 and $3'$, say. In just the same way as we did with A_5, we can calculate from first principles that the elements which extend this group to 7:6 by inverting B and centralizing C, are exactly those of shape

$$\begin{pmatrix} \alpha&\beta&\beta&\gamma&\beta&\gamma&\gamma\\ \beta&\beta&\gamma&\beta&\gamma&\gamma&\alpha\\ \beta&\gamma&\beta&\gamma&\gamma&\alpha&\beta\\ \gamma&\beta&\gamma&\gamma&\alpha&\beta&\beta\\ \beta&\gamma&\gamma&\alpha&\beta&\beta&\gamma\\ \gamma&\gamma&\alpha&\beta&\beta&\gamma&\beta\\ \gamma&\alpha&\beta&\beta&\gamma&\beta&\gamma \end{pmatrix}$$

The fact that this is an involution yields the equation

$$\alpha^2 + 3\beta^2 + 3\gamma^2 = 1,$$

as well as implying that this element is conjugate to A so has trace -1, that is

$$\alpha + 3\beta + 3\gamma = -1.$$

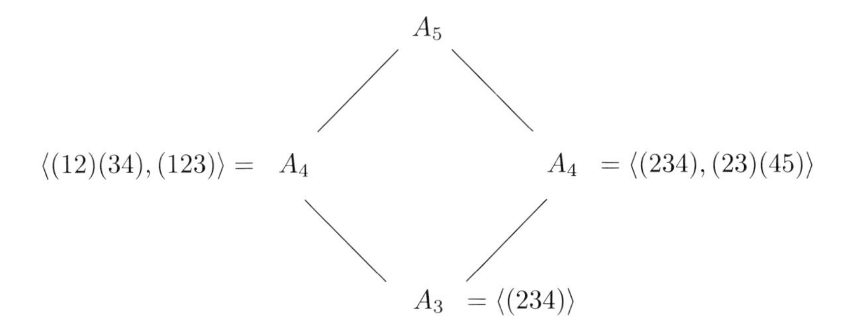

Figure 3.3: Generating A_5 from subgroups A_4

As before, this yields a 1-parameter family of solutions (in fact, ten solutions), and we have a choice of methods for determining the (unique) case which gives J_1, according to how much knowledge of the group structure we are prepared to use.

If we assume for example that we know that the centralizer of C (an element of order 3) is $3 \times D_{10}$, then this implies that the product of A with our new element has order 5, and therefore (from the 11-modular character computed by Janko) has trace 7 or 3. This gives $-\alpha + 3\beta - 3\gamma = 7$ or 3, so either $\beta = 1$ and $\alpha + 3\gamma = 7$, or $\beta = 4$ and $\alpha + 3\gamma = 9$, Using the remaining equation $\alpha^2 + 3\beta^2 + 3\gamma^2 = 1$, we reduce to the two cases $(\alpha, \beta, \gamma) = (0, 1, 6)$or $(9, 1, 3)$. We can then test these two cases further, and show that the last one is the correct one. (The condition stated, that the given element has order 5, is enough to eliminate the other case, but is a trifle tedious to check by hand.)

3.5 A Generalization

The method described above is asymmetrical, in that the two generating subgroups H and $N_G(L)$ are treated differently. This method, going from H to L and then to $N_G(L)$, is often the easiest in practice, but there is a more general method, in which $N_G(L)$ is replaced by a subgroup K which does not necessarily normalize L. Then we must start with representations of *both* H and K, and find appropriate subgroups of H and K which are isomorphic to L. Then we look at all possible ways of identifying these two subgroups.

To illustrate the method, we again consider constructing the 3-dimensional representation of A_5, using two subgroups A_4 intersecting in A_3, as shown in Figure 3.3.

For each of the groups H and K, isomorphic to A_4, we may take the representation given by the matrices

$$\begin{pmatrix} 1 & 0 & 0 \\ 0 & -1 & 0 \\ 0 & 0 & -1 \end{pmatrix} \text{ and } \begin{pmatrix} 0 & 1 & 0 \\ 0 & 0 & 1 \\ 1 & 0 & 0 \end{pmatrix}.$$

In both cases the subgroup A_3 is generated by the second matrix given. (In general we would need at this stage to find a matrix conjugating one copy of the subgroup to the other.)

The next step is to find *all* matrices which conjugate one copy of A_3 to the other, which in this case means all matrices which centralize A_3. This is the same calculation we have done already (this is no coincidence!), and we obtain all matrices of the form

$$\begin{pmatrix} \alpha & \beta & \gamma \\ \gamma & \alpha & \beta \\ \beta & \gamma & \alpha \end{pmatrix}.$$

Since conjugating by a scalar has no effect, this is really a 2-parameter family of cases to consider.

For ease of calculation, we first change basis so that the subgroup A_3 is diagonalized (this is a kind of 'standard basis' that we will see more of later). Thus we conjugate both H and K by $\begin{pmatrix} 1 & 1 & 1 \\ 1 & \omega & \omega^2 \\ 1 & \omega^2 & \omega \end{pmatrix}$ to obtain

$$\frac{1}{3}\begin{pmatrix} 1 & 1 & 1 \\ 1 & \omega^2 & \omega \\ 1 & \omega & \omega^2 \end{pmatrix}\begin{pmatrix} 1 & 0 & 0 \\ 0 & -1 & 0 \\ 0 & 0 & -1 \end{pmatrix}\begin{pmatrix} 1 & 1 & 1 \\ 1 & \omega & \omega^2 \\ 1 & \omega^2 & \omega \end{pmatrix} = \frac{1}{3}\begin{pmatrix} -1 & 2 & 2 \\ 2 & -1 & 2 \\ 2 & 2 & -1 \end{pmatrix}$$

and $\begin{pmatrix} 1 & 0 & 0 \\ 0 & \omega & 0 \\ 0 & 0 & \omega^2 \end{pmatrix}$. To obtain A_5 we now take H to be generated by these two matrices, and L to be generated by the latter of them, and K to be generated by L together with some conjugate of the first matrix by an element $\begin{pmatrix} \alpha & 0 & 0 \\ 0 & \beta & 0 \\ 0 & 0 & \gamma \end{pmatrix}$ which centralizes L. We may assume that $\alpha = 1$, which means that our second generator for K looks like

$$\frac{1}{3}\begin{pmatrix} \alpha^{-1} & 0 & 0 \\ 0 & \beta^{-1} & 0 \\ 0 & 0 & \gamma^{-1} \end{pmatrix}\begin{pmatrix} -1 & 2 & 2 \\ 2 & -1 & 2 \\ 2 & 2 & -1 \end{pmatrix}\begin{pmatrix} \alpha & 0 & 0 \\ 0 & \beta & 0 \\ 0 & 0 & \gamma \end{pmatrix}$$

$$= \frac{1}{3}\begin{pmatrix} -1 & 2\beta^{-1} & 2\gamma^{-1} \\ 2\beta & -1 & 2\beta\gamma^{-1} \\ 2\gamma & 2\gamma\beta^{-1} & -1 \end{pmatrix}.$$

We can now look at the product of this involution with the original involution from H, and proceed in any of the various ways we have already discussed, to determine which values of β and γ are consistent with the group generated being A_5.

3.6 Thompson's Theorem

At this point it is worth pausing to consider in general how many cases we are going to be left with. A theoretical analysis of this will help us to decide in any given case what choices of subgroups H, K and L are most suitable for the construction we have in mind. Thompson's theorem [26] tells us the answer to this question. His original paper refers only to complex representations, but it is easy to see that his argument nowhere depends on the field.

Theorem 3.1. *Suppose that H and K are two subgroups of $GL(V)$, intersecting in L. Then the $GL(V)$-orbits on pairs (H^x, K^y) with $L \subseteq H^x \cap K^y$, and with L^x conjugate to L in H, and L^y conjugate to L in K, are in one-to-one correspondence with the double cosets of $C_{GL(V)}(H)$ and $C_{GL(V)}(K)$ in $C_{GL(V)}(L)$, via the map $(H^x, K^y) \mapsto Hxy^{-1}K$.*

Proof. The condition that L^x is conjugate to L in H means that we may assume $x \in C_{GL(V)}(L)$, and similarly $y \in C_{GL(V)}(L)$. Conversely, if $x, y \in C_{GL(V)}(L)$, then $L^x = L$ and $L^y = L$, and $L = L^x \cap L^y \subseteq H^x \cap K^y$. Now if (H^a, K^b) is conjugate to (H^x, K^y), then there is an element $g \in GL(V)$ such that $H^x = H^{ag}$ and $K^y = K^{bg}$, and $L^g = L$. Again we may assume that $g \in C_{GL(V)}(L)$, and $agx^{-1} \in H$, and $bgy^{-1} \in K$, so that

$$Hab^{-1} = Hxg^{-1}.gy^{-1}K = Hxy^{-1}K.$$

Conversely, if $Hab^{-1}K = Hxy^{-1}K$, then there are elements $h \in H$ and $k \in K$ with $hab^{-1}k = xy^{-1}$, so that $x^{-1}ha = y^{-1}k^{-1}b = g$, say. Therefore

$$(H^x, K^y)^g = (H^{xg}, K^{yg}) = (H^{xx^{-1}ha}, K^{yy^{-1}k^{-1}b}) = (H^a, K^b).$$

□

To apply this result, we need to understand the centralizers of H and K in $GL(V)$. Over an algebraically closed field F of characteristic zero, such as $\mathbb{C}$, this is easy, as Schur's Lemma tells us that if the representation of H on V has r distinct constituents, with multiplicities $e_1, \ldots, e_r$, then

$$C_{GL(V)}(H) \cong GL_{e_1}(F) \times \cdots \times GL_{e_r}(F).$$

Over finite fields it can be a little more difficult, since Maschke's Theorem no longer holds.

Let us see how this works in the examples we have already looked at. For A_5 in dimension 3 over $\mathbb{C}$, generated by $H \cong A_4$ and $K \cong S_3$, the representation of H is absolutely irreducible, while the representation of K is the direct sum of the alternating representation and the degree 2 representation. The representation of $L \cong A_3$ is the direct sum of the three distinct irreducible representations, so $C_{GL(V)}(L) \cong (\mathbb{C}^\times)^3$. Similarly, $C_{GL(V)}(H) \cong \mathbb{C}^\times$ (that is, just scalars), and $C_{GL(V)}(K) \cong (\mathbb{C}^\times)^2$. Since all these groups obviously contain the scalar matrices, we can quotient them out, and obtain $\overline{C(L)} \cong (\mathbb{C}^\times)^2$, $\overline{C(H)} \cong 1$, and

$\overline{C(K)} \cong \mathbb{C}^\times$. Thus the double cosets of $C(H)$ and $C(K)$ in $C(L)$ correspond to the single cosets of $\overline{C(K)}$ in $\overline{C(L)}$, so to the elements of $\overline{C(L)}/\overline{C(K)} \cong \mathbb{C}^\times$. This is reflected in the fact that after using all the available information about H, K and L separately, we were left with a single unknown to solve for.

If, as in this case, the three centralizers are all abelian (which is equivalent in characteristic zero to the representation being a direct sum of distinct irreducible representations), then the double coset condition can be replaced by a single coset condition. Specifically, the double cosets are in one-to-one correspondence with the elements of $C(L)/C(H).C(K)$. But

$$C(H).C(K) \cong C(K)/(C(H) \cap C(K)),$$

and if the underlying field is $\mathbb{C}$ then all these groups are direct products of copies of $\mathbb{C}^\times$, so are characterized by their dimension. So the number of unknowns equals the dimension of $C(L)/C(H).C(K)$, which is

$$\begin{aligned} &\dim C(L) - \dim(C(H).C(K)) \\ &= \dim C(L) - (\dim C(H) + \dim C(K) - \dim(C(K) \cap C(H)) \\ &= \dim C(L) - \dim C(H) - \dim C(K) + \dim(C(K) \cap C(H)). \end{aligned}$$

Now there is no loss in assuming we are trying to construct an irreducible representation, where $C(H) \cap C(K) = C(G)$ consists only of scalars, that is, has dimension 1. In the above example we have $\dim C(L) = 3$, $\dim C(H) = 1$ and $\dim C(K) = 2$, so the number of unknowns is $3 - 1 - 2 + 1 = 1$.

In the construction of J_1, we had $H \cong 2^7{:}7{:}3$ acting irreducibly, $L \cong 7{:}3$ acting as $1+3+3'$ over $GF(11)$, and $K \cong 7{:}6$ acting as $1+6$. (Here and elsewhere we denote representations by their degrees, with $'$ or letters to distinguish different representations of the same degree.) Thus we have $\dim C(H) = 1$, $\dim C(L) = 3$ and $\dim C(K) = 2$, so the number of unknowns is $3 - 1 - 2 + 1 = 1$ again.

Finally we mention the example which motivated Thompson's statement of the theorem, namely the uniqueness of the Monster. He assumed the existence of the degree 196883 character, and calculated its values. Then restricting to the maximal subgroup $2^{1+24}Co_1$, the character becomes

$$299 + 98280 + 98304,$$

and restricting further to $2^{1+24}2^{11}M_{24}$ it becomes

$$23 + 276 + 552 + 48576 + 49152 + 98304.$$

The latter group is the intersection of the former with another maximal subgroup, of shape $2^{2+11+22}(M_{24} \times S_3)$, with character

$$23 + 828 + 48576 + 147456.$$

Thompson's formula now gives the number of unknowns as

$$\dim C(L) - \dim C(H) - \dim C(K) + 1 = 6 - 3 - 4 + 1 = 0.$$

In other words, there is no freedom at all, the situation is uniquely determined, and so there is at most one Monster with the given properties.

3.7 The Meataxe

If we are to construct much larger representations than those described above, then we are going to need some computational assistance. What is required is more or less exactly what was provided by the first versions of Richard Parker's Meataxe package [16], namely some elementary linear algebra operations such as matrix multiplication and Gaussian elimination, together with a "split" algorithm for chopping representations into irreducible constituents, and a "standard basis" algorithm for finding equivalences between irreducibles. Let us take the elementary operations as read, and describe briefly the operations of "split" and "standard basis".

Both of these algorithms require a "seed vector" which can be obtained in a variety of ways. The original implementation of the Meataxe assumed that the underlying group was generated by two elements A and B, and calculated the nullspaces of a fixed sequence of group algebra elements ($A + B + AB$, $A + B + AB + ABB$, ...). For "split", one can take any non-zero null vector of any of these "words", while for standard basis one takes the first one where the nullity is 1. Later implementations by Ringe [19] and Holt and Rees [5] have more general techniques which work even in situations where Parker's original implementation will fail.

The "split" algorithm takes the seed vector, and repeatedly applies the group generators, building up a basis for a subspace in echelon form, until the subspace is invariant under the action of the group. The action of the group on the subspace can then be readily calculated, by taking the images of the echelonized basis under the group generators, and expressing these as linear combinations of the basis vectors. The action on the quotient space is obtained by a similar exercise in Gaussian elimination.

By repeated application of this technique, a representation can be chopped up into its irreducible constituents. Moreover, Norton's irreducibility criterion [16] can be used to prove that the pieces so obtained really are irreducible, so cannot be split further. In its simplest form, this criterion states that if an element of nullity 1 fails to split a representation (a) regarded as acting on row vectors from the right, *and* (b) regarded as acting on column vectors from the left, then the representation is irreducible. (Strictly speaking, one of these representations is for the group algebra itself, while the other is for the opposite algebra, but since these are isomorphic, the distinction is not important.)

Finally, we describe the "standard basis" algorithm. There are a number of variants of the algorithm, but they all produce a basis with the property that the generating matrices assume a standard form when written with respect to this basis. Thus two equivalent representations of the same group can both be put into canonical form, or one can be conjugated to the other inside the ambient general linear group. There are two essential steps to this. First we must find a basis which respects a composition series for the module. This can be achieved by keeping track of all the basis changes involved in the repeated splitting of the module, or by more sophisticated techniques developed in later implementations of the Meataxe. Second, we must find a "standard basis" for each irreducible constituent.

The first step generalizes finding the eigenspaces in the abelian case considered earlier. The second step was not necessary in that case, as all vectors in a given eigenspace are equivalent from this point of view.

The second step requires an algorithm like that described and implemented by Parker [16]. He first defined a suitable seed vector v in a basis-independent way, as above. Then he applied the group generators g_1 and g_2 to produce a list of (row) vectors in canonical order

$$v, vg_1, vg_2, vg_1g_1, vg_1g_2, vg_2g_1, vg_2g_2, \ldots$$

Only those vectors which are linearly independent of all the previous ones are kept. Since the space is finite-dimensional, we obtain after a finite number of steps a "standard basis" for the underlying vector space. If S denotes the matrix whose rows form this basis, then the conjugated generators Sg_iS^{-1} are in canonical form.

There is one other significant issue, which was not important in Parker's original applications of the Meataxe, where one only deals with a single group, with a fixed set of generators. In our applications, however, we need the idea of "standard generators". In our generalized method (Section 3.5), we obtain two copies of a group L, and need to identify them. Computationally this is only possible if we have corresponding sets of generators in the two groups. In practice, one needs to define a set of generators sufficiently precisely that they are determined up to isomorphism, and they also need to be reasonably easy to find in practice (see [28]).

3.8 Some more Applications

As mentioned above, the original motivation for the development of these techniques came from the need to produce existence and uniqueness proofs for the sporadic simple groups. These ranged from Janko's construction of J_1, which can be repeated by hand as indicated above, to the Cambridge team's construction of J_4 (see [15]), where Benson, Conway, Norton, Parker and Thackray proved existence and uniqueness of J_4 using 112×112 matrices over $GF(2)$ which had been produced by Norton and Parker.

More recently, different motivations have become important, for example, the desire to study properties of the group in question which are not amenable to theoretical analysis. I am thinking mainly of (a) the complete determination of the maximal subgroups, which was at least part of the motivation behind my construction of the Baby Monster [29], which led to [30], [31], [32], etc., and (b) modular characters and representations, which was again part of the motivation behind [8], [9], [10], [3], and led to solutions of many such problems in the modular atlas [7].

We next describe briefly the Norton–Parker construction of J_4, as it was historically the first example where the computational Meataxe techniques were of necessity used.

This is the first example we have considered in which the representation does not restrict to completely reducible representations of all the subgroups used. Moreover, the representation itself was only conjectured to exist, and even one of the subgroups used was not known to be a subgroup. These facts could only be proved *after* the group constructed was shown to be J_4.

The conjecture was that J_4 had a 112-dimensional representation over $F = GF(2)$, and that it had a subgroup isomorphic to $U_3(11){:}2$. If so, then the representation would restrict to this subgroup as a uniserial module with structure $\begin{smallmatrix}1\\110\\1\end{smallmatrix}$. Restricting further to the Borel subgroup, or Sylow 11-normalizer, $11^{1+2}{:}(5 \times SD_{16})$, the constituents remain irreducible, but the representation becomes a direct sum $\begin{smallmatrix}1\\1\end{smallmatrix} \oplus 110$ of a 2-dimensional uniserial module and a 110-dimensional irreducible. Now the Sylow 11-normalizer in J_4 is $11^{1+2}{:}(5 \times 2S_4)$, which contains the above group to index 3, and has representation $2 \oplus 110$.

We have $C(H) = C(L) \cong C_2$, and $C(K) = 1$, so again there is a unique possibility. Half of the difficulty in the construction is making the appropriate representation of $U_3(11){:}2$ (or indeed in proving that it has this structure in the first place!). It turns out that a uniserial module of structure $\begin{smallmatrix}1\\110\end{smallmatrix}$ can be obtained as a quotient module of the permutation module of degree 1332 (obtained by permuting the isotropic 1-spaces in the natural module). Similarly, a module with structure $\begin{smallmatrix}110\\1\end{smallmatrix}$ can be obtained as a submodule of this permutation module.

Now we can use a standard basis method to obtain generators g_i for the group in these two representations in the forms

$$g_i \mapsto \begin{pmatrix} 1 & 0 \\ v_i & M_i \end{pmatrix}, g_i \mapsto \begin{pmatrix} M_i & 0 \\ w_i & 1 \end{pmatrix},$$

where the M_i are 110×110 matrices, v_i are column vectors, and w_i are row vectors. This enables us to write down matrices

$$\begin{pmatrix} 1 & 0 & 0 \\ v_i & M_i & 0 \\ \lambda_i & w_i & 1 \end{pmatrix},$$

and all we need to do is determine which of the two possible values $0, 1$ to use for each λ_i. A process of trial and error soon tells us which possibilities give $U_3(11){:}2$ and which give $2 \times U_3(11){:}2$.

Now the original construction proceeded by finding a subgroup $11^{1+2}{:}5$, on which the representation restricts as the direct sum of a 2-dimensional trivial module, and a 55-dimensional irreducible $GF(4)$-module (regarded as a 110-dimensional irreducible over $GF(2)$). Then an element of order 3 was adjoined which normalizes this group in the correct way. In effect, we have the direct sum of two irreducible $GF(4)$-modules, and the usual calculation shows that there are just three possibilities to consider, and it is easy to show by computer calculation of the order of some suitable elements that only one of these cases is consistent with generating J_4.

More details of this construction, including a full construction by hand of the representation of $U_3(11){:}2$, can be found in Norton's article [15].

3.9 The Baby Monster

With practice, the methods described above become fairly routine to apply, and the size of the group, or even the representation, is not necessarily an obstacle. This is demonstrated by the computer construction of the 4370-dimensional representation of the Baby Monster group over $GF(2)$, described in [29].

In this case, the chosen strategy was to start with the subgroup Fi_{23}, which turns out to act on the space as the direct sum of two irreducible modules, $782 \oplus 3588$, and then restrict to $O_8^+(3)$, under which the 782 breaks up as $1 \oplus 1 \oplus 260abc$, that is, the direct sum of two copies of the trivial representation and three distinct 260-dimensional representations. The 3588 similarly breaks up as the direct sum $260def \oplus 2808$ of three more 26-dimensional representations, and a 2808-dimensional one.

The extra element required must act on $O_8^+(3)$ as a specified outer automorphism, in the normal 2^2 subgroup of the outer automorphism group S_4. Therefore it acts trivially on the 2-dimensional fixed space, and permutes the 260-dimensional constituents in a specified manner. Once the standard basis calculation is done (which was the major part of the computation), there is a unique possibility, which can be written down.

The only other problem was constructing the 3588-dimensional representation of Fi_{23}. It turned out, by a lucky chance, that this was contained in the permutation representation on 31671 points. While chopping a representation of this size was a challenge in 1991, it may be considered routine today.

3.10 An Analogue for Permutation Groups

So far we have considered methods for constructing matrix groups over finite fields, and matrix groups over fields of characteristic 0. All these methods are very similar, and one is naturally inclined to ask, what other types of representations can be constructed by the same sort of procedure? The two main requirements, beyond basic manipulations, are a "standardization" algorithm of some sort (essentially, a constructive isomorphism test), and a method of eliminating all the unwanted cases (of which there may be infinitely many).

In the case of permutation groups, the latter is not a problem: both in principle, because permutation groups are finite, and in practice. The analogue of the "standard basis" procedure is a procedure for producing a "standard ordering" of the points. To obtain such an ordering we need firstly a "seed point", and secondly an algorithm for generating the remaining points of the given orbit in a standard order.

The obvious way to obtain a seed point is to take a fixed point of a subgroup. Indeed, we can find words in the generators of our group L, which give generators for the stabilizer of a point. Then we want to take the fixed point of this "standard subgroup" as our seed point—however, if the point stabilizer is not self-normalizing in L, there will not be a unique fixed point. On the other hand, the centralizer of L in the full symmetric group acts regularly on the fixed points of L. In particular, the order of this centralizer equals the number of fixed points of L. Thus for many of our purposes it does not matter which fixed point we take, while for other purposes we will need to consider all of them in order to determine which is the right one.

Having obtained a seed point, we need an algorithm to obtain the rest of the points in a specified order. This is essentially the same as for matrix groups, except that it is much easier: instead of adding a new vector to our standard basis when it is linearly independent of all the previous vectors, we simply add a new point to the list when it is distinct from all previous points.

Let us consider an example. Suppose we wish to construct the permutation representation of A_5 on 6 points, using the same subgroups as before: that is, we start with A_4, restrict to A_3, and finally extend to S_3. The representation restricts to a transitive representation of A_4, which is uniquely determined and may be taken to be generated by the permutations $(12)(34)$ and $(135)(246)$. Restricting to A_3, generated by $(135)(246)$, we need a standard ordering of the points with respect to the given generator of this subgroup. Here there are two orbits, and the point stabilizer is trivial in each case. Thus we have three possible seeds for each orbit, giving 9 possibilities for a choice of standard ordering, as well as two possibilities for which orbit to take first. Let us choose arbitrarily the standard ordering $(1,3,5,2,4,6)$ for this first case.

Next we must find all the possible standard orderings for the subgroup A_3 with respect to the inverse generator $(153)(264)$. We may choose the seed point for the first orbit arbitrarily, but then we have three choices for the seed point in the second orbit. Moreover, we do not know whether or not the two orbits are interchanged by the extra generator. Thus we have six possibilities for the second standard ordering of the points, as follows:

$$(1,5,3,2,6,4),(1,5,3,6,4,2),(1,5,3,4,2,6),$$

$$(2,6,4,1,5,3),(2,6,4,5,3,1),(2,6,4,3,1,5).$$

These give rise to the following six possibilities for our extra generator, given by mapping $(1,3,5,2,4,6)$ to one of the above six orderings:

$$(35)(46),(35)(26),(35)(42),(12)(36)(45),(125436),(123654).$$

We could now, for example, eliminate the last three as they are odd permutations, and A_5 is a simple group. In the first three cases, we multiply our new element by one of the original elements of the A_4, and obtain:

$$(12)(34).(35)(46) \quad = \quad (12)(3645)$$

$$\begin{array}{rcl}(12)(34).(35)(26) & = & (162)(345)\\ (12)(34).(35)(24) & = & (14532)(6).\end{array}$$

This enables us to eliminate the first case, as A_5 contains no elements of order 4. The other two cases are interchanged by the involution which centralizes the A_4, namely (12)(34)(56), so are equivalent. We conclude that the 6-point representation of A_5 may be generated by the three elements

$$(12)(34), (135)(246), (35)(26).$$

The alert reader may have noticed that the "standard ordering" part of the argument in this case involves nothing more nor less than finding the centralizer of the subgroup A_3 in the full symmetric group S_6. This is exactly analogous to the matrix group case, where the standard basis argument is what allows us to write down the centralizer of the subgroup in the corresponding general linear group.

A similar but more interesting example is the construction of $L_2(11)$ on 11 points (as opposed to its natural action which is on 12 points). We know that $L_2(11)$ may be generated by a Frobenius group 11:5 (the image modulo scalars of the group of invertible upper triangular matrices), and a dihedral group D_{10} (the image of the monomial subgroup), intersecting in a 5 (the image of the diagonal subgroup).

As permutations on 11 points $\{0,1,2,3,4,5,6,7,8,9,X\}$ we may take the group 11:5 to be generated by $(0123456789X)$ and $(13954)(267X8)$. There are just ten ways of extending the 5 to D_{10}, half of which interchange the two 5-cycles, so involve odd permutations. The other five are $(34)(59)(68)(7X)$, $(34)(59)(27)(8X)$, $(34)(59)(28)(6X)$, $(34)(59)(26)(78)$, and $(34)(59)(2X)(67)$.

We calculate the five products of these with our original element of order 11, and obtain the following permutations:

$$(01235X87)(4)(69),$$

$$(0128)(35X967)(4),$$

$$(01296)(35X78)(4),$$

$$(01279635X)(4)(8),$$

$$(012)(35X)(4)(689)(7).$$

We thus eliminate the first, second, and fourth, since $L_2(11)$ contains no elements of order 8, 12 or 9. The other two are interchanged by the involution $(1X)(29)(38)(47)(56)$ which normalizes the original Frobenius group, and centralizes the 5. This means that we actually get two distinct actions of $L_2(11)$ on 11 points, interchanged by the outer automorphism of $L_2(11)$.

3.11 Some Applications of the Permutation Group Construction

The method just described has been used in some form or other for proving the existence of many of the sporadic simple groups. One case which is quite easy to describe is the Higman–Sims group. This group first came to light as a rank 3 extension of M_{22}, acting on 100 points, and was constructed by Donald Higman and Charles Sims [4]. Specifically, they constructed a graph of valence 22 on 100 points, defined as a special point $*$, the 22 points permuted by M_{22}, and the 77 hexads defined on those points, by joining $*$ to the 22 points, joining a hexad to the six points it contains, and joining two hexads if they are disjoint as sets. This graph is clearly invariant under M_{22}.

Now restrict to the point stabilizer $M_{21} \cong L_3(4)$ in M_{22}. This has orbits $1+21$ on the 22 points, and orbits $21+56$ on the 77 hexads. The point stabilizers in the three non-trivial orbits are $2^4{:}A_5$, the other conjugacy class of $2^4{:}A_5$, and one of the classes of A_6. In terms of the group $L_3(4)$, the vertices of the graph are 21 "points" and 21 "lines" of the projective plane, and 56 "ovals" consisting of 6 points, no three of which lie on a line. Now all of these point stabilizers are self-normalizing in $L_3(4)$, so by our construction it follows that there is a unique extension to $L_3(4){:}2$ in which the outer automorphism exchanges the two fixed points and the two orbits of size 21. The only question is, what is the group generated by M_{22} and this extra element? Clearly there are just two possibilities, namely a 2-transitive group which is easily seen to be A_{100}, or a group G in which the point stabilizer is M_{22}, so that G has order $44,352,000$.

Now we can describe the edges of the graph purely in terms of the group $L_3(4)$: a point is joined to a line if and only if they are incident in the projective plane; an oval is joined to an oval if and only if they do not meet in the projective plane; an oval is joined to the six points it contains, and to the six lines which do not meet it. The duality automorphism which we adjoin interchanges points and lines, but preserves incidence between them. It also preserves the ovals: an oval considered as a set of six points is mapped to an oval considered as the six points which lie on only one of a set of six lines. Thus it is an elementary exercise to show that the edges of the graph are preserved by duality, which implies that the group generated by M_{22} and this duality automorphism of $L_3(4)$ is a simple group of order $44,352,000$.

A rather harder construction of this type was Livingstone's construction of J_1 as a permutation group on 266 points [13]. This was obtained entirely by hand. The idea here is take a group $L_2(11)$, and a subgroup A_5, and extend the latter to $A_5 \times 2$. By showing that there is such a group satisfying Janko's hypotheses, Livingstone obtained an independent existence proof of J_1, as well as a proof that $L_2(11)$ is a subgroup. The details are messy, so we only give a rough outline of the construction.

We start with $L_2(11)$ acting on 266 points, with orbits of size 1, 11, 110, 132, and 12, and corresponding point stabilizers $L_2(11)$, A_5, S_3, 5, and 11:5. We construct a graph of valence 11 on these points: we label the 110 vertices by ordered pairs of the 11 points, and we label the 132 vertices by ordered pairs of the 12 points, and join each point to each of the pairs that it lies in.

The remaining edges join the various pairs, and are harder to describe. We now restrict to the subgroup A_5, which has orbits $1+10$ on the 11, $10+10+30+60$ on the 110, $12+60+60$ on the 132, and is transitive on the 12. Now we need to calculate the index of each point stabilizer in its normalizer, in A_5. This is 1 for the orbits of size 1 and 10, 2 for the orbits of size 12 and 30, and 60 for the regular orbits. It is easy to see that the extra involution exchanges the two fixed points, the two orbits of size 12, two of the orbits of size 10, and two of the orbits of size 60. Therefore the total number of such involutions in S_{266} is 14400. Remarkably, Livingstone was able to use the graph to determine exactly which of these 14400 cases gives rise to J_1, and to deduce the existence of J_1.

3.12 The Monster

When we come to the largest of the sporadic simple groups, the so-called 'Monster', it has to be admitted that the sheer size of the group and the representation do play a role. The smallest representation is of degree 196882 over $GF(2)$, and simply to write down two generating matrices would require approximately 10GB of memory. While this is not beyond the resources of modern computational facilities, it is beyond the resources of the average desktop workstation, and some improvements to the basic methods seem desirable. Moreover, a naive matrix multiplication might take some months of processor time, which we would wish to avoid if at all possible.

The process by which a computer construction of the Monster was achieved is described in some detail in [12]. We remark first that all our improvements are technical ones, and that there is no difference in the basic method between our construction of the Monster, and the construction of A_5 in Section 3.2 above.

One of these technical improvements concerns calculations with induced representations. The examples we use are induced from the 2-dimensional representation of S_3, but the same principles can be applied to any induced representation. In essence we are trying to produce a "standard basis" for induced representations which respects the block structure of the matrices. Thus we first use a matrix standard basis on the representation from which we induce up. This corresponds to writing the individual blocks with respect to a standard basis. Then we use a permutation standard ordering on the blocks themselves to produce a full standard basis for the induced representation.

There are very great advantages in using standard bases of this form. The most obvious is the potentially enormous saving of time and space in keeping to what may be very sparse matrices. However, there are occasions, which arise in the Monster, where things are not quite as pleasant as one might hope. For example, a representation may be an induced representation in several inequivalent ways. This results in our final standard basis matrix, which extends $3^{1+12}{\cdot}3{\cdot}Suz$ to the Monster, containing some 324×324 blocks, in addition to the large number of 2×2 blocks which we expected. Nevertheless, this element of the Monster may be encoded as a permutation on 87480 points, a list of 87480 elements of S_3, two 324×324 matrices and a 538×538 matrix over $GF(2)$, and can therefore be stored in less than 350 kB, which is a massive improvement on 5 GB!

3.13 A World-Wide-Web Atlas of Group Representations

Now that constructing matrix representations of simple groups, and others, has become relatively routine, it seems appropriate to approach the problem systematically, and construct all "reasonable" representations in a wholesale manner. The ever-decreasing price of computer memory means that large numbers of representations can be stored in a database for retrieval by interested researchers. Some first steps in that direction have been taken in our world-wide-web atlas of group representations [35], which contains some 700 representations of the sporadic groups and other interesting groups. A start has been made on providing some other information about the groups. For example, we have programs which produce lists of conjugacy class representatives for all the sporadic simple groups of order up to 10^{16}, and other programs which produce representatives of the conjugacy classes of maximal subgroups, for many of these groups.

Acknowledgements

I would like to thank Professor Gerhard Michler for inviting me to give the two expository lectures at this conference, which are revised and extended in the present paper. I am very grateful to Richard Parker, who taught me almost everything that is described here.

Bibliography

[1] J. H. Conway, R. T. Curtis, S. P. Norton, R. A. Parker and R. A. Wilson: *An* ATLAS *of Finite Groups*, Clarendon Press, Oxford, 1985.

[2] G. Cooperman, L. Finkelstein, M. Tselman and B. York: *Constructing permutation representations for large matrix groups*, in: Proc. ISSAC-94. ACM Press, 1994.

[3] H. Gollan: *Über die Konstruktion modularer Darstellungen am Beispiel der einfachen Tits-Gruppe* ${}^2F_4(2)'$ *in Charakteristik* 5, Dissertation, Essen, 1988.

[4] D. G. Higman and C. C. Sims: *A simple group of order* 44, 352, 000, Math. Z. **105**, (1968), 110–113.

[5] D. F. Holt and S. Rees: *Testing modules for irreducibility*, J. Austral. Math. Soc. Ser. A **57**, (1994), 1–16.

[6] Z. Janko: *A new finite simple group with Abelian Sylow 2-subgroups, and its characterization*, J. Algebra **3**, (1966), 147–186.

[7] C. Jansen, K. Lux, R. A. Parker and R. A. Wilson: *An Atlas of Brauer Characters*, Clarendon Press, Oxford, 1995.

[8] C. Jansen and R. A. Wilson: *Two new constructions of the O'Nan group*, J. London Math. Soc. **56**, (1997), 579–583.

[9] C. Jansen and R. A. Wilson: *The 2- and 3-modular characters of the O'Nan group*, J. London Math. Soc. **57**, (1998), 71–90.

[10] C. Jansen and R. A. Wilson: *The minimal faithful 3-modular representation for the Lyons group*, Comm. Algebra **24**, (1996), 873–879.

[11] D. L. Johnson: *Topics in the theory of group presentations*, LMS Lecture Note Series **42**, Cambridge University Press, 1980.

[12] S. A. Linton, R. A. Parker, P. G. Walsh and R. A. Wilson: *Computer construction of the Monster*, J. Group Theory, to appear.

[13] D. Livingstone: *On a permutation representation of the Janko group*, J. Algebra **6**, (1967), 43–55.

[14] W. Meyer, W. Neutsch and R. Parker: *The minimal 5-representation of Lyons' sporadic group*, Math. Ann. **272**, (1985), 29–39.

[15] S. P. Norton: *The construction of J_4*, in: *The Santa Cruz conference on finite groups (eds. B. Cooperstein and G. Mason)*, pp. 271–277. Proceedings of Symposia in Pure Mathematics, Vol. 37, American Mathematical Society, 1980.

[16] R. A. Parker: *The computer calculation of modular characters (The 'Meataxe')*, in: *Computational Group Theory (ed. M. D. Atkinson)*, Academic Press, 1984, pp. 267–274.

[17] R. A. Parker: *An integral Meat-axe*, in: *The Atlas 10 years on (eds. R. T. Curtis and R. A. Wilson)*, pp. 215–228. LMS Lecture Note Series **249**, Cambridge University Press, 1998.

[18] R. A. Parker and R. A. Wilson: *Computer construction of matrix representations of finite groups over finite fields*, J. Symbolic Comput. **9**, (1990), 583–590.

[19] M. Ringe: *The C Meataxe 2.3, documentation,*, RWTH Aachen, 1995.

[20] A. Ryba and R. Wilson: *Matrix generators for the Harada-Norton group*, Experimental Math. **3**, (1994), 137–145.

[21] M. Schönert et al.: *GAP 3.4 Manual (Groups, Algorithms, and Programming)*, RWTH Aachen, 1994.

[22] I. A. Suleiman and R. A. Wilson: *Computer construction of matrix representations of the covering group of the Higman–Sims group*, J. Algebra **148**, (1992), 219–224.

[23] I. A. Suleiman and R. A. Wilson: *Construction of the fourfold cover of the Mathieu group M_{22}*, Experimental Math. **2**, (1993), 11–14.

[24] I. A. Suleiman and R. A. Wilson: *Covering and automorphism groups of* $U_6(2)$, Quart. J. Math. (Oxford) **48**, (1998), 511–517.

[25] I. A. Suleiman and R. A. Wilson: *Construction of exceptional covers of generic groups*, Math. Proc. Cambridge Philos. Soc., to appear (1998).

[26] J. G. Thompson: *Finite-dimensional representations of free products with an amalgamated subgroup*, J. Algebra **69**, (1981), 146–149.

[27] P. G. Walsh: *Computational study of the Monster and other sporadic groups*, Ph.D. thesis, Birmingham, 1996.

[28] R. A. Wilson: *Standard generators for sporadic simple groups*, J. Algebra **184**, (1996), 505–515.

[29] R. A. Wilson: *A new construction of the Baby Monster, and its applications*, Bull. London Math. Soc. **25** (1993), 431–437.

[30] R. A. Wilson: *Some new subgroups of the Baby Monster*, Bull. London Math. Soc. **25**, (1993), 23–28.

[31] R. A. Wilson: *More on the maximal subgroups of the Baby Monster*, Arch. Math. (Basel) **61**, (1993), 497–507.

[32] R. A. Wilson: *The maximal subgroups of the Baby Monster, I*, J. Algebra, to appear.

[33] R. A. Wilson: *Matrix generators for Fischer's group Fi_{24}*, Math. Proc. Cambridge Philos. Soc. **113**, (1993), 5–8.

[34] R. A. Wilson: *A construction of the Lyons group in $GL_{2480}(4)$ and a new uniqueness proof*, Arch. Math. (Basel) **70**, (1998), 11–15.

[35] R. A. Wilson et al.: *A world-wide-web atlas of group representations*, `http://www.mat.bham.ac.uk/atlas/`

Part II

Keynote Articles

Progress in Mathematics, Vol. 173,

Chapter 4

Derived Tubularity: a Computational Approach

M. Barot and J. A. de la Peña

Let A be a finite dimensional algebra over an algebraically closed field. We denote by mod_A the category of finite dimensional left A-modules and by $D^b(A)$ the derived category of mod_A (see for example [17]] for definitions). We say that two algebras A and B are *derived equivalent* if their derived categories $D^b(A)$ and $D^b(B)$ are equivalent as triangulated categories.

An important problem in the Representation Theory of Algebras has been to characterize those algebras A which are derived equivalent to well understood classes of algebras, for instance to representation-finite hereditary algebras [16], [17], to tame hereditary algebras [16] or to tubular algebras [18], [3]. The aim of this work is to discuss the above mentioned characterizations from a computational point of view. In particular, we present an *algorithm* to decide whether or not certain classes of algebras are derived equivalent to a tubular algebra. This algorithm has been implemented in a C++ program which will be available as part of the CREP package, see [12].

Let A be an algebra as above with finite global dimension. Let $K_0(A) \xrightarrow{\sim} \mathbb{Z}^n$ be its Grothendieck group which we will consider equipped with a (non-symmetric) bilinear form $\langle -, - \rangle_A$ such that for two modules $X, Y \in \mathrm{mod}_A$ we have,

$$\langle [X], [Y] \rangle_A = \sum_{i=0}^{\infty} (-1)^i \dim_k \mathrm{Ext}_A^i(X, Y).$$

The associated quadratic form $\chi_A(v) = \langle v, v \rangle_A$ is called the *Euler form* of A. For two derived equivalent algebras A and B, the Euler forms χ_A and χ_B are equivalent.

If χ_A is non-negative, then $\mathrm{rad}\, \chi_A = \{v \in \chi_A : \chi_A(v) = 0\}$ is a subgroup of $\mathbb{Z}^n$ and there is an induced positive definite form $\bar{\chi}_A : \mathbb{Z}^n / \mathrm{rad}\, \chi_A \to \mathbb{Z}$ (see (4.4)). The form $\bar{\chi}_A$ accepts only finitely many *positive roots* (a root $v \in \mathbb{Z}^n / \mathrm{rad}\, \chi_A \cong \mathbb{Z}^{n-s}$ satisfies $\bar{\chi}_A(v) = 1$), see [33]. The main result behind our algorithm is the following (for the definition of strong simple connectedness see Section 4.1.2).

Theorem 4.1. *(Main Theorem) Let $A = kQ/I$ be a strongly simply connected algebra and assume that its Grothendieck group is $K_0(A) \cong \mathbb{Z}^n$ with $n > 6$.*

Then A is derived equivalent to a tubular algebra if and only if the following conditions are satisfied:

(a) χ_A *is non-negative of* corank $\chi_A = 2$*;*

(b) *there exists a source or a sink* a *of* Q *such that* $Q \setminus \{a\}$ *is connected and the convex subcategory* B *of* A *with vertices* $Q \setminus \{a\}$ *has a non-negative Euler form of corank one.*

(c) $\overline{\chi}_A$ *has* 36, 63 *or* 120 *positive roots.*

We briefly describe the contents of the paper. In Section 4.1 we recall the relevant characterizations of derived equivalences mentioned above. In Section 4.2 we survey some results related to algorithmic procedures for integral quadratic forms. Moreover we show in (4.2.10) (see also [24]):

Theorem 4.2. *Each connected non-negative unit form* q *determines uniquely a Dynkin graph* Δ *such that* $\overline{q}$ *is* $\mathbb{Z}$*-equivalent to the form* p_Δ *asociated with* Δ*.*

Using this theorem we prove in Section 4.3 the main result. Finally, in Section 4.4 we present the list of all algebras whose underlying quiver is a simple oriented line and which are derived equivalent to a tubular algebra. This list was calculated using our computer program.

The present work is an extended version of the report presented in the meeting 'Computational Methods for Representations of Groups and Algebras' held in Essen in April 1997. We thank SFB 343 Bielefeld for supporting our participation in the meeting. The first author thankfully acknowledges support from PADEP, UNAM and the second from DGAPA, UNAM and CONACyT.

4.1 Derived Equivalence for Algebras

4.1.1 Let A be a basic, finite-dimensional k-algebra over a fixed algebraically closed field k. It is well-known that $A = kQ/I$ where Q is a finite quiver and I is an admissible ideal of the path algebra kQ. We shall consider A as a *spectroid*, that is, as a k-category whose objects are the vertices Q_0 of Q and the space of maps from x to y is $A(x,y) = e_y A e_x$, where e_x denotes the primitive idempotent associated to the vertex x, see [14].

We assume that $Q_0 = \{1, \ldots, n\}$ and that Q is connected and has no oriented cycle. In particular, the global dimension of A is finite.

4.1.2 We recall that a vertex $a \in Q_0$ is *separated* in A if any two different direct summands of the radical rad P_a of the indecomposable projective module P_a (= projective cover of the simple S_a) have their supports in different connected components of the quiver $Q^{(x)} = Q \setminus \{y \in Q_0\colon$ there is a path from y to $x\}$. The algebra A is *separated* if every vertex $a \in Q_0$ is separated.

Finally, A is *strongly simply connected* if every full and convex (= path closed) subcategory B of A is separated. See [35] for equivalent properties.

If A is representation-finite and separated, then A is strongly simply connected [11]. In any case, 'separation' is a condition which may be easily checked.

4.1.3 A basic, finite-dimensional and hereditary algebra H is a path algebra $k\Delta$ for Δ a finite connected quiver without oriented cycle. Recall that H is representation-finite (resp. tame) if and only if Δ is of Dynkin (resp. extended-Dynkin) type. In [16] it was shown that A is derived equivalent to H if and only if A is *tilting-cotilting equivalent* to H, that is, there is a sequence of algebras $A = A_0, A_1, \dots, A_m = H$ and a sequence of modules ${}_{A_i}T^{(i)}$ $(0 \le i < m)$ such that $A_{i+1} = \mathrm{End}_{A_i} T^{(i)}$ and $T^{(i)}$ is either a tilting or a cotilting module. Then we may reformulate a result in [4, (5.1)], in the following way.

Theorem 4.3. [4] *Let $A = kQ/I$ be a separated algebra. Then A is derived equivalent to $k\Delta$ with Δ a quiver of Dynkin type if and only if χ_A is positive definite.*

4.1.4 That the positivity of the Euler form is preserved by derived equivalence is a consequence of the following general argument: if $F\colon D^b(A) \to D^b(B)$ is an equivalence of triangulated categories, then there is an induced isomorphism $f\colon K_0(A) \to K_0(B)$ such that $f([X^{\cdot}]) = [FX^{\cdot}]$, for any object $X^{\cdot} \in D^b(A)$, where $[X^{\cdot}] = \sum_{i\in\mathbb{Z}}(-1)^i[X^i] \in K_0(A)$. Then f commutes with the corresponding bilinear forms (we say that f is an *isometry*). In particular, χ_A is non-negative if and only if so is χ_B, and in this case, $\mathrm{corank}_{\cdot}\chi_A = \mathrm{corank}_{\cdot}\chi_B$.

The next natural step after (4.1.3) is the following:

Theorem 4.4. [4], [7] *Let A be a strongly simply connected algebra. Then A is derived equivalent to $k\Delta$ with Δ of extended-Dynkin type if and only if χ_A is non-negative and* $\mathrm{corank}\,\chi_A = 1$.

The result is shown in [4] in case A is representation-finite and extended in [7] to the representation-infinite situation.

4.1.5 Let $t \ge 3$ and $p = (p_1, \dots, p_t)$ be a sequence of numbers $2 \le p_i$. Consider the following quiver $Q(p)$:

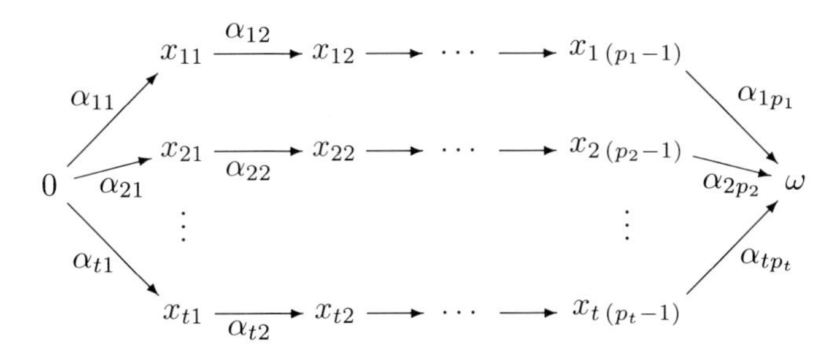

Let $\lambda = (\lambda_3, \dots, \lambda_t)$ be a sequence of pairwise different elements of $k \setminus \{0\}$. Then the *canonical algebra* $C(p, \lambda)$ is defined as the quotient of $kQ(p)$ by the ideal generated by the relations: $\alpha^{(i)} = \alpha^{(1)} + \lambda_i\alpha^{(2)}$, $i = 3, \dots, t$, where $\alpha^{(i)} = \alpha_{ip_i}\cdots\alpha_{i1}$. Canonical algebras have been extensively studied, see for instance [33], [15]. Of particular interest are *canonical tubular algebras* obtained when $p = (2, 2, 2, 2)$, $(3, 3, 3)$, $(2, 4, 4)$ or $(2, 3, 6)$. These algebras are *tame* and the module category is described in [33]. The class of *tubular algebras* is obtained in [33] as certain tilted algebras of canonical tubular algebras.

We say that A is *derived tubular* if it is derived equivalent to a canonical tubular algebra. In this case χ_A is non-negative and corank $\chi_A = 2$.

4.1.6 The *Tits form* $q_A\colon \mathbb{Z}^n \to \mathbb{Z}$ of A is obtained as a 'truncation' of χ_A in the following way: $q_A(v) = \sum_{i=0}^{2}(-1)^i \sum_{a,b\in Q_0} v(a)v(b) \dim_k \operatorname{Ext}_A^i(S_a, S_b)$. We recall that for a representation-finite algebra A, the Tits form is *weakly positive* (that is, $q_A(v) > 0$ for $0 \neq v \in \mathbb{N}^n$). If A is separated, the converse holds [9]. Meanwhile, for a *tame* algebra, the Tits form is *weakly non-negative* (that is, $q_A(v) \geq 0$ for $v \in \mathbb{N}^n$) [27]. If A is strongly simply connected, it is conjectured that the converse holds [28].

Theorem 4.5. [6] *The algebra A is representation-finite and derived tubular if and only if the following conditions are satisfied:*

(0) *A has more than six vertices ($n > 6$);*

(1) *χ_A is non-negative and* corank $\chi_A = 2$*;*

(2) $\chi_A^{-1}(1) \cap (\operatorname{rad}\chi_A)^{\perp} = \emptyset$*;*

(3) *q_A is weakly positive;*

(4) *A is separated.*

Where $V^{\perp} = \{w \in \mathrm{K}_\circ(A) : \langle w, v\rangle = 0$ for every $v \in V\}$. Conditions (0), (1) and (4) are easy to check (also by a computer). For condition (3) there are efficient algorithms as we shall recall in Section 4.2. Condition (2) is the hardest to check. Our work in Section 4.3 is devoted to substitute this condition for another easier to check (by computer).

4.1.7 We shall consider also the following extension of (4.1.6) to representation-infinite algebras.

Theorem 4.6. [7] *Let $A = kQ/I$ be a strongly simply connected algebra with $n = |Q_0| > 6$. Then A is derived tubular if and only if conditions (1) and (2) above hold.*

Let us observe that it is impossible to have a criterion for derived tubularity in the case $n = 6$, which uses only properties of the Euler form. Indeed, for $p = (2,2,2,2)$ and $\lambda = (1,\rho)$ with $\rho \neq 0,1$, the algebra $C = C(p,\lambda)$ is canonical tubular, while the algebra $C' = kQ(p)/\langle \alpha^{(3)} - \alpha^{(2)} - \alpha^{(1)}, \alpha^{(4)} - \alpha^{(2)} - \alpha^{(1)}\rangle$ is not derived tubular, but $\chi_C = \chi_{C'}$. See [7].

In the case $n = 6$, a complete list of all derived tubular algebras (strongly simply connected or not) was obtained in [7]. We reproduce in Fig. 1 this list.

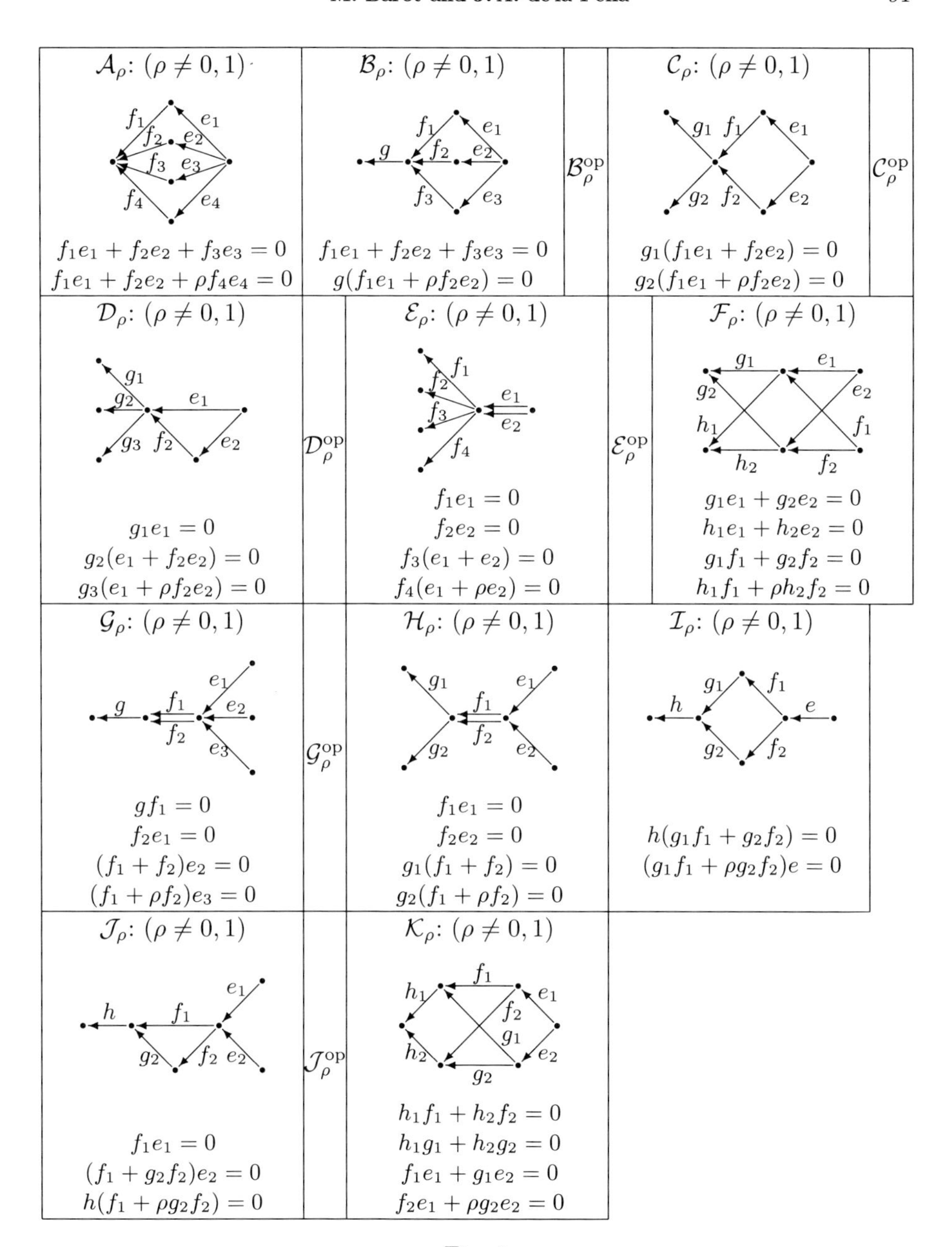

Fig. 1

4.1.8 We need to recall some central steps of the *proof of Theorems* (4.1.6) and (4.1.7) with the purpose of sketching how condition (2) is used. We consider only the sufficiency of the conditions in Theorem (4.1.7).

We assume that $A = kQ/I$ is strongly simply connected and such that Q has $n > 6$ vertices. Assume moreover that conditions (1) and (2) are satisfied. Choose a a source or sink in Q such that $Q \setminus \{a\}$ is connected, say a is a source.

Then $A = B[M]$ is a one-point extension for $B = A/(a)$ and $M = \operatorname{rad} P_a$. Since A is strongly simply connected, M is indecomposable and B is again strongly simply connected. Clearly, χ_B is non-negative and $\operatorname{corank} \chi_B \geq 1$ (indeed, if v and w are generators of $\operatorname{rad} \chi_A$, we may always choose w satisfying $w(a) = 0$). Consider $p = [P_a] \in K_0(A)$ which is a root of χ_A. Since $\langle p, v \rangle_A = v(a)$ for any vector $v \in K_0(A)$ and by assumption (2) we have $\langle p, \operatorname{rad} \chi_A \rangle_A \neq 0$, thus $\operatorname{corank} \chi_B = 1$. By (4.1.4), there is a hereditary algebra $H = k\Delta$ of extended-Dynkin type which is derived equivalent to B. The argument [6, (5.4)] shows that the non-negativity of χ_A implies the existence of a derived equivalence $F\colon D^b(B) \to D^b(H)$ such that $FM \in \operatorname{mod}_H$ and F extends to an equivalence $\hat{F}\colon D^b(B[M]) \to D^b(H[FM])$. In particular $\chi_{H[FM]}$ is non-negative. By [27] and condition (1), only the following situations are possible:

(a) FM is a simple regular module and $H[FM]$ is a tubular algebra;

(b) Δ is of type $\tilde{\mathrm{D}}_{n-2}$, FM is a regular module of regular length 2 in the tube of rank $n-4$ (≥ 3) in the Auslander-Reiten quiver Γ_H. In this case $H[FM]$ is said to be 2-tubular.

Finally, condition (2) is not satisfied by 2-tubular algebras (see below) and the proof is completed.

As illustration of the last step above, we consider a 'typical' 2-tubular algebra A:

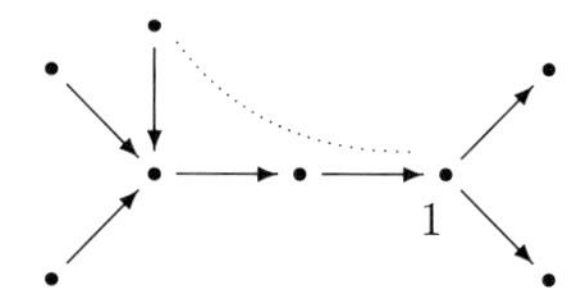

It is easy to see that χ_A is non-negative of $\operatorname{corank} \chi_A = 2$ but $e_1 \in \chi_A^{-1} \cap (\operatorname{rad} \chi_A)^{\perp}$.

4.2 Reduction of Integral Quadratic Forms

4.2.1 An *integral quadratic form* q is a map $q\colon \mathbb{Z}^n \to \mathbb{Z}$ of the shape

$$q(x_1, \dots, x_n) = \sum_{i=1}^{n} q_i x_i^2 + \sum_{i<j} q_{ij} x_i x_j$$

where $q_i \in \mathbb{Z}$ and $q_{ij} \in \mathbb{Z}$ for all $i, j \in \{1, \dots, n\}$. We say that q is a *semi-unit form* if $q_i \in \{0, 1\}$ for all $i \in \{1, \dots, n\}$. Sometimes we also will consider the special case where $q_i = 1$ for all $i \in \{1, \dots, n\}$ and then say that q is a *unit form.*

With a semi-unit form q we associate a bigraph G_q (see for example [21], [24]). We illustrate this in the following example:

$$q(x_1, x_2, x_3) = x_1^2 + x_2^2 - 2x_1x_3 + x_2x_3 \qquad G_q :$$

In particular, we say that q is connected if the graph G_q is connected. Conversely, any bigraph G defines a quadratic form $p_G : \mathbb{Z}^{G_0} \to \mathbb{Z}$, where G_0 denotes the set of vertices of G.

In the following observations we will deal with semi-unit forms which are weakly non-negative (or even weakly positive). Observe that this implies $q_{ij} \in \{-1, -2\}$ whenever $q_{ij} < 0$.

4.2.2 In [21], a reduction procedure for semi-unit forms was introduced which allows to verify the weak positivity or weak non-negativity of forms.

Let $q\colon \mathbb{Z}^n \to \mathbb{Z}$ be a semi-unit form and $i \neq j$ be indices such that $q_{ij} < 0$. Define the form $q'\colon \mathbb{Z}^{n+1} \to \mathbb{Z}$ by

$$q'(y) = q\rho(y) + y_i y_j$$

where $\rho\colon \mathbb{Z}^{n+1} \to \mathbb{Z}^n$ denotes the linear map sending $e_s \mapsto e_s$ if $1 \leq s \leq n$ and $e_{n+1} \mapsto e_i + e_j$. Then q' is said to be obtained from q by *edge reduction* with respect to i and j.

Proposition 4.1. [21], [23] *Let $q\colon \mathbb{Z}^n \to \mathbb{Z}$ be a semi-unit form and q' be obtained from q by edge reduction. Then*

(a) *q is weakly positive if and only if q' is weakly positive. In this case, ρ induces a bijection between the set of positive roots $\sum^1(q) = q^{-1}(1) \cap \mathbb{N}^n$ and $\sum^1(q')$.*

(b) *q is weakly non-negative if and only if q' is weakly non-negative. In this case, ρ induces a bijection between the set of positive isotropic vectors $\sum^0(q) = q^{-1}(0) \cap \mathbb{N}^n$ and $\sum^0(q')$.*

4.2.3 We say that an iterated edge reduction of a semi-unit form $q\colon \mathbb{Z}^n \to \mathbb{Z}$ is *exhaustive* if every reduction step only involves reductions with respect to indices $\leq n$ and the resulting semi-unit form q' satisfies $q'_{ij} \geq 0$ for all $1 \leq i < j \leq n$. For instance the unit form q in the example below admits an exhaustive reduction with respect to the sequence (of couples of vertices) $\{2,3\}$, $\{1,2\}$, $\{1,2\}$, the result is indicated as $q^{(1)}$:

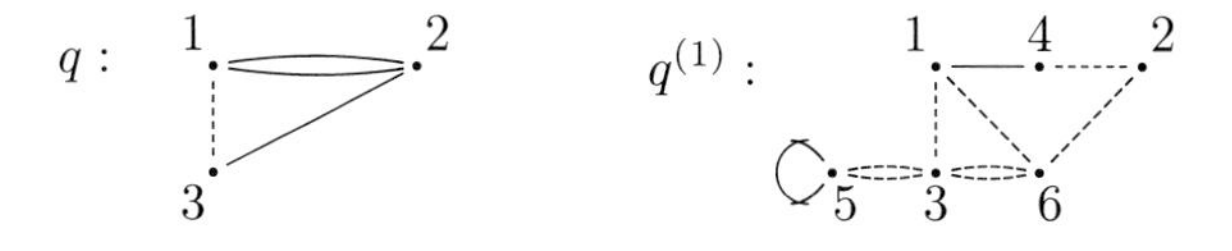

Theorem 4.7. *[23] Let* $q\colon \mathbb{Z}^n \to \mathbb{Z}$ *be a semi-unit form and* $q^{(k)}\colon \mathbb{Z}^{n_k} \to \mathbb{Z}$, $k = 0, 1, 2 \ldots$ *be a sequence of semi-unit forms obtained by iterated exhaustive reductions of* q.

(a) q *is weakly positive if and only if there is some* $k \leq 30$ *such that* $q_{ij}^{(k)} \geq 0$ *for all* $1 \leq i < j \leq n_k$

(b) q *is weakly non-negative if and only if* $q_i^{(k)} \geq 0$ *for all* $k \leq 31$ *and* $1 \leq i \leq n_k$.

4.2.4 We recall that a semi-unit form $q\colon \mathbb{Z}^n \to \mathbb{Z}$ is called *critical* if either $n = 1$ and $q(x_1) = 0$, $n = 2$ and $q(x_1, x_2) = (x_1 - x_2)^2$ or $n \geq 3$ and q is not weakly positive but every restriction $q^{(i)} = q(x_1, \ldots, x_{i-1}, 0, x_{i+1}, \ldots, x_n)$ is weakly positive. In the latter case, q is non-negative and there is a vector $z_0 \in \mathbb{N}^n$ such that $\operatorname{rad} q = \mathbb{Z}z_0$ (for $n = 1$, $z_0 = (1)$; for $n = 2$, $z_0 = (1, 1)$). See [33].

Similarly, the semi-unit form $q\colon \mathbb{Z}^n \to \mathbb{Z}$ with $n \geq 3$ is called *hypercritical* if q is not weakly non-negative but every restriction $q^{(i)}$ is weakly non-negative. Clearly a semi-unit form q is weakly positive (resp. weakly non-negative) if and only if there is no restriction $q|_I$ of q which is critical (resp. hypercritical). Observe that a critical semi-unit form is already a unit form. Critical and hypercritical unit forms have been classified [20], [22].

A strongly simply connected algebra A is said to be a *critical algebra* (resp. *hypercritical algebra*) if its Tits form q_A is critical (resp. hypercritical). These algebras have also been classified [19], [37]. Their importance is due to the following:

Theorem 4.8. *Let A be a strongly simply connected algebra. Then*

(a) [10] q_A *is weakly positive (equivalently, A is representation-finite) if and only if A does not contain a convex critical subcategory.*

(b) [28] q_A *is weakly non-negative if and only if A does not contain a convex hypercritical subcategory.*

For other results related to these problems the reader may see [31]. Criteria as the above Theorem are normally well-adapted for eye-checking but are very slow when run in a computer. The implementations of (4.2.3) run much faster.

4.2.5 Clearly, if $\operatorname{gl.dim} A \leq 2$, then the quadratic forms q_A and χ_A coincide. Most of the applications of these forms reduces to this case when the homological information contained in χ_A may be compared with the combinatorial information related to q_A. In the general case there are still some relations.

Lemma 4.1. [32] *Let A be a strongly simply connected algebra. If χ_A is weakly positive (resp. weakly non-negative), then q_A is also weakly positive (resp. weakly non-negative).*

Proof. Assume q_A is not weakly positive (resp. not weakly non-negative). Since A is strongly simply connected, then by (4.2.4), A contains a convex subcategory B which is critical (resp. hypercritical). Since $\operatorname{gl.dim} B \leq 2$, then $q_B = \chi_B$ is not weakly positive (resp. not weakly non-negative) which implies that χ_A is not weakly positive (resp. not weakly non-negative). □

4.2.6 Let $q\colon \mathbb{Z}^n \to \mathbb{Z}$ be a semi-unit form. A vector $v \in \mathbb{N}^n$ is called *critical* for q if the restriction q^v of q to the support $\operatorname{supp} v = \{i\colon v(i) \neq 0\}$ is a critical form and $\operatorname{rad} q^v = \mathbb{Z}v$. The set of critical vectors for q is denoted by C_q.

Lemma 4.2. *[30] Let $q\colon \mathbb{Z}^n \to \mathbb{Z}$ be a semi-unit form.*

(a) *If q is weakly non-negative, then every vector $w \in \sum^0(q)$ may be written as $w = \sum_{v \in C_q} \lambda_v v$ for some numbers $\lambda_v \in \mathbb{Z}^+$, where $\mathbb{Z}^+$ denotes the non-negative rational numbers.*

(b) *If q is non-negative, then the dimension of the space generated by $\sum^0(q)$ is the maximal number of linearly independent vectors in C_q (this number is called the* positive corank *of q and it is denoted by* $\operatorname{corank}^+ q$*).*

(c) *If q is non-negative and there is a sincere vector $v \in \sum^0(q)$, then* $\operatorname{corank} q = \operatorname{corank}^+ q$.

4.2.7 We introduce some operations on quadratic forms, the so called deflations and inflations which have been successfully applied for various problems [26], [20], [24].

Let $q\colon \mathbb{Z}^n \to \mathbb{Z}$ be a semi-unit form with $q_{ij} \neq 0$ and set $\varepsilon = -$ (resp. $\varepsilon = +$) if $q_{ij} < 0$ (resp. $q_{ij} > 0$). Then we define the invertible $\mathbb{Z}$-matrix $T_{ij}^{\varepsilon}\colon \mathbb{Z}^n \to \mathbb{Z}$ by the rule

$$T_{ij}^{\varepsilon}(e_s) = \begin{cases} e_s & \text{if } s \neq i \\ e_i - \varepsilon e_j & \text{if } s = i \end{cases}$$

We say that T_{ij}^- is a *deflation* (resp. T_{ij}^+ is an *inflation*) for q.

In comparison to the edge reductions which preserve weak positivity and weak non-negativity, deflations and inflations preserve positivity and non-negativity. The following result shown in [24] is a generalization of $\operatorname{corank} q = 1$, considered in [26].

Theorem 4.9. *[24] Let $q\colon \mathbb{Z}^n \to \mathbb{Z}$ be a non-negative semi-unit quadratic form. Then* $\operatorname{corank} q = \operatorname{corank}^+ q = s$ *if and only if there is an iteration of deflations with composition T such that the form $q' = qT$ is the orthogonal sum $q^0 \oplus q^1$ of two forms such that q^0 is the zero form in s variables and q^1 is a positive definite form in $n - s$ variables.*

Example 4.1. We consider the quadratic form p_Δ associated to the following (extended Dynkin) diagram

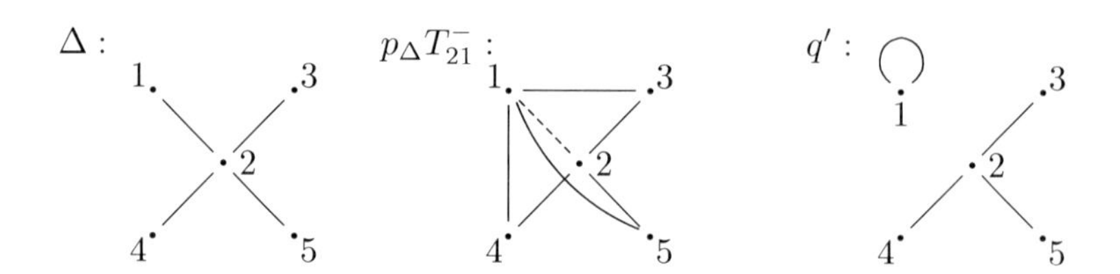

Applying to p_Δ the transformation T_{21}^- we get the quadratic form associated to the bigraph in the middle. The sequence $(T_{21}^-, T_{31}^-, T_{41}^-, T_{51}^-, T_{21}^-)$ applied to p_Δ yields the form $q' = qT_{21}^- T_{31}^- T_{41}^- T_{51}^- T_{21}^-$ in the variables x_1, x_2, x_3, x_4, x_5 but since x_1 does not appear explicitely, we have $q' = q^0 \oplus q^1$ where $q^1(x_2, x_3, x_4, x_5) = q'(0, x_2, x_3, x_4, x_5)$ and $q^0 = 0$. The form q^1 is positive definite and $\operatorname{corank} q = \operatorname{corank}^+ q = 1$.

4.2.8 We denote by ζ the zero form in one variable.

Lemma 4.3. *Let $q \neq 0$ be a connected non-negative semi-unit form, T be a deflation or an inflation and $q' = qT$. If q' decomposes properly, say $q' = q^0 \oplus q^1$ with $q^1 \neq 0$ then $q^0 = \zeta$ and q^1 is a connected non-negative semi-unit form.*

Proof. Let $T = T_{ij}^\varepsilon$ be deflation or an inflation. Let $G_{q'}$ be the bigraph associated to q' and assume it is not connected. Since G_q is connected, then $G_{q'}$ has exactly two components, C_0 containing j and C_1 containing i. In particular, $q'_{ij} = 0$.

An obvious calculation yields:

$$0 = q'_{ij} = q_{ij} + 2\varepsilon \quad \text{and} \quad q'_j = q_i\varepsilon^2 + q_{ij}\varepsilon + q_j = q_i + q_j - 2$$

As observed in [24], q' is still a semi-unit form and hence $q_i = 1 = q_j$ and $q'_j = 0$. For any $s \neq j$ we should have $q'_{sj} = 0$, otherwise $q'(2e_j - q'_{sj}e_s) < 0$ which contradicts the non-negativity of q'. Therefore, C_0 contains only the vertex j and $q' = q^0 \oplus q^1$, where q^i is the form associated to C_i, $i = 0, 1$. Clearly, q^0 is the zero form in one variable. □

4.2.9 We present now a generalization of (4.2.7) which will be central in the proof of our main result (compare also with [38]).

Theorem 4.10. *Let q be a connected, non-negative semi-unit form of corank s. Then there exists an iteration of deflations and inflations with composition T such that qT is the orthogonal sum $\zeta^s \oplus p_\Delta$, where ζ^s is the zero form in s variables and p_Δ is the quadratic form associated to a (connected) Dynkin diagram Δ (hence p_Δ is positive definite). Moreover, Δ is uniquely determined by q.*

Proof. We proceed by induction on $s = \operatorname{corank} q$. If $s = 0$ the form q is already positive definite. Hence by [26], there exists a sequence of inflations with composition T such that $qT = p_\Delta$ for some Dynkin diagram Δ. By (4.2.8), Δ is connected.

For the induction step we first show in (a) that there is a sequence of inflations with composition T_a such that the radical of the form $q^a = qT_a$ contains a positive vector v_a and then in (b) we prove that there exists a sequence of deflations with composition T_b such that $q^a T_b$ is the orthogonal sum of the zero form in one variable with a non-negative connected semi-unit form of corank $s-1$.

For any $v \in \mathbb{Z}^n$ we define $|v| = \sum_{i=1}^n |v(i)|$, $\operatorname{supp} v = \{i : v(i) \neq 0\}$ and $v^+, v^- \in \mathbb{Z}^n$ by

$$v^\varepsilon(i) = \begin{cases} \varepsilon v(i) & \text{if } \varepsilon v(i) > 0 \\ 0 & \text{else} \end{cases} \qquad (\varepsilon = \pm)$$

(a) Choose $v \in \operatorname{rad} q$, $v \neq 0$. If there are $i \in \operatorname{supp} v^+$ and $j \in \operatorname{supp} v^-$ such that $q_{ij} > 0$ then we apply T_{ji}^+ to q obtaining $\tilde{q} = qT_{ji}^+$ and $\tilde{v} = (T_{ji}^+)^{-1} v$. Observe that $\tilde{v} = v + v(j)e_i$, thus $|\tilde{v}^-| < |v^-|$. We repeat this procedure until this is no longer possible and obtain a quadratic form q^a and a radical vector v'_a. Now we have

$$0 = q^a(v'_a) = q^a({v'_a}^+) + q^a({v'_a}^-) + \sum_{(i,j)} (q^a)_{ij} v'_a(i) v'_a(j)$$

where the sum runs over $\operatorname{supp} {v'_a}^+ \times \operatorname{supp} {v'_a}^-$. Any summand on the right hand side is at least zero hence equals zero. So we get step (a) by setting $v_a = {v'_a}^+ + {v'_a}^-$.

(b) If there exist $i, j \in \operatorname{supp} v_a$ with $(q^a)_{ij} < 0$ we apply $T = T_{ij}^-$ if $v_a(i) \geq v_a(j)$ (or $T = T_{ji}^-$ if $v_a(i) < v_a(j)$). Observe that for $\tilde{q}_a = q^a T$ and $\tilde{v}_a = T^{-1} v_a$ we have that $\tilde{v}_a$ is positive again with $|\tilde{v}_a| < |v_a|$. We repeat this procedure as long as possible and end up with a quadratic form q^b and a positive radical vector v_b. Then we have

$$0 = q^b(v_b) = \sum_{i=1}^n (q^b)_i v_b(i)^2 + \sum_{i,j \in \operatorname{supp} v_b} (q^b)_{ij} v_b(i) v_b(j)$$

where any summand on the right hand side is non-negative, hence zero. In particular $(q^b)_i = 0$ whenever $v_b(i) \neq 0$. This proves step (b). □

Lemma 4.4. *Let $q\colon \mathbb{Z}^n \to \mathbb{Z}$ be a non-negative semi-unit form. Let $S, T\colon \mathbb{Z}^n \to \mathbb{Z}^n$ be $\mathbb{Z}$-invertible such that $qS = \zeta^s \oplus p_\Sigma$ and $qT = \zeta^t \oplus p_\Delta$ where $\Sigma = \cup_{i=1}^a \Sigma_i$ and $\Delta = \cup_{j=1}^b \Delta_j$ are disjoint unions of Dynkin graphs, then we have $s = t$, $a = b$ and $\Sigma_i = \Delta_{\pi(i)}$ for $i = 1, \dots, a$ where π is a permutation.*

Proof. Since p_Σ is positive definite, s is the corank of qS hence $s = t$.

Let $C_\Sigma = q_\Sigma^{-1}(1) \subset \mathbb{Z}^{n-s}$ and $G(C_\Sigma)$ be the graph having as points the elements of C_Σ and an edge $x \,\text{———}\, y$ if $x \pm y \in C_\Sigma \cup \{0\}$ for $x, y \in C_\Sigma$. The connected components of $G(C_\Sigma)$ are exactly the graphs $G(C_{\Sigma_i})$ for $i = 1, \dots, a$. The map $T \circ S^{-1}$ induces a bijection $C_\Sigma \xrightarrow{\sim} C_\Delta$ which respects the associated graph structure, i.e. $T \circ S^{-1}$ induces an isomorphism $G(C_\Sigma) \xrightarrow{\sim} G(C_\Delta)$, hence $a = b$ and there exists a permutation π such that $G(C_{\Sigma_i}) \xrightarrow{\sim} G(C_{\Delta_{\pi(i)}})$. Therefore $\Sigma_i \xrightarrow{\sim} \Delta_{\pi(i)}$ holds for every i. □

4.2.10 Let $q\colon \mathbb{Z}^n \to \mathbb{Z}$ be a non-negative semi-unit form. Then $\operatorname{rad} q$ is a pure subgroup of $\mathbb{Z}^n$ (that is, if $n \in \mathbb{Z} \setminus \{0\}$ and $nv \in \operatorname{rad} q$, then $v \in \operatorname{rad} q$). Hence $\mathbb{Z}^n / \operatorname{rad} q \xrightarrow{\sim} \mathbb{Z}^{n-s}$ as $\mathbb{Z}$-modules where $s = \operatorname{corank} q$. Thus we may consider the induced map

$$\bar{q}\colon \mathbb{Z}^n / \operatorname{rad} q \to \mathbb{Z}, v + \operatorname{rad} q \mapsto q(v)$$

which is, by choosing a $\mathbb{Z}$-base of $\mathbb{Z}^n / \operatorname{rad} q$, a positive definite form.
The following result, which (surprisingly) seems to be new, answers the question whether there is a basis of $\mathbb{Z}^n / \operatorname{rad} q$, which makes $\bar{q}$ a unit form, affirmatively.

Theorem 4.11. *Let q be a non-negative semi-unit form. Then there exists a disjoint union Δ of Dynkin graphs Δ_i $(i = 1, \dots, n)$ such that $\bar{q}$ is $\mathbb{Z}$-equivalent to p_Δ. Moreover n is the number of connected components of q and Δ is (up to the order of the Δ_i) uniquely determined by q.*

Proof. By (4.2.9), there is an invertible $\mathbb{Z}$-matrix $T\colon \mathbb{Z}^n \to \mathbb{Z}^n$ such that $qT^{-1} = \zeta^s \oplus p_\Delta \colon \mathbb{Z}^s \times \mathbb{Z}^{n-s} \to \mathbb{Z}$ where Δ is the disjoint union of Dynkin diagrams Δ_i $(i = 1, \dots, n)$ and n is the number of connected components of q. Since $\mathbb{Z}^s = T(\operatorname{rad} q) = \operatorname{rad}(qT^{-1})$ we get an induced $\mathbb{Z}$-isomorphism $\bar{T}\colon \mathbb{Z}^n / \operatorname{rad} q \to \mathbb{Z}^s$ making the following diagram commutative

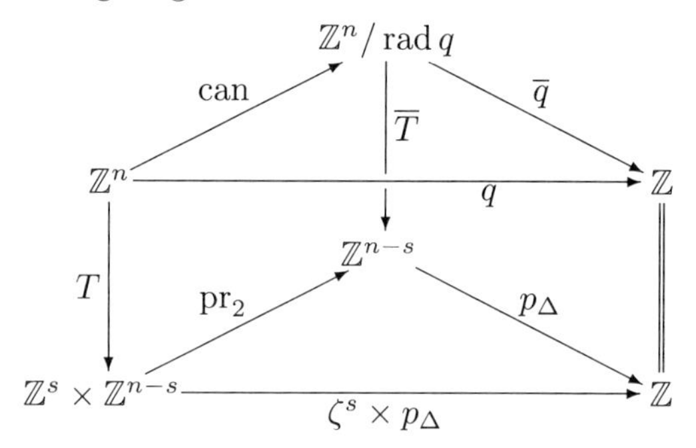

hence $\bar{q} = p_\Delta \bar{T}$ which is the desired equivalence. The assertion about the uniqueness follows by (4.4). □

Corollary 4.1. *Each connected non-negative unit form q determines uniquely a Dynkin graph Δ such that $\bar{q}$ and p_Δ are $\mathbb{Z}$-equivalent.*

4.3 Derived Tubularity: an Algorithm

Lemma 4.5.

(a) *Let C be a canonical tubular algebra of type $(3,3,3)$, $(2,4,4)$ or $(2,3,6)$. The Dynkin diagram $\Delta(\chi_C)$ associated to the Euler form χ_C as in (4.2.9) is of the form $\mathbb{E}_p, p = 6, 7, 8$, respectively.*

(b) *Let C_0 be a representation-infinite tame concealed algebra of type $\tilde{\mathrm{D}}_n$ and let M be an indecomposable C_0-module regular of regular length 2 lying in a tube of rank $n-2$. Consider the one-point extension $B = C_0[M]$. Then the Dynkin diagram $\Delta(\chi_B)$ is of the form D_n.*

Proof. It is enough to apply deflations and inflations in the way indicated by (4.2.9).

(a) is left as an easy exercise.

(b) Any algebra of the type $B = C_0[M]$ as indicated is tilted (and therefore derived equivalent) to an algebra E whose Euler form χ_E has the following bigraph (we draw the case $n = 6$)

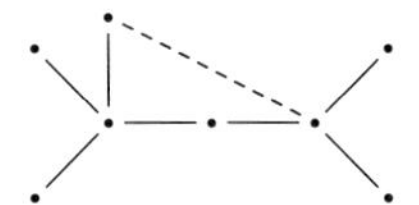

Again, it is an easy exercise to transform this graph to a Dynkin diagram D_5 by means of deflations. □

4.3.1 Proof of the Main Theorem

Proof. Let A be a strongly simply connected algebra of the form $A = kQ/I$ with $Q_0 = \{1, \ldots, n\}$ and $n > 6$.

Assume first that A is derived tubular. Then we know that χ_A is non-negative of corank $\chi_A = 2$ and $A = B[M]$ for some representation-infinite tilted connected algebra B such that corank $\chi_B = 1$, see for example [7]. Moreover we may assume that A is derived equivalent to a canonical tubular algebra C of type $(3,3,3)$, $(2,4,4)$ or $(2,3,6)$. Then by (4.2.9) and (4.5), the Dynkin graph $\Delta(\chi_A)$ is equal to $\Delta(\chi_C)$ of type $\mathbb{E}_p$, $p = 6,7,8$, respectively. The number of positive roots of $\bar{\chi}_A \simeq p_{\Delta(\chi_A)}$ is $36, 63, 120$, respectively.

Conversely, assume that the conditions (a), (b) and (c) hold. Choose a sink or source a in Q such that $Q \setminus \{a\}$ is connected and χ_B is non-negative with corank one, where $B = A/(a)$. We assume that a is a source (the case where a is sink, is handled dually), so the B-module $M = \operatorname{rad} P_a$ is indecomposable.

By (4.1.4), we get that B is derived equivalent to a hereditary algebra $H = k\Delta$, where Δ is an extended-Dynkin diagram of type $\tilde{\mathrm{D}}_n$ or $\tilde{\mathbb{E}}_p$ $(p = 6,7,8)$. As in (4.1.8), we get a derived equivalence between $A = B[M]$ and $H[N]$ for some regular H-module N. There are two possibilities: either $H[N]$ is a tubular algebra or Δ is of type $\tilde{\mathrm{D}}_m$ and N lies on a tube of rank $n-4$ (≥ 3) in Γ_H and has regular length 2. In the latter case, (4.5) implies that $\Delta(\chi_{H[N]}) = \Delta(\chi_A)$ is a Dynkin diagram of type D_n. But then the number of positive roots of $\bar{\chi}_A = p_{\Delta(\chi_A)}$ is $n(n-1)$ which cannot be 36, 63 or 120. A contradiction showing that $H[N]$ is a tubular algebra and A is derived tubular. □

4.3.2 We remark that the 'strong simple connectedness' hypothesis is necessary in the main theorem. Namely, consider the algebra $A = kQ/I$ given by the following quiver with commutative relations

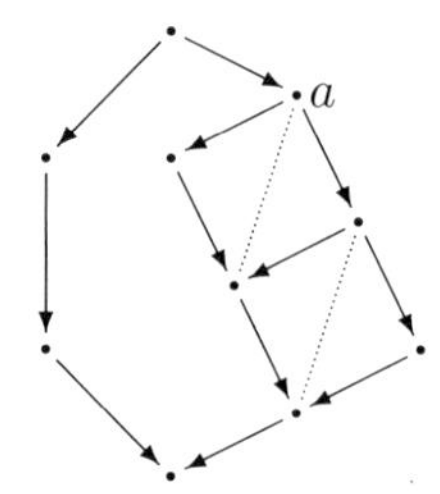

Then we have the following facts:

- χ_A is non-negative of corank $\chi_A = 2$;
- the algebra $B = A/(a)$ is tame concealed of type $\tilde{\mathbb{E}}_8$;
- $\Delta(\chi_A)$ is of type $\mathbb{E}_8$ and hence $\bar{\chi}_A$ has 120 positive roots;
- the algebra A is wild since the universal covering $\tilde{A}$ of A contains a convex subcategory of the following form

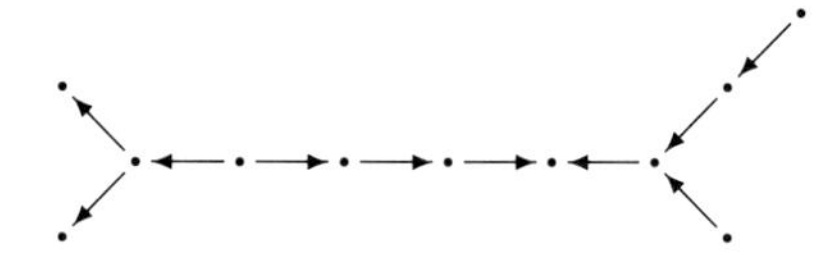

therefore, A is not derived tubular (as observed in [32], every derived tubular algebra is tame).

4.3.3 In a preliminary stage of our research we suggested the following *algorithm* to check whether or not $\chi_A^{-1}(1) \cap \chi_A^{-1}(0)^{\perp} = \emptyset$ for a non-negative form χ_A with corank $\chi_A = 2$.

First construct a $\mathbb{Z}$-bases (v_1, v_2) of rad χ_A.

A subset $\mathcal{S}$ of $\chi_A^{-1}(1)$ is said to be *reduced* if $s_1, s_2 \in \mathcal{S}$ and $s_1 - s_2 \in \chi_A^{-1}(0)$ or $s_1 + s_2 \in \chi_A^{-1}(0)$ imply that $s_1 = s_2$. Any reduced set is finite.

Construct a maximal reduced subset $\mathcal{S}$ of $\chi_A^{-1}(0)$. Then the following hold:

- If $\langle v_1, v_2 \rangle_A = 0$, then $\chi_A^{-1}(1) \cap \chi_A^{-1}(0)^{\perp} = \emptyset$ if and only if $\langle v_1, s \rangle_A \neq 0$ or $\langle v_2, s \rangle_A \neq 0$ for every $s \in \mathcal{S}$
- If $\langle v_1, v_2 \rangle_A \neq 0$, then $\chi_A^{-1}(1) \cap \chi_A^{-1}(0)^{\perp} = \emptyset$ if and only if $\varepsilon_1 \frac{\langle v_2, s \rangle}{\langle v_2, v_1 \rangle} v_1 + \varepsilon_2 \frac{\langle v_1, s \rangle}{\langle v_1, v_2 \rangle} v_2$ is not in $\mathbb{Z}^n$ for any $s \in \mathcal{S}$ and $\varepsilon_1, \varepsilon_2 \in \{1, -1\}$.

Clearly, the cardinality of $\mathcal{S}$ is the number of positive roots of $\bar{\chi}_A$. It was the implementation of the above algorithm which suggested the result presented in the main theorem.

4.4 Examples

We append the list of all derived tubular algebras whose underlying quiver is linearly ordered. The list was calculated by our C++ program.

Each figure represents a class of algebras which is obtained from the given algebra in the picture by a sequence of the following two operations: add a relation $\beta\alpha$, if $\beta\alpha \neq 0$, and change the orientation of all arrows simultaneously. In this way, picture (6) defines 16 non-isomorphic algebras.

For each algebra we give at the right hand side generators of the radical of the Euler form. In order to obtain the generators of the radical of the Euler form of an algebra which is obtained by adding a relation of length two one calculates starting with the given vectors $[v_1 \ v_2 \ \cdots \ v_n]$ the new ones in the following way:

Add a relation from 2 to 4: $[(v_1 - v_2) \ \ -v_2 \ \ (v_3 - v_2) \ \ v_4 \ \cdots \ v_n]$.
Add a relation from 1 to 3: $[-v_1 \ \ (v_2 - v_1) \ \ v_3 \ \ v_4 \ \cdots \ v_n]$.
Add a relation from 2 to 4 and a relation from 1 to 3:

$$[(-v_1 + v_2) \ \ -v_1 \ \ (v_3 - v_2) \ \ v_4 \ \cdots \ v_n].$$

Similarily one calculates on the side of the sink of the algebra.

4.4.1 Algebras with 9 Points (8 algebras)

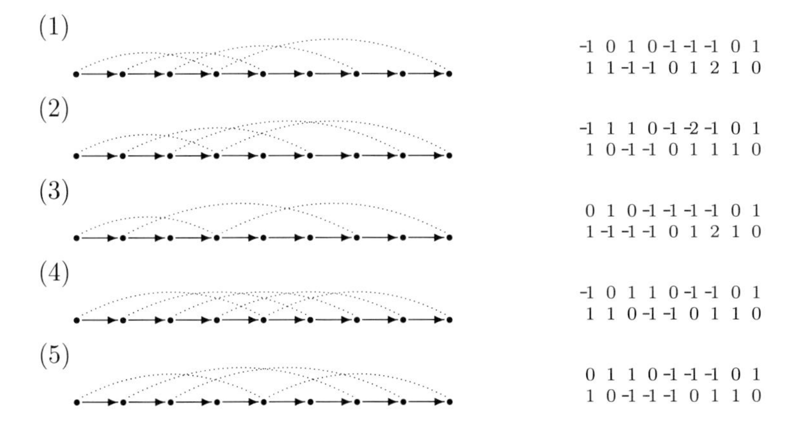

4.4.2 Algebras with 10 Points (123 algebras)

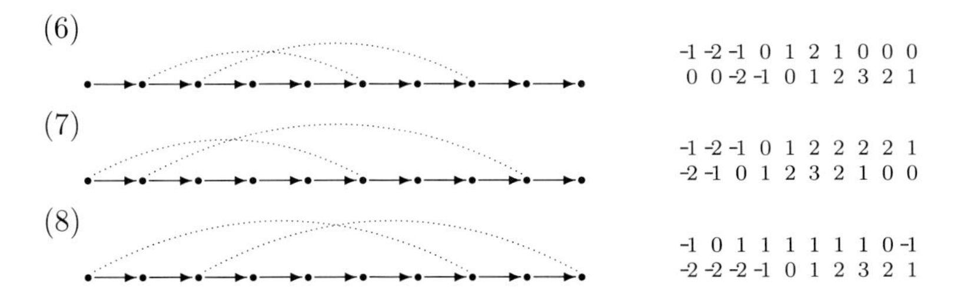

(9)

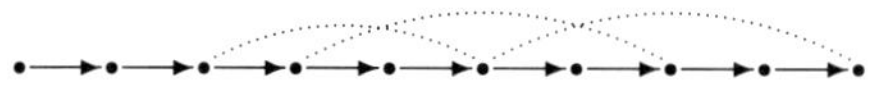

0 0 0 1 1 0 -1 -1 0 1
1 2 3 0 -1 -2 0 2 1 0

(10)

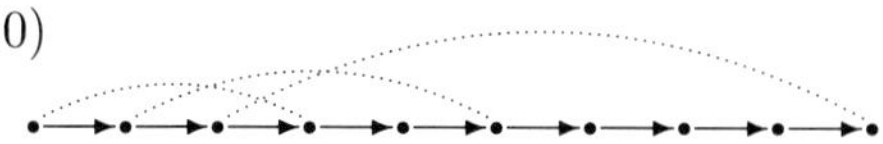

0 1 0 -2 -1 0 2 3 2 1
1 0 -1 -1 0 1 1 1 1 1

(11)

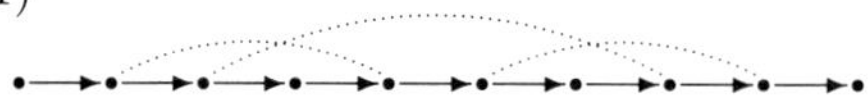

0 0 1 0 -1 -2 -1 0 2 1
1 2 0 -1 -2 -1 0 1 0 0

(12)

0 -1 0 1 2 3 2 1 0 -2
-1 0 1 2 2 2 1 0 -1 -1

(13)

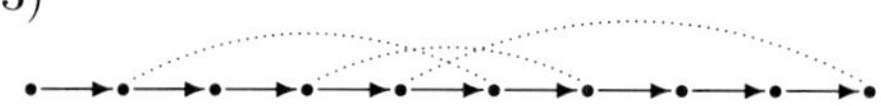

0 0 1 2 0 -1 -2 -1 0 1
1 2 0 -2 -1 0 3 2 1 0

(14)

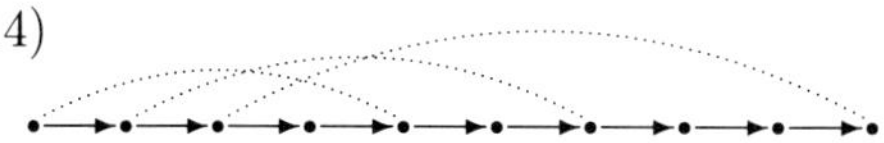

1 0 -1 -1 -1 0 1 1 1 1
0 -2 -1 0 1 2 3 2 1 0

(15)

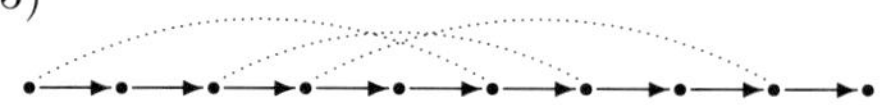

1 0 -1 -2 -1 0 2 2 2 1
0 -1 -2 -1 0 1 2 1 0 0

(16)

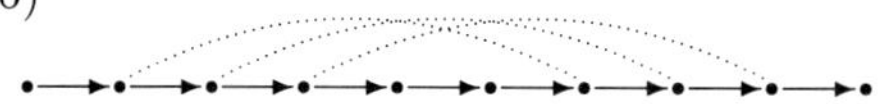

0 0 -1 -2 -1 0 1 2 2 1
1 2 2 1 0 -1 -2 -1 0 0

(17)

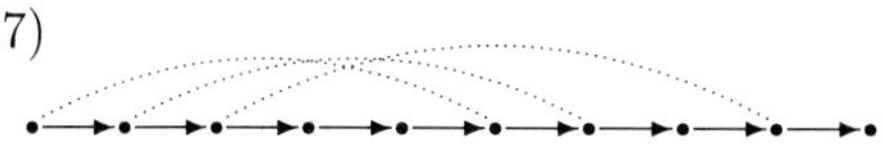

0 -1 -2 -1 0 1 2 2 2 1
-1 -2 -1 0 1 2 2 1 0 0

(18)

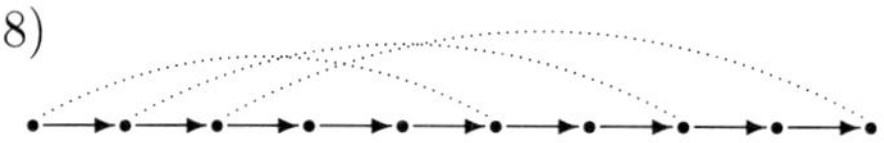

1 0 -1 -1 -1 -1 0 1 1 1
-1 -2 -1 0 1 2 2 2 1 0

(19)

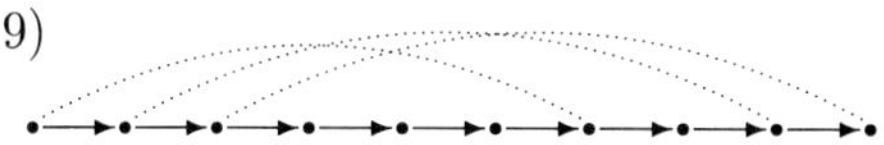

-1 0 1 1 1 1 1 0 -1 -1
-2 -2 -1 0 1 2 3 2 1 0

(20)

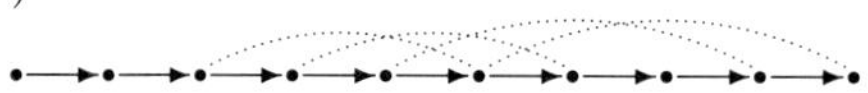

1 2 3 0 -1 0 2 1 0 -2
0 0 0 1 0 -1 -1 0 1 1

(21)

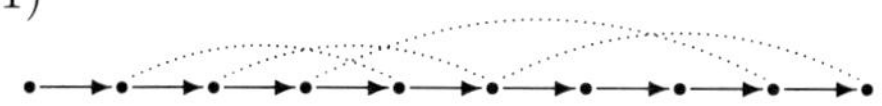

1 2 0 -1 0 3 2 1 0 -2
0 0 1 0 -1 -2 -1 0 1 1

(22)

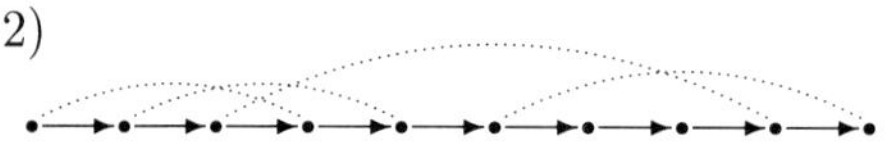

-1 0 1 0 -2 -2 -1 0 2 1
0 1 0 -1 -2 -1 0 1 1 0

(23)

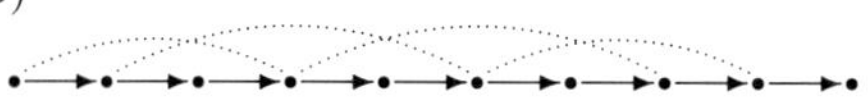

-1 0 1 2 0 -2 -1 0 2 1
0 -1 0 1 1 1 0 -1 0 0

(24)

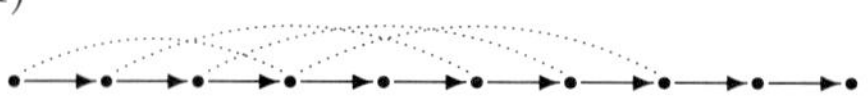

0 1 0 -2 -1 0 2 3 2 1
1 0 -1 -1 0 1 1 0 0 0

(25)

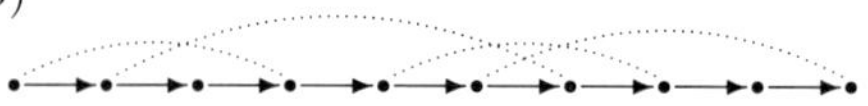

-1 0 1 2 2 0 -1 -1 0 1
0 1 0 -1 -2 -1 0 2 1 0

(26)

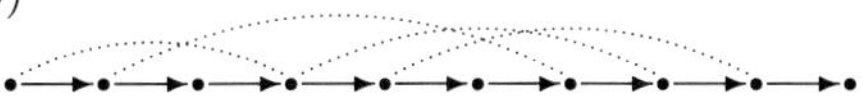

$$\begin{array}{cccccccccc} 0 & 1 & 0 & -1 & -2 & -1 & 0 & 2 & 2 & 1 \\ 1 & 0 & -1 & -2 & -1 & 0 & 1 & 1 & 0 & 0 \end{array}$$

(27)

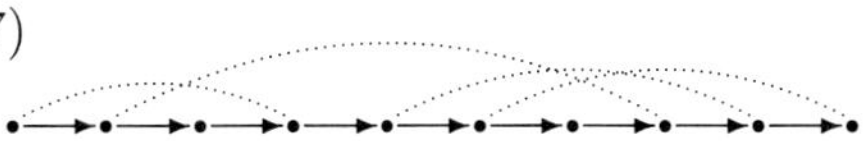

$$\begin{array}{cccccccccc} 0 & -1 & 0 & 1 & 2 & 2 & 1 & 0 & -2 & -1 \\ -1 & 0 & 1 & 2 & 2 & 1 & 0 & -1 & -1 & 0 \end{array}$$

(28)

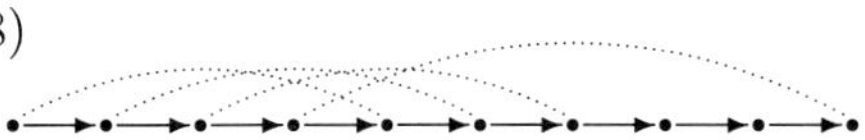

$$\begin{array}{cccccccccc} 0 & -1 & -2 & 0 & 1 & 2 & 2 & 1 & 0 & -1 \\ 1 & 0 & -2 & -1 & 0 & 2 & 3 & 2 & 1 & 0 \end{array}$$

(29)

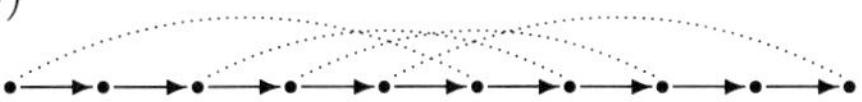

$$\begin{array}{cccccccccc} 0 & 1 & 2 & 2 & 0 & -1 & -2 & -1 & 0 & 1 \\ 1 & 0 & -1 & -2 & -1 & 0 & 2 & 2 & 1 & 0 \end{array}$$

(30)

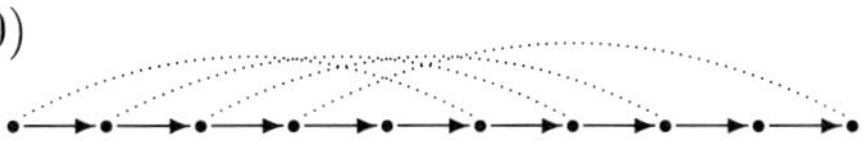

$$\begin{array}{cccccccccc} 1 & 1 & 0 & -1 & -1 & -1 & 0 & 1 & 1 & 1 \\ 0 & -1 & -2 & -1 & 0 & 1 & 2 & 2 & 1 & 0 \end{array}$$

(31)

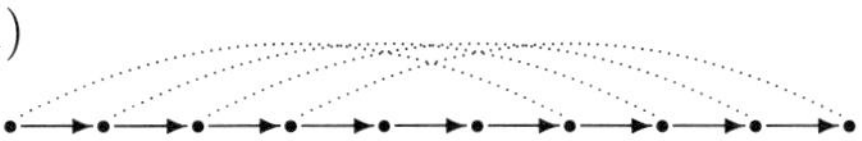

$$\begin{array}{cccccccccc} 0 & 1 & 2 & 2 & 1 & 0 & -1 & -2 & -2 & -1 \\ -1 & -2 & -2 & -1 & 0 & 1 & 2 & 2 & 1 & 0 \end{array}$$

(32)

$$\begin{array}{cccccccccc} 0 & 0 & 1 & 0 & -1 & 0 & 1 & 1 & 0 & -1 \\ -1 & -2 & 0 & 1 & 0 & -1 & 0 & 2 & 1 & 0 \end{array}$$

(33)

$$\begin{array}{cccccccccc} 0 & -1 & 0 & 1 & & -2 & -1 & 0 & 2 & 1 \\ -1 & 0 & 1 & 0 & -1 & -1 & 0 & 1 & 0 & 0 \end{array}$$

(34)

$$\begin{array}{cccccccccc} 0 & -1 & 0 & 1 & 2 & 0 & -1 & -1 & 0 & 1 \\ -1 & 0 & 1 & 0 & -2 & -1 & 0 & 2 & 1 & 0 \end{array}$$

(35)

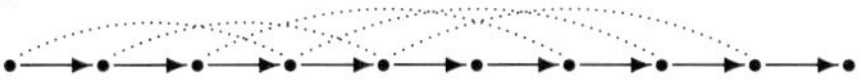

$$\begin{array}{cccccccccc} -1 & -2 & 1 & 2 & 0 & -1 & -2 & 0 & 2 & 1 \\ 0 & 1 & 0 & -1 & -1 & 0 & 1 & 1 & 0 & 0 \end{array}$$

(36)

$$\begin{array}{cccccccccc} 1 & 2 & 0 & -1 & 0 & 2 & 1 & 0 & -2 & -1 \\ 0 & 0 & 1 & 0 & -1 & -1 & 0 & 1 & 1 & 0 \end{array}$$

(37)

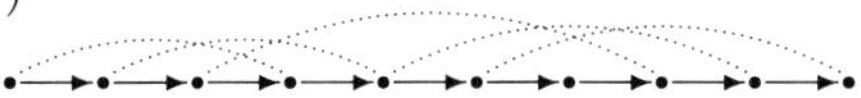

$$\begin{array}{cccccccccc} -1 & 0 & 1 & 0 & -2 & -2 & -1 & 0 & 2 & 1 \\ 0 & 1 & 0 & -1 & -2 & -1 & 0 & 1 & 1 & 0 \end{array}$$

(38)

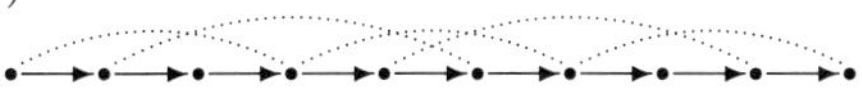

$$\begin{array}{cccccccccc} 0 & 1 & 0 & -1 & -1 & 0 & 2 & 1 & 0 & -1 \\ -1 & 0 & 1 & 2 & 0 & -1 & -1 & 0 & 1 & 0 \end{array}$$

(39)

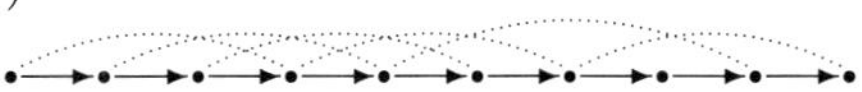

$$\begin{array}{cccccccccc} 1 & 0 & -1 & 0 & 1 & 0 & -2 & -1 & 0 & 1 \\ 0 & 1 & 0 & -1 & 0 & 1 & 1 & 0 & -1 & 0 \end{array}$$

(40)

$$\begin{array}{cccccccccc} 0 & 1 & 0 & -1 & 0 & 1 & 0 & -1 & 0 & 1 \\ 1 & 0 & -1 & 0 & 1 & 0 & -1 & 0 & 1 & 0 \end{array}$$

Bibliography

[1] I. Ágoston, E. Lukács and C. M. Ringel *Frobenius functions on translation quivers*, Conf. Proc. Can. Math. Soc **18**, pp. 17–37, 1996.

[2] I. Assem and A. Skowroński: *On some classes of simply connected algebras*, Proc. London Math. Soc. (3) **56**, pp. 417–450, 1988.

[3] I. Assem and A. Skowroński: *Algebras with cycle-finite derived categories*, Math. Ann. **280**, pp. 441–463, 1988.

[4] I. Assem and A. Skowroński: *Quadratic forms and iterated tilted algebras*, Journal of Algebra **128**, pp. 55–85, 1990.

[5] I. Assem and A. Skowroński: *On tame repetitive algebras*, Fundamenta Math. **142**, pp. 59–84, 1993.

[6] M. Barot: *Representation-finite derived tubular algebras*, (Preprint, Mexico 1996)

[7] M. Barot and J. A. de la Peña: *Derived tubular strongly simply connected algebras*, (To appear in Proc. Amer. Math. Soc.)

[8] R. Bautista, F. Larrión and L. Salmerón: *On simply connected algebras*, Journal of London Math. Soc. (2) **27**, pp. 212–220, 1983.

[9] K. Bongartz: *Algebras and quadratic forms*, J. London Math. Soc (2) **28**, pp. 461–469, 1983.

[10] K. Bongartz: *Critical simply connected algebras*, Manuscr. Math. **46**, pp. 117–136, 1984.

[11] O. Bretscher and P. Gabriel: *The standard form of a representation-finite algebra*, Bull. Soc. Math. France **111**, pp. 21–40, 1983.

[12] P. Dräxler and R. Nörenberg: *CREP Manual, Version 1.0, Using Maple as Surface*, Sonderforschungsbereich 343, Ergänzungsreihe 96–002 1996.

[13] P. Gabriel: *Unzerlegbare Darstellungen I*, *Manuscripta Math.*, **6**, pp. 71–103, 1972.

[14] P. Gabriel, B. Keller and A. V. Roiter: *Representations of finite-dimensional algebras*, Encycl. Math. Sc., Algebra VIII, **73**, 1992.

[15] W. Geigle and H. Lenzing: *A class of weighted projective curves arising in representation theory of finite dimensional algebras*, in: Singularities, Representations of Algebras and Vector Bundles, Springer Lecture Notes in Mathem atics, **1273**, pp. 265–297, 1985.

[16] D. Happel: *On the derived category of a finite-dimensional algebra*, Comment. Math. Helvetici **62**, pp. 339–389, 1987.

[17] D.Happel: *Triangulated Categories in the Representation Theory of Finite Dimensional Algebras*, , London Mathematical Society, Lecture Notes Series **119** , Cambridge University Press 1988.

[18] D. Happel and C. M. Ringel: *The derived category of a tubular algebra*, In: Representation Theory I, Springer Lecture Notes in Mathematics **1177**, pp. 156–180, 1984.

[19] D. Happel and D. Vossieck: *Minimal algebras of infinite representation type with preprojective component*, Manuscripta Math. **42**, pp. 221–243, 1983.

[20] H. J. von Höhne: *On weakly positive unit forms*, Commentari Math. Helv. **63**, pp. 312–336, 1988.

[21] H. J. von Höhne: *Edge reduction for unit forms*, Arch. Math. **65**, pp. 300–302, 1995.

[22] H. J. von Höhne: *On weakly non-negative unit forms*, Proc. London Math. Soc. (3) **73**, pp. 47–67, 1996.

[23] H. J. von Höhne and J. A. de la Peña: *Edge reductions for weakly non-negative unit forms*, Archiv Math. **70**, pp. 270–277, 1998.

[24] H. J. von Höhne and J. A. de la Peña: *Isotropic vectors of non-negative integral quadratic forms*, European J. Comb. 19 No. **5**, pp. 621–638, 1998.

[25] D. Hughes and J. Waschbüsch: *Trivial extensions of tilted algebras*, Proc. London Math. Soc. (3) **46**, pp. 346–364, 1982.

[26] S. A. Ovsienko: *Integer weakly positive forms*, in: Schurian Matrix problems and quadratic forms, Kiev, pp. 3–17, 1978.

[27] J. A. de la Peña: *On the representation type of one point extensions of tame concealed algebras*, Manuscr. Math. **61**, pp. 183–194, 1988.

[28] J. A. de la Peña: *Algebras with hypercritical Tits form*, Banach Center Publ. Part 1 **26**, pp. 353–396, 1990.

[29] J. A. de la Peña: *On the dimension of the module varieties of tame and wild algebras*, Communications in Algebra (6) **19**, pp. 1795–1807, 1991.

[30] J. A. de la Peña: *On the corank of the Tits form of a tame algebra*, Journal of Pure and Applied Algebra **107**, pp. 89–105, 1996.

[31] J. A. de la Peña: *The Tits form of a Tame Algebra*, in: Representation Theory of Algebras, Can. Math. Soc. Conference Proc., **19**, pp. 159–183, 1996.

[32] J.A. de la Peña: *Derived tame algebras*, to appear in Proceedings of AMS Conference on Representation Theory, Seattle 1997.

[33] C. M. Ringel: *Tame algebras and integral quadratic forms*, Springer Lecture Notes in Mathematics **1099**, Berlin, Heidelberg, New York, Springer Verlag 1984.

[34] A. Skowroński: *Selfinjective algebras of polynomial growth*, Math. Ann. **285**, pp. 177–199, 1989.

[35] A. Skowroński: *Simply connected algebras and Hochschild cohomologies*, Canadian Math. Soc., Conference Proceedings **14**, pp. 431–447, 1993.

[36] A. Skowroński: *Simply connected algebras of polynomial growth*, Compositio Math. **109**, pp. 99–133, 1997.

[37] L. Unger: *The concealed algebras of the minimal wild hereditary algebras*, Bayreuther Math. Schr. **31**, pp. 145–154, 1990.

[38] M. Zeldich: *Sincere weakly positive non-negative unit forms*, in: Quadratic forms in Representation Theory of finite dimensional algebras, Bielefeld 1996.

Progress in Mathematics, Vol. 173, © 1999 Birkhäuser Verlag Basel/Switzerland

Chapter 5

Problems in the Calculation of Group Cohomology[1]

Jon F. Carlson

5.1 Introduction

For several years now, I have occupied myself with a project to compute the mod-p cohomology of finite p-groups using computer algebra. To date, programs have been written to build minimal projective resolutions for modules over finite p-groups and create chain maps between resolutions representing cohomology classes. The cup products can be computed by treating them as compositions of chain maps. With this machinery it is possible to find a set of generators and relations for a cohomology ring $H^*(G, k)$ in the case that G is a finite p-group and k is a finite field of characteristic p. There are, of course, practical problems that arise if the group or the field is too big. However even for groups and fields of reasonable size, we encounter the problems of producing an output in a form that is usable for interpretation in theoretical investigations and of verifying that the output gives a complete description of the cohomology ring of the group. The description of the algorithms for the actual computation of the cohomology is given in the literature [9, 10, 11]. In this paper, I will focus on some of the problems which still need work.

The problem with such calulations is that they require techniques and methods (algorithms) from a wide variety of algebraic areas. In the beginning the computations are almost all linear algebra though, it is helpful to have an adept implementation of groups and of modules for group algebras available. In the later stages, particularly the verification of the complete calculation, the computation requires reasonably sophisticated applications from commutative algebra. One approach that could be taken is to write all of the code from scratch. But this does not seem practical given the wide range of techniques which are used in the advanced stage of some of the applications that are needed. My personal choice has been to write all of my programs in MAGMA [7], which offers both the module theory and the commutative algebra. I owe a great debt of gratitude to John Cannon and his group in Sydney (particularly Allen Steel) for all of the help that they have given me. At the same time I have had to write many of my

[1]Partly supported by grants from NSF and the EPSRC

own routines, even for the commutative algebra portion of the calculations, and it is clear to me that there are many improvements that can be made both in my program and in the general state of what is available for a project like this one. Hence I would be grateful for any help or suggestions from the readers.

It should also be mentioned that David Green has written stand-alone versions of some of the programs [15].

So far, the programs have played a crucial role in the calculation of the mod-2 cohomology of several groups, namely the Janko groups J_2 and J_3 [12], and the Higman-Sims group HS [3]. The latter calculation is still in progress but some resolution is expected soon. In addition, Dave Benson and I are working on a project to compute the cohomology of the double cover $2M_{12}$ of the Mathieu group M_{12}. The scheme for this project would be to identify the elements of the computed cohomology of the Sylow 2-subgroup of M_{12} with the generators in the presentation of [2] and then use the inflation map routines which have been recently included in the cohomology package. Then it should be possible to complete the calculation using the hypercohomology spectral sequence.

One of the major aims of the project has been the computation of the mod-2 cohomology rings for the groups of order 64. The rings for the 51 groups of order 32 were computed by Rusin [17] several years ago and they have served as test cases and general inspiration for several investigations (e.g. see [1, 6]). It is hoped that the new calculations will play a similar role. As of the writing of this report and the EuroConference in Essen, more than 90% of the cohomology rings of the 267 groups of order 64 had been successfully computed. Only 20 or so problem cases remained to be done. The obstructions to finishing the last few cases are varied and mostly involve technical questions in computational commutative algebra. The actual calculation of the cohomology ring – as far as it goes – is reasonably straightforward, and breaks down in only a few cases because the Groebner basis calculations of the relations ideal is too complicated. More often the problems arise in the verification that the computation is complete, that all generators and relations for the cohomology ring have been found in the course of the calculation. The test for completion is rather tortured and is based on obtaining positive answers to some questions which are only conjectural, but seem to hold for all of the examples that have been computed. The tests invoke mostly computer applications that deal only with the output in the form of generators and relations for a quotient of a polynomial ring.

There are three main aims to this note. First I want to give some explanation and discussion of the conjectures (questions) on which the test for the completion of the calculation is based. Most of these can be verified using Groebner basis techniques on the output of the calculation. However in Section 5.3 we note that there are some problems with a few of the verifications. The second aim is to give some discussion of the computational problems. In some cases it may be a matter of finding a correct and efficient algorithm for making the calculation. In other cases it is possible to circumvent the difficulties by employing a different test for the completion of the calculation. So the third aim of the note is to outline a couple of tests for the completion which are not already recorded in the literature. This we do in Section 5.4.

It should be emphasized that the output of the calculation is being made public as it is completed. It is posted on my web page

http://www.math.uga.edu/~jfc/groups/cohomology.html

Some explanation of the project and the problems is given there. In addition, anyone who is interested is welcome to get the programs, written in MAGMA code, which run the calculation.

For what follows, the reader is urged to consult the general reference books [4, 14] for basic information on modules and cohomology and the book [16] for the details of the commutative algebra.

5.2 Questions and Conjectures

For notation assume that G is a finite p-group and that k is a field of characteristic p. We also use k to denote the one-dimensional trivial kG-module. Let $H^*(G,k)$ be the cohomology ring of G. It is known that $H^*(G,k)$ is a finitely generated algebra over k and that its Krull dimension is equal to the p-rank of G. More specifically a theorem of Quillen (see [4]) says that the minimal prime ideals of $H^*(G,k)$ are precisely the radicals of the kernels of the restrictions to the maximal elementary abelian p-subgroups of G. Moreover we know by Duflot's theorem [13] that the depth of the cohomology ring is at least equal to the p-rank of the center of G. The depth of a ring is defined to be the length of the longest regular sequence of elements of the ring. Indeed we know (see Proposition 5.2 at the end of this section) that if r is the p-rank of the center of G and if $\zeta_1, \dots, \zeta_r$ is a sequence of elements of $H^*(G,k)$ whose restriction to the center of G is a system of parameters then $\zeta_1, \dots, \zeta_r$ is a regular sequence for the cohomology ring. In the case that the cohomology ring has a regular sequence of length equal to its Krull dimension then, by definition, the cohomology is Cohen-Macaulay.

A great deal of what we wish to discuss in this section is concerned with the existence of sequences of parameters with special properties. A homogeneous system of parameters for $H^*(G,k)$ is a sequence of homogeneous elements $\zeta_1, \dots, \zeta_n$ with the property that the subring of $H^*(G,k)$ generated by the elements is a polynomial subring and the algebra $H^*(G,k)$ is a finitely generated module over the polynomial subring $k[\zeta_1, \dots, \zeta_n]$. In particular it is required that the length of any homogeneous system of parameters be equal to the Krull dimension of the cohomology ring. If $V_G(k)$ is the maximal ideal spectrum of $H^*(G,k)$ and if $V_G(\zeta_i)$ is the set of all maximal ideals containing the ideal generated by ζ_i then it can be seen that $\zeta_1, \dots, \zeta_n$ is a homogeneous system of parameters if and only if n is the Krull dimension and $\cap_1^n V_G(\zeta_i) = \{0\}$, where 0 denotes the maximal ideal of all elements of positive degree.

Question 5.1. *Does the cohomology ring $H^*(G,k)$ have a quasi-regular sequence of length equal to the Krull dimension? [5,10]*

A sequence $\zeta_1, \dots, \zeta_n$ of homogeneous elements is a quasi-regular sequence provided

(i) ζ_1 is a regular element, i.e. multiplication by ζ_1 is an injective map from $H^*(G,k)$ to itself and

(ii) for every $i > 1$ multiplication by ζ_i

$$\zeta_i : (H^*(G,k)/(\zeta_1, \dots, \zeta_{i-1}))_m \longrightarrow (H^*(G,k)/(\zeta_1, \dots, \zeta_{i-1}))_{m+deg(\zeta_i)}$$

is injective whenever the degree m is at least equal to $deg(\zeta_1) + \cdots + deg(\zeta_{i-1})$.

From the standpoint of performing computer calculations the advantage of having a quasi-regular sequence is clear from Proposition 10.3 of [5]. Specifically, if $H^*(G,k)$ has a quasi-regular sequence $\zeta_1, \dots, \zeta_n$ then all of the ring is generated by elements in degrees at most $\sum_1^n deg(\zeta_i)$. This holds because the Koszul complex described in the Proposition has no homology and, in particular the last term which is $H^m(G,k)/\sum \zeta_i \cdot H^{m-deg(\zeta_i)}(G,k)$ is zero provided m is large enough. So the only problem with finding the entire cohomology ring is making certain that a basis for the ideal of relations is computed. A different proof of this fact can be reasonably easily deduced from the discussion in Section 5.4. Note that in the paper [10], the Condition G (3.4) is essentially the existence of a quasi-regular sequence.

The essential cohomology of G is defined to be intersection of the kernels of the restriction maps $res_{G,H} : H^*(G,k) \longrightarrow H^*(H,k)$ for all maximal subgroups H of G. So it is the ideal of the cohomology ring consisting of all elements which are not detected on any subgroup of G. Let $\mathcal{E} = \mathcal{E}(G,k)$ denote the essential cohomology of G. It should be clear that it is a graded ideal.

Proposition 5.1. *(See Theorem 2.3 of [9]) If the essential cohomology is nonzero, then its dimension is equal to the p-rank of the center of G.*

One motivation for wanting to know the proposition is clear. If the essential cohomology is zero, then the cohomology ring $H^*(G,k)$ is determined entirely by its restrictions to the maximal subgroups of G. Thus the essential cohomology is the part that is not determined by the restrictions. By an induction, we can assume the cohomology of the maximal subgroups is known.

The dimension of $\mathcal{E}$ is the same as the dimension of the variety $V_G(J)$ or the Krull dimension of the ring $H^*(G,k)/J$ where J is the annihilator of $\mathcal{E}$. The Proposition concerns the existence of an associated prime $\mathbf{p}$ in $H^*(G,k)$ such that the dimension of $H^*(G,k)/\mathbf{p}$ is equal to the depth of the cohomology ring. That is, if $\mathcal{E} \neq 0$, then $\mathcal{E}$ has an element whose annihilator $\mathbf{p}$ is prime and has dimension equal to the p-rank of the center of G, which is the depth. In some sense this is part of a more general question.

Question 5.2. *Does the cohomology ring always have an associated prime* $\mathbf{p}$ *such that the dimension* $H^*(G,k)/\mathbf{p}$ *is equal to the depth of* $H^*(G,k)$*? [9]*

For arbitrary commutative rings it is easy to find examples for which the last question does not have an affirmative answer. For example, suppose that

$R = k[x,y,u,v]/(u^2, v^2, uv, xu - yv)$. Then the ideal generated by u and v as a $k[x,y]$-module looks like the augmentation ideal of $k[x,y]$. Thus the ring has depth one, although the Krull dimension is two and every nonzero element of $k[x,y]$ is a regular element.

It is not hard to see that minimum dimension $H^*(G,k)/\mathbf{p}$ for $\mathbf{p}$ any of the associated prime is less than the depth. So the question (5.2) is really about the converse. In the case that the depth d of $H^*(G,k)$ is equal to the p-rank r of G then the cohomology ring is Cohen-Macaulay and there is no question at all. As to the ideal $\mathcal{E}$, Adem and Karagueuzian [1] have shown that in this case the essential cohomology is nonzero if and only if the group has a unique central maximal elementary abelian p-subgroup. But this does not address the question of how we determine the depth.

Another way of expressing this is to say that the real issue is detection. That is, we say that a collection of subgroups of G detects the cohomology of G if no element of $H^*(G,k)$ is in the kernel of the restricton to all of the subgroups in the collection. The theorem of [1] affirms in a roundabout way that the cohomology of centralizers of the maximal elementary abelian subgroups are precisely the detectors of the cohomology of G in the Cohen-Macaulay case. In general, we know [9] that if the depth of the cohomology ring is d then the cohomology is detected on the centralizers of the elementary abelian subgroups of rank d. The question again is the converse.

Question 5.3. *Suppose the depth of G is d and $d < r$ where r is the p-rank of G. Is it necessary that the intersection of the kernels of the restrictions to the centralizers of the elementary abelian p-subgroups of rank $d+1$ is not zero? Must that intersection have dimension d?*

An affirmative answer to (Quesetion 5.3) would also imply an affirmative answer to (Quesetion 5.2). The answer is yes if the group has a unique maximal elementary abelian subgroup which is not central. This is because there is a nonzero element of

$$H^1(G,k) \cong k \otimes H^1(G, \mathbb{Z}/p) \cong k \otimes \operatorname{Hom}(G, \mathbb{Z}/p)$$

corresponding a maximal subgroup (the kernel of the homomorphism) which contains the centralizer of of the maximal elementary abelian. That element restricts to zero on the centralizer. The same argument will work in the case of a semi-dihedral group where there is a maximal subgroup which contains the centralizers of all of the maximal elementary abelian subgroups. However the reader should be careful of what this means. For example, the group of order 32 ($p = 2$) with Hall-Senior number 33 has two maximal elementary abelian subgroups of orders 8 and 16. So the Krull dimension of the cohomology is 4, but the depth is 3. The cohomology is actually detected on the elementary abelians, but not on the centralizer of the only elementary abelian subgroup of rank 4.

Affirmative answers to these questions would be a great aid in the computational problems. However at least until the issues are settled, it is necessary to ask another question.

Question 5.4. *Assume that $\mathcal{E} \neq 0$ and the dimension of $\mathcal{E}$ is d, the p-rank of the center of G. Let $\zeta_1, \dots, \zeta_d$ be a regular sequence of maximal length, then is $\mathcal{E}$ a free module over the polynomial subring $k[\zeta_1, \dots, \zeta_d]$?*

An affirmative answer here allows us to establish limits on the maximum degrees of a minimal set of relations among the generators of the cohomology ring. In particlular, we can show the following.

Theorem 5.1. *[10] Suppose that the generators and relations for the cohomology ring $H^*(G, k)$ have been calculated out to degree N and that the following conditions hold.*

i) Questions 5.1 and 5.4 have affirmative answers for the quotient of the computed polynomial ring by the ideal of computed generators.

ii) N is larger than the sum of the degrees of the elements in a homogeneous system of parameters for the computed cohomology.

iii) N is larger than the degrees of the generators of the kernels of the restriction of the computed cohomology to the maximal subgroups of G.

Then the computation is complete and all generators of the cohomology ring and the ideal of relations have been computed.

In the later sections of the paper we want to consider some of the problems involved in verifying the hypotheses of the theorem. However first we should clarify one point which does not seem to appear anywhere else in the literature. Namely we sketch a proof of the following.

Proposition 5.2. *Suppose that $\zeta_1, \dots, \zeta_t$ is a sequence of homogeneous elements of $H^*(G, k)$ whose restriction to the center $Z(P)$ of a Sylow p-subgroup P of G is a regular sequence. Then $\zeta_1, \dots, \zeta_t$ is a regular sequence on $H^*(G, k)$.*

Proof. First notice that there is a transfer map $tr_P^G : H^*(P, k) \longrightarrow H^*(G, k)$ and a restriction map $res_{G,P} : H^*(G, k) \longrightarrow H^*(P, k)$ such that the compositon $tr_P^G res_{G,P}$ is multiplication by the index $|G : P|$. If $\zeta_1, \dots, \zeta_t$ is a sequence of elements in $H^*(G, k)$ such that $res_{G,P}(\zeta_1), \dots, res_{G,P}(\zeta_t)$ is regular then $\zeta_1, \dots, \zeta_t$ is regular because $H^*(G, k)$ is a direct summand of $H^*(P, k)$ as a module over the polynomial subring generated by $\zeta_1, \dots, \zeta_t$. Consequently we may assume that $G = P$ is a p-group.

From here we follow the construction of [8]. Let C denote the center of G and suppose that $res_{G,C}(\zeta_1), \dots, res_{G,C}(\zeta_t)$ is a regular sequence. Then by Theorem 10.3.4 of [14] we know that $H^*(C, k)$ is a free module, by restriction, over the polynomial ring $R = k[\zeta_1, \dots, \zeta_t]$. Additionally there is a multiplication map $\mu : C \times G \longrightarrow G$ and an inclusion $\iota : G \longrightarrow C \times G$. These maps induce algebra homomorphisms

$$\mu^* : H^*(G, k) \longrightarrow H^*(C \times G, k) \cong H^*(C, k) \otimes H^*(G, k)$$

and

$$\iota^* : H^*(C \times G, k) \longrightarrow H^*(G, k)$$

with the property that $\iota^* \mu^*$ is the identity on $H^*(G, k)$. We view both of these homomorphisms as homomorphisms of R-modules. As such $H^*(G, k)$ is a direct summand of $H^*(C \times G, k)$. Then the proof is completed by the observation that $H^*(C \times G, k)$ is a free R-module. This last fact is most easily seen by observing that for all $i \geq 0$ the quotient module

$$(H^*(C, k) \otimes \sum_{n \geq i} H^n(G, k)) \quad / \quad (H^*(C, k) \otimes \sum_{n \geq i+1} H^n(G, k))$$

is isomorphic as an R-module to a direct sum of copies of $H^*(C, k)$, the number of copies being the dimension of $H^i(G, k)$. □

5.3 Some Problems in the Computations

As we said in the introduction, a major purpose of this note is the discussion of some of the difficulties which arise in the computation of the group cohomology. In several cases the problems have become most obvious in the effort to compute all of the cohomology rings of the groups of order 64. For this project a few of the difficult situations have occurred in several different cases and there has been the opportunity to compare methods and results. Occasionally, the problems have been solved in the sense that the computation has been successful even though it might be slow.

The method of computation is outlined in [11] and [10] and we will not discuss it much here. Suffice it to say that there are two main types of operations used. Initially the calculation is almost all linear algebra. This part includes the construction of the projective resolution, the chain maps for the minimal generators of the cohomology ring, and the set of all relations among the generators in degrees within the range of the computation. It also includes the constructions of the restriction maps to subgroups. The linear algebra portion can be very memory intensive because of the size of the modules and spaces and can also be time consuming when the resolutions get large. On the other hand it is very reliable and reasonably predictable. As long as the size of the computation is not too large for the machine, then it always finishes and it usually finishes in a reasonable amount of time.

The real difficulties in the computations come up in the second phase, which is mostly commutative algebra applications. I must admit that this is the area that I know least about and some of the problem that have arisen have clearly been due to my ignorance of the proper use of the methods. I am still trying to separate the problems which can be solved by intelligent use of available methods from those that are simply hard computations or for which no really good method has been discovered.

It is a fact that some Groebner basis computations are very difficult. Still a few of the successes seem quite amazing. Consider the example of the group number 153 of order 64. It is isomorphic to the Sylow 2-subgroup of the Suzuki group $Sz(8)$. The cohomology ring has Krull dimension 3 and is the quotient of a polynomial ring on 20 generators by an ideal that is minimally generated by 136 relations. All computations including the kernels of the restrictions to all maximal subgroups and the unique maximal elementary abelian subgroup and the nilradical and essential cohomology were performed in approximately 9 days on a SUN ULTRA 2200. However other cases with fewer generators and relations have not gone as well. The search for a solution for the last few groups of order 64 may finally degenerate into a search for any Groebner basis technique or trick that can do the job. The running times are extremely sensitive to the ordering on the variables and it is also possible that the right ordering might make a big difference.

In the rest of the section let me just mention a few computational difficulties which, I hope, have reasonable solutions. The ultimate aim is to develop a collection of programs which could run through the cohomology rings of most or all of a set of groups without a glitch. At the present time the program runs until something goes wrong and it crashes. The information which is posted on my web page is the result of care in adjusting and restarting the computation. However this is certainly not the first time that we will run the cohomology of the groups of order 64. As we figure out more questions and conjectures about the cohomology rings we will find new ideas to test and add more pieces to the computation. It would be very desirable to have it run smoothly and not require a lot of monitoring.

5.3.1 Systems of Parameters and Regular Sequences The problem is to find an efficient method of computing a system of parameters in low degrees for a ring which is the quotient of a polynomial ring P by an ideal I. Strangely the problem is made more difficult by our desire to keep the field small, and to have the system begin with a regular sequence of maximal length. The last condition is important in the verification that the computed cohomology ring has a quasi-regular sequence (see Question 5.1). The method that is currently in use in the project is a search that begins by looking at the generators of the cohomology and then looks at more complicated expressions in the generators. At each step of the search we simply look for elements which have the property that factoring out the element decreases the Krull dimension by one. In this way we get a system of parameters. However we may not have a regular sequence of maximal length at the beginning of the system.

The problem is illustrated by the following examples. Suppose that $P = k[x, y, u]$ for a field k and that $I = (u^2, xyu)$. Then it is easy to see that P/I has Krull dimension 2 and that the sequence x, y is a system of parameters for P/I. However neither x nor y is a regular element since x annihilates yu and y annihilates xu. Still the depth of the ring is one because $x + y$ is a regular element. In the case that $k = GF(2)$, then there is an example that is even worse. Suppose that $P = k[x, y, u]$ and $I = (u^2, xy(x + y)u)$. Again x, y is a system of parameters, but no element of degree one is a regular element. To find a regular element we must go to degree 2 (that is, $x^2 + xy + y^2$ is regular).

So we see that to find regular elements it is necessary to avoid the associated primes in the ring. It would be nice to have an implementation of an algorithm to find associated primes. However it might also be expensive to recompute the associated primes at each stage. The algorithm currently in use in the project involves extending the field and checking random elements when a problem with the regular sequence is encountered. It works well in small cases, but is slow and probably inefficient in others.

5.3.2 Free Generators for the essential Cohomology One of the checks for the completion of the cohomology ring calculation is a verification that the essential cohomology $\mathcal{E}$ is a finitely generated free module over the polynomial ring $k[\zeta_1, \dots, \zeta_d]$ where d is the depth and $\zeta_1, \dots, \zeta_d$ is a regular sequence of maximal length in the computed cohomology. Magma does not currently have an algorithm that would test the freeness of a module over a polynomial ring that is presented in this fashion. The program which is being used in the cohomology project is rather naive and doomed to failure in certain circumstances. Basically it is the following.

Suppose that R is the computed cohomology ring. Then we know that R is itself a free module over the polynomial ring $k[\zeta_1, \dots, \zeta_d]$ (see Theorem 10.3.4 of [14]). Any k-basis of $R/(\zeta_1, \dots, \zeta_d)$ is a free $k[\zeta_1, \dots, \zeta_d]$-basis of R. Likewise any k-basis of $\mathcal{E}/(\zeta_1, \dots, \zeta_d)$ is a set of generators of $\mathcal{E}$ as a $k[\zeta_1, \dots, \zeta_d]$-module. Hence if the k-basis of $\mathcal{E}/(\zeta_1, \dots, \zeta_d)$ is linearly independent in $R/(\zeta_1, \dots, \zeta_d)$ then $\mathcal{E}$ is a free module as desired.

The test outlined above is reasonably easy to implement, but it tells you nothing when it fails. Moreover in some cases it is certain to fail. For suppose that $p = 2$ and $G \cong \mathbb{Z}/2 \times H$ where H is a 2-group. Then $H^*(G, k) \cong H^*(\mathbb{Z}/2, k) \otimes H^*(H, k)$ and $H^*(\mathbb{Z}/2, k) \cong k[\eta]$ is a polynomial ring in one variable. Now suppose that there is an element β in the essential cohomology of H. Then $1 \otimes \beta$ is not in the essential cohomology of G because H is one of the maximal subgroups. However $\eta \otimes \beta$ is in the essential cohomology of G. So the problem is that η is a regular element and the essential cohomology is all contained in $\eta \otimes H^*(H, k)$.

The programs that run at the moment simply ignore the failures of the test. I have been checking the cases of failure by hand. In the example of the last paragraph, it can be seen that if the answer to (Question 5.4) is yes for H, then it is the same for G. Indeed, if enough steps have been computed to get the cohomology of H then the calculation is complete. However it would be much better to have an efficient implementation of an algorithm for testing when a module is free.

In some other cases, the completion of the calculation can be checked by other tests such as those in the next section.

5.4 Two Exceptional Cases

We just noticed in the last section that there are problems sometimes in verifying the hypothesis of Question 5.4. Fortunately the aforementioned problems seem only to have occurred in situations where there exist alternate tests for the completion of the calculations. However the alternate tests are not well exposed in the literature. The aim of this section is to state the requirements of the alternate tests and sketch the proofs. The proofs rely on some module theoretic methods, so we need to introduce some notation. As a general reference for the homological algebra we refer the reader to the book by Benson [4].

As before assume that G is a p-group but now let k be an algebraically closed field of characteristic p. If M is a kG-module let $\phi : P \longrightarrow M$ be a projective cover of M. That is, P is a minimal projective kG-module such that there exists such a homomorphism ϕ which is surjective. The kernel of ϕ is denoted $\Omega(M)$. For $n > 1$ we define inductively $\Omega^n(M) = \Omega(\Omega^{n-1}(M))$. Similarly we let $\Omega^{-1}(M)$ be the cokernel of an injective hull $\theta : M \longrightarrow Q$ and for $n > 1$ let $\Omega^{-n}(M) = \Omega^{-1}(\Omega^{-n+1}(M))$. It should be remembered that projective kG-modules are also injective.

In the case that $M = k$ the trivial kG-module, then we have an exact sequence

$$0 \longrightarrow \Omega^n(k) \longrightarrow P_{n-1} \longrightarrow \dots \longrightarrow P_0 \longrightarrow k \longrightarrow 0$$

where P_* is a part of a minimal projective resolution of k. Then it can be seen that, because of the minimality, any element ζ of $H^*(G,k) \cong \mathrm{Ext}^*(k,k)$ is uniquely represented by a homomorphism (cocycle) $\zeta : \Omega^n(k) \longrightarrow k$. So if ζ is not zero then we have an exact sequence

$$0 \longrightarrow L_\zeta \longrightarrow \Omega^n(k) \xrightarrow{\zeta} k \longrightarrow 0 \tag{5.1}$$

Throughout the section we assume that G has the property that the p-rank of the center $Z(G)$ of G is at least $r-1$, where r is the p-rank of G. So there exists a homogeneous system of parameters $\zeta_1, \dots, \zeta_r$ for $H^*(G,k)$ such that the sequence $\zeta_1, \dots, \zeta_{r-1}$ is regular. Let n_i be the degree of ζ_i for all i. Notice that if p is an odd prime then n_i must be even for all i since otherwise $\zeta_i^2 = 0$. For technical reasons we assume that $n_i \geq 2$ even when $p = 2$.

Our hypothesis that ζ_1 is regular means that in the long exact sequence

$$\dots \longrightarrow \mathrm{Ext}^n_{kG}(k,k) \xrightarrow{\hat{\zeta}_1} \mathrm{Ext}^n_{kG}(\Omega^{n_1}(k),k) \longrightarrow \mathrm{Ext}^n_{kG}(L_{\zeta_1},k)$$
$$\longrightarrow \mathrm{Ext}^{n+1}_{kG}(k,k) \longrightarrow \dots$$

the module $\mathrm{Ext}^n_{kG}(\Omega^{n_1}(k),k) \cong \mathrm{Ext}^{n+n_1}_{kG}(k,k) \cong H^{n+n_1}(G,k)$ and the map $\hat{\zeta}_1$ is injective for all $n \geq 0$. It follows that

$$\mathrm{Ext}^n_{kG}(L_{\zeta_1},k) \cong H^{n+n_1}(G,k)/\zeta_1 \cdot H^n(G,k)$$

for all $n \geq 0$. In fact we can do the same thing for Tate cohomology and get that $\widehat{\mathrm{Ext}}^{-1}_{kG}(L_{\zeta_1},k) \cong H^{n_1-1}(G,k)$ since $\zeta_1 \cdot H^{-1}(G,k) \cong 0$ [6]. For convenience of notation let $U_1 = \Omega^{-n_1}(L_{\zeta_1})$. By degree shifting in the cohomology we have that

$$\operatorname{Ext}^m_{kG}(U_1, k) \cong \operatorname{Ext}^m_{kG}(\Omega^{-n_1}(L_{\zeta_1}), k) \cong$$

$$\operatorname{Ext}^{m-n_1}(L_{\zeta_1}, k) \cong H^m(G,k)/\zeta_1 \cdot H^{m-n_1}(G,k)$$

for $m \geq n_1 - 1$. Now consider ζ_2. Assume that $r > 2$ so that ζ_2 is the second element of a regular sequence. If we tensor the exact sequence of modules (Theorem 5.2) corresponding to ζ_2 with L_{ζ_1} we get a sequence

$$0 \longrightarrow L_{\zeta_1} \otimes L_{\zeta_2} \longrightarrow L_{\zeta_1} \otimes \Omega^{n_2}(k) \stackrel{1\otimes\zeta_2}{\longrightarrow} L_{\zeta_1} \otimes k \longrightarrow 0$$

Now notice that $L_{\zeta_1} \otimes \Omega^{n_2}(k) \cong \Omega^{n_2}(L_{\zeta_1}) \oplus P$, for some projective module P. So with careful degree shifting, and by letting $U_2 = \Omega^{-n_1}(L_{\zeta_1}) \otimes \Omega^{-n_2}(L_{\zeta_2})$, we have that

$$\operatorname{Ext}^m_{kG}(U_2, k) \cong H^m(G,k)/(\zeta_1 \cdot H^{m-n_1}(G,k) + \zeta_2 \cdot H^{m-n_2}(G,k))$$

for all $m \geq n_1 + n_2 - 2$. Obviously we can continue this process as long as the elements in the sequence are regular. Our first result in this direction is the following.

Theorem 5.2. *Suppose that the p-rank of $Z(G)$ is equal to the p-rank, r, of G and that $r > 1$. Then the cohomology ring $H^*(G,k)$ is generated by elements in degrees at most $s = \sum_1^r (n_i - 1)$. Moreover if t is the largest degree of any generator then the ideal of relations for the cohomology ring is generated in degrees at most $s + t$.*

Proof. The proof relies on the varieties of the modules in question. Recall that $V_G(k)$ is the maximal ideal spectrum of the cohomology ring $H^*(G,k)$, and if M is a kG-module then $V_G(M)$ is the set of maximal ideals which contain the annihilator $J(M)$ of $\operatorname{Ext}^*_{kG}(M,M)$ in $H^*(G,k)$. The most relevant facts are that $V_G(M) = \{0\}$ if and only if M is projective and the Tensor Product Theorem which says that the variety of the tensor product of two modules is the intersection of their varieties. Another important result is that $V_G(L_\zeta) = V_G(\Omega^m(L_\zeta)) = V_G(\zeta)$, the set of all maximal ideals which contain $\zeta \in H^*(G,k)$, for all m. See [4] or [14] for details.

For the case in question, the important thing to notice is that the cohomology ring $H^*(G,k)$ is Cohen-Macaulay [13] and the sequence $\zeta_1, \ldots, \zeta_r$ is a regular sequence. Also

$$U_r = \bigotimes_{i=1}^{r} \Omega^{-n_i}(L_{\zeta_i}).$$

Consequently $V_G(U_r) = \cap_1^r V_G(\zeta_i) = \{0\}$, since $\zeta_1, \ldots, \zeta_r$ is a system of parameters for the cohomology ring. Hence U_r is projective, and $\operatorname{Ext}^m_{kG}(U_r, k) = \{0\}$ for all $m > 0$. So from the above argument we conclude that

$$H^m(G,k) = \zeta_1 \cdot H^{m-n_1}(G,k) + \cdots + \zeta_r \cdot H^{m-n_r}(G,k)$$

whenever $m > s$. This proves the first part of the theorem.

For the rest we need to recall that the cohomology ring is a free module over the polynomial subring $k[\zeta_1, \dots, \zeta_r]$ (see Theorem 10.3.4 of [14]). So if $\alpha_1, \dots, \alpha_t$ is a set of generators for this module, then a minimal set of relations can be given in the form

$$\alpha_i \cdot \alpha_j = \sum_1^t \eta_{ijk} \alpha_k$$

where the coefficients η_{ijk} are in $k[\zeta_1, \dots, \zeta_r]$. □

The second case that we want to consider is the case in which the codepth of $H^*(G, k)$ is known to be at most one. That is, suppose that the p-rank of $Z(G)$ is $r - 1$. Then we know that

$$\operatorname{Ext}^m_{kG}(U_{r-1}, k) \cong H^m(G, k) / \sum_1^{r-1} \zeta_i \cdot H^{m-n_i}(G, k)$$

for all $m \geq \sum_1^{r-1}(n_i - 1)$. This time the thing to notice is that $V_G(U_{r-1}) = \cap_1^{r-1} V_G(\zeta_i)$ is a variety of dimension one because $\zeta_1, \dots, \zeta_r$ is a system of parameters. As a consequence U_{r-1} is a periodic kG-module and because $V_G(\zeta_r) \cap V_G(U_{r-1}) = \{0\}$ multiplication by ζ_r induces an isomorphism

$$\zeta_r : \operatorname{Ext}^m_{kG}(U_{r-1}, k) \longrightarrow \operatorname{Ext}^{m+n_r}_{kG}(U_{r-1}, k)$$

for all $m > 0$. All of this is made reasonably clear if we tensor the sequence (Theorem 5.2) for $\zeta = \zeta_r$ with the module U_{r-1}. That is, $U_{r-1} \otimes L_{\zeta_r}$ is projective, hence injective, because its variety is $\{0\}$, and the sequence splits. It follows that

$$\zeta_r : H^m(G, k) / \sum_1^{r-1} \zeta_i \cdot H^{m-n_i}(G, k)$$

$$\longrightarrow H^{m+n_r}(G, k) / \sum_1^{r-1} \zeta_i \cdot H^{m+n_r-n_i}(G, k)$$

is an isomorphism for $m \geq s' = \sum_1^{r-1}(n_i - 1)$. Hence we have proved the following

Proposition 5.3. *[5] Suppose that $\zeta_1, \dots, \zeta_{r-1}$ is a regular sequence for $H^*(G, k)$ where $r =$ p-rank(G). Then $\zeta_1, \dots, \zeta_r$ is a quasi-regular sequence.*

Our aim is to prove the following.

Theorem 5.3. *Suppose that the p-rank of the center of G is $r-1$ where r is the p-rank of G. Let $\zeta_1, \dots, \zeta_r$ be a system of parameters for the computed cohomology such that $\zeta_1, \dots, \zeta_{r-1}$ is a regular sequence and the restrictions of $\zeta_1, \dots, \zeta_{r-1}$ to the center of G is a regular sequence for $H^*(Z(G), k)$. Assume also that we have an isomorphism*

$$\zeta_r : R^m / \sum_1^{r-1} \zeta_i \cdot R^{m-i} \longrightarrow R^{m+n_r} / \sum_1^{r-1} \zeta_i \cdot R^{m+n_r-n_i}$$

Let $s = \sum_1^r (n_i - 1)$ *where* $n_i = deg(\zeta_i)$. *Then if the cohomology of* G *has been calculated to degree* $s + w$ *where* w *is the maximum of the degrees of the generators of the cohomology, then the computation is complete.*

Proof. It should be clear that if the hypothesis of (Proposition 5.3) is satisfied then all of the generators of $H^*(G,k)$ can be taken to lie in degrees at most s. What about the relations? For convenience of notation let R denoted the computed cohomology ring. So we have a ring homomorphism $\theta : R \longrightarrow H^*(G,k)$.

The thing to notice is that both $H^*(G,k)$ and R are free modules over the polynomial subring $Q = k[\zeta_1, \dots, \zeta_{r-1}]$. So if $\{\beta_i\}$ is a (necessarily countable) set of free generators of R as a module over Q then we have a complete, though infinite, set of relations of the form

$$\beta_i \cdot \beta_j = \sum_k \mu_{ijk}\beta_k \tag{5.2}$$

for $\mu_{ijk} \in Q$. Now because of the isomorphism we may assume that $\beta_i = \zeta_r\beta_j$ whenever $i > s$. Hence by the hypothesis and the isomorphism (Proposition 5.3) the elements $\{\theta(\beta_i)\}$ form a basis for $H^*(G,k)$ as a module over Q. Then θ is an isomorphism of free Q-modules. So it is also injective and a ring isomorphism. □

Remark 5.1. The hypothesis for the last theorem may seem a bit complicated, but all of the items in it are easy to check in the computer calculation. Moreover they are built into the programs that are currently being run.

Bibliography

[1] A. Adem and D. Karagueuzian: *Essential cohomology of finite groups*, Comm. Math. Helv. **72**, (1997), 101–109.

[2] A. Adem, R. J. Milgram and J. Maginnis: *The geometry and cohomology of the Mathieu group, M*12, J. Algebra **139**, (1991), 90–133.

[3] A. Adem, J. F. Carlson and R. J. Milgram: *The cohomology of the Higman-Sims group*, (in preparation).

[4] D. J. Benson: *Representations and Cohomology*, II: *Cohomology of Groups and Modules*, Cambridge Studies in Advanced Mathematics, Cambridge University Press, Cambridge **31**, (1991).

[5] D. J. Benson and J. F. Carlson: *Projective resolutions and Poincaré duality complexes*, Trans. Amer. Math. Soc. **342**, (1994), 447–488.

[6] ______: *Products in negative cohomology*, J. Pure Appl. Algebra **82**, (1992), 107–129.

[7] W. Bosma and J. Cannon: *Handbook of Magma Functions*, Magma Computer Algebra, Sydney, 1996

[8] C. Broto and H.-W. Henn: *Some remarks on central elementary abelian p-subgroups and cohomology of classifying spaces*, Quart. J. Math **44**, (1993), 155–163.

[9] J. F. Carlson: *Depth and transfer maps in the cohomology of groups*, Math Zeit. **218**, (1995), 461–468.

[10] ______: *Calculating group cohomology: Tests for completion*, J. Sym. Comp., to appear.

[11] J. F. Carlson, E. Green and G. J. A. Schneider: *Computing the ext algebras for the group algebras of finite groups*, J. Sym. Comp.

[12] J. F. Carlson, J. Maginnis and R. J. Milgram: *The cohomology of the sporadic groups J_2 and J_3*, J. Algebra, to appear.

[13] J. Duflot: *Depth and equivariant cohomology*, Comm. Math. Helvetici **56**, (1981), 627–637.

[14] L. Evens: *The Cohomology of Groups*, Oxford University Press , New York (1991).

[15] D. Green: *Private correspondence.*

[16] H. Matsumura: *Commutative Ring Theory*, Cambridge University Press, Cambridge, 1986.

[17] D. Rusin: *The cohomology of groups of order 32*, Math. Comp. **53**, (1989), 359–385.

Progress in Mathematics, Vol. 173, © 1999 Birkhäuser Verlag Basel/Switzerland

Chapter 6

On a Tensor Category for the Exceptional Lie Groups

Arjeh M. Cohen and Ronald de Man

6.1 Introduction

The construction of representations of Lie groups is intertwined with combinatorics. For instance the combinatorial Littlewood-Richardson rule tells us how the tensor product of two irreducible GL_n-representations W_1 and W_2 decomposes into irreducibles, provided n is big with respect to the size of the Young diagrams corresponding to W_1 and W_2. Moreover, each irreducible GL_n-representation occurs in a tensor power $V^{\otimes d}$ of the natural representation V and can be isolated by use of combinatorics of the symmetric group S_d acting on $V^{\otimes d}$ by permuting the d factors. The intriguing part of the latter two combinatorial involvements is that they hardly depend on the series parameter n. For instance, once the primitive idempotents of the enveloping algebra of S_d inside $\mathrm{End}(V^{\otimes d})$ are figured out, the irreducible components of $V^{\otimes d}$ for GL_n follow (again, at least for n sufficiently large).

The tensor category captures the essence of the combinatorics of these decompositions, which enables a simultaneous treatment of representations for a series of Lie groups parameterized by a parameter such as n above. In Section 6.2 of this paper we introduce tensor categories in a somewhat informal way, giving the basic machinery and terminology for this paper. In Section 6.3 we treat an example based on GL_n, in which both V and its dual are taken as starting point (rather than just V), using diagrams as building blocks. We describe a diagrammatic notation which was introduced by Penrose (see [8]) as a substitute for overly indexed tensor notation. This notation is well known to physicists (cf. Cvitanović [2]). But the main goal of this paper is to discuss Deligne's conjecture ([3]) about the existence of a tensor category for the exceptional Lie groups. We describe the conjecture in Section 6.4. Prior to Deligne, work on these representation categories was conducted by El Houari ([6]) and Vogel ([9]). In Section 6.5, a generic tensor category for automorphism groups of complex finite-dimensional Lie algebras is introduced. Finally, in Section 6.6, we show some implications of Deligne's conjecture for low degree representations and some of the problems arising in an attempt to understand the structure of the conjectured tensor category.

The authors would like to thank Deligne for his valuable suggestions for improvement of this paper.

6.2 Tensor Categories

This section will give the necessary definitions and properties of tensor categories. The main purpose is to show how concepts like duality are paraphrased in categorical language. To prevent it from growing too big, we sometimes omit the details. These can be found in for example the first section of [5]. We assume the reader is familiar with the notions of category and functor, as can be found in [7].

In this paper, a *tensor category* $\mathcal{C}$ will be a category $\mathcal{C}$ together with a tensor functor $\otimes : \mathcal{C} \times \mathcal{C} \to \mathcal{C}$ satisfying certain associativity and commutativity conditions, and an identity object 1 satisfying $1 \otimes X = X \otimes 1 = X$ for all objects X. One important consequence of the associativity and commutativity conditions is that they permit us to define, for any finite set I, a functor

$$\bigotimes_{i \in I} : \mathcal{C}^I \to \mathcal{C}$$

in a 'consistent' (functorial) way. A *tensor functor* $F : \mathcal{C} \to \mathcal{D}$ will be a functor preserving the tensor product with its associativity and commutativity conditions. In this vein, there is the notion of *tensor equivalence* for tensor categories.

In the categories that we consider, the Hom-sets are abelian groups with morphism composition being bilinear (in other words, they are Ab-categories in the terminology of [7]); moreover all functors are supposed to respect addition. Then $\mathrm{End}_{\mathcal{C}}(X)$ is a ring with multiplication coming from morphism composition. The ring $\mathrm{End}_{\mathcal{C}}(1)$ can be seen to be commutative and its multiplication can be seen to coincide with the tensor product of morphisms. All Hom-sets are modules over this ring with the action given by $f \cdot g = f \otimes g$ for $f : 1 \to 1$, $g : X \to Y$ where we use the identifications $1 \otimes X = X$ and $1 \otimes Y = Y$. We speak of a tensor category *over the ring* R if $\mathrm{End}_{\mathcal{C}}(1) \cong R$.

The fundamental example of a tensor category is the category of finite-dimensional vector spaces over a field k equipped with the usual tensor product $\otimes$. Another example is the category having finite sets as objects, bijections as morphisms, and disjoint union as tensor product.

Let G be a group and k a field. Taking the finite-dimensional representations of G over k as objects and equivariant maps as morphisms gives what we will call the *representation category* of G (over k), denoted $\mathbf{Rep}_G$. We can make this into a tensor category by adding the usual tensor product $\otimes$ of representations. An important property of this category is that, for objects X and Y, the vector space $\mathrm{Hom}_k(X, Y)$ with the usual G-action is again an object of the category. In categorical sense $\mathrm{Hom}_k(X, Y)$ distinguishes itself as the object that gives natural isomorphisms

$$\mathrm{Hom}_G(T, \mathrm{Hom}_k(X,Y)) \to \mathrm{Hom}_G(T \otimes X, Y)$$

for all objects T (by mapping $f : T \to \mathrm{Hom}_k(X,Y)$ to the map $t \otimes x \mapsto f(t)(x)$). In other words, $\mathrm{Hom}_k(X,Y)$ represents the functor $T \mapsto \mathrm{Hom}_G(T \otimes X, Y)$.

Now we consider an arbitrary tensor category $\mathcal{C}$. Let X, Y be two objects. If the functor $T \mapsto \mathrm{Hom}_{\mathcal{C}}(T \otimes X, Y)$ is representable, we let $\underline{\mathrm{Hom}}(X,Y)$ denote the representing object. Taking $T = \underline{\mathrm{Hom}}(X,Y)$, the natural isomorphism gives

$$\mathrm{Hom}_{\mathcal{C}}(\underline{\mathrm{Hom}}(X,Y), \underline{\mathrm{Hom}}(X,Y)) \cong \mathrm{Hom}_{\mathcal{C}}(\underline{\mathrm{Hom}}(X,Y) \otimes X, Y).$$

We let $\mathrm{ev}_{X,Y} : \underline{\mathrm{Hom}}(X,Y) \otimes X \to Y$ denote the morphism corresponding to $\mathrm{id}_{\underline{\mathrm{Hom}}(X,Y)}$. The isomorphisms $\eta_T : \mathrm{Hom}_{\mathcal{C}}(T, \underline{\mathrm{Hom}}(X,Y)) \to \mathrm{Hom}_{\mathcal{C}}(T \otimes X, Y)$ are then given by

$$\eta_T(f) = \mathrm{ev}_{X,Y} \circ (f \otimes \mathrm{id}_X).$$

The *dual* $X^\vee$ of an object X is defined to be $\underline{\mathrm{Hom}}(X, 1)$. We let $\mathrm{ev}_X : X^\vee \otimes X \to 1$ denote the map $\mathrm{ev}_{X,1}$. Let $f : X \to Y$ be a morphism and suppose both X and Y have duals. The dual morphism ${}^\top f : Y^\vee \to X^\vee$ is defined to be $\eta_{Y^\vee}^{-1}(\mathrm{ev}_Y \circ (\mathrm{id}_{Y^\vee} \otimes f))$: the unique morphism satisfying $\mathrm{ev}_X \circ ({}^\top f \otimes \mathrm{id}_X) = \mathrm{ev}_Y \circ (\mathrm{id}_{Y^\vee} \otimes f)$.

Composing $\mathrm{ev}_X : X^\vee \otimes X \to 1$ with the natural isomorphism $\psi_X : X \otimes X^\vee \to X^\vee \otimes X$ (obtained from the axioms) we find a morphism $X \otimes X^\vee \to 1$ that, using $\mathrm{Hom}_{\mathcal{C}}(T, X^{\vee\vee}) \cong \mathrm{Hom}_{\mathcal{C}}(T \otimes X^\vee, 1)$, corresponds to a morphism $X \to X^{\vee\vee}$. Objects X for which this morphism is an isomorphism are called *reflexive.*

We say that a tensor category $\mathcal{C}$ is *rigid* if $\underline{\mathrm{Hom}}(X,Y)$ exists for any two objects of $\mathcal{C}$, all objects are reflexive, and certain naturally defined morphisms are isomorphisms (among which are morphisms $Y \otimes X^\vee \to \underline{\mathrm{Hom}}(X,Y)$). For such categories we define the *trace* $\mathrm{Tr}(f)$ of a morphism $f : X \to X$ as follows. Dualizing ev_X gives a morphism $\delta_X : 1 \to X \otimes X^\vee$. We set

$$\mathrm{Tr}(f) = \mathrm{ev}_X \circ \psi_X \circ (f \otimes \mathrm{id}_{X^\vee}) \circ \delta_X.$$

Note that $\mathrm{Tr}(f)$ is an element of $\mathrm{End}_{\mathcal{C}}(1)$.

These definitions coincide with the usual ones for $\mathbf{Rep}_G$ via the isomorphism $\mathrm{End}_{\mathcal{C}}(1) \cong k$. For an object X of $\mathbf{Rep}_G$, the evaluation map ev_X assigns to $f \otimes v$ the value $f(v)$. Given a basis $\{e_i\}$ of X and dual basis $\{e^i\}$ in $X^\vee$, we have $\delta_X(1) = \sum_i e_i \otimes e^i$. The *dimension* $\dim(X)$ of an object X is defined to be $\mathrm{Tr}(\mathrm{id}_X)$. For $\mathbf{Rep}_G$ this is the usual dimension of X modulo the characteristic of k.

Finally, suppose $\mathcal{C}$ is an abelian tensor category over a field k (that is, the underlying category is abelian in the sense of [7]). Then $\mathcal{C}$ is called *semisimple* if, for each object X, the k-algebra $\mathrm{End}_{\mathcal{C}}(X)$ is semisimple.

6.3 A Representation Category for the General Linear Groups

The groups under consideration in this section are the complex general linear groups $\mathrm{GL}_n = \mathrm{GL}_n(\mathbb{C})$ $(n \in \mathbb{N})$. Let t be an indeterminate. We define a tensor category $\mathcal{C}$ over $\mathbb{Z}[t]$ that essentially describes GL_n-representations. For the objects we take finite sets A endowed with an orientation $\epsilon_A : A \to \{+,-\}$. We define an *admissible diagram* from A to B to be a set of arrows going from the positive points of A and negative points of B to the negative points of A and positive points of B, such that every point of $A \cup B$ is a begin point or end point of exactly one arrow. We also allow the diagram to have a finite number of closed (oriented) circles.

Note that we can compose such diagrams by putting them next to each other and connecting arrows. For $\mathrm{Hom}_{\mathcal{C}}(A, B)$ we take the free $\mathbb{Z}[t]$-module on the admissible diagrams, modulo the relation 'adding a circle' = 'multiplication by t'. Composition of these morphisms is induced by the composition of diagrams. For the tensor product of A and B we take the disjoint union of A and B.

The rigidity of this category is easy to verify: the dual of an object is obtained by changing the orientation of its points.

In drawings, we denote positive points by a white dot and negative points by a black dot. Connecting line segments denote the arrows. The direction of the arrow is clear once A and B are known, and is hence omitted. With composition in mind, we will depict a diagram $A \to B$ with A on the right side and B on the left side (to make composition easier).

Example 6.1. The ring $\mathrm{End}_{\mathcal{C}}(\{+,+,-\})$ is a free $\mathbb{Z}[t]$-module with basis

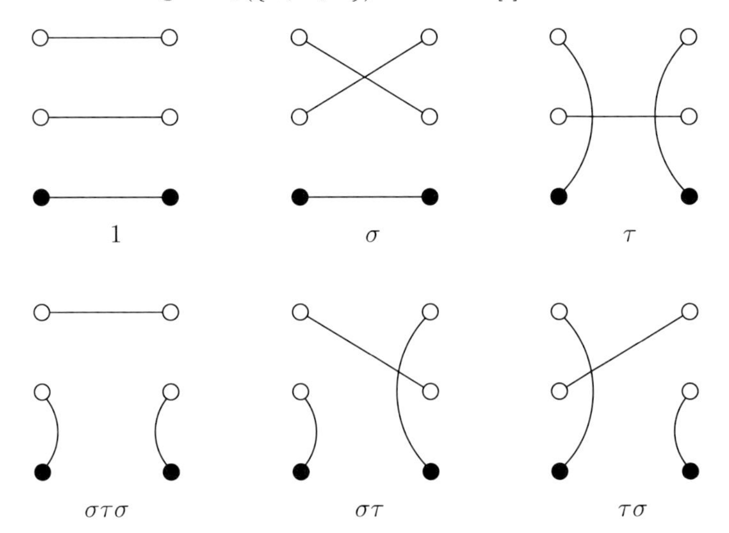

Multiplication is given by

	1	σ	τ	$\sigma\tau\sigma$	$\sigma\tau$	$\tau\sigma$
1	1	σ	τ	$\sigma\tau\sigma$	$\sigma\tau$	$\tau\sigma$
σ	σ	1	$\sigma\tau$	$\tau\sigma$	τ	$\sigma\tau\sigma$
τ	τ	$\tau\sigma$	$t\tau$	$\tau\sigma$	τ	$t\tau\sigma$
$\sigma\tau\sigma$	$\sigma\tau\sigma$	$\sigma\tau$	$\sigma\tau$	$t\sigma\tau\sigma$	$t\sigma\tau$	$\sigma\tau\sigma$
$\sigma\tau$	$\sigma\tau$	$\sigma\tau\sigma$	$t\sigma\tau$	$\sigma\tau\sigma$	$\sigma\tau$	$t\sigma\tau\sigma$
$\tau\sigma$	$\tau\sigma$	τ	τ	$t\tau\sigma$	$t\tau$	$\tau\sigma$

Traces are computed by the formula of Section 6.2. Graphically, this gives:

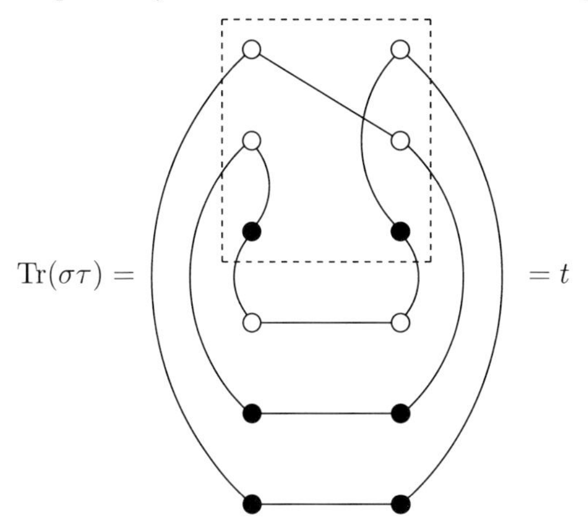

Similarly, $\mathrm{Tr}(1) = t^3$, $\mathrm{Tr}(\sigma) = \mathrm{Tr}(\tau) = \mathrm{Tr}(\sigma\tau\sigma) = t^2$ and $\mathrm{Tr}(\tau\sigma) = t$.

We now show how to obtain $\mathbf{Rep}_{\mathrm{GL}_n}$ from $\mathcal{C}$. The first step is to substitute t by the integer n. So we consider the category $\mathcal{C}_n$ having the same objects as $\mathcal{C}$ and with morphisms

$$\mathrm{Hom}_{\mathcal{C}_n}(X, Y) = \mathrm{Hom}_{\mathcal{C}}(X, Y) \otimes_{\mathbb{Z}[t]} \mathbb{C},$$

where $\mathbb{Z}[t] \to \mathbb{C}$ is determined by the specialization $t \mapsto n$.

We define a functor $F : \mathcal{C}_n \to \mathbf{Rep}_{\mathrm{GL}_n}$ as follows. Let V denote the natural representation of GL_n. For S an object of $\mathcal{C}_n$, we set $F(S) = \bigotimes_{i \in S} V^{\epsilon(i)}$, where $V^+ = V$ and $V^- = V^\vee$. Note that any diagram without circles can be obtained by forming tensor products of copies of the four morphisms

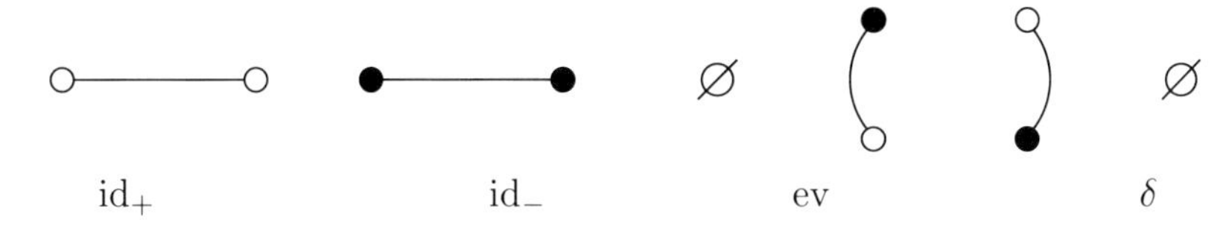

Therefore we set

$$\begin{aligned} F(\mathrm{id}_+) &= \mathrm{id}_V : V \to V, \\ F(\mathrm{id}_-) &= \mathrm{id}_{V^\vee} : V^\vee \to V^\vee, \\ F(\mathrm{ev}) &= \mathrm{ev}_V : V^\vee \otimes V \to 1, \\ F(\delta) &= \delta_V : 1 \to V \otimes V^\vee, \end{aligned}$$

and extend this first to all diagrams by forming tensor products, and then to all morphisms by linearity. This is clearly well defined, but we need to check that it commutes with composition. This follows from standard duality arguments. For example, the relation ‘adding a circle’ = ‘multiplication by n’ is respected since F maps circles to the morphism in $\mathbf{Rep}_{\mathrm{GL}_n}$ representing the dimension of V, which is n.

Now by the first fundamental theorem for GL_n (see [10]), the map from $\mathrm{Hom}_{\mathcal{C}_n}(X,Y)$ to $\mathrm{Hom}_{\mathrm{GL}_n}(F(X),F(Y))$ induced by F is surjective for all objects X,Y. To see this, consider the case where $X=\emptyset$ and $Y=\{1,2,\ldots,2k\}$ with $\epsilon_Y(i)=+$ for $1\leq i\leq k$ and $\epsilon_Y(i)=-$ for $k+1\leq i\leq 2k$. Let $\{e_i\}$ be a basis of V and let $\{e^j\}$ denote its dual basis (satisfying $e^j(e_i)=\delta_{ij}$). Then

$$\begin{aligned}\mathrm{Hom}_{\mathrm{GL}_n}(F(X),F(Y)) &= \mathrm{Hom}_{\mathrm{GL}_n}(\mathbb{C}, V^{\otimes k}\otimes V^{\vee\otimes k})\\ &\cong \left(V^{\otimes k}\otimes V^{\vee\otimes k}\right)^{\mathrm{GL}_n},\end{aligned}$$

and a spanning set for this space is given by the $k!$ elements

$$\sum_{i_1,\ldots,i_k} e_{i_1}\otimes\cdots\otimes e_{i_k}\otimes e^{i_{\sigma(1)}}\otimes\cdots\otimes e^{i_{\sigma(k)}},$$

with σ running over S_k. These elements clearly correspond to the $k!$ elements formed by tensoring k copies of δ_V in all possible ways.

For arbitrary X and Y we either can reduce to the case above by dualizing, or both $\mathrm{Hom}_{\mathcal{C}_n}(X,Y)$ and $\mathrm{Hom}_{\mathrm{GL}_n}(F(X),F(Y))$ are 0 (this happens when the total number of positive points in $X\otimes Y^\vee$ does not equal the number of negative points).

In general the map $\mathrm{Hom}_{\mathcal{C}_n}(X,Y)\to\mathrm{Hom}_{\mathrm{GL}_n}(F(X),F(Y))$ need not be injective. Let $I_n(X,Y)$ denote its kernel. The elements of $I_n(X,Y)$ can be identified without the help of $\mathbf{Rep}_{\mathrm{GL}_n}$:

Lemma 6.1. *The elements of $I_n(X,Y)$ are exactly the morphisms $f:X\to Y$ having $\mathrm{Tr}_{\mathcal{C}_n}(g\circ f)=0$ for all $g:Y\to X$.*

Proof. Consider morphisms $f:X\to Y$ and $g:Y\to X$ in $\mathcal{C}_n$. Then

$$F(\mathrm{Tr}_{\mathcal{C}_n}(g\circ f))=\mathrm{Tr}_{\mathrm{GL}_n}(F(g)\circ F(f)).$$

Since the map $F:\mathrm{End}_{\mathcal{C}_n}(1)\to\mathrm{End}_{\mathrm{GL}_n}(1)$ corresponds to the identity map $\mathbb{C}\to\mathbb{C}$, it follows that $\mathrm{Tr}_{\mathcal{C}_n}(g\circ f)=0$ if and only if $\mathrm{Tr}_{\mathrm{GL}_n}(F(g)\circ F(f))=0$. Therefore it suffices to see that a morphism $f:X\to Y$ in $\mathbf{Rep}_{\mathrm{GL}_n}$ is the zero map if and only if $\mathrm{Tr}_{\mathrm{GL}_n}(g\circ f)=0$ for all $g:Y\to X$. This follows easily from the reductivity of GL_n. □

The collection of $I_n(X,Y)$ can be considered as an ideal of $\mathcal{C}_n$: if $f:X\to Y$ and $g:Y\to Z$, then $g\circ f\in I_n(X,Z)$ whenever $f\in I_n(X,Y)$ or $g\in I_n(Y,Z)$, and if $f:X_1\to Y_1$ and $g:X_2\to Y_2$ then $f\otimes g\in I_n(X_1\otimes X_2,Y_1\otimes Y_2)$ whenever $f\in I_n(X_1,Y_1)$ or $g\in I_n(X_2,Y_2)$. This observation allows us to define a category $\mathcal{C}_n/I_n$ having the same objects as $\mathcal{C}_n$, with morphisms

$$\mathrm{Hom}_{\mathcal{C}_n/I_n}(X,Y)=\mathrm{Hom}_{\mathcal{C}_n}(X,Y)/I_n(X,Y),$$

and composition $\circ$ and tensor $\otimes$ induced by those of $\mathcal{C}_n$.

Now the functor F induces a functor $\overline{F} : \mathcal{C}_n/I_n \to \mathbf{Rep}_{\mathrm{GL}_n}$ defining a full embedding of $\mathcal{C}_n/I_n$ in $\mathbf{Rep}_{\mathrm{GL}_n}$. Of course, $\mathbf{Rep}_{\mathrm{GL}_n}$ has more objects. Since every GL_n-module is isomorphic to a submodule of some $V^{\otimes n} \otimes V^{\vee \otimes m}$, we can extend $\mathcal{C}_n/I_n$ to a category tensor equivalent to $\mathbf{Rep}_{\mathrm{GL}_n}$ by formally adding direct sums and images of idempotents. The result of this process is the *Karoubian envelope* $(\mathcal{C}_n/I_n)^{\mathrm{kar}}$.

Theorem 6.1. *For each $n \in \mathbb{N}$, the functor $\overline{F}$ extends to a tensor equivalence between the categories $(\mathcal{C}_n/I_n)^{\mathrm{kar}}$ and $\mathbf{Rep}_{\mathrm{GL}_n}$.*

Before we turn our attention to $\mathcal{C}$ again, we note that a consequence of the second fundamental theorem for GL_n (cf. [10]) is that for X, Y fixed, the map

$$\mathrm{Hom}_{\mathcal{C}_n}(X, Y) \to \mathrm{Hom}_{\mathrm{GL}_n}(F(X), F(Y))$$

is actually an isomorphism for n sufficiently large. So

$$\mathrm{Hom}_{\mathcal{C}}(X, X) \otimes_{\mathbb{Z}[t]} \mathbb{C} = \mathrm{End}_{\mathcal{C}}(X) \otimes_{\mathbb{Z}[t]} \mathbb{C}$$

with $t \mapsto n$ is a semisimple finite-dimensional $\mathbb{C}$-algebra for $n \gg 0$. This implies that

$$\mathrm{End}_{\mathcal{C}}(X) \otimes_{\mathbb{Z}[t]} \mathbb{Q}(t)$$

is a semisimple finite-dimensional $\mathbb{Q}(t)$-algebra for all objects X. Therefore we define a tensor category $\mathcal{C}_t$ over $\mathbb{Q}(t)$ by taking the objects of $\mathcal{C}$, with morphisms

$$\mathrm{Hom}_{\mathcal{C}_t}(X, Y) = \mathrm{Hom}_{\mathcal{C}}(X, Y) \otimes_{\mathbb{Z}[t]} \mathbb{Q}(t).$$

Since $\mathrm{End}_{\mathcal{C}_t}(X)$ is now a semisimple algebra for all objects X, the Karoubian envelope $\mathcal{C}_t^{\mathrm{kar}}$ is a semisimple category.

Computations performed in $\mathcal{C}_t$, such as decomposition of a tensor product, will have significance in $\mathbf{Rep}_{\mathrm{GL}_n}$ as long as n is sufficiently large. So $\mathcal{C}_t$ basically describes $\mathbf{Rep}_{\mathrm{GL}_n}$ for large n.

Example 6.2. The isotypic components of $\{+,+,-\}$ in $\mathcal{C}_t^{\mathrm{kar}}$ can be found by computing the primitive central idempotents of the endomorphism algebra $A = \mathrm{End}_{\mathcal{C}_t}(\{+,+,-\})$ of which the multiplication table was given in Example 6.1. These turn out to be

$$\begin{aligned}
e_1 &= \frac{1}{t^2-1}(t(\tau + \sigma\tau\sigma) - (\sigma\tau + \tau\sigma)),\\
e_2 &= \frac{1}{2}(1+\sigma) + \frac{1}{2(t+1)}(\sigma\tau + \tau\sigma + \tau + \sigma\tau\sigma),\\
e_3 &= \frac{1}{2}(1-\sigma) + \frac{1}{2(t-1)}(\sigma\tau + \tau\sigma - \tau - \sigma\tau\sigma).
\end{aligned}$$

Since $e_1 A$ is a 4-dimensional subspace, the isotypic component $\mathrm{Im}(e_1)$ is the direct sum of two isomorphic irreducible representations. Likewise, we see that $\mathrm{Im}(e_2)$ and $\mathrm{Im}(e_3)$ are both irreducible. The dimensions of the components are given as traces of the idempotents, which can be computed by linearity from the traces of the basis elements in Example 6.1:

$$\begin{aligned}\mathrm{Tr}(e_1) &= 2t,\\ \mathrm{Tr}(e_2) &= t(t+2)(t-1)/2,\\ \mathrm{Tr}(e_3) &= t(t+1)(t-2)/2.\end{aligned}$$

For GL_n ($n \geq 3$), this reflects the fact that $V \otimes V \otimes V^\vee$ decomposes into three isotypic components of dimensions $2n$, $n(n+2)(n-1)/2$, $n(n+1)(n-2)/2$, with the first component being a sum of two irreducibles.

6.4 The Exceptional Series

Let $\mathfrak{g}$ denote one of the exceptional Lie algebras G_2, F_4, E_6, E_7, E_8, and let $G = \mathrm{Aut}(\mathfrak{g})$. Then G is an exceptional group of adjoint type, and can be written as $G \rtimes \Gamma$, the semidirect product of the group G and the group Γ of automorphisms of the corresponding Dynkin diagram.

It turns out that we can decompose $\mathfrak{g}^{\otimes 2}$ in a 'uniform' way:

$$\begin{aligned}\mathrm{Sym}^2\,\mathfrak{g} &= 1 + Y_2 + Y_2^*,\\ \textstyle\bigwedge^2 \mathfrak{g} &= \mathfrak{g} + X_2,\end{aligned}$$

with

$$\begin{aligned}\dim \mathfrak{g} &= -2\frac{(\mu-5)(\mu+6)}{\mu(\mu+1)}\\ \dim X_2 &= 5\frac{(\mu-5)(\mu+6)(\mu-3)(\mu+4)}{\mu^2(\mu+1)^2}\\ \dim Y_2 &= -90\frac{(\mu+4)(\mu-5)}{\mu^2(\mu+1)(2\mu+1)}\\ \dim Y_2^* &= -90\frac{(\mu-3)(\mu+6)}{\mu(\mu+1)^2(2\mu+1)}\end{aligned}$$

and $\mu = 3/2, 2/3, 1/2, 1/3, 1/5$, for $G = G_2, F_4, E_6, E_7, E_8$, respectively. Note that $\dim Y_2$ and $\dim Y_2^*$ are related by the involution $*$ sending μ to $-1-\mu$, and that both $\dim \mathfrak{g}$ and $\dim X_2$ stay invariant under $*$.

This was observed in [9] where it was deduced from the fact that $(\mathrm{Sym}^4\,\mathfrak{g})^G$ is one-dimensional for these Lie algebras. We remark that similar observations can be found in [2] and [6].

It soon turned out that the trivial Lie algebra (i.e., the one-dimensional one), those of types A_1, A_2, D_4, and also the super Lie algebra $SOSp(1,2)$ can be included in the list (corresponding to $\mu = 5, 3, 2, 1, 4$, respectively). Computations [1] gave similar results for $\mathfrak{g}^{\otimes 3}$ and $\mathfrak{g}^{\otimes 4}$. This led Deligne [3] to conjecture the existence of a semisimple tensor category $\mathcal{E}_t$ having specializations to the various $\mathbf{Rep}_G$ by $t \mapsto \mu$.

In the next sections we will investigate a possible combinatorial structure behind $\mathcal{E}_t$. First, we consider an arbitrary simple Lie algebra $\mathfrak{g}$ and construct a kind of universal category $\mathcal{L}$. Then we specialize to the exceptional algebras and try to find relations describing the exceptional series.

6.5 The Category $\mathcal{L}$

Let $\mathfrak{g}$ be a finite-dimensional semisimple Lie algebra over $\mathbb{C}$ and set $G = \mathrm{Aut}(\mathfrak{g})$. We denote the Killing form of $\mathfrak{g}$ by $(\cdot,\cdot) : \mathfrak{g} \otimes \mathfrak{g} \to \mathbb{C}$. Since this form is non-degenerate, we can use it to identify $\mathfrak{g}$ with its dual. Hence

$$\mathrm{Hom}_G(\mathfrak{g}^{\otimes n}, \mathfrak{g}^{\otimes m}) \;\cong\; \left(\mathfrak{g}^{\otimes(n+m)}\right)^G.$$

Therefore we know all morphisms once we know the invariants of the spaces $\mathfrak{g}^{\otimes n}$. Independently of the type of $\mathfrak{g}$, we can already give two of these invariants: those induced by $\mathrm{id}_{\mathfrak{g}} : \mathfrak{g} \to \mathfrak{g}$ and $[\cdot,\cdot] : \mathfrak{g} \otimes \mathfrak{g} \to \mathfrak{g}$. To be more precise, $\mathrm{id}_{\mathfrak{g}}$ corresponds to the element $\sum_i e_i \otimes e^i \in \mathfrak{g}^{\otimes 2}$, with $\{e_i\}$ a basis of $\mathfrak{g}$ and $\{e^j\}$ its dual basis. Likewise, $[\cdot,\cdot]$ gives the element

$$\sum_{i,j,k} c^{ijk} e_i \otimes e_j \otimes e_k$$

of $\mathfrak{g}^{\otimes 3}$, where $c^{ijk} = (e^k, [e^i, e^j])$, so that they are the coefficients defined by $[e^i, e^j] = \sum_k c^{ijk} e_k$.

As in the previous sections, we would like a graphical representation of these elements. Letting dots correspond to $\mathfrak{g}$, we can denote the tensor element $\sum_i e_i \otimes e^i$ by the diagram

and the element $\sum c^{ijk} e_i \otimes e_j \otimes e_k$ by

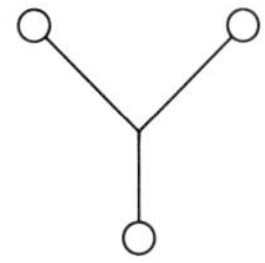

Note that, since this element is alternating, we have

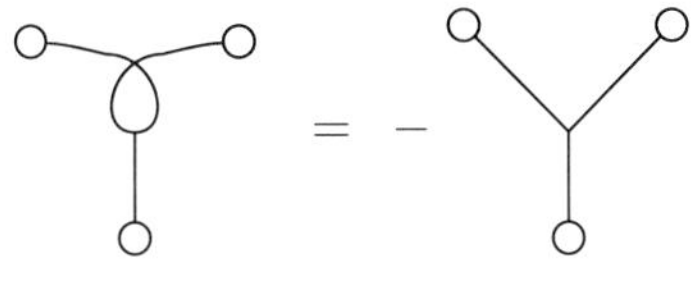

so points of valency 3 should have a fixed cyclic orientation (cyclic order of the three edges) and changing orientation corresponds to multiplication by -1.

With these, we build more invariants and diagrams. First, an invariant formed by tensoring copies of $\sum_i e_i \otimes e^i$ and $\sum c^{ijk} e_i \otimes e_j \otimes e_k$ is denoted by a disjoint union of copies of the two diagrams above. Applying contraction maps $\mathfrak{g}^{\otimes n} \to \mathfrak{g}^{\otimes(n-2)}$, that is, replacing two tensor factors by their value under the Killing form, we obtain more invariants. Appropriate diagrams are formed by connecting the points corresponding to the contracted components.

The diagrams obtained this way are finite graphs for which each vertex has valency 1, 2, or 3, and where any trivalent vertex has a fixed cyclic orientation assigned to it. Any such diagram can be dissected into copies of the two basic diagrams in an essentially unique way (up to order of contraction).

Note that contracting a tensor element with $\sum_i e_i \otimes e^i$ gives the same element again. For diagrams, this has as consequence that we can divide any edge into two edges connected by a point of valency 2 without changing the corresponding invariant. Therefore we can neglect points of valency 2 and consider graphs with vertices of valency 1 and cyclically oriented vertices of valency 3 only. We call these *trivalent diagrams* and usually omit the dots at the points of valency 1.

By way of example, we indicate how to write down the tensor element corresponding to the diagram

We interpret the cyclic orientation as being given by 'clockwise is positive'. It is formed by contracting two oriented trivalent nodes, say

$$\sum_{i,j,k} c^{ijk} e_i \otimes e_j \otimes e_k \quad \text{and} \quad \sum_{p,q,r} c^{pqr} e_p \otimes e_q \otimes e_r,$$

along k and p, giving

$$\sum_{i,j,k,p,q,r} c^{ijk} c^{pqr} (e_k, e_p) e_i \otimes e_j \otimes e_q \otimes e_r.$$

The Jacobi identity can easily be translated into a relation between diagrams. We consider the three elements of $\mathfrak{g}^{\otimes 4}$

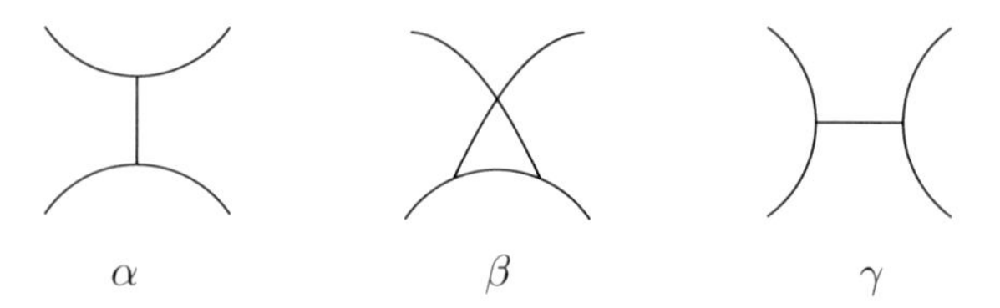

As before the orientation of the trivalent nodes is assumed to be clockwise. To compare these elements, it is easiest to look at the induced functionals $\mathfrak{g}^{\otimes 4} \to 1$.

Labeling the endpoints anti-clockwise by x, y, z, w, with x denoting the lower-left endpoint, we have

$$(\alpha, x \otimes y \otimes z \otimes w) = ([x, y], [z, w]),$$
$$(\beta, x \otimes y \otimes z \otimes w) = ([z, x], [y, w]),$$

and

$$\begin{aligned}(\gamma, x \otimes y \otimes z \otimes w) &= ([w, x], [y, z]) = (w, [x, [y, z]]) \\ &= (w, -[y, [z, x]] - [z, [x, y]]) \\ &= (-([w, y], [z, x]) - ([w, z], [x, y]) \\ &= ([z, x], [y, w]) + ([x, y], [z, w]) \\ &= (\alpha + \beta, x \otimes y \otimes z \otimes w).\end{aligned}$$

Hence $\gamma = \alpha + \beta$:

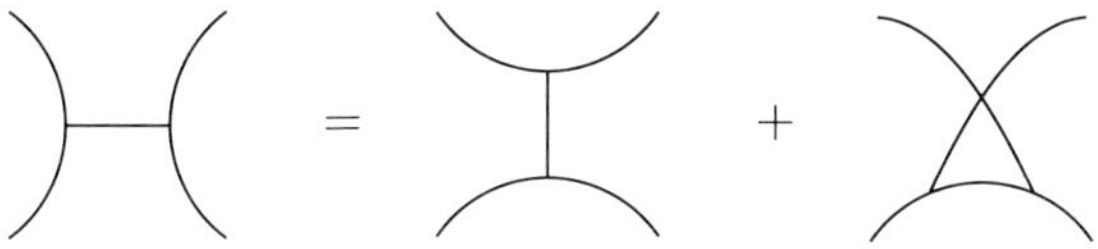

$$= \qquad + $$

This diagrammatic formula is known as the IHX relation (cf. [6, 9]). As we saw before, anti-commutativity of $[\cdot, \cdot]$ is expressed by

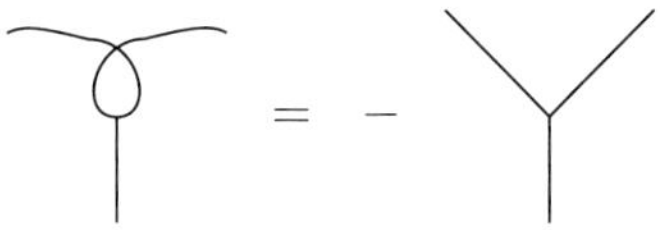

$$= \; -$$

We use these diagrams to construct a category $\mathcal{L}$. In this construction, the diagrams are viewed as combinatorial objects, without reference to the Lie algebra $\mathfrak{g}$. As objects we take finite sets (without orientation). For finite sets X, Y we define $\mathrm{Hom}_{\mathcal{L}}(X, Y)$ to be the free $\mathbb{Z}$-module with basis the trivalent diagrams that have as set of monovalent points the disjoint union of X and Y, modulo the two relations above. It is clear how to define composition on diagrams. We extend this by linearity to obtain composition in $\mathcal{L}$. As usual, we take disjoint unions for tensor products.

It can easily be seen that any object in $\mathcal{L}$ is self-dual and that $\mathcal{L}$ is rigid. We summarize this section in the following theorem.

Theorem 6.2. *Let $\mathfrak{g}$ denote a finite-dimensional complex Lie algebra having a non-degenerate Killing form $(\cdot, \cdot) : \mathfrak{g}^{\otimes 2} \to \mathbb{C}$ and let $G = \mathrm{Aut}(\mathfrak{g})$. There exists a tensor functor $F : \mathcal{L} \to \mathbf{Rep}_G$, unique up to unique isomorphism, mapping singletons to $\mathfrak{g}$ and satisfying*

$$F(\text{———}) = (1 \mapsto \sum e_i \otimes e^i),$$

$$F(\,\text{Y}\,) = (1 \mapsto \sum c^{ijk} e_i \otimes e_j \otimes e_k).$$

The diagrams are interpreted here as morphisms $1 \to \mathfrak{g}^{\otimes n}$; all other interpretations follow by dualizations of tensor factors.

6.6 The Exceptional Categories

Let $\mathfrak{g}$ be one of the exceptional simple Lie algebras G_2, F_4, E_6, E_7, E_8, and let d denote its dimension. Set $G = \mathrm{Aut}(\mathfrak{g})$, and let $F : \mathcal{L} \to \mathbf{Rep}_G$ be as in Theorem 6.2. In Section 6.6.1, we look for linear combinations of morphisms in the kernel of $F : \mathrm{Hom}_{\mathcal{L}}(X, Y) \to \mathrm{Hom}_G(F(X), F(Y))$, where X, Y are small tensor powers of $\mathfrak{g}$. In Section 6.6.2 this is used to compute $\mathrm{End}(\mathfrak{g}^{\otimes 2})$ and the corresponding decomposition of $\mathfrak{g}^{\otimes 2}$. Based on the relations found so far, a tensor category quotient of $\mathcal{L}$ can be found through which F factors. This can be seen as a first step towards the conjectured category $\mathcal{E}_t$. In Section 6.6.3, we observe that more relations are needed to describe the conjectured category $\mathcal{E}_t$ fully.

6.6.1 Finding Relations All of the relations we will obtain involve diagrams consisting of a circle and a number of legs. These diagrams have an easy interpretation as elements of $(\mathfrak{g}^{\otimes n})^\vee$: the circle with n legs corresponds to the linear map sending $x_1 \otimes \cdots \otimes x_n$ to $\mathrm{Tr}(\mathrm{ad}\, x_1 \cdots \mathrm{ad}\, x_n)$.

One obvious relation is given by the fact that the dimension d of $\mathfrak{g}$ is equal to the trace of $\mathrm{id}_{\mathfrak{g}}$:

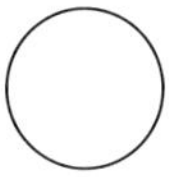

$$\bigcirc \quad = \quad d$$

To obtain further relations, one way to proceed is to count the number of invariants in certain degrees.

First, since $\mathfrak{g}$ is simple, there is no non-zero morphism $\mathfrak{g} \to 1$. This translates to the fact that any diagram with exactly one monovalent vertex vanishes.

Next, by the decomposition in Section 6.4 of $\mathfrak{g}^{\otimes 2}$, there is only one independent morphism $\mathfrak{g}^{\otimes 2} \to 1$. Of course, by Schur's Lemma this also follows from the simplicity of $\mathfrak{g}$. So any diagram with exactly two monovalent vertices maps to a scalar multiple of the diagram having two vertices and one connecting edge. One important example is the Casimir endomorphism of $\mathfrak{g}$:

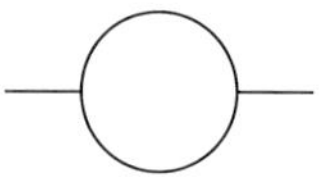

$$-\bigcirc- \quad = \quad C \;\;\text{———}$$

with C a scalar. Since we can interpret the lefthand side as the morphism sending $x \otimes y$ to $\mathrm{Tr}(\mathrm{ad}\ x\ \mathrm{ad}\ y)$, we find that both sides represent the Killing form. Hence $C = 1$.

From the same decomposition of $\mathfrak{g}^{\otimes 2}$, we find that there is only one independent morphism $\mathfrak{g}^{\otimes 2} \to \mathfrak{g}$. Of course, this is the Lie bracket

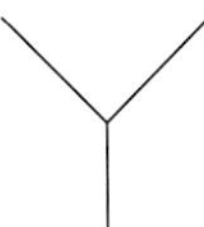

Hence any diagram with three monovalent vertices is a scalar multiple of the Lie bracket. In particular, the diagram

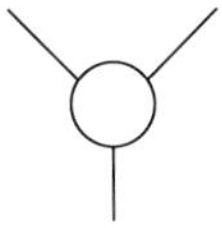

is anti-commutative:

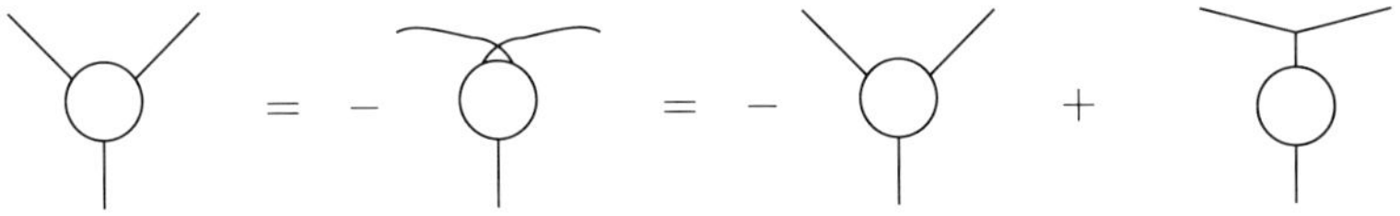

$$= - \qquad = - \qquad +$$

Therefore,

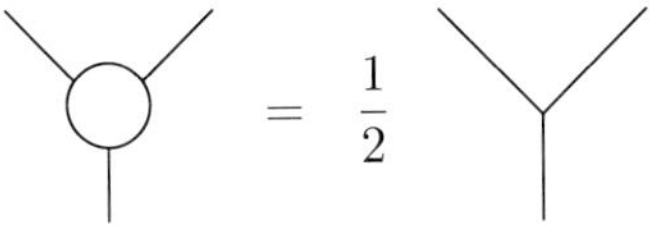

$$= \frac{1}{2}$$

The next relation depends on the fact that $\left(\mathrm{Sym}^4\, \mathfrak{g}\right)^G$ is one-dimensional for exceptional Lie algebras. To make use of this fact, note that

$$+ \qquad +$$

is invariant under the action of S_4 (permuting vertices), and that the same holds for

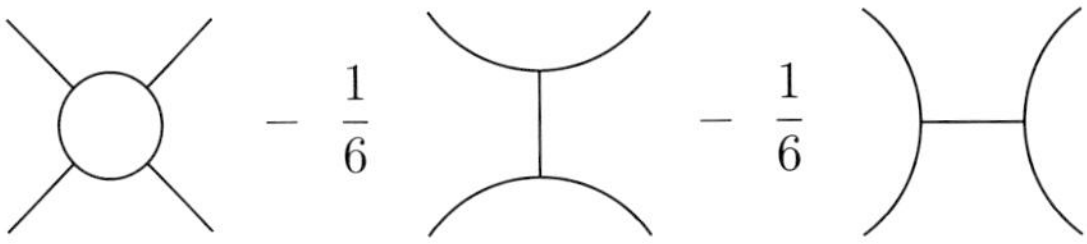

$$- \frac{1}{6} \qquad - \frac{1}{6}$$

So we know that for some scalar ξ we have

$$= \frac{1}{6} \qquad + \frac{1}{6}$$

$$+ \xi \; (\qquad + \qquad + \qquad)$$

To find ξ, we simply contract two pairs of vertices to get $d = d/6 + \xi(d^2 + 2d)$, and hence $\xi = 5/(6(d+2))$.

6.6.2 Endomorphisms of the Second Tensor Power The relations found in the preceding section already are sufficient to describe the endomorphism algebra of $\mathfrak{g}^{\otimes 2}$. From the decomposition in Section 6.4, we know this will be a five-dimensional commutative algebra. As a basis we take the morphisms

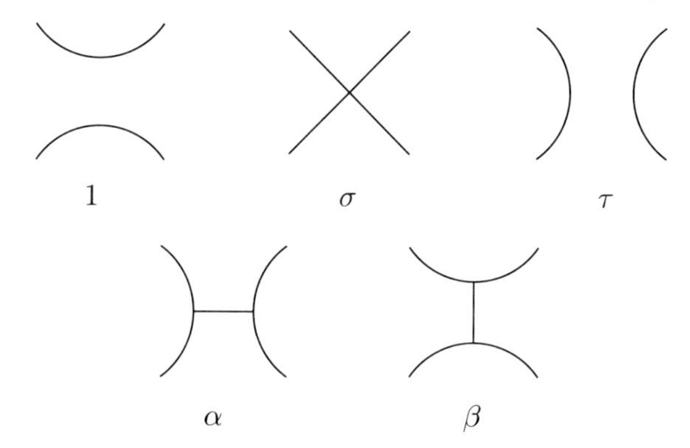

Using the relations it is not difficult to find the multiplication table:

	1	σ	τ	α	β
1	1	σ	τ	α	β
σ	σ	1	τ	$-\alpha$	$\beta - \alpha$
τ	τ	τ	$d\tau$	0	τ
α	α	$-\alpha$	0	α	$\alpha/2$
β	β	$\beta - \alpha$	τ	$\alpha/2$	$\frac{1}{6}(\alpha + \beta) + \frac{5}{6(d+2)}(1 + \sigma + \tau)$

We can use the endomorphism algebra to verify the dimension formulae in Section 6.4. The representations occurring in the decomposition of $\mathfrak{g}^{\otimes 2}$ correspond to the primitive idempotents in the commutative algebra $\mathrm{End}(\mathfrak{g}^{\otimes 2})$. These idempotents turn out to be

$$\alpha, \frac{1}{2}(1 - \sigma) - \alpha, \frac{1}{d}\tau,$$

$$B = \frac{1}{1+2\mu}\left(6\beta - 3\alpha + \frac{\mu}{2}(1+\sigma) - \frac{\mu+6}{d}\tau\right), \frac{1}{2}(1+\sigma) - \frac{1}{d}\tau - B,$$

where $\mu = -\frac{1}{2} + \frac{1}{2}\sqrt{1+\frac{240}{d+2}}$, so that $d = -2\frac{(\mu-5)(\mu+6)}{\mu(\mu+1)}$. The dimension of a representation is equal to the trace of the corresponding idempotent and an easy calculation gives the formulae of Section 6.4.

6.6.3 The Category $\mathcal{E}_t$ We now consider the conjectural tensor category $\mathcal{E}_t$, whose 'specialization' $t \mapsto \mu$ should give the relevant data for the corresponding exceptional Lie algebra $\mathfrak{g}$. It will be a tensor category over $\mathbb{Q}(t)$, whose objects are those of $\mathcal{L}$, and whose morphisms are of the form

$$\mathrm{Hom}_{\mathcal{E}_t}(X,Y) = (\mathrm{Hom}_{\mathcal{L}}(X,Y) \otimes_{\mathbb{Z}} \mathbb{Q}(t))/I(X,Y),$$

for an ideal I containing the relations found in section 6.6.1, where we replace d by $-2\frac{(t-5)(t+6)}{t(t+1)}$. This is justified by the fact that the linear combinations found in Section 6.6.1 have coefficients depending rationally on d. Note also that the idempotents of Section 6.6.2 have meaning over $\mathbb{Q}(t)$.

Although we do not know whether it exists, we can already describe $\mathcal{E}_t$ in 'low degrees': the decomposition formulae tell us what the irreducible objects are occurring in $\mathfrak{g}^{\otimes n}$ for n small, and using diagrams we can describe the morphisms between them. So $\mathrm{End}_{\mathcal{E}_t}(\mathfrak{g}^{\otimes 2})$ is the five-dimensional algebra with multiplication table as given in Section 6.6.2.

The ad-hoc arguments of the preceding sections have been used to describe $\mathrm{Hom}_{\mathcal{E}_t}(1,\mathfrak{g}^{\otimes n})$ for $n \le 4$. To proceed further, we will use a more systematic approach. It depends on 'guessing' a basis for the space of morphisms $\mathrm{Hom}_{\mathcal{E}_t}(1,\mathfrak{g}^{\otimes n})$, and then using a naturally defined bilinear form to express other elements in this basis.

The bilinear form $\langle\cdot,\cdot\rangle : \mathrm{Hom}_{\mathcal{E}_t}(1,\mathfrak{g}^{\otimes n}) \times \mathrm{Hom}_{\mathcal{E}_t}(1,\mathfrak{g}^{\otimes n}) \to \mathbb{Q}(t)$ is defined as follows. Let $f, g : 1 \to \mathfrak{g}^{\otimes n}$ be morphisms. Then ${}^{\top}f \circ g : 1 \to 1$ corresponds to multiplication by a scalar, and we let $\langle f, g\rangle$ denote this scalar. It is easily verified that $\langle\cdot,\cdot\rangle$ is a symmetric bilinear form. If we have f and g in the form of diagrams, we can try and evaluate this diagram using the relations already found.

The Gram matrix of the $\mathrm{Hom}_{\mathcal{E}_t}(1,\mathfrak{g}^{\otimes 4})$-basis of section 6.6.2 with respect to $\langle\cdot,\cdot\rangle$ is

$$\begin{pmatrix} d^2 & d & d & 0 & d \\ d & d^2 & d & -d & -d \\ d & d & d^2 & d & 0 \\ 0 & -d & d & d & d/2 \\ d & -d & 0 & d/2 & d \end{pmatrix}$$

The determinant is $3d^5(d+2)(d-3)^2/4$, confirming that the chosen basis elements are linearly independent.

To attack $\mathrm{Hom}_{\mathcal{E}_t}(1,\mathfrak{g}^{\otimes 5}) = \mathrm{Hom}_{\mathcal{E}_t}(\mathfrak{g}^{\otimes 2},\mathfrak{g}^{\otimes 3})$, we first use the decompositions of $\mathfrak{g}^{\otimes 2}$ and $\mathfrak{g}^{\otimes 3}$ to compute its dimension. This gives a number of 16 expected basis elements. As candidate basis elements we take the diagrams on 5 points having no cycles (since many diagrams having cycles can be reduced by use of the relations we have). Since this gives more than 16 elements, there will be linear dependencies. Assuming that this set of diagrams contains a basis and that $\langle \cdot, \cdot \rangle$ is nondegenerate, we can use this bilinear form to remove redundant elements. Doing this, we end up with 16 linearly independent diagrams, hence a basis. Any diagram can be expressed in this basis, and we can effectively compute the coefficients provided we know how to evaluate the relevant closed diagrams (i.e., without monovalent vertices) occurring as values of the bilinear form. For example,

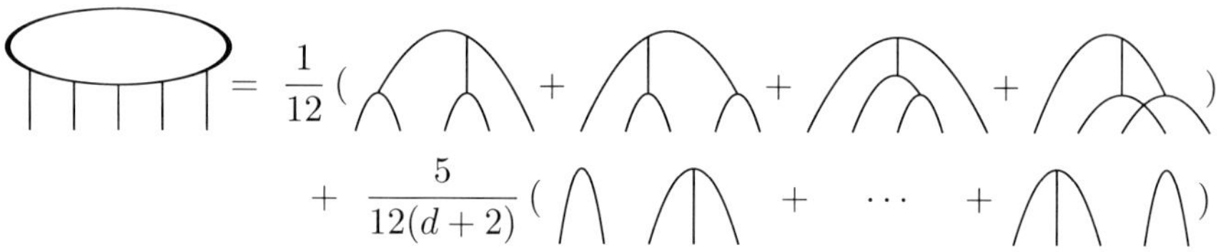

$$= \frac{1}{12}\left(\quad + \quad + \quad + \quad \right) + \frac{5}{12(d+2)}\left(\quad + \quad \cdots \quad + \quad \right)$$

In a similar fashion, a basis of $\mathrm{Hom}_{\mathcal{E}_t}(1, \mathfrak{g}^{\otimes 6})$ can be found. The decomposition of $\mathfrak{g}^{\otimes 3}$ leads us to expect 80 elements. However, in this case diagrams having no cycles do not suffice. We explain why not. The subspace of $\mathrm{Hom}_{\mathcal{E}_t}(1, \mathfrak{g}^{\otimes 6})$ invariant under the action of S_6 is of dimension two (which we derive from the decomposition tables, suggesting that 1 occurs with multiplicity two in $\mathrm{Sym}^6(\mathfrak{g})$). But symmetrizing (taking the sum over all permutations) any diagram with no cycles and at least one trivalent vertex, we get zero since a transposition of two endpoints meeting in the same trivalent vertex changes the sign of the diagram. Therefore the only symmetric morphism obtained from diagrams without cycles is the symmetrization of the morphism representing three tensor copies of the evaluation morphism. As the second symmetric morphism, it turns out we can take the symmetrization of the circle with six legs. This shows that the circle with six legs is linearly independent of the diagrams with no cycles. Here we have a potential obstruction for our method when applied to $\mathrm{Hom}_{\mathcal{E}_t}(1, \mathfrak{g}^{\otimes n})$ for higher n: we do not know how to rewrite closed diagrams having no cycles of length at most 5.

We conclude by making some remarks about the present state of knowledge regarding $\mathcal{E}_t$. Decomposition tables have been found for $\mathfrak{g}^{\otimes n}$, $n \leq 5$. All objects that occur have a dimension formula factoring completely in linear forms, with one exception. It is expected (or at least hoped) that this exception is in fact the sum of two or more irreducible objects of $\mathcal{E}_t$. There turns out to be an additive structure on the set of irreducible objects, much like the addition of heighest weights in the representation theory of semisimple algebraic groups. In [4], a 7-parameter family of irreducible objects is identified, including for each element a set of affine linear forms in t (each form having a multiplicity in $\mathbb{Z}$). These forms play the role of the factors occurring in the Weyl dimension formula, explaining the factorization property of the dimensions of these irreducible objects. The family should extend to infinitely many parameters. Currently work is being done to find the linear forms for larger families.

Bibliography

[1] A.M. Cohen and R. de Man: *Computational evidence for Deligne's conjecture regarding exceptional Lie groups*, C.R. Acad. Sci. Paris 322 (1996), série I, 427–432.

[2] P. Cvitanović: *Group Theory, Part II: Exceptional Lie groups*, preprint, 1984–1997.

[3] P. Deligne: *La série exceptionnelle de groupes de Lie*, C.R. Acad. Sci. Paris 322 (1996), série I, 321–326.

[4] P. Deligne and R. de Man: *La série exceptionnelle de groupes de Lie II*, C.R. Acad. Sci. Paris 323 (1996), série I, 577–582.

[5] P. Deligne and J.S. Milne: *Tannakian categories*, in: Hodge cycles, Motives and Shimura Varieties, volume 900 of Lecture Notes in Math., 101–228, Springer-Verlag, Berlin, 1982.

[6] M. El Houari: *Nouvelle classification des (super) algèbres de Lie simples par le biais de leurs invariants tensoriels*, Thesis, Université de Paris Sud, Centre d'Orsay, 1994.

[7] S. Mac Lane: *Categories for the Working Mathematician*, volume 5 of Graduate Texts in Math., Springer-Verlag, New York, 1971.

[8] R. Penrose: *Applications of negative dimensional tensors*, in: Combinatorial mathematics and its applications, ed. D.J.A. Welsh, Academic Press, New York, 1971.

[9] P. Vogel: *Algebraic structures on modules of diagrams*, preprint, 1995.

[10] H. Weyl: *The Classical Groups*, Princeton Math. Ser. 1, Princeton University Press, 1946.

Progress in Mathematics, Vol. 173, © 1999 Birkhäuser Verlag Basel/Switzerland

Chapter 7

Non-Commutative Gröbner Bases and Anick's Resolution

Svetlana Cojocaru, Alexander Podoplelov and Victor Ufnarovski

Abstract

Noncommutative algebras, defined by the generators and relations, are considered. The definition and main results connected with the Gröbner bases, Hilbert series and Anick's resolution are formulated. Special attention is paid to universal enveloping algebras. Four main examples illustrate the main concepts and ideas. Algorithmic problems arising in the calculation of the Hilbert series are investigated. The existence of finite state automata, defining the behavior of the Hilbert series is discussed.

Two programs for calculating Gröbner bases and Anick's resolution are considered.

This article is a short and elementary introduction into the theory of noncommutative Gröbner bases and its applications. It contains only known results and sufficiently many examples. We restricted ourselves to the case of graded algebras and excluded part of possible material, contained in the other articles of this issue.

Besides that it contains the description of two software tools to deal with noncommutative graded algebras.

BERGMAN is en effective program for calculating the Gröbner bases (both for commutative and non-commutative case) and monomial Poincaré series, developed in Stockholm university (J.Backelin) for SUN-station (and some other types of computers). In our implementation on IBM-PC we used the original source of J.Backelin (and his valuable help) ([4]) and added some new additional functions especially for the noncommutative algebras.

ANICK is a program, developed by authors (source code - A.Podoplelov) for calculating Anick's resolution and (through it) Gröbner basis and Yoneda product in the noncommutative graded algebras. The last version of BERGMAN contains the corresponding version of ANICK. To illustrate their possibilities we give all necessary definitions and several examples.

7.1 Main Examples

Let $A = \langle X \mid R\rangle$ be a finitely presented (noncommutative) associative algebra over field K. Here are the main examples of algebras that will help us to illustrate some following definitions (later we will refer to them as main examples).

Example 7.1. $A = \langle x, y \mid x^2 = 0, xy^2 = 0\rangle$

Example 7.2. $A = \langle x, y \mid x^2 = y^2\rangle$

Example 7.3. $A = \langle x, y \mid x^2 - xy\rangle$

Example 7.4. $A = \langle e_1, e_2, e_3, \cdots \mid [e_i, e_j] = (i-j)e_{i+j}\rangle$, where char $K = 0$ and $[a, b] = ab - ba$.

Though the last example does not appear to be finitely presented, it is evident that e_1 and e_2 generate the algebra ($e_{k+1} = \frac{[e_k, e_1]}{k-1}$ for $k \geq 2$). Slightly less trivial is the fact that it is sufficient to have only two relations: $[e_3, e_2] = e_5$; $[e_4, e_3] = e_7$ (see [8]). Nevertheless, we will use this example in this infinite presentation to give the reader a opportunity to see how definitions work in the infinite case.

This is also an example of an universal enveloping algebra: $A = U(L)$, where L is a Lie algebra with the same sets of generators and relations except that the commutator $[a, b]$ is now interpreted as a Lie product. If we interpret it as a graded commutator $[a, b] = ab - (-1)^{|a||b|}ba$, the second example can be also considered as an universal enveloping algebra of 4-dimensional Lie superalgebra: $L = \langle x, y \mid [x, x] = [y, y]\rangle$.

7.2 Hilbert Series and global Dimension

Note that all algebras in our examples are graded algebras: $A = \oplus A_n$,where all components A_n are finite dimensional and $A_nA_m \subseteq A_{m+n}$. For the last example grading is less trivial: $e_n \in A_n$.

We restrict our attention to graded algebras and introduce the following

Definition 7.1. The generating function $H_A = H_A(t) = \sum_0^\infty (\dim A_n)t^n$ is called the Hilbert series of an algebra A.

The Hilbert series of a graded algebra is one of our main objects of interest. It is a very useful invariant in the commutative case, but in the noncommutative case it also contains a lot of important information about the algebra. First of all, it plays the role of generalized dimension of an algebra. For example, it has the following trivial properties:

- $H_{A\oplus B} = H_A + H_B$; $H_{A\otimes B} = H_AH_B$, $\frac{1}{H_{A*B}} = \frac{1}{H_A} + \frac{1}{H_B} - 1$,
- If L is a graded Lie algebra (superalgebra), $H_L = \sum_1^\infty a_nt^n$ and $A = U(L)$, then

$$H_A = \prod_1^\infty \frac{1}{(1-t^n)^{a_n}} \quad (H_A = \prod_1^\infty \frac{(1+t^{2n-1})^{a_{2n-1}}}{(1-t^{2n})^{a_{2n}}}).$$

If L is a Lie algebra from our main example 4, then

$$H_L = t + t^2 + t^3 + \cdots \Rightarrow H_A = \prod_1^{\infty} \frac{1}{1-t^n} = \sum p(n)t^n,$$

where $p(n)$ is the number of partitions. So it is the example with non-rational Hilbert series. In the second main example we have

$$H_L = 2t + 2t^2 \Rightarrow H_A = \frac{(1+t)^2}{(1-t^2)^2} = \frac{1}{1-2t+t^2}.$$

In our main examples 1,3 the Hilbert series is equal to $(1-2t+t^2)^{-1}$. It can be directly checked from the first example, but the last one will be discussed in the sequel.

Secondly, the following two theorems of D. Anick ([1, 2]) show non-trivial properties of Hilbert series:

Theorem 7.1. *Let $H_X (H_R)$ be the generating function of the number of the generators (relations, both minimal) of given degree. Then*

$$gl.\dim A \leq 2 \Leftrightarrow H_A = (1 - H_X + H_R)^{-1}.$$

(inequality for free algebras only.)

In our main examples the value of $(1 - H_X + H_R)$ is equal

1	2,3	4
$1-2t+t^2+t^3$	$1-2t+t^2$	$1-t-t^2+t^5+t^7$

so only the second and third examples are algebras of global dimension 2.

Theorem 7.2. *For every system of diophantine equations $S = 0$, there exists a finitely presented algebra A (which can be constructively expressed in terms of the coefficients of the S) such that A has global dimension 2 if and only if the system $S = 0$ has no solutions.*

Moreover this algebra is an universal enveloping algebra of a Lie superalgebra, defined by quadratic relations only. This theorem has important, though unpleasant:

Corollary 7.1. *One cannot*

- *find an algorithm that calculates the global dimension of an arbitrary finitely-presented algebra.*
- *find an algorithm that takes relations as input and gives the Hilbert series as output. (Even more, one cannot detect in general if the Hilbert series of a given algebra is equal some fixed series)*

- *predict in general the behavior of a Hilbert series, knowing only finitely many of its coefficients.*

Instead of Hilbert series one can calculate such invariants as growth or GK-dimension and they are defined in non-graded algebras too. We recall the definition and some facts (see [9] for more details).

Definition 7.2. Let V be a finite dimensional subspace of an algebra A and $d_V(n) = \dim(V + V^2 + \cdots V^n)$. Then

$$GK\text{-}Dim(A) = \sup_V \overline{\lim_{n\to\infty}} \frac{\ln d_V(n)}{\ln n}$$

If A is generated by V, then the function $d_V(n)$ up to a certain eqvivalence is the *growth* of A. For example, polynomial growth means finite GK-dimension and exponential means $d_V(n) > a^n$ for some $a > 1$.

Theoretically there exists an algorithm, checking for graded finitely-presented algebras that the growth is linear or constant (more exactly, in the case when all generators have degree 1: for fixed number of generators, relations and their degrees there exists a constant n such that $\dim A_n \leq n$ iff A has a linear growth and $\dim A_n = 0$ iff A is finite dimensional).

It would be interesting to know if there exists an algorithm, checking that A has polynomial growth.

7.3 Normal Words and Gröbner Basis

Despite pessimistic conclusions in the end of the previous chapter, there is some hope to find the Hilbert series in some important cases. Let us introduce some important definitions.

Let S be the set of the all words in the alphabet X (identifying 1 with the empty word). Consider the following ordering on S:

$f > g$ if either the length of word f is greater than that of g or they have the same length, but f is greater then g lexicographically. (More ingenious, the so called admissible ordering may be considered too, but we restrict our attention only to this case.)

Definition 7.3. A word $f \in S$ is called normal (for A) if it cannot be written in A as a linear combination of words that are less than f.

In our first main example the words $1, x, y, xy, yx, y^2$ are normal, but x^2, xy^2 are not. The same is true for the second main example if $x > y$. Why xy^2 is not normal? Because $xy^2 = x^3 = y^2x$.

In the last main example the words of the form $e_1^{k_1} e_2^{k_2} \cdots e_m^{k_m}$ are normal according the PBW-theorem, if alphabet and ordering is $e_1 < e_2 < e_3 \cdots$ It is much more complicated to express normal words in the alphabet e_1, e_2 only (see the end of article).

The following evident theorem explains how normal words can be used for the calculation of the Hilbert series:

Theorem 7.3. *The set N, consisting of all normal words, forms a basis for the algebra A. Its Hilbert series can be calculated as $H_A = \sum_0^\infty d_n t^n$, where d_n is the number of normal words of degree n.*

Following D. Anick, let us introduce

Definition 7.4. A word $f \in S$ is called an obstruction if f is not normal itself, but every proper subword is normal.

Note that, other expressions (such as "tips", for example) are used instead of "obstruction". We denote the set of all obstructions as F. Because obstructions are not normal words every $f_i \in F$ can be written as a linear combination of normal words: $f_i = u_i$.

Definition 7.5. The set $G = \{f_i - u_i\}$ is called a (reduced) Gröbner basis (for A).

In our main examples:

1. $G = \{x^2, xy^2\}$ (evidently)

2. $G = \{x^2 - y^2, xy^2 - y^2x\}$.

3. $G = \{xy^kx - xy^{k+1} \mid k = 0, 1, 2, \dots\}$

4. $G = \{e_je_i - e_ie_j - (j - i)e_{i+j} \mid j > i\}$

All those examples can be easy checked by hand (see [9]). BERGMAN is a powerful tool to calculate Gröbner bases in the more complicated cases.

Of course, knowing an obstruction set we can easy reconstruct normal words:

Theorem 7.4. *Let $B = \langle X \mid F \rangle$. Then algebras A and B have the same sets of normal words and in particular, $H_A = H_B$.*

Definition 7.6. Algebra B from previous theorem is called the monomial algebra, associated with A.

So, the algebra from our first main example is the associated monomial algebra for our second example. Note also, that all universal enveloping algebras for Lie algebras with the same Hilbert series have, according the PBW theorem, the same associated monomial algebra (in some alphabet). Note also that definitions depend on the choice of the generator set (alphabet) and ordering.

More general a Gröbner basis for any ideal I is its subset G, such that the set of highest terms of elements from G is exactly the set of obstructions for $A = K\langle X \rangle / I$. Note that the reduced Gröbner basis may be easily obtained from an arbitrary Gröbner basis (by selfreduction) and is determined uniquely for a given ordering.

7.4 Graphs, Languages and Monomial Algebras

In this section we try to understand how to calculate Hilbert series and growth in the monomial algebras $A = \langle X \mid F \rangle$, where F is a set of words in the alphabet X.

The set of normal words can be described here directly: a word is normal iff it does not contain any of word from F as a subword. So, the problem does not look so difficult. Nevertheless, from the computational point of view, it appears to be less trivial problem that one might expect. To appreciate this, the reader can try to invent an *efficient* algorithm, answering the follows easy question: is a given monomial algebra finite dimensional?

One of possible ways to deal with monomial algebras is to use different kinds of graphs. We start from the easiest case: finitely-presented monomial algebras, i.e. algebras of form $A = \langle X \mid F \rangle$, where F is a finite set of words in the alphabet X. We can, of course, restrict ourself by the case when no one of the words from F is a subword of another (by the way, it means that F is both the reduced Gröbner basis and the obstructions set).

Let $n+1$ be the maximum of length of words from F. Let as construct a graph $U(F) = (V, E)$, where the set of vertices V consist of all normal words of length n and edges E are described as follows: $f \longrightarrow g$ if and only if there exists generators $x, y \in X$ such that $fx = yg$ and this word is normal.

Example 7.5. For $A = \langle x, y \mid xy \rangle$ we have $n = 1$ and $U(A)$ looks like this.

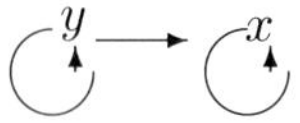

$$y \circlearrowleft \longrightarrow x \circlearrowleft$$

In our first (and second) main examples the graph is

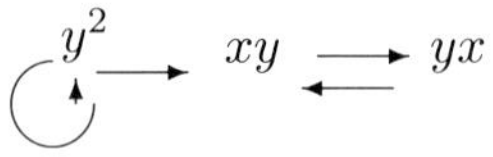

$$y^2 \circlearrowleft \longrightarrow xy \rightleftarrows yx$$

The following theorem ([11]) solves the problem of growth:

Theorem 7.5.

- *The growth of a monomial finitely-presented algebra $A = \langle X \mid F \rangle$ is exponential if and only if there are two different cycles in the graph $U(F)$ with a common vertex.*
- *Otherwise it is polynomial of degree d, where d is the maximal possible number of cycles through which one path can pass.*
- *There is a bijective correspondence between the set of normal words of length $\geq n$ and paths in the graph.*
- *The Hilbert series H_A is a rational function.*

For example, the growth of algebra in the both our last examples is polynomial of degree 2. Using this theorem the reader can easy check that the growth of algebra $A\langle x, y \mid x^2 \rangle$ is exponential. The rather efficient algorithm for calculating the growth, based on this theorem is presented in [10]

The graph $U(F)$ has found several theoretical applications. For example, Jacobson radical $J(A)$, PI-property, primarity, left and right Noetherian property can be easy expressed and checked using this graph. See [9] and references inside about more details.

Another approach is to consider the obstruction set F as a language and to use the corresponding ideas (here F may be infinite). Especially important are so called regular languages. Recall, that *regular language* is a set of words, obtained from the finite number of words using finite number of the following operations:

$$A \cup B; AB; A^* = \cup_0^\infty A^n$$

They can be described by a finite state automaton. It appears that F is regular if and only if the set N of normal words is regular. For example, in our third main example the regular set F can be described as $x(y^*)x$ and the automaton to recognize it looks like this.

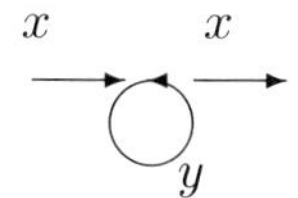

The automaton for normal words can be used for calculating both growth (the same criteria as above) and Hilbert series. The reader can find more details about this and other types of graphs for normal words in [9].

7.5 $n-$Chains and Poincaré series

The next step is to introduce some homological algebra. Let us consider a graph $\Gamma = (V, E)$, where the set of vertices V consist of union of the unit 1, alphabet X and all proper suffices of the obstructions. Edges E are defined as follows: $1 \to x$ for every $x \in X$ and in other cases $f \to g$ if and only if the word fg contains exactly one obstruction and this obstruction is its suffix (maybe coinciding with fg).

In our main Examples 7.1 and 7.2 the graph Γ looks like:

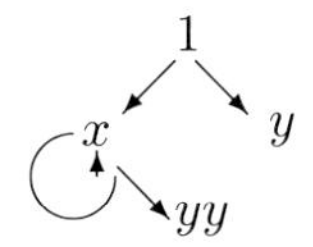

In the third example vertices (except 1 and y) have form $y^n x$ and are all connected to each other (including itself).

It the fourth main example vertices are e_i and (considering 1 as e_∞) every e_i is connected with every e_j with $i > j$.

Definition 7.7. An $n-chain$ is a word, that can be read in the graph Γ during a path of length $(n+1)$, starting from 1. Let C_n be a set of all $n-$chains.

So,

The only $-1-$chain is 1 itself: $C_{-1} = 1$.

The only $0-$chains are letters from the alphabet: $C_0 = X$.

The only $1-$chains are obstructions: $C_1 = F$.

Let us enumerate $n-$chains for $n \geq 2$ in our examples. In the first two

$$C_n = \{x^{n+1}, x^n y^2\}$$

In the third

$$C_n = \{xy^{k_1}xy^{k_2}x \cdots xy^{k_n}x \mid k_i \geq 0\}.$$

In the last main example

$$C_n = \{e_{i_1}e_{i_2}\dots e_{i_{n+1}} \mid i_1 > i_2 > \cdots i_{n+1}\}.$$

Definition 7.8. The monomial Poincaré series for an algebra A is defined as

$$P_A^{mon}(s,t) = \sum c_{m,n} t^m s^n,$$

where $c_{m,n}$ is the number of $(n-1)-$chains of degree m.

In our first two main examples

$$P_A^{mon}(s,t) = 1 + 2ts + (t^2+t^3)s^2 + (t^3+t^4)s^3 + \cdots.$$

In the third

$$P_A^{mon}(s,t) = 1 + 2ts + \sum_{m \geq n \geq 2} \binom{m-2}{n-2} t^m s^n =$$

$$= 1 + 2ts + t^2 s^2 \left(\sum_{k=0}^{\infty} (st+t)^k \right) s = 1 + 2ts + \frac{t^2 s^2}{1-t-st}.$$

In the fourth main example $c_{m,n}$ is equal the number of partitions of m to n distinct summands.

Let us recall the definition of classical Poincaré series

Definition 7.9. The double Poincaré series for an algebra A is defined as the generating function

$$P_A(s,t) = \sum_{m,n} \dim(\mathrm{Tor}^A_{n,m}(K,K)) t^m s^n,$$

where $Tor_n^A(K,K)$ is considered as a graded module.

From the point of view of calculation of the Hilbert series we can restrict our attention to monomial Poincaré series:

Theorem 7.6.

$$H_A^{-1} = P_A(-1,t) = P_A^{mon}(-1,t)$$

If A is monomial algebra, then $P_A^{mon}(s,t) = P_A(s,t)$, so $n-$chains corresponds to homology of associated monomial algebra.

In our first two main examples we have

$$H_A^{-1} = 1 - 2t + t^2 + t^3 - t^3 - t^4 + t^4 + t^5 - \cdots = 1 - 2t + t^2$$

In the third one we have $H_A^{-1} = 1-2t+t^2$ too. The reader can simply interpret himself the connections between partitions, that we have got as a sequence in the last main example.

7.6 Anick's Resolution

In order to calculate the Poincaré series in general case we construct Anick's resolution ([3]):

$$C_n \otimes A \to C_{n-1} \otimes A \to \cdots C_{-1} \otimes A \to K \to 0$$

It is sufficient to define module homomorphisms $d_n : C_n \otimes A \to C_{n-1} \otimes A$ only for terms $f \otimes 1$. It is convenient to identify $C_n \otimes N$ with $C_n N$. Then the map d_n is defined by induction as

$$d_{n+1}(f) = f - i_n d_n(f)$$

and $i_n : \ker d_{n-1} \to KC_nN$ (here KC_nN is the linear hull of C_nN) is defined recursively:

$$i_n(u) = \alpha\hat{u} + i_n(u - \alpha d_n(\hat{u})),$$

where $\hat{u}$ is highest term of u and α is its coefficient although $\hat{u}$ belongs both $C_{n-1}N$ and C_nN (otherwise u cannot be reduced and does not belong to the kernel.)

Note, that:

- d_0 calculates, for every non empty word f, its normal form $\bar{f}$, i_0 acts identically.
- d_1 calculates, for any obstruction f an element $f - \bar{f}$, i.e. recovers the element of Gröbner basis from its obstruction. To apply d_1 for arbitrary word of form fs one needs to be more careful: Let R_0 be a linear map that, applied to non-empty word g, does not change first letter, but reduce the remaining part of the word to its normal form. Then $d_1(fs) = R_0(fs - \bar{f}s)$. (In the general case one needs to use the map $R_n : C_{n+1}N \to C_nN$, that fixes $n-$chain in the beginning and reduce the remaining part to normal form, for example $R-1$ is reduction to the normal form).

In our main examples:

1. $d_n(f) = f$ for $f \in C_n$ and $d_n(fs) = R_{n-1}(fs)$ in general. (Those formulas are valid for every monomial algebra).

2. In tensor language:

$$d_2 : x^3 \to x^2 \otimes y - xy^2 \otimes 1,$$

$$d_2 : x^2y^2 \otimes 1 \to x^2 \otimes y^2 - xy^2 \otimes x$$

(see also the end of the article).

7.7 Finite State Automata and Lie Algebras

The main problem in noncommutative case is that Gröbner basis is usually infinite. Nevertheless, using finite state automata we can try to predict the infinite behavior of our Gröbner basis or at least the infinite obstruction set. The main idea of this approach was described in [10] and can be illustrated here by our third main example: having sufficiently many terms from obstruction set, for example, $x^2, xyx, xy^2x, xy^3x, xy^4x$ one can predict the whole family: xy^nx. This kind of prediction can be formalized in the terms of regular languages (or equivalently, finite state automata). Rather often this prediction gives the correct answer, that can be proved using another arguments. Nevertheless the possibility of prediction are restricted. First of all it is impossible in general case, as we have mentioned in the second section. Second, algebras that have regular obstruction set have also rational Hilbert series and either polynomial or exponential growth. So, the Hilbert series for our fourth main example could not be predicted in this manner after finitely many calculations in terms of only two generators e_1, e_2. But even in a quite nice class of universal enveloping algebras those predictions are impossible, as we see:

Theorem 7.7. *[5] Let L be a free solvable Lie algebra of solvability length k, $U(L)$ be its universal enveloping algebra. If $k > 2$ then the growth of L and $U(L)$ is almost exponential (it means, that it is less than exponential growth $[2^m]$ but greater than growth $[2^{m^\alpha}]$ for any $\alpha\langle 1$).*

Much more advanced result and interesting generalization of GK-dimension can be found in [6].

7.8 Bergman Package under MS-DOS

The package BERGMAN was developed by J.Backelin (Stockholm university) originally for SUN/Sparc stations, but, being written on PSL (a dialect of LISP, used in REDUCE), was successfully transfered to IBM PC in the Institute of Mathematics of Moldova. It can be used both under REDUCE and (more comfortably) under specially written shell (implemented by A. Colesnicov and L. Malahova).

Main possibilities of the original version:

- commutative Gröbner bases calculations in two different strategies
- noncommutative Gröbner basis calculations
- calculations of Hilbert series and Poincaré series of the associated monomial algebra
- arbitrary long integer or Z_p coefficients of relations.
- input and output both in LISP and Maple notations.

From the shell are also accessible several separate programs, specially written for the noncommutative case that permit, after calculating Gröbner bases

- predict the behavior of the highest terms of the infinite Gröbner basis using its finite part.
- calculate growth and Hilbert series.

To estimate the possibilities that BERGMAN gives let us consider one hard

Example 7.6.

$$A = \langle x, y, z, t \mid xx - xy - yx, xz + zx - yt - ty,$$

$$yz + zy, zt + tz, zz, tt\rangle$$

Its Hilbert series can be calculated both from the noncommutative and commutative point of view (considering homology of the related commutative algebra – see [7] for details). So one can compare this with one of the best commutative programs - MACAULAY. (But maybe the best results could be obtained by the considering this example from the point of view of Lie superalgebras. See Figure 1.)

Sparcserver 690 MP (40 MHZ Cypress Sparc, 192 MB int. memory) Macintosh SE/30 (68030, 16.67MHZ, 8M) and 486SX 25 MHz IBM PC compatible computer with 4 MB RAM were used.

Pure commutative Gröbner bases calculations can be performed rather fast too. For example the well-known 6-cyclic system of equations takes from 1 till 3 minutes (depending on the ordering) and these calculations cannot be performed on the same computer both by MAPLE and MATHEMATICA).

7.9 ANICK – Program for calculating Betti Numbers and Gröbner Bases

The recursive construction for Anick's resolution, described above, was implemented by authors both as a part of the program BERGMAN (in LISP) and as a separate program ANICK (C++) for different computers. It permits to calculate the resolution and Betti numbers for sufficiently small examples (4-5 generators and several quadratic relations usually permit to calculate Betti numbers up to

Deg $\leq$	Time for BERGMAN Sparcserver 40 MHZ 192 MB	Time for MACAULAY Sparcserver 40 MHZ 192 MB	Time for MACAULAY Mac 16.7 MHZ 8 MB	Time for BERGMAN IBM-PC 25 MHZ 4 MB
2	0.32	<1	<1	1
3	0.37	<1	1	1
4	0.44	<1	5	2
5	0.53	1	16	3
6	0.61	4	75	3
7	0.73	35	535	4
8	0.94	372	5020	5
9	1.21	3416	–	6
10	1.56	31742	–	8
11	2.04	259647	–	9
12	2.84	–	–	12
13	4.47	–	–	18
14	7.49	–	–	31
15	13.9	–	–	116
16	27.8	–	–	477
17	59.3	–	–	2443
18	167	–	–	11601
19	411	–	–	–
20	1061	–	–	–

Figure 1

degree 15, sometimes more, depending on example and computer). We plan to discuss this implementation in a separate article and here restrict our attention by one aspect of implementation only.

According to the construction, the Gröbner basis should be known (at least up to given degree) to calculate the resolution. Let us suppose we give as input data not fully tasked Gröbner basis, hence the program will have at its disposal only part of the Gröbner basis elements. This means that when the graph of chains is constructed not all the obstructions participated in this process and hence some of the chains could not be obtained from this graph. If in this situation we try to calculate the resolution we will see that there exist such chains to be calculated in the graph that requires these absent chains. So, consequently we get the fact that if we want to have the whole resolution calculated till the degree k, it is necessary to have all the chains till this degree k. But this fact would not be very interesting if it answered only whether we have a complete Gröbner basis or not.

Using the information on the way of the calculations it turned out it is possible to find in this critical situation a new Gröbner basis element.

Let us consider this on the example. Let A be our second main example.

$$A = \langle x, y \mid x^2 - y^2 \rangle$$

The corresponding (preliminary) graph of chains looks now like

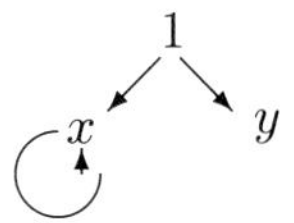

Let us try to calculate the differential for 1-chain x^2. We know that that $d_0(x) = x, d_0(y) = y$ and $d_0(f) = \overline{f}$, though we do not know yet how to calculate in general the reduction $\overline{u}$ (normal form) of arbitrary element u. Nevertheless, we can do it in the degree 2 (here we already know all the elements of our Gröbner basis).

$$d_1(x^2) = x^2 - i_0(d_0(x^2)) = x^2 - i_0(y^2) =$$
$$x^2 - y^2 - i_0(y^2 - d_0(y^2)) = x^2 - y^2 - i_0(y^2 - y^2) = x^2 - y^2.$$

And we see that no critical situations occurred while calculating all the i_0. But let us see on the process of the calculation for the 2-chain x^3:

$$d_2(x^3) = x^3 - i_1(d_1(x^2 * x)) = x^3 - i_1(R_0((x^2 - y^2) * x)) =$$
$$x^3 - i_1(R_0(x^3 - y^2x)) = x^3 - i_1(x \cdot \overline{x^2} - y \cdot \overline{yx}) =$$
$$x^3 - i_1(xy^2 - y^2x)$$

To calculate the integral i_1 it is necessary to have an 1-chain xy^2 which for a moment cannot be got from the graph of chains.

In this situation let us add the argument of i_1 to the Gröbner basis, reconstructing the graph of chains. In our case it is:

$$xy^2 - y^2x$$

Hence our new basis will consist already of two elements:

$$x^2 - y^2, \qquad xy^2 - y^2x.$$

In the graph there will appear the new vertex y^2 and the edge $x \to y^2$, i.e.

$$d_2(x^3) = x^3 - xy^2.$$

As we have seen the idea of constructing the Gröbner basis is to calculate in the fixed degree all the 2-chains. After computing these chains in the given degree we get the complete Gröbner basis in this degree. But it is not necessary reduced, so the final stage is to reduce the new Gröbner basis elements to their normal form.

We conclude the article with the calculations of our four main example by ANICK. To save space we restrict ourselves to degree 6 in three of them. In the first main example the input file looks like this.

```
% setinvars y,x;
% (setq maxdeg 6)
 x*x,
 x*y*y,

Done
.
```

The command file is as follows (semicolon starts comments)

```
   ConstructBasis             ; Calculate Groebner basis
   Alphabet                   ; Printing the alphabet
   Graph                      ; Printing the graph
   Resolution                 ; Printing the Anick's resolution
   Betti                      ; Printing the Betti numbers
                              ; (except trivial B(0,0)=1
   MacaulayBetti              ; Printing the Betti numbers
                              ; in Macaulay form (A(i,j)=B(j,i+j))
   Basis                      ; Printing the Groebner's basis

; The End
```

The output file looks like this.

```
Alphabet : y,x;

Groebner basis is :
xx,
xyy;

The Graph is :
N   Vertex      Edge(s)
-----------------------
0   1           1,2;
1   y           ;
2   x           2,3;
3   yy          ;
-----------------------

Resolution :
D(-1,1) = [1.1],
```

```
D(0,y) = [1.y],
D(0,x) = [1.x],
D(1,xx) = [x.x],
D(1,xyy) = [x.yy],
D(2,xxx) = [xx.x],
D(2,xxyy) = [xx.yy],
D(3,xxxx) = [xxx.x],
D(3,xxxyy) = [xxx.yy],
D(4,xxxxx) = [xxxx.x],
D(4,xxxxyy) = [xxxx.yy],
D(5,xxxxxx) = [xxxxx.x];
Total : 12


Betti numbers B(n+1,m) are:
B(1,1)=2
B(2,2)=1
B(2,3)=1
B(3,3)=1
B(3,4)=1
B(4,4)=1
B(4,5)=1
B(5,5)=1
B(5,6)=1
B(6,6)=1


Betti numbers are:
    0  1  2  3  4  5  6
  +---------------------------
0 | 1  2  1  1  1  1  1
1 | -  -  1  1  1  1
2 | -  -  -  -  -
3 | -  -  -  -
4 | -  -  -
5 | -  -
6 | -

Groebner basis is :
xx,
xyy;
The total working time is 1 s.
```

For the second main example output looks like this.

```
Alphabet : y,x;

Groebner basis is :
xx-yy;
```

```
The Graph is :
N   Vertex       Edge(s)
-----------------------
0   1            1,2;
1   y            ;
2   x            2;
-----------------------


Resolution :
D(-1,1) = [1.1],
D(0,y) = [1.y],
D(0,x) = [1.x],
D(1,xx) = [x.x]-[y.y],
D(1,xyy) = [x.yy]-[y.yx],
D(2,xxx) = [xx.x]-[xyy.1],
D(2,xxyy) = [xx.yy]-[xyy.x],
D(3,xxxx) = [xxx.x]-[xxyy.1],
D(3,xxxyy) = [xxx.yy]-[xxyy.x],
D(4,xxxxx) = [xxxx.x]-[xxxyy.1],
D(4,xxxxyy) = [xxxx.yy]-[xxxyy.x],
D(5,xxxxxx) = [xxxxx.x]-[xxxxyy.1];
Total : 12

Betti numbers B(n+1,m) are:
B(1,1)=2
B(2,2)=1

Betti numbers are:
    0   1   2   3   4   5   6
  +----------------------------
0 | 1   2   1   -   -   -   -
1 | -   -   -   -   -   -
2 | -   -   -   -   -
3 | -   -   -   -
4 | -   -   -
5 | -   -
6 | -

Groebner basis is :
xx-yy,
xyy-yyx;
The total working time is  1 s.
```

In the third main example we can see the main disadvantage of Anick's resolution: it might be too huge, though in reality this algebra is isomorphic to monomial algebra $A = \langle x, y \mid xy = 0 \rangle$ with a very short resolution.

Input:

```
% setinvars y,x;
% (setq maxdeg 6)
 x*x - x*y,

Done
.
```

Output:

```
Alphabet : y,x;

Groebner basis is :
xx-xy;

The Graph is :
N   Vertex      Edge(s)
-----------------------
0   1           1,2;
1   y           ;
2   x           2;
-----------------------

Resolution :
D(-1,1) = [1.1],
D(0,y) = [1.y],
D(0,x) = [1.x],
D(1,xx) = [x.x]-[x.y],
D(1,xyx) = [x.yx]-[x.yy],
D(2,xxx) = [xx.x]-[xx.y]+[xyx.1],
D(1,xyyx) = [x.yyx]-[x.yyy],
D(2,xyxx) = [xyx.x]-[xyx.y]+[xyyx.1],
D(2,xxyx) = [xx.yx]-[xx.yy]+[xyyx.1],
D(3,xxxx) = [xxx.x]-[xxx.y]+[xxyx.1]-[xyxx.1],
D(1,xyyyx) = [x.yyyx]-[x.yyyy],
D(2,xyyxx) = [xyyx.x]-[xyyx.y]+[xyyyx.1],
D(2,xyxyx) = [xyx.yx]-[xyx.yy]+[xyyyx.1],
D(3,xyxxx) = [xyxx.x]-[xyxx.y]+[xyxyx.1]-[xyyxx.1],
D(2,xxyyx) = [xx.yyx]-[xx.yyy]+[xyyyx.1],
D(3,xxyxx) = [xxyx.x]-[xxyx.y]+[xxyyx.1]-[xyyxx.1],
D(3,xxxyx) = [xxx.yx]-[xxx.yy]+[xxyyx.1]-[xyxyx.1],
D(4,xxxxx) = [xxxx.x]-[xxxx.y]+[xxxyx.1]-[xxyxx.1]+[xyxxx.1],
D(1,xyyyyx) = [x.yyyyx]-[x.yyyyy],
D(2,xyyyxx) = [xyyyx.x]-[xyyyx.y]+[xyyyyx.1],
D(2,xyyxyx) = [xyyx.yx]-[xyyx.yy]+[xyyyyx.1],
D(3,xyyxxx) = [xyyxx.x]-[xyyxx.y]+[xyyxyx.1]-[xyyyxx.1],
D(2,xyxyyx) = [xyx.yyx]-[xyx.yyy]+[xyyyyx.1],
D(3,xyxyxx) = [xyxyx.x]-[xyxyx.y]+[xyxyyx.1]-[xyyyxx.1],
```

```
D(3,xyxxyx) = [xyxx.yx]-[xyxx.yy]+[xyxyyx.1]-[xyyxyx.1],
D(4,xyxxxx) = [xyxxx.x]-[xyxxx.y]+[xyxxyx.1]-[xyxyxx.1]+[xyyxxx.1],
D(2,xxyyyx) = [xx.yyyx]-[xx.yyyy]+[xyyyyx.1],
D(3,xxyyxx) = [xxyyx.x]-[xxyyx.y]+[xxyyyx.1]-[xyyyxx.1],
D(3,xxyxyx) = [xxyx.yx]-[xxyx.yy]+[xxyyyx.1]-[xyyxyx.1],
D(4,xxyxxx) = [xxyxx.x]-[xxyxx.y]+[xxyxyx.1]-[xxyyxx.1]+[xyyxxx.1],
D(3,xxxyyx) = [xxx.yyx]-[xxx.yyy]+[xxyyyx.1]-[xyxyyx.1],
D(4,xxxyxx) = [xxxyx.x]-[xxxyx.y]+[xxxyyx.1]-[xxyyxx.1]+[xyxyxx.1],
D(4,xxxxyx) = [xxxx.yx]-[xxxx.yy]+[xxxyyx.1]-[xxyxyx.1]+[xyxxyx.1],
D(5,xxxxxx) = [xxxxx.x]-[xxxxx.y]+[xxxxyx.1]-[xxxyxx.1]+[xxyxxx.1]
             -[xyxxxx.1];
Total : 34

Betti numbers B(n+1,m) are:
B(1,1)=2
B(2,2)=1

Betti numbers are:
    0   1   2   3   4   5   6
  +---------------------------
0 | 1   2   1   -   -   -   -
1 | -   -   -   -   -   -
2 | -   -   -   -   -
3 | -   -   -   -
4 | -   -   -
5 | -   -
6 | -

Groebner basis is :
xx-xy,
xyx-xyy,
xyyx-xyyy,
xyyyx-xyyyy,
xyyyyx-xyyyyy;
The total working time is 1 s.
```

The input file for the last main example is as follows.

```
% setinvars x5(5),x4(4),x3(3),x2(2),x1(1);
% (setq maxdeg 15)

 x2*x1-x1*x2-x3,
 x3*x1-x1*x3-2*x4,
 x4*x1-x1*x4-3*x5,
 x3*x2-x2*x3-x5,
 3*x4*x3-3*x3*x4-x5*x2+x2*x5,

Done
 .
```

The command file looks like this (here we do not print the resolution).

```
  ConstructBasis          ; Calculate Groebner basis
  Alphabet                ; Printing the alphabet
  Graph                   ; Printing the graph
; Resolution              ; Printing the Anick's resolution
  Betti                   ; Printing the Betti numbers
                          ; (except trivial B(0,0)=1
  MacaulayBetti           ; Printing the Betti numbers
                          ; in Macaulay form (A(i,j)=B(j,i+j))
  Basis                   ; Printing the Groebner's basis
```

The output is:

```
Alphabet : x5(5),x4(4),x3(3),x2(2),x1;

Betti numbers B(n+1,m) are:
B(1,1)=1
B(1,2)=1
B(2,5)=1
B(2,7)=1
B(3,12)=1
B(3,15)=1


Betti numbers are:
    0   1   2   3   4   5   6   7   8   9   10  11  12  13  14  15
  +------------------------------------------------------------------
0 | 1   1   -   -   -   -   -   -   -   -   -   -   -   -   -   -
1 | -   1   -   -   -   -   -   -   -   -   -   -   -   -   -
2 | -   -   -   -   -   -   -   -   -   -   -   -   -   -
3 | -   -   1   -   -   -   -   -   -   -   -   -   -
4 | -   -   -   -   -   -   -   -   -   -   -   -
5 | -   -   1   -   -   -   -   -   -   -   -
6 | -   -   -   -   -   -   -   -   -   -
7 | -   -   -   -   -   -   -   -   -
8 | -   -   -   -   -   -   -   -
9 | -   -   -   1   -   -   -
10| -   -   -   -   -   -
11| -   -   -   -   -
12| -   -   -   1
13| -   -   -
14| -   -
15| -

Groebner basis is :
x1x2-x2x1+x3,
x1x3-x3x1+2x4,
x1x4-x4x1+3x5,
x2x3-x3x2+x5,
x2x5-3x3x4+3x4x3-x5x2,
x1x5-2x2x4+2x4x2-x5x1,
```

```
x2x2x4-2x2x4x2+4x3x5+x4x2x2-4x5x3,
x2x4x3-x3x2x4+x3x4x2-x4x3x2-5x4x5+6x5x4,
x2x4x4+x3x3x4-2x3x4x3-2x4x2x4+x4x3x3+x4x4x2,
x2x4x5-4/7x3x3x5-18/7x3x4x4+8/7x3x5x3+15/7x4x3x4+
3/7x4x4x3-x4x5x2-x5x2x4-4/7x5x
3x3+x5x4x2,
x3x3x5+10/3x3x4x4-2x3x5x3-20/3x4x3x4+10/3x4x4x3+x5x3x3,
x3x4x5-1/4x3x5x4-3/4x4x3x5-1/4x4x5x3-3/4x5x3x4+x5x4x3,
x3x3x3x4-3x3x3x4x3+3x3x4x3x3-x4x3x3x3+28/5x4x4x5-56/5x4x5x4+28/5x5x4x4,
x3x5x5+6/5x4x4x5-12/5x4x5x4-2x5x3x5+6/5x5x4x4+x5x5x3,
x3x3x4x4-2x3x4x3x4+2x4x3x4x3-x4x4x3x3+6x4x5x5-12x5x4x5+6x5x5x4,
x3x4x3x5-2/21x3x4x4x4-x3x5x3x4+3/4x3x5x4x3+76/21x4x3x4x4-7/4x4x3x5x3-
146/21x4x4
x3x4+24/7x4x4x4x3+3/4x4x5x3x3+x5x3x3x4-7/4x5x3x4x3+x5x4x3x3;
The Graph is :
N   Vertex      Edge(s)
-----------------------
0   1           1,2,3,4,5;
1   x5          ;
2   x4          ;
3   x3          10,9,11,13,14,15;
4   x2          1,3,6,7,8,9;
5   x1          2,3,4,1;
6   x2x4        3,2,1;
7   x4x3        10,9,11,13,14,15;
8   x4x4        ;
9   x4x5        ;
10  x3x5        1;
11  x3x3x4      1,2,10;
12  x3x4        1,10;
13  x5x5        ;
14  x3x4x4      ;
15  x4x3x5      1;
-----------------------
The total working time is 5 s.
```

One can use the obtained Poincare series to find a known Euler's identity, which we can describe here as:

$$H_A^{-1} = 1 + \sum_1^{\infty} (-1)^n \left(t^{\frac{3n^2-n}{2}} + t^{\frac{3n^2+n}{2}} \right)$$

Bibliography

[1] Anick, D.: *Non-commutative graded algebras and their Hilbert series*, J.Algebra, **78**, No 1, (1982) 120–140.

[2] Anick, D.: *Diophantine equations Hilbert series and undecidable spaces*, Ann. Math.,II Ser. **122**, (1985) 87–112.

[3] Anick, D.: *On the homology of associative algebras*, Trans. Am. Math. Soc., **296**, No 2, (1986), 641–659.

[4] Backelin, J., Bergman: *Computer algebra program*, available by ftp to `ftp.matematik.su.se`

[5] Lichtman, A., Ufnarovski V.: *On Growth of Lie Algebras*, Algebra Colloquium **2**,1 (1985), 45–49.

[6] Petrogradsky V.M.: *On some types of intermediate growth in Lie algebras*, Uspechi Mat.Nauk, 1993, **48**,5, 181–182.

[7] Roos J.E.: *A computer-aided study of the graded Lie algebra of a local commutative noetherian ring*, Journal of Pure and Applied Algebra **91**, (1994).

[8] Ufnarovskij V.: *Poincaré series of graded algebras*, Mat.Zametki **27**, N0 1 (1980), 21–32. English transl: Math. Notes **27**, (1980), 12–18.

[9] Ufnarovski, V.: *Combinatorial and Asymptotic Methods of Algebra*, in: ‘Algebra-VI’ (A.I.Kostrikin and I.R.Shafarevich, Eds), Encyclopaedia of Mathematical Sciences, Vol. **57**, Springer,(1995), 5–196.

[10] Ufnarovski V.: *Calculations of growth and Hilbert series by computer*, Lect. Notes Pure Appl. Math, **151**, (1993) 247–256

[11] Ufnarovskij V.: *A growth criterion for graphs and algebras defined by words*, Mat Zametki 31, N0 3 (1980), 465–472. English transl: Math. Notes **31**, (1982), 238–241.

Progress in Mathematics, Vol. 173, © 1999 Birkhäuser Verlag Basel/Switzerland

Chapter 8

A new Existence Proof of Janko's Simple Group J_4

G. D. Cooperman, W. Lempken, G. O. Michler and M. Weller

8.1 Introduction

Janko's large simple sporadic group J_4 was originally constructed by Benson, Conway, Norton, Parker and Thackray as a subgroup of the general linear group $GL_{112}(2)$ of all invertible 112×112-matrices over the field $GF(2)$ with 2 elements, see [1] and [13]. So far the construction of the 112-dimensional 2-modular irreducible representation of J_4 is only described in Benson's thesis [1] at Cambridge University. Furthermore, its proof is very involved.

In his paper [12] Lempken has constructed two matrices $x, y \in GL_{1333}(11)$ of order $o(x) = 42$, $o(y) = 10$, respectively, which describe a 1333-dimensional 11-modular irreducible representation of J_4. These two matrices are the building blocks for the new existence proof for J_4 given in this article.

In [17] the fourth author has used this linear representation of the finite group $G = \langle x, y \rangle$ to construct a permutation representation of G of degree 173 067 389 with stabilizer $M = \langle x^3, y, (x^{14})^t \rangle$, where $t = (x^{14}y^5)^2$. His main result is described in Section 8.3. It is based on a high performance computation on the supercomputers of the Theory Center of Cornell University and the University of Karlsruhe.

Using Weller's permutation representation we show in Theorem 8.3 of this article that the group $G = \langle x, y \rangle$ is simple and has order

$$|G| = 2^{21} \cdot 3^3 \cdot 5 \cdot 7 \cdot 11^3 \cdot 23 \cdot 29 \cdot 31 \cdot 37 \cdot 43.$$

Furthermore, we construct an involution $u_1 \neq 1$ of G and an element a_1 of order 3 as words in x and y such that $H = C_G(u_1)$ has the following properties:

(a) The subgroup $Q = O_2(H)$ is an extra-special group of order $|Q| = 2^{13}$ such that $C_H(Q) = \langle u_1 \rangle$.

(b) $P = \langle a_1 \rangle$ is a Sylow 3-subgroup of $O_{2,3}(H)$, and $C_Q(P) = \langle u_1 \rangle$.

(c) $H/O_{2,3}(H) \cong Aut(M_{22})$, the automorphism group of the Mathieu group M_{22}, $N_H(P) \neq C_H(P) \cong 6M_{22}$, the sixfold cover of M_{22}.

Hence $G \cong J_4$ by Theorem A of Janko's article [11].

In fact we give generators of these subgroups of H in terms of short words in x and y, see Theorem 8.3. Therefore all the assertions of this result can easily be checked by means of the computer algebra systems GAP or MAGMA without using the programs of [17].

In Section 8.2 we determine the group structure of the subgroup $M = \langle x^3, y, (x^{14})^t \rangle$, where $t = (x^{14}y^5)^2$. Proposition 8.1 asserts that M is the split extension of an elementary abelian group E of order 2^{11} by the simple Mathieu group M_{24}. By Proposition 8.2 the restriction of Lempken's 1333-dimensional 11-modular representation V of $G = \langle x, y \rangle$ to M decomposes into two irreducible 11-modular representations W and S of dimensions $\dim_F W = 45$ and $\dim_F S = 1288$. From these data the fourth author has constructed the above mentioned permutation representation of G having degree 173067389 in [17].

Section 8.4 is devoted to determine the group structure of $H = \langle x^7, y^5, (x^{14})^a, (r_1)^b \rangle$ of G, where a and b are suitably chosen elements of G described in Lemma 8.3. In Lemma 8.2 we construct an involution $u_1 \neq 1$ of G such that $H \leq C_G(u_1)$. In fact, we show in Proposition 8.3 that the group H has all the properties stated in assertions (a), (b) and (c) above.

In Section 8.5 we study the fusion of the involutions of the subgroup M in G. Proposition 8.4 asserts that G has 2 conjugacy classes of involutions $(u_1)^G$ and $(w_1)^G$. Using this result and another high performance computation determining the number of fixed points of the involution u_1 on the permutation module of degree 173 067 389 we prove in Proposition 8.5 that $H = C_G(u_1)$. In the final section it is shown that G is a simple group. This is done in Theorem 8.3, which completes our existence proof of Janko's simple group J_4.

Concerning our notation and terminology we refer to the Atlas [4] and the books by Butler [3], Gorenstein [7], Gorenstein, Lyons, Solomon [8] and Isaacs [10].

8.2 Lempken's Subgroup $G = \langle x, y \rangle$ of $GL_{1333}(11)$

Throughout this paper F denotes the prime field $GF(11)$ of characteristic 11. Let V be the canonical 1333-dimensional vector space over F. In Theorem 3.16 and Remark 3.21 of [12] Lempken describes the construction of two 1333×1333-matrices $x, y \in GL_{1333}(11)$ of order $o(x) = 42$ and $o(y) = 10$, which will become the starting data for the construction and new existence proof of Janko's group J_4. Because of their size these matrices cannot be restated here, but they can be received by e-mail from eowmob@@exp-math.uni-essen.de.

Throughout this paper $G = \langle x, y \rangle$ is the subgroup of $GL_{1333}(11)$ generated by the matrices x and y of orders 42 and 10, respectively. The following notations are taken from Lempken's article [12]. There he considers the subgroup $M = \langle x^3, y, (x^{14})^t \rangle$ with $t = (x^{14}y^5)^2$ as well. However we cannot quote any result of [12] on the structure of the subgroup M, because Lempken assumes the existence of the simple Janko group J_4.

In this section we show that the subgroup M is a split extension of an elementary abelian normal subgroup E of order $|E| = 2^{11}$ by the simple Mathieu group M_{24}. Furthermore, the module structure of the restriction of the 1333-dimensional representation of G to the subgroup M is determined.

The following notations are kept throughout the remainder of this article.

Notation 8.1. In $G = \langle x, y\rangle \leq GL_{1333}(11)$ define the following elements:
$r_0 = yx^{21}y^{-1}$
$r_1 = x^{14}yx^{21}y^{-1}x^{-14} = (r_0)^{x^{28}}$
$r_2 = y^3x^{21}y^7$
$r_3 = x^{14}y^3x^{21}y^7x^{-14} = (r_2)^{x^{28}}$
$v_1 = y^6x^{21}y^4$
$v_2 = y^8x^{21}y^2$
$v_3 = y^4x^{21}y^6$
$v_4 = x^{21}$
$w_1 = [x^6, y^5]$
$u_1 = [x^{-6}w_1x^6, r_1] = (v_1r_2)^2 = [x^{-12}(y^5x^6)^2(x^{21})^{y^{-1}x^{28}}]^2$
$u_2 = (v_3r_0)^2$
$u_3 = u_1(v_4r_0)^2$
$u_4 = (v_3r_2)^2$
$u_5 = u_4(v_4r_2)^2$
$u_6 = [x^{21}(x^{21})^y]^2$
$s_1 = y^2r_1y^{-2} = (r_1)^{y^8}$
$s_2 = (x^{21})^{yx^{28}} = x^{14}y^{-1}x^{21}yx^{-14}$
$d_1 = [(r_1)^b, s_1]$, where $b = y^{-2}x^{-6}$
$d_2 = (x^{21})^y$
$a_1 = d_1x^6d_1x^{24}d_1$
$t_1 = s_1(r_1)^bs_1$
$t_2 = (x^{21})^{y^5}$
$q_0 = r_1r_3d_1s_1(x^6y^2)^4$
$a_3 = s_1(q_0)^3s_1(q_0)^4s_1s_2$
$a_6 = t_1(x^{14}y^5)^{-2}x^{14}(x^{14}y^5)^2t_1$
$z = (x^{14})^t$, where $t = (x^{14}y^5)^2$

Observe that the elements $a_1, a_3, a_6, z \in G$ have order 3, and $q_0 \in G$ has order 7. All other elements are involutions of G.

Lemma 8.1. *Let $L = \langle x^6, y^2, z\rangle \leq G = \langle x, y\rangle$. Let $T = \langle x^6, y^2\rangle$, $j = (x^6y^2x^{12})^3$, and $s = y^6(y^2x^{18})^4$. Then the following assertions hold:*

(a) $T = \langle x^6, y^2\rangle \cong GL_4(2)$.

(b) $r_1 = (j)^s \in T$.

(c) $r_0 = (r_1)^{z^2} \in L$.

(d) $t_2 = (r_0)^{y^6}, r_2 = (r_0)^{y^8}, d_2 = (r_0)^{y^2} \in L$.

(e) *$E_2 = \langle r_0, r_2, d_2, t_2 \rangle$ is an elementary abelian normal subgroup of $N = \langle T, E_2 \rangle$ with order $|E_2| = 2^4$.*

(f) *$N = E_2T$, $E_2 \cap T = 1$, and N is perfect.*

(g) *$|L : N| = 759$, and $|L| = 2^{10} \cdot 3^3 \cdot 5 \cdot 7 \cdot 11 \cdot 23$.*

(h) *$L \cong M_{24}$, the simple Mathieu group.*

Proof.

(a) By MAGMA the simple group $GL_4(2) \cong A_8$ has the following presentation with respect to the generators a and b of orders 5 and 7, respectively:

$$(*)\quad \begin{aligned} & a^5 = b^7 = (ba^3)^4 = 1\,, \\ & (b^2a)^2 \cdot b^{-1}ab^{-1}a^3b^{-1}a^{-1} = 1\,, \\ & b^3aba^3b^{-3}ab^{-1}a^2 = 1\,, \\ & b^3a^{-1}b^{-1}a^3b^{-2}a^{-1}b^{-1}aba^{-1} = 1\,, \\ & b^2aba^{-1}b^{-1}a^3b^{-1}ab^{-1}aba = 1\,, \\ & (b^2ab^{-1}a^{-1}b^{-1}a^{-1})^2 = 1. \end{aligned}$$

Choosing $a = y^2$ and $b = x^6$ it follows by means of MAGMA that all the relations of $(*)$ are satisfied. Hence $T = \langle x^6, y^2 \rangle \cong GL_4(2)$.

(b) Certainly $j = (x^6y^2x^{12})^3$, $s = y^6(y^2x^{18})^4 \in T$. Hence $(j)^s \in T$. Using MAGMA one checks that $r_1 = (j)^s$.

(c) From $r_1 \in T \leq L$ we obtain $(r_1)^{z^2} \in L$. Using MAGMA again one gets that $r_0 = (r_1)^{z^2}$.

(d) As y^2 has order 5 it is easily checked that conjugation by y^2 yields the following orbit:

$$y^2 : t_2 \to r_2 \to r_0 \to d_2 \to r_0r_2d_2t_2 \to t_2.$$

Hence all assertions of (d) hold.

(e) Similarily x^6 has the following conjugation action:

$$x^6 : t_2 \to t_2, \text{ and } r_0 \to r_2 \to d_2 \to r_0d_2$$

Thus $E_2 = \langle r_0, r_2, d_2, t_2 \rangle$ is a normal elementary abelian subgroup of $N = \langle T, E_2 \rangle$ of order $|E_2| = 2^4$.

(f) As T is a simple group by (a) we now get $N = E_2T$, and $E_2 \cap T = 1$. Since E_2 is a simple 2-modular representation of T, it follows that N is perfect.

(g) Using the coset enumeration algorithm of MAGMA we see that $|L : N| = 759$. Hence $|L| = 2^{10} \cdot 3^3 \cdot 5 \cdot 7 \cdot 11 \cdot 23$.

(h) A matrix computation inside $GL_{1333}(11)$ shows that $[z, t_2] = z$. Therefore $L = \langle N, z\rangle$ is perfect by (f). As $|L : N|$ is odd, any Sylow 2-subgroup of L is isomorphic to a Sylow 2-subgroup of N, and therfore to those of $L_5(2)$ or M_{24}. Applying now Theorem 1 of Schoenwaelder [16] we get $L \cong M_{24}$. □

Proposition 8.1. *Let* $M = \langle x^3, y, z\rangle \leq G = \langle x, y\rangle$. *Then the following assertions hold:*

(*a*) $E = \langle u_1, u_2, u_3, u_4, u_5, u_6, v_1r_0, v_2r_2, v_3d_2, v_4t_2, y^5\rangle$ *is an elementary abelian normal subgroup of* M *with order* $|E| = 2^{11}$.

(*b*) $L = \langle x^6, y^2, z\rangle$ *is a subgroup of* M *such that* $M = EL$, $E \cap L = 1$, *and* L *is isomorphic to the simple Mathieu group* M_{24} *acting irreducibly on* E.

(*c*) $|M| = 2^{21} \cdot 3^3 \cdot 5 \cdot 7 \cdot 11 \cdot 23$.

(*d*) M *is perfect.*

(*e*) M *has six conjugacy classes of involutions with representatives:* u_1, r_0, r_1r_2, w_1, u_4r_0 *and* $u_3r_1r_2$.

Proof.

(a) By Lemma 8.1 (b) the elements t_2, d_2, r_0, r_2 are contained in $L = \langle x^6, y^2, z\rangle \leq M = \langle x^3, y, z\rangle$. The following equations are verified by means of MAGMA:

$$\begin{aligned}
[y^5, t_2] &= v_4t_2, [y^5, d_2] = v_3d_2, [y^5, r_2] = v_2r_2, [y^5, r_0] = v_1r_0, [v_2r_2, r_0] \\
&= u_1, \\
[v_3d_2, r_0] &= u_2, [v_4t_2, r_0] = u_1u_3, [v_3d_2, r_2] = u_4, [v_4t_2, r_2] = u_4u_5, \text{ and} \\
[v_4t_2, d_2] &= u_6. \\
\text{Hence } E &= \langle u_1, u_2, u_3, u_4, u_5, u_6, v_1r_0, v_2r_2, v_3d_2, v_4t_2, y^5\rangle \\
&\leq M.
\end{aligned}$$

Using the computer and MAGMA it is checked that the 11 involutions $u_1, u_2, u_3, u_4, u_5, u_6, v_1r_0, v_2r_2, v_3d_2, v_4t_2$ and y^5 commute pairwise, and that they generate an elementary abelian normal subgroup of M with order $|E| = 2^{11}$.

(b) $L = \langle x^6, y^2, z\rangle$ is a simple subgroup of M by Lemma 8.1. Thus $E \cap L = 1$, and EL is a subgroup of M. We claim that $M = EL$. Certainly, $y = y^5 \cdot (y^2)^3 \in EL$. Lemma 8.1 (d) asserts that $t_2 \in L$. Hence

$$x^3 = (x^6)^4 \cdot x^{21} = (x^6)^4 \cdot v_4 = (x^6)^4 \cdot (v_4t_2) \cdot t_2 \in EL$$

Thus $M = \langle x^3, y, z\rangle = EL$. It is well known that the smallest, non-trivial, irreducible 2-modular representation of $L \cong M_{24}$ is of degree 11. Hence L acts irreducibly on E, because $1 \neq w_1 = [x^6, y^5] \in [L, E]$.

(c) By (a), (b) and Lemma 8.1 (g) we have

$$|M| = |E \cdot L| = 2^{11} \cdot 2^{10} \cdot 3^3 \cdot 5 \cdot 7 \cdot 11 \cdot 23 = 2^{21} \cdot 3^3 \cdot 5 \cdot 7 \cdot 11 \cdot 23.$$

(d) As L is a simple group, $L = L' \leq M'$. Since L acts irreducibly on E by (b), we have $E = [E, L] \leq M'$. Therefore $M = EL = M'$.

(e) It is well known that the simple Mathieu group M_{24} has 2 non-isomorphic simple 2-modular representations of degree 11. They are dual to each other. Hence there are 2 non-isomorphic split extensions $2^{11}M_{24}$. By GAP [15] the character tables of these groups are both known. It follows that M has six conjugacy classes. Certainly, u_1^M, w_1^M, r_0^M, $(u_4r_0)^M$ and $(r_1r_2)^M$ are 5 different conjugacy classes of involutions of M, because their lengths $|u_1^M| = 7 \cdot 11 \cdot 23$, $|w_1^M| = 2^2 \cdot 3 \cdot 23$, $|r_0^M| = 2^4 \cdot 3^2 \cdot 5 \cdot 11 \cdot 23$, $|(u_4r_0)^M| = 2^4 \cdot 3^2 \cdot 5 \cdot 7 \cdot 11 \cdot 23$, and $|(r_1r_2)^M| = 2^6 \cdot 3^2 \cdot 7 \cdot 11 \cdot 23$ are all distinct. Furthermore, $|(u_3r_1r_2)^M| = 2^6 \cdot 3^2 \cdot 7 \cdot 11 \cdot 23$, but the matrices r_1r_2 and $u_3r_1r_2$ are not conjugate in M, because they have different traces $tr(r_1r_2) = 9$ and $tr(u_3r_1r_2) = 0$ in $GF(11)$ as is checked by means of MAGMA.

□

The following result is due to Lempken [12].

Proposition 8.2. *Let $G = \langle x, y \rangle$, and $M = \langle x^3, y, z \rangle$. Then the following assertions hold:*

(a) $V = F^{1333}$ is a simple FG-module.

(b) M is a subgroup of G such that the restriction $V_{|M} = W \bigoplus S$, where W and S are simple FM-modules with dimensions $dim_F W = 45$ and $dim_F S = 1288$.

Proof. Assertion (a) is a restatement of Theorem 3.20 of [12]. (b) is checked by means of Parker's Meat-Axe algorithm contained in GAP, see [15]. □

8.3 Transformation of G into a Permutation Group

In [5] Cooperman, Finkelstein, York and Tselman have described a method to construct a permutation representation $\pi : X \to S_m$ of a finite group X with a given subgroup U of index $m = |X : U|$ from a linear representation $\kappa : X \to GL_n(K)$ of X over a finite field K.

This transformation is an important idea, because most of the efficient algorithms in computational group theory deal with permutation groups, see [3]. In particular, there is a membership test for a permutation $\sigma \in S_m$ to belong to the subgroup $\pi(X)$.

Using Algorithm 2.3.1 of [6] M. Weller [17] strengthened the results of Cooperman et al. [5] as follows:

Theorem 8.1. *Let K be a finite field of characteristic $p > 0$. Let U be a subgroup of a finite group G, and let V be a simple KG-module such that its restriction $V_{|U}$ contains a proper non-zero KU-submodule W. Then there is an algorithm to construct:*

(a) *The stabilizer $\hat{U} = Stab_G(W) = \{g \in G | Wg = W \leq V\}$,*

(b) *a full set of double coset representatives $x_i, 1 \leq i \leq k$, of $\hat{U}$ in G, i. e.*

$$G = \bigcup_{i=1}^{k} \hat{U} x_i \hat{U},$$

(c) *a base $[\beta_1, \beta_2, \cdots, \beta_j]$ and strong generating set $\{g_s | 1 \leq s \leq q\}$ of G with respect to the action of G on the cosets of $\hat{U}$, which coincides with the given operation of G on the KU-submodule W of V.*

Using an efficient implementation of this algorithm on the supercomputers of the Theory Center at Cornell University and of the computer center of Karlsruhe University M. Weller [17] has obtained the following result.

Theorem 8.2. *Let $G = \langle x, y \rangle \leq GL_{1333}(11)$ and $M = \langle x^3, y, z \rangle$. Then the following assertions hold:*

(a) $|G : M| = 173067389$

(b) $|G| = 2^{21} \cdot 3^3 \cdot 5 \cdot 7 \cdot 11^3 \cdot 23 \cdot 29 \cdot 31 \cdot 37 \cdot 43$

(c) *If Ω denotes the index set of the cosets Mg_i of M in G, then G induces on $\Omega = \{1, 2, \cdots, 173067389\}$ a faithful permutation action with stabilizer $stab_G(1) = M$.*

(d) *$G = \bigcup_{i=1}^{7} Mx_iM$, where the double coset representatives x_i of M are given by the following words:*

$$\begin{array}{ll}
x_1 = 1 & (|Mx_1M| = 1) \\
x_2 = x^{14}c^{12} & (|Mx_2M| = 15180) \\
x_3 = x_6 y^2 c^{16} y c^{13} y c^4 y c^{10} y c^{18} x c^{21} & (|Mx_3M| = 28336) \\
x_4 = x_3 c^3 x^{-1} c^{11} & (|Mx_4M| = 3400320) \\
x_5 = x_6 x c & (|Mx_5M| = 32643072) \\
x_6 = x c^{12} x c^3 x c^2 & (|Mx_6M| = 54405120) \\
x_7 = x_5 x c^9 x c^{16} x^2 c^{16} & (|Mx_7M| = 82575360)
\end{array}$$

where $c := (x^{14})^t y^4 (x^{14})^t y^{-1} (x^{14})^t$ has order 23, and $t = (x^{14}y^5)^2$.

8.4 Group Structure of the approximate Centralizer H

Lempken [12] determines a suitable involution $u_1 \in G$ and an approximation H of the centralizer $C_G(u_1)$. It is now defined by means of the Notation 8.1.

Lemma 8.2. *In $G = \langle x, y\rangle$ let $a = r_1 y^{-4} x^6 y^4$ and $b = y^{-2}x^{-6}$. The subgroup $H = \langle x^7, y^5, (x^{14})^a, (r_1)^b\rangle$ of $G = \langle x, y\rangle$ contains the involution $u_1 \neq 1$, and*

$$H \leq C_G(u_1).$$

Proof. The subgroup H of G is defined in Lemma 2.5 of [12]. Using GAP [15] it can easily be checked that $u_1^2 = 1$, and that u_1 commutes with the given generators of H. □

The remainder of this section is devoted to determine the group structure of H.

Lemma 8.3. *Let $H = \langle x^7, y^5, (x^{14})^a, (r_1)^b\rangle$, $M = \langle x^3, y, z\rangle$, where $a = r_1(x^6)^{y^4}$ and $b = y^{-2}x^{-6}$. Let $W = H \cap M$. Then the following assertions hold:*

(a) $|H : W| = 77$

(b) $Q = \langle u_1, u_2, u_3, u_4, u_5, v_1, v_2, r_0, r_1, r_2, r_3, d_1, s_1\rangle$ *is a normal extra-special 2-subgroup of H with $|Q| = 2^{13}$.*

(c) $K = \langle d_2, s_2, t_2, a_1, a_3, a_6\rangle$ *is a subgroup of W with center $Z(K) = \langle a_1\rangle$ such that $K \cong 3A_6$.*

(d) $A = \langle u_1, u_6, v_3d_2, v_4t_2, y^5\rangle$ *is an elementary abelian subgroup of W with order $|A| = 2^5$ normalized by $K\langle t_1\rangle$, and $A \cap K\langle t_1\rangle = 1$.*

(e) $H \cap M = QAK\langle t_1\rangle$, *and* $K\langle t_1\rangle \cong 3S_6$.

(f) $|H| = 2^{21} \cdot 3^3 \cdot 5 \cdot 7 \cdot 11$

Proof.

(a) By Theorem 8.2 $G = \langle x, y\rangle$ has a faithful permutation action on the 173067389 cosets Mg of the subgroup $M = \langle x^3, y, z\rangle$. Using the computer we restrict this permutation representation to the subgroup H. It follows that

$$|H : H \cap M| = 77.$$

(b) Let $Q = \langle u_1, u_2, u_3, u_4, u_5, v_1, v_2, r_0, r_1, r_2, r_3, d_1, s_1\rangle$. Using again the computer and the permutation representation of G described in Theorem 8.2 it follows that $Q \leq H \cap M$. Furthermore, we get the following relations:

$$(u_2)^{s_1} = u_1u_2, (u_3)^{r_3} = u_1u_3, (u_4)^{d_1} = u_1u_4$$
$$(u_1u_4u_5)^{r_1} = u_1(u_1u_4u_5) = u_4u_5, (v_1)^{r_2} = u_1v_1, \quad (v_2)^{r_0} = u_1.$$

Since u_1 commutes with all the generators it follows that $\langle s_1, u_2\rangle$, $\langle r_3, u_3\rangle$, $\langle d_1, u_4\rangle$, $\langle r_1, u_1u_4u_5\rangle$, $\langle r_2, v_1\rangle$ and $\langle r_0, v_2\rangle$ are six dihedral subgroups of order 8 with almagamated subgroup $\langle u_1\rangle$, which commute pairwise as subgroups. Hence Q is their central product. In particular, Q is an extra-special 2-group of order $|Q| = 2^{13}$. Another matrix computation shows that Q is invariant under conjugation by the 4 given generators of H. Thus Q is normal in H.

(c) Let $K := \langle a_1, d_2, s_2, t_2, a_3, a_6 \rangle$. Using the computer again we see that K is a subgroup of $W = H \cap M$. Define $a := t_2 a_1^2 d_2 a_3^2 a_6 a_3^2 a_6^2 s_2 a_1 s_2 a_3$, $b := s_2 t_2$, $o(a) = 2$, $o(b) = 4$ and $o(ab) = 15$.

Let

$$
\begin{aligned}
x_1 &:= b^2ab^2abab^2abab, \\
x_2 &:= a, \\
x_3 &:= b^3abab^3ab^2abab^2ab^2ab, \\
x_4 &:= b^3ab^3ab^3ab^2ab^2abababa, \text{ and} \\
c &:= x_3^2.
\end{aligned}
$$

Then the following relations hold in $B = \langle x_1, x_2, x_3, x_4 \rangle \leq K$:

$$
\begin{aligned}
c^3 &= 1, x_1^3 = 1, x_2^2 = 1, x_3^2 = c, x_4^2 = 1, (x_1x_2)^3 = 1, (x_2x_3)^3 = 1, (x_3x_4)^3 \\
&= 1, \\
(x_1x_3)^2 &= 1, (x_1x_4)^2 = 1, (x_2x_4)^2 = 1, c^{x_1} = c, c^{x_2} = c, c^{x_3} = c, c^{x_4} \\
&= c.
\end{aligned}
$$

By Huppert [9], p.138 $B/\langle c \rangle$ is isomorphic to the alternating group A_6. Thus $|B| = 3|A_6|$. Now $|\langle a, b \rangle| = 3|A_6|$ by MAGMA. Furthermore we have $a = x_2$, and

$$
b = \\
x_1^2x_2x_1^2x_3^4x_1^2x_2x_1^2x_3^4x_2x_4x_1x_4x_3^5x_2x_4x_3^2x_1x_3^5x_2x_3x_4x_1^2x_4x_3^2x_1x_2x_1x_4x_3^2x_4x_1.
$$

Hence $B = \langle x_1, x_2, x_3, x_4 \rangle = \langle a, b \rangle$. We claim that $K = \langle a, b \rangle$. This follows immediately from the following equations:

$$
\begin{aligned}
a_1 &= (bab^2ab^3abab^3ab^3)^8 \\
d_2 &= b^2 \\
s_2 &= (bab^2ab^3abab^3ab^3)^6 \\
t_2 &= bab^2ab^3(ababab^3)^5abab^3a \\
a_3 &= bab^2ab^3(ababab^3)^2abab^3ab^3ababab^3 \\
a_6 &= bab^2ab^3(ababab^3)^2abab^3ab^2 \ .
\end{aligned}
$$

Let $p_1 = b^2ab^2abab^2ab^3abab^3$. Then p_1 has order 3 and commutes with a_1. Furthermore, $[p_1, a_3] = a_1$. Hence $D = \langle p_1, a_3 \rangle$ is a Sylow 3-subgroup of K. It is extra-special. Therefore its center $Z(D) = \langle a_1 \rangle = \langle c \rangle$ does not split off. Hence $K = \langle a, b \rangle \cong 3A_6$, the non-split 3-fold cover $3A_6$ of A_6.

(d) Another application of the permutation representation of G described in Theorem 8.2 on the computer shows that $A = \langle u_1, u_6, v_3d_2, v_4t_2, y^5 \rangle \leq W$, and that $a_1, a_3, a_6 \in W$. Then A is an elementary 2-subgroup of W of order $|A| = 2^5$. Another computation shows that A is normalized by $K\langle t_1 \rangle$, and $A \cap K\langle t_1 \rangle = 1$.

(e) and (f) As $(a_1)^{t_1} = (a_1)^2, (AK)^{t_1} = AK$ and $K\langle t_1\rangle \cong 3S_6$ by (c).

Certainly $H \cap M \le C_M(u_1)$ by Lemma 8.2. Using MAGMA we see that $|(u_1)^M| = 1771 = 7{\cdot}11{\cdot}23$. Therefore $|C_M(u_1)| = |M| : |(u_1)^M| = 2^{21}{\cdot}3^3{\cdot}5$ by Proposition 8.1.

Thus $H \cap M = C_M(u_1)$, because $|H \cap M| \ge |QAK\langle t_1\rangle| = 2^{13}\cdot 2^4 \cdot 2^3 \cdot 3^3 \cdot 5 \cdot 2 = 2^{21}\cdot 3^3 \cdot 5$. Hence $|H| = 2^{21}\cdot 3^3\cdot 5\cdot 7\cdot 11$ by (a), and $W = QAK\langle t_1\rangle$.

□

Proposition 8.3. *Let* $H = \langle x^7, y^5, (x^{14})^a, (r_1)^b\rangle$, *where* $a = r_1(x^6)^{y^4}$, $b = y^{-2}x^{-6}$. *Then the following assertions hold:*

(a) $U_0 = \langle u_6, v_3, v_4, d_2, s_2, t_2, a_1, a_3, a_6, x^{14}, y^5\rangle$ *is a subgroup of* H *with center* $Z(U_0) = \langle u_1 a_1\rangle$

(b) $U_0/Z(U_0) \cong M_{22}$, *and* $U_0 \cong 6M_{22}$

(c) $Q \cap Z(U_0) = \langle u_1\rangle$.

(d) *The element* a_1 *of order 3 generates a Sylow 3-subgroup of* $O_{2,3}(H)$, *and* $C_Q(a_1) = Z(Q) = \langle u_1\rangle$.

(e) $U = U_0 : \langle t_1\rangle = N_H(\langle a_1\rangle)$.

(f) $H = QU$, $U \cap Q = U_0 \cap Q = Z(U) = \langle u_1\rangle$, *and* $U_0 = C_H(a_1)$

(g) $U/Z(U_0) \cong Aut(M_{22})$, *the automorphism group of the simple Mathieu group* M_{22}.

Proof. By Lemma 8.3 we know that u_6, v_3, v_4, s_2, t_2 and $d_2 = (s_2t_2)^2$ belong to $W = H \cap M$. Certainly $x^{14} \in H$.

By Lemma 8.3 (e) we have $a_1, a_3 \in W$. From $[y^5, s_2] = (y^5 s_2)^2 = u_1 u_6 \in U_0$ follows that $u_1 a_1 \in U_0$ and has order 6. By Lemma 8.3

$$U_0 = \langle u_6, v_3, v_4, d_2, s_2, t_2, a_1, a_3, a_6, x^{14}, y^5\rangle$$

is a subgroup of H. Using MAGMA it is checked that the matrix $u_1 a_1$ commutes with the 11 generators of U_0. Thus $\langle u_1 a_1\rangle \le Z(U_0)$.

By Lemma 8.3 (d) $A = \langle u_1, u_6, v_3 d_2, v_4 t_2, y^5\rangle$ is an elementary abelian subgroup of U_0 with $|A| = 2^5$. By Lemma 8.3 (c) and (d) it is normalized by the perfect subgroup $K = \langle d_2, s_2, t_2, a_1, a_3, a_6\rangle$ of U_0. Using MAGMA it can be checked that $A = [A, K]$ is a uniserial $GF(2)K$-module. Hence

$$AK = AK' \le (U_0)'.$$

Lemma 8.3 (a), (e) and (f) assert that $|U_0 : AK| = 77$. Since x^{14} has order 3, and $U_0 = \langle AK, x^{14}\rangle$ we get $x^{14} \in U_0'$. Hence U_0 is perfect.

Let $\bar{U}_0 = U_0/\langle u_1 a_1 \rangle$. Then $\bar{U}_0$ is perfect. Let $\bar{H} = H/Q\langle a_1 \rangle$. From Lemma 8.3 we get $|\bar{H} : \bar{U}_0| = 2$, and $|\bar{U}_0| = 2^7 \cdot 3^2 \cdot 5 \cdot 7 \cdot 11$.

Now we claim that $\bar{U}_0$ is simple. Since x^{14} does not normalize A, it follows that $\langle u_1 a_1 \rangle$ is the largest normal subgroup of U_0 contained in AK. Thus $O_2(\bar{U}_0) = O_3(\bar{U}_0) = O_5(\bar{U}_0) = 1$. Suppose that Y is a minimal normal subgroup of $\bar{U}_0$. If $|Y|$ is odd, then $|Y| \in \{7, 11\}$, and $\bar{U}_0$ splits over Y. Furthermore, $Aut(Y)$ is cyclic. As $\bar{U}_0$ is perfect we get $Y \leq Z(\bar{U}_0)$. Hence $\bar{U}_0'$ is a proper subgroup of $\bar{U}_0$, a contradiction.

Therefore $|Y|$ is even, and $Y \cap (\bar{A} : \bar{K}) \neq 1$. As $\bar{A}$ is not normal in $\bar{U}_0$, we get $\bar{A} : \bar{K} \leq Y$. Thus $Y = \bar{U}_0$, because $|\bar{U}_0 : Y|$ is odd and $\bar{U}_0$ is perfect. Hence $\bar{U}_0$ is a simple group of order $|\bar{U}_0| = 2^7 \cdot 3 \cdot 5 \cdot 7 \cdot 11$. Now Theorem A of Parrott [14] asserts that $\bar{U}_0 \cong M_{22}$, the simple Mathieu group M_{22}. Therefore (b) holds. Assertions (a) and (c) are immediately clear.

(d) Using MAGMA it can be seen that $(a_1)^{t_1} = (a_1)^2$. By Lemma 8.3 and (b) we have $H = QU_0\langle t_1 \rangle, |H : QU_0| = 2$ and $O_{2,3}(QU_0) = Q\langle a_1 \rangle$. Hence $O_{2,3}(H) = Q : \langle a_1 \rangle$. Another computation with MAGMA yields that $C_Q(a_1) = Z(Q) = \langle u_1 \rangle$.

(e) Certainly $U = U_0 : \langle t_1 \rangle \leq N_H(\langle a_1 \rangle)$. In fact, $U = N_H(\langle a_1 \rangle)$ by (d) and the Frattini argument applied to the Sylow 3-subgroup $\langle a_1 \rangle$ of $O_{2,3}(H)$.

(f) is now obvious.

(g) By Lemma 8.3 and (f) we know that $(H \cap M)/Q \cong 2^4 : \hat{3}S_6$. Therefore $(H \cap M)/O_{2,3}(H) \cong 2^4 : S_6$. Hence $H/O_{2,3}(H) \cong Aut(M_{22})$.

□

8.5 The Order of $C_G(u_1)$

In this section the order of the centralizer $C_G(u_1)$ of the involution $u_1 \in G = \langle x, y \rangle$ is determined. From Proposition 8.3 we then get: $H = C_G(u_1)$.

Proposition 8.4. *The group $G = \langle x, y \rangle$ has two conjugacy classes of involutions with representatives u_1, $w_1 \in GL_{1333}(11)$ having traces $tr(u_1) = 9$, $tr(w_1) = 0 \in F$.*

Proof. Certainly the matrices u_1 and w_1 are not conjugate in G, because they have different traces $tr(u_1) = 9$, $tr(w_1) = 0$ in $F = GF(11)$. By Proposition 8.4 the following elements u_1, w_1, r_0, $u_4 r_0$, $r_1 r_2$, and $u_3 r_1 r_2$ of M yield a complete set of representatives of all six conjugacy classes of M.

Let $q_0 = r_1 r_3 d_1 s_1 (x^6 y^2)^4$, and $a_7 = (x^{14})^{y^5 x^{14}}$. Then in G the following fusion takes place:

$$u_1 \sim r_0 \sim (r_1 r_2) \text{ and } w_1 \sim (u_4 r_0) \sim (u_3 r_1 r_2),$$

because $r_0^{y^5 x^{14} r_3 (q_0)^6} = u_1 = (r_1 r_2)^{y^5 x^{14} t_2 r_2 s_2 (q_0)^6 (a_7)^2 (q_0)^6}$, $(u_4 r_0)^{y^5 x^{14} r_3} = w_1 = (u_3 r_1 r_2)^{y^5 x^{14} s_2 (q_0)^6 y^5 x^{14} r_3}$

By Theorem 8.2 (b) the index of M in G is odd. Therefore each involution i of G has a G-conjugate $i^g \in M$. Hence i^g is contained in one of the six conjugacy classes of M. Since they are G-fused to u_1^G or w_1^G it follows that either $i \in u_1^G$ or $i \in w_1^G$. □

Proposition 8.5. $H = C_G(u_1)$

Proof. Let f be the number of fixed points of the permutation afforded by u_1 on the 173067389 cosets of M in G. As $|G : M| = 173067389$ is odd, each involution i of u_1^G is contained in $f > 0$ different conjugates M^g of M for some $g \in G = \langle x, y \rangle$. By the proof of Proposition 8.4 the group G fuses the conjugacy classes u_1^M, $(r_0)^M$ and $(r_1 r_2)^M$. Furthermore, $|u_1^M| = 7 \cdot 11 \cdot 23$, $|r_0^M| = 2^4 \cdot 3^2 \cdot 5 \cdot 11 \cdot 23$, and $|(r_1 r_2)^M| = 2^6 \cdot 3^2 \cdot 7 \cdot 11 \cdot 23$. Hence

$$\begin{aligned} |u_1^G| &= \frac{|G : M|(|u_1^M| + |r_0^M| + |(r_1 r_2)^M|)}{f} \\ &= \frac{173067389 \cdot 11 \cdot 23(7 + 2^4 \cdot 3^2 \cdot 5 + 2^6 \cdot 3^2 \cdot 7)}{f} \\ &= \frac{173067389 \cdot 11 \cdot 23 \cdot 4759}{f} \end{aligned}$$

Using now the computer again, we see that u_1 has $f = 52349$ fixed points. Therefore

$$|u_1^G| = |G : C_G(u_1)| = 11^2 \cdot 23 \cdot 29 \cdot 31 \cdot 37 \cdot 43,$$

and $|C_G(u_1)| = 2^{21} \cdot 3^2 \cdot 5 \cdot 7 \cdot 11$ by Theorem 8.2. Now Lemma 8.3 and Proposition 8.3 assert that $H = C_G(u_1)$. □

8.6 The Main Result

In this section we show that $G = \langle x, y \rangle$ is a simple group. As $C_G(u_1)$ satisfies the hypothesis of Janko's theorem A [11] by Propositions 8.3 and 8.5 our existence proof for Janko's simple group J_4 then is complete.

Theorem 8.3. *Let $G = \langle x, y \rangle$ where $x, y \in GL_{1333}(11)$ are matrices constructed in [12] of orders $o(x) = 42$ and $o(y) = 10$. Then G is a simple group of order*

$$|G| = 2^{21} \cdot 3^3 \cdot 5 \cdot 7 \cdot 11^3 \cdot 23 \cdot 29 \cdot 31 \cdot 37 \cdot 43$$

such that $u_1 = [x^{-12}(y^5 x^6)^2 (x^{21})^{y^{-1}x^{28}}]^2 \neq 1$ is an involution of G with centralizer $H = C_G(u_1) = \langle x^7, y^5, (x^{14})^a, (r_1)^b \rangle$, where $a = r_1 (x^6)^{y^4}$, and $b = y^{-2} x^{-6}$. Furthermore, with the Notation 8.1 the following assertions hold:

(a) *$Q = O_2(H) = \langle u_1, u_2, u_3, u_4, u_5, v_1, v_2, r_0, r_1, r_2, r_3, d_1, s_1 \rangle$ is an extra-special normal subgroup of H with $|Q| = 2^{13}$.*

(b) *The element $a_1 = d_1 x^6 d_1 y^{24} d_1$ of order 3 generates a Sylow 3-subgroup of $O_{2,3}(H)$, and $C_Q(a_1) = Z(Q) = \langle u_1 \rangle$.*

(c) $U_0 = C_H(a_1) = \langle u_6, v_3, v_4, d_2, s_2, t_2, a_1, a_3, a_6, x^{14}, y^5 \rangle \cong 6M_{22}$, *the sixfold cover of the Mathieu group* M_{22} *with center* $Z(U_0) = \langle u_1 a_1 \rangle$.

(d) $U = N_H(\langle a_1 \rangle) = U_0 : \langle t_1 \rangle$ *is a subgroup of* H *with* $U/Z(U_0) \cong Aut(M_{22})$, *and center* $Z(U) = \langle u_1 \rangle$.

(e) $H = QU$, *and* $Q \cap U = \langle u_1 \rangle = C_Q(a_1)$.

In particular, G *is isomorphic to Janko's simple group* J_4.

Proof. In view of Proposition 8.5, Theorem 8.2, Lemma 8.3 and Proposition 8.3 it remains to show that G is a simple.

Proposition 8.1 asserts that $M = \langle x^3, y, (x^{14})^t \rangle = M'$,where $t = (x^{14}y^5)^2$. Hence $x^3, y \in M' \leq G'$. Furthermore, $(x^{14})^t \in M' \leq G'$. As G' is normal in G we see that $x^{14} \in G'$. But $gcd(3, 14) = 1$ and so $\langle x \rangle = \langle x^3, x^{14} \rangle$. Therefore $G = \langle x, y \rangle = G'$, and G is perfect.

Let N be any normal subgroup of G. If $|N|$ is even, then there is an involution $y \neq 1$ in N. By Proposition 8.4 it is either conjugate to u_1 or to w_1 in G. Using Proposition 8.1 (b) and the fusion of the conjugacy classes u_1^M, r_0^M, $(r_1 r_2)^M$ of involutions of M in G it follows that

$$\langle y^G \cap M \rangle = M \leq N.$$

By Theorem 8.2 the index $|G : M|$ is odd. Hence G/N is a solvable group by the Feit-Thompson theorem. As G is perfect, we get $N = G$.

Therefore we may assume that $|N|$ is odd. As u_1 and u_2 are two commuting involutions $W = \langle u_1, u_2 \rangle$ is a Klein four-group acting on the normal subgroup N. Using the computer it follows that the matrix $u_1 u_2 \in GL_{1333}(11)$ has trace $tr(u_1 u_2) = 9$. Since $tr(u_1 u_2) = tr(u_1)$ Proposition 8.4 implies that all three involutions of W belong to u_1^G. Now the Brauer-Wielandt formula of [8], p. 198 asserts that

$$|N||C_N(W)|^2 = |C_N(u_1)|^3.$$

By Proposition 8.3 $O_{2'}(H) = 1$, because $H = C_G(u_1)$ by Proposition 8.5. Hence $C_N(u_1) = 1$. Thus $N = 1$. Therefore G is a simple group, and $G \cong J_4$ by Theorem A of Janko [11]. □

Acknowledgements

The authors of this paper have been supported by the DFG research project "Algorithmic Number Theory and Algebra".

A substantial part of the high performance computations proving Theorem 8.2 were conducted using the resources of the Cornell Theory Center, which receives major funding from the National Science Foundation and New York State with additional support from the Research Resources at the National Institutes of Health, IBM Cooperation and members of the Corporate Research Institute. The total computing time on all the involved nodes was 28137 CPU-h.

We owe special thanks to Professor J. Guckenheimer and Dr. A. Hoisie for their support.

The authors also would like to thank the Computer Center of Karlsruhe University for providing 38785 CPU-h on their supercomputer IBM RS/6000 SP with 256 knots. This help was necessary to complete the above mentioned computations. We are very grateful to Professor W. Schönauer for his assistance.

Bibliography

[1] D. J. Benson: *The Simple Group* J_4, PhD. Thesis, Trinity College, Cambridge (1980).

[2] W. Bosma, J. Cannon: *MAGMA Handbook*, Sydney (1993)

[3] G. Butler: *Fundamental algorithms for permutation groups*, Lect. Notes in Computer Science, Springer Verlag, Heidelberg (1991)

[4] J. H. Conway, R. T. Curtis, S. P. Norton, R. A. Parker, R. A. Wilson: *Atlas of finite groups*, Clarendon Press, Oxford (1985).

[5] G. Cooperman, L. Finkelstein, M. Tselman, B. York: *Constructing permutation representations for large matrix groups*, J. Symbolic Computation **24**, (1997), 471-488.

[6] H. W. Gollan: *A new existence proof for Ly, the sporadic group of R. Lyons*, Preprint IEM Essen (1995).

[7] D. Gorenstein: *Finite simple groups*, Plenum Press, New York (1982).

[8] D. Gorenstein, R. Lyons, R. Solomon: *The classification of the finite simple groups*, Mathematical Surveys and Monographs **40**, No. 2, American Mathematical Society, Providence, Rhode Island (1996).

[9] B. Huppert: *Endliche Gruppen I*, Springer Verlag, Heidelberg (1983).

[10] I. M. Isaacs: *Character theory of finite groups*, Academic Press, New York (1972).

[11] Z. Janko: *A new finite simple group of order* $86 \cdot 775 \cdot 571 \cdot 046 \cdot 077 \cdot 562 \cdot 880$ *which possesses* M_{24} *and the full covering group of* M_{22} *as subgroups*, J. Algebra **42**, (1972), 564–596.

[12] W. Lempken: *Constructing* J_4 *in* $GL(1333, 11)$, Communications in Algebra **21**, (1993), 4311–4351.

[13] S. Norton: *The construction of* J_4, Proceedings of Symposia in Mathematics AMS **37**, (1980), 271–277.

[14] D. Parrott: *On the Mathieu groups* M_{22} *and* M_{11}, J. Austr. Math. Soc. **11**, (1970), 69–81.

[15] M. Schönert et al.: *GAP-Groups, Algorithms, and Programming*, 3rd ed., Lehrstuhl D für Mathematik, RWTH Aachen (1993).

[16] U. Schoenwaelder: *Finite groups with a Sylow 2-subgroup of type M_{24}*, II. J. Algebra **28**, (1974), 46–56.

[17] M. Weller: *Construction of large permutation representations for matrix groups*, Preprint IEM Essen (1997).

Progress in Mathematics, Vol. 173, © 1999 Birkhäuser Verlag Basel/Switzerland

Chapter 9

The Normalization: a new Algorithm, Implementation and Comparisons

Wolfram Decker, Theo de Jong, Gert-Martin Greuel and Gerhard Pfister

9.1 Introduction

We present a new algorithm for computing the normalization $\bar{R}$ of a reduced affine ring R, together with some remarks on efficiency based on our experience with an implementation of this algorithm in SINGULAR (cf. [2]).

Our method to compute $\bar{R}$ (the integral closure of R in its total ring of fractions) is based on a criterion for normality, due to Grauert and Remmert [3], which was rediscovered in [5]. The criterion states that $R = \bar{R}$ if and only if the canonical map $R \longrightarrow \operatorname{Hom}_R(J, J)$ is an isomorphism, where J denotes a reduced ideal such that its zero set contains the non–normal locus of $\operatorname{Spec} R$. In general this map is only injective and we obtain an inclusion of rings,

$$R \subset \operatorname{Hom}_R(J, J) \subset \bar{R}.$$

Our method is to present $\operatorname{Hom}_R(J, J)$ as an affine ring R_1, which is of type $R[T_1, \ldots, T_s]$ modulo an ideal generated by linear and quadratic relations in the T_i.

We continue in the same manner with R_1 and obtain a sequence of rings

$$R = R_0 \subsetneqq R_1 \subsetneqq \cdots \subsetneqq R_k = \operatorname{Hom}_{R_k}(J_k, J_k),$$

such that $R_k = \bar{R}$ by the criterion of Grauert and Remmert (the algorithm must stop, since $\bar{R}$ is finite over R).

At the end of this paper we describe several special cases which allow us to do some steps in the algorithm more efficiently. Examples show that these refinements may be essential to the ability to compute the normalization.

The normalization provides a decomposition of R into the normalization of the prime components, in particular, it computes the number of irreducible components. Our algorithm does not need any prime decomposition, but we may, of course, first make such a decomposition and then normalize the components. The last table shows that this is sometimes useful, but there are also examples where the normalization is easily computed, the prime decomposition, however, is not computable in a reasonable time.

It is clear that this algorithm applies to the case where R is the localization of an affine ring with respect to a general monomial ordering (for example, the localization at a maximal ideal) as described in [1].

9.2 Criterion for Normality

Here we describe the algorithm, mentioned in the introduction. Other algorithms were given, for example, by Seidenberg [6], Stolzenberg [7], Gianni, Trager [4] and Vasconcelos [8].

The algorithm is based on the following criterion for normality due to Grauert and Remmert [3]:

Proposition 9.1. *Let R be a noetherian reduced ring and J be a radical ideal containing a non–zero divisor such that the zero set of J, $V(J)$ contains the non– normal locus of Spec(R). Then R is normal if and only if $R = \mathrm{Hom}_R(J, J)$.*

Proof. Let $\bar{R}$ be the normalization of R. Then we have the canonical inclusions

$$R \subset \mathrm{Hom}_R(J, J) \subset \bar{R}$$

$$r \rightsquigarrow \varphi_r, \qquad \varphi \rightsquigarrow \frac{\varphi(x)}{x}$$

where φ_r is the map defined by the multiplication with r and x is a non–zero divisor of J. It is easy to see that $\frac{\varphi(x)}{x}$ is independent of the choice of x. Since J is finitely generated, the characteristic polynomial of φ defines an integral relation and, therefore, $\frac{\varphi(x)}{x} \in \bar{R}$.

Now we claim that $\mathrm{Hom}_R(J, J) = \mathrm{Hom}_R(J, R) \cap \bar{R}$. Indeed, let $h \in \bar{R}$ and $h^n + a_{n-1}h^{k-1} + \cdots + a_0 = 0$, $a_i \in R$. If $h \in \mathrm{Hom}_R(J, R)$, that is $hJ \subset R$, then for all $g \in J$ we have $hg \in R$ and

$$(gh)^n + ga_{n-1}(gh)^{n-1} + \cdots + g^n a_0 = 0$$

which implies $(gh)^n \in J$. But J is a radical ideal and, therefore, $gh \in J$, which implies $h \in \mathrm{Hom}_R(J, J)$.

Now we are prepared for the proof of the proposition.

One implication is trivial. To prove the other, assume $R = \mathrm{Hom}_R(J, J)$. This implies $R = \mathrm{Hom}_R(J, R) \cap \bar{R}$.

Let $h = \frac{f}{g} \in \bar{R}$, it suffices to show that $hJ \subset R$. Let $\Delta = \{P \in \mathrm{Spec}(R) | h \notin R_P\}$, then obviously, Δ is contained in the non–normal locus of Spec(R) and, therefore, by assumption $\Delta \subseteq V(J)$. Δ can be defined by the ideal $C = \{u \in R | hu \in R\}$. By the abstract Nullstellensatz, we obtain $\sqrt{C} \supseteq J$. Now, by definition $h \cdot C \subseteq R$ and we may choose an integer d such that $R \supseteq h\sqrt{C}^d \supseteq hJ^d$. Assume that d is minimal such that $R \supseteq hJ^d$. If $d > 1$, then there exists an $a \in J^{d-1}$ such that $ha \notin R$ but $ha \in \bar{R}$ and $haJ \subset R$. This implies, because of $\mathrm{Hom}_R(J, J) = \mathrm{Hom}_R(J, R) \cap \bar{R}$, that $ha \in J \subseteq R$, which gives a contradiction. Therefore, $d = 1$ and $hJ \subseteq R$, which proves the proposition. □

Remark 9.1. Let J, R be as in the proposition and x a non–zero divisor of J. We saw in the beginning of the proof that

1)
$$xJ : J = x \cdot \mathrm{Hom}_R(J, J)$$

and, consequently,

2)
$$R = \mathrm{Hom}_R(J, J)$$

if and only if $xJ : J \subseteq \langle x \rangle$.

3) Let $u_0 = x, u_1, \ldots, u_s$ be generators of $xJ : J$ as R–module. Because $\mathrm{Hom}_R(J, J)$ is a ring we have $\frac{s(s+1)}{2}$ relations

$$\frac{u_i}{x} \cdot \frac{u_j}{x} = \sum_{k=0}^{s} \xi_k^{ij} \frac{u_k}{x}, s \geq i \geq j \geq 1, \xi_k^{ij} \in R \text{ in } \tfrac{1}{x}(xJ : J).$$

Together with the linear relations, the syzygies between $u_0, \ldots, u_s$, they define the ring structure of $\mathrm{Hom}_R(J, J)$:

$$\begin{array}{ccc} R[T_1, \ldots, T_s] & \twoheadrightarrow & \mathrm{Hom}_R(J, J) \\ T_i & \rightsquigarrow & \dfrac{u_i}{x}. \end{array}$$

The kernel of this map is the ideal generated by

$$T_i T_j - \sum_{k=0}^{s} \xi_k^{ij} T_k, \quad (T_0 = 1) \text{ and } \sum_{k=0}^{s} \eta_k T_k \text{ such that } \sum_{k=0}^{s} \eta_k u_k = 0.$$

9.3 The Normalization Algorithm

Now we are prepared to give the normalization algorithm:

Input: a radical ideal $I \subseteq K[x_1, \ldots, x_n]$.

Output: s polynomial rings $R_1, \ldots, R_s$ and s prime ideals $I_1 \subset R_1, \ldots, I_s \subset R_s$ and s maps $\pi_i : R \to R_i$, such that the induced map $\pi : K[x_1, \ldots, x_n]/I \to R_1/I_1 \times \cdots \times R_s/I_s$ is the normalization of $K[x_1, \ldots, x_n]/I$.

9.3.1 normal(I[, inform]) Additional information by the user (respectively by the algorithm) can be given in the optional list inform, as for instance,

- I defines an an isolated singularity,
- some elements of the radical of the non–normal locus, which are already known.

Result $= \emptyset$
compute idempotents of $K[x_1, \ldots, x_n]/I$.
This is optional and gives just the information about the normalization as splitting into a direct sum of rings.
Assume $K[x_1, \ldots, x_n]/I = K[x_1, \ldots, x_n]/I_1 \times \cdots \times K[x_1, \ldots, x_n]/I_s$.
For $i = 1$ **to** s **do**
compute $J =$ singular locus of I_i
choose $f \in J \setminus I_i$ and compute $I_i : f$ **to**
check whether f is a zero divisor
if $I_i : f \supsetneqq I_i$
Result $=$ Result $\cup$ normal$(I_i : (I_i : f)) \cup$ normal$(I_i : f)$
(Notice that $\sqrt{I_i, f} = I_i : (I_i : f)$ in this situation.)
else
If we have an isolated singularity at $0 \in K^n$ **then**

$$J = (x_1, \ldots, x_n)$$

In general, if J_0 is the radical of the singular locus of a normalization loop before, given by the list inform, **then**

$$J = \sqrt{I_i, f + J_0}$$

else
$J = \sqrt{I_i, f}$
$H = fJ : J$
if $H = \langle f \rangle$
Result $=$ Result $\cup \{K[x_1, \ldots, x_n], I_i, id\}$
else
assume $H = fJ : J = \langle f, u_1, \ldots, u_s \rangle$
then
compute an ideal L,
$L \subseteq K[x_1, \ldots, x_n, T_1, \ldots, T_s]$
(as described in the remark above) such that
$K[x_1, \ldots, x_n, T_1, \ldots, T_s]/L \xrightarrow{\sim} \mathrm{Hom}(J, J)$
$T_i \rightsquigarrow \frac{u_i}{f}$,
let
$\varphi : K[x_1, \ldots, x_n] \to K[x_1, \ldots, ex_n, T_1, \ldots, T_s]$
be the inclusion.
$S =$ normal (L),
compose the maps of S with φ.
Result $=$ Result $\cup S$
return Result

It remains to give an algorithm to compute the idempotents.

We shall explain this for the case when the input ideal I is (weighted) homogeneous with strictly positive weights.

An idempotent e, that is, $e^2 - e \in I$, has to be homogeneous of degree 0. Therefore it will not occur in the first loop.

It may occur after one normalization loop in $\mathrm{Hom}(J, J) \simeq K[x_1, \ldots, x_n, T_1, \ldots, T_s]/L$ because some of the generators may have the same degree.

Let $T \subseteq \{T_1, \ldots, T_s\}$ be the subset of variables of degree 0.

Then $L \cap K[T]$ is zero–dimensional because $T_j^2 - \sum \xi_k^{jj} T_k \in L \cap K[T]$ for all $T_j \in T$ (the weights are ≥ 0 and, therefore, $\xi_k^{jj} \in K$, $T_k \in T$).

For this situation there is an easy algorithm:

Input: $I \subseteq K[x_1, \ldots, x_n]$ a (weighted) homogeneous radical ideal, $\deg(x_1) = \cdots = \deg(x_k) = 0, \deg(x_i) > 0$ for $i > k$, $I \cap K[x_1, \ldots, x_k]$ being 0–dimensional.

Output: ideals $I_1, \ldots, I_s$ such that $K[x_1, \ldots, x_n]/I = K[x_1, \ldots, x_n]/I_1 \times \cdots \times K[x_1, \ldots, x_n]/I_s$ and $I \cap K[x_1, \ldots, x_k] = \cap(I_v \cap K[x_1, \ldots, x_k])$ is the prime decomposition

9.3.2 Idempotents (I)

Result $= \emptyset$
compute $J = I \cap K[x_1, \ldots, x_k]$
compute $J = P_1 \cap \cdots \cap P_s$ the (0–dimensional) prime decomposition.
For $i = 1$ **to** s **do**
choose $g_i \neq 0$ in $\bigcap_{v \neq i} P_v$
Result $=$ Result $\cup \{I : g_i\}$
return Result

9.4 Implementation and Comparisons

The above algorithm has been implemented by the authors. The implementation in SINGULAR is available as SINGULAR library `normal.lib`.

To have an efficient version of the normalization algorithm, we had to take care of several special cases and tricks for the implementation:

1. If the variety defined in a certain normalization loop has an isolated singularity (let us say at the origin), the varieties arising in the following loops will have an isolated singularity, too. Therefore, there is no longer any need to compute the singular locus: its radical is $(x_1, \ldots, x_n)$.

2. If we have computed some radical containing the non–normal locus as zero–set, then we add it in the next step to the corresponding ideal. The "old" radical elements turned out to be very helpful in speeding up the computations.

3. Similar to the property 1), equidimensionality is kept. Then we only have to compute the equidimensional radical of the ideal containing the non–normal locus, which is faster.

4. In some examples it is faster to make a primary decomposition (for example, à la Gianni, Trager, Zacharias) prior to normalizing. In other examples, a direct computation of the normalization is faster.

5. Similar to the property 1), irreducibility is kept. Then a check for non–zero divisors is not necessary.

6. Similar to the property 1), Cohen–Macaulayness is kept. Then a test for regularity in codimension 2 as criterion for normality is very useful.

7. It is also useful to try to reduce the number of variables after having computed the ring structure of $\mathrm{Hom}(J, J)$ by substituting those which can be expressed by other variables.

We illustrate the algorithm by computing the normalization of the cuspidal plane cubic:

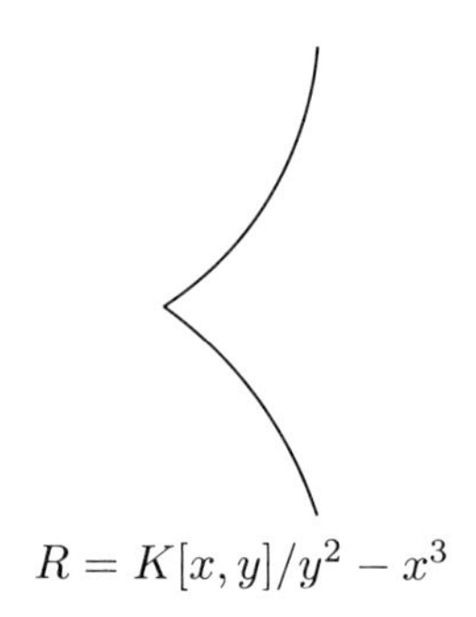

$$R = K[x, y]/y^2 - x^3$$

- Radical of the singular locus : $J = (x, y)$

- $R \subsetneq \mathrm{Hom}_R(J, J) = (1, \frac{y}{x})$

- the linear relations are $x^2 - yT_1, y - xT_1$ the quadratic relation is $T_1^2 - x$ and, therefore

$$\mathrm{Hom}_R(J, J) = R[T_1]/(x^2 - yT_1, y - xT_1, T_1^2 - x)$$

- reducing the number of variables by $y = xT_1,\ x = T_1^2$ we obtain $\bar{R} = K[T_1]$ and as map

$$\begin{array}{ccc} R & \to & K[t_1], \\ x & \rightsquigarrow & T_1^2, \\ y & \rightsquigarrow & T_1^3. \end{array}$$

9.5 Examples

1. Example of Huneke (cf. [8])

$$
\begin{aligned}
&5abcde - a^5 - b^5 - c^5 - d^5 - e^5,\\
&ab^3c + bc^3d + a^3be + cd^3e + ade^3,\\
&a^2bc^2 + b^2cd^2 + a^2d^2e + ab^2e^2 + c^2de^2,\\
&abc^5 - b^4c^2d - 2a^2b^2cde + ac^3d^2e - a^4de^2 + bcd^2e^3 + abe^5,\\
&ab^2c^4 - b^5cd - a^2b^3de + 2abc^2d^2e + ad^4e^2 - a^2bce^3 - cde^5,\\
&a^3b^2cd - bc^2d^4 + ab^2c^3e - b^5de - d^6e + 3abcd^2e^2 - a^2be^4 - de^6,\\
&a^4b^2c - abc^2d^3 - ab^5e - b^3c^2de - ad^5e + 2a^2bcde^2 + cd^2e^4,\\
&b^6c + bc^6 + a^2b^4e - 3ab^2c^2de + c^4d^2e - a^3cde^2 - abd^3e^2 + bce^5
\end{aligned}
$$

   This is a non–minimal abelian surface of degree 15 in $\mathbb{P}^4$ which is linked $(5,5)$ to a Horrocks–Mumford surface.

2. $x^2y^2 + x^2z^2 + y^2z^2$

   This is a 3–nodal quartic:

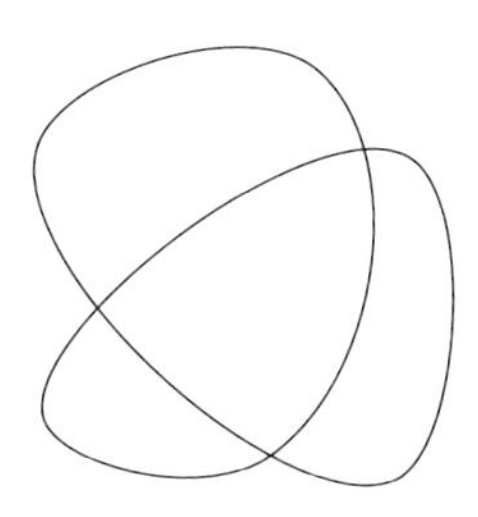

3. Example of Sturmfels

   The radical of the ideal generated by the 2×2 permanents of a generic 3×3–matrix.

$$
\begin{aligned}
&bv + su,\\
&bw + tu,\\
&sw + tv,\\
&by + sx,\\
&bz + tx,\\
&sz + ty,\\
&uy + vx,\\
&uz + wx,\\
&vz + wy,\\
&bvz.
\end{aligned}
$$

4. $wy - vz$,
   $vx - uy$,

$tv - sw,$
$su - bv,$
$tuy - bvz.$

5. Example of Riemenschneider

   The radical of the base space of the versal deformation of the first "general" cyclic quotient singularity of embedding dimension 6.

   $xz,$
   $vx,$
   $ux,$
   $su,$
   $qu,$
   $txy,$
   $stx,$
   $qtx,$
   $uv^2z - uwz,$
   $uv^3 - uvw,$
   $puv^2 - puw.$

6. The intersection of the ideals of example 2) and 4).

The examples illustrate the remarks made at the beginning of this section:

Example 1) takes several hours if the information that it is an isolated singularity is not given to the next loop.

Example 2) takes several minutes if the "old" radical is not used in the next loop.

Example 3) and 5) are much faster if a primary decomposition is performed prior to the normalization (using the algorithm of Gianni, Trager, Zacharias).

We obtained the following timings for our implementation in SINGULAR (cf. [2]) (in seconds) on an HP740, where $*$ means that the computation took more than one hour:

	computation of the radical	computation of the minimal associated primes	prime decomposition	prime decomposition and normalization	normalization	number of components
1	42	*	*	*	7	1
2	1	1	1	1	1	1
3	3	7	9	11	*	15
4	1	2	2	30	28	3
5	1	2	9	3	33	4
6	2	4	4	32	22	4

Bibliography

[1] Greuel, G.-M., Pfister, G.: *Advances and improvements in the theory of standard bases and syzygies*, Arch. Math. **66**, 163–176 (1996).

[2] Greuel, G.-M., Pfister, G., Schönemann, H.: *SINGULAR Reference Manual*, Reports On Computer Algebra Number 12, May 1997, Centre for Computer Algebra, University of Kaiserslautern, www.mathematik.uni-kl.de/ zca/Singular.

[3] Grauert, H., Remmert, R.: *Analytische Stellenalgebren*, Springer 1971.

[4] Gianni, P., Trager, B.: *Integral closure of noetherian rings*, Proceedings ISSAC 97.

[5] de Jong, T.: *An algorithm for computing the integral closure*, J. of Symbolic Comp. 26, 273–277 (1998).

[6] Seidenberg, A.: *Construction of the integral closure of a finite integral domain II*, Proc. Amer. Math. Soc. 52, 368–372 (1975).

[7] Stolzenberg, G.: *Constructive normalization of an algebraic variety*, Bull. Amer. Math. Soc. 74, 595–599 (1968).

[8] Vasconcelos, W.: *Computing the integral closure of an affine domain*, Proc. AMS 113 (3), 633–638 (1991).

Progress in Mathematics, Vol. 173, © 1999 Birkhäuser Verlag Basel/Switzerland

Chapter 10

A Computer Algebra Approach to sheaves over Weighted Projective Lines

Piotr Dowbor and Thomas Hübner

Abstract

The theoretical background of the TUBULAR package, developed for a computer algebra treatment of sheaves over weighted projective (tubular) lines is presented. The package is also of interest for classifying indecomposable modules over finite dimensional tubular algbras, resp. G-equivariant sheaves over elliptic curves with respect to certain actions of finite groups. The encoding of indecomposable sheaves by a set of discrete data (using the tubular structure of $\operatorname{coh}\mathbb{X}$) and the effective method of determining the class $[\mathcal{F}]$ of indecomposable $\mathcal{F} \in \operatorname{coh}\mathbb{X}$ in the Grothendieck group $\mathrm{K}_0\,\mathbb{X}$ (based on the telescoping functor technique) are explained. Certain additional facts improving efficiency of this method are discussed.

10.1 Introduction

10.1.1 The Category $\operatorname{coh}\mathbb{X}$ **of Coherent Sheaves** We can present only a short summary of the basic facts needed for the purpose of this article. For further details, also for the link to representations of finite dimensional algebras (in particular to Ringel's classifications of modules over tubular algebras [5]), resp. sheaves on elliptic curves (the connections between the Atyiah's classification of vector bundles on smooth elliptic curves [1] and the results concearning weighted projective lines of the weight type $(2,2,2,2)$) we refer the reader to [2].

For a *weight sequence* $\underline{\mathbf{p}} = (p_1, \dots, p_t)$ of (nondecreasing) integers $p_i \geq 1$ let

$$\mathbb{L}(\underline{\mathbf{p}}) := \bigoplus_{i=1}^{t} \mathbb{Z}\vec{x_i} \; / \; \left(\{p_i\vec{x_i} - p_j\vec{x_j}\}_{i,j=1,\dots,t}\right)$$

be the abelian group of rank 1 generated by $\vec{x_1}, \dots, \vec{x_t}$ subject to the relations

$$p_1\vec{x_1} = p_2\vec{x_2} = \dots = p_t\vec{x_t} \; (=: \vec{c}).$$

Each element $\vec{x} \in \mathbb{L}(\underline{\mathbf{p}})$ can be uniquely written in a *normal form*; i.e.

$$\vec{x} \;=\; \sum_{i=1}^{t} \alpha_i\vec{x_i} \;+\; \alpha_0\vec{c}$$

with integers $\alpha_0 \in \mathbb{Z}$ and $0 \le \alpha_i < p_i$. Setting

$$\vec{x} \ge \vec{0} \quad :\Longleftrightarrow \quad \vec{x} \in \sum_{i=1}^{t} \mathbb{N}\vec{x_i}$$

defines a partial *order* on the set of elements of $\mathbb{L}(\underline{\mathbf{p}})$. We denote by $\vec{\omega} = -\sum_{i=1}^{t} \vec{x_i} + (t-2)\vec{c}$ the *dualizing element* of $\mathbb{L}(\underline{\mathbf{p}})$.

Throughout, let k denote an algebraically closed field and $k[X_1, \dots, X_t]$ the polynomial k-algebra in commuting variables $X_1, \dots, X_t$. For a sequence $\underline{\lambda} = (\lambda_1, \dots, \lambda_t)$ of *pairwise different* points $\lambda_i \in \mathbb{P}^1(k)$ (normalized such that $\lambda_1 = \infty, \lambda_2 = 0$) let

$$S = S(\underline{\mathbf{p}}, \underline{\lambda}) \;:=\; k[X_1, \dots, X_t] \;/\; (\{X_2^{p_2} - \lambda_i X_1^{p_1} - X_i^{p_i}\}_{i=3,\dots,t})$$

be the $\mathbb{L}(\underline{\mathbf{p}})$-graded ($\deg x_i = \vec{x_i}$, where x_i denotes the residue-class of X_i) factorial ring of graded Krull dimension 2. The *projective spectrum*

$$\mathbb{X} = \mathbb{X}(\underline{\mathbf{p}}, \underline{\lambda}) := \mathrm{Proj}^{\mathbb{L}(\underline{\mathbf{p}})}(S)$$

will be called a *weighted projective line* of type $(\underline{\mathbf{p}}, \underline{\lambda})$. Geometrically, $\mathbb{X}(\underline{\mathbf{p}}, \underline{\lambda})$ consists of

- the *exceptional points* $x_1, \dots, x_t$ of multiplicities
$$p(x_1) = p_1, \dots, p(x_t) = p_t;$$
- the *ordinary points* $x = x^{p_2} - \mu x^{p_1}$ of multiplicity $p(x) = 1$ (for $\mu \notin \underline{\lambda}$),
- and the *generic point* $\xi = 0$.

We denote by $\mathcal{O} = \mathcal{O}_{\mathbb{X}}$ the *structure sheaf* of $\mathbb{L}(\underline{\mathbf{p}})$-graded algebras over $\mathbb{X}(\underline{\mathbf{p}}, \underline{\lambda})$ produced by S. $\mathbb{X}(\underline{\mathbf{p}}, \underline{\lambda})$ is said to be of *tubular type* if and only if $\underline{\mathbf{p}}$ is one of $(2,4,4), (2,3,6), (3,3,3), (2,2,2,2)$. Defining $\mathcal{O}$-modules $\mathcal{F}$ (such that $\mathcal{F}(U)$ is an $\mathbb{L}(\underline{\mathbf{p}})$-graded module over $\mathcal{O}(U)$) allows to introduce the category $\mathrm{coh}\,\mathbb{X}$ of $\mathbb{L}(\underline{\mathbf{p}})$-graded *coherent sheaves* over $\mathbb{X}$. The category $\mathrm{coh}\,\mathbb{X}$ is an abelian *hereditary* k-category (i.e. $\mathrm{Ext}^2(-,-) \equiv 0$), which neither contains projective nor injective objects. Due to heredity, the *derived category* $\mathcal{D}^b\mathrm{coh}\,\mathbb{X}$ of $\mathrm{coh}\,\mathbb{X}$ is equivalent to its full subcategory $\mathrm{add} \bigcup_{n\in\mathbb{Z}} (\mathrm{coh}\,\mathbb{X})[n]$ and

$$\mathrm{Hom}_{\mathcal{D}^b\mathrm{coh}\,\mathbb{X}}(\mathcal{F}[n], \mathcal{G}[m]) \;=\; \mathrm{Ext}^{m-n}_{\mathrm{coh}\,\mathbb{X}}(\mathcal{F}, \mathcal{G})$$

holds for all $\mathcal{F}, \mathcal{G} \in \mathrm{coh}\,\mathbb{X}$. The *Euler characteristic*

$$\langle \mathcal{F}, \mathcal{G} \rangle \;=\; \dim_k \mathrm{Hom}(\mathcal{F}, \mathcal{G}) - \dim_k \mathrm{Ext}^1(\mathcal{F}, \mathcal{G})$$

induces a *bilinear form* on the *Grothendieck group* $\mathrm{K}_0\,\mathbb{X}$ of $\mathrm{coh}\,\mathbb{X}$ (which is always finitely generated free).

On $\mathrm{coh}\,\mathbb{X}$ (and $\mathcal{D}^b\mathrm{coh}\,\mathbb{X}$), $\mathbb{L}(\underline{\mathbf{p}})$-shift by an arbitrary element of $\mathbb{L}(\underline{\mathbf{p}})$ yields an automorphism. In particular, shift by the dualizing element $\vec{\omega}$ serves for the *Auslander-Reiten-translation* on $\mathrm{coh}\,\mathbb{X}$, since the *Serre-Duality* formula

$$\mathrm{DExt}^1(\mathcal{F}, \mathcal{G}) \;\cong\; \mathrm{Hom}\left(\mathcal{G}, \mathcal{F}(\vec{\omega})\right)$$

holds for any $\mathcal{F}, \mathcal{G} \in \mathrm{coh}\,\mathbb{X}$ (here $D = \mathrm{Hom}_k(-, k)$ denotes the standard duality).

10.1.2 Canonical Algebras: tilting from Sheaves to Modules

Theorem 10.1. *(Cf. [2])* $T := \bigoplus_{\vec{0}\leq\vec{x}\leq\vec{c}}\mathcal{O}(\vec{x})$ *is a* tilting object *in* $\operatorname{coh}\mathbb{X}$ *with the endomorphism algebra* $\operatorname{End}(T)$ *naturally isomorphic to the* canonical algebra $\Lambda = \Lambda(\underline{\mathbf{p}}, \underline{\lambda})$, *which is given by the quiver*

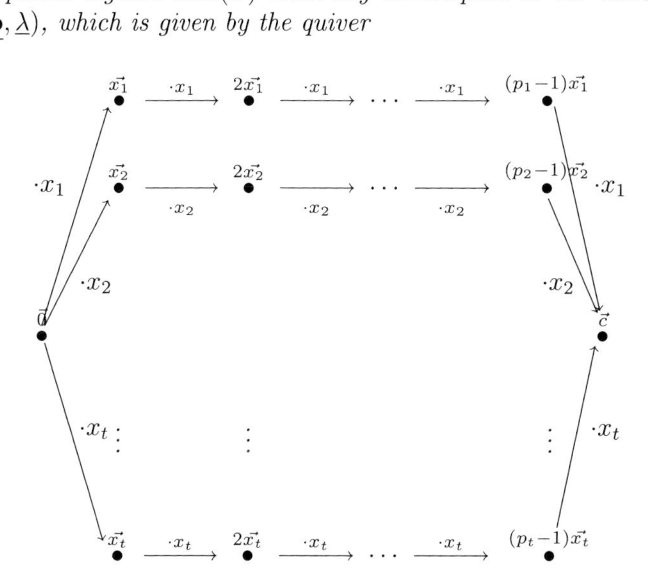

subject to relations $x_i^{p_i} \;=\; x_2^{p_2} - \lambda_i x_1^{p_1} \quad (i = 3, \dots, t)$. *(Cf. [5]).*

Therefore, there exists a *triangle equivalence* $\mathcal{D}^b\operatorname{coh}\mathbb{X} \cong \mathcal{D}^b \operatorname{mod}\Lambda$ inducing an isomorphism of the corresponding Grothendieck groups $\mathrm{K}_0\,\mathbb{X} \cong \mathrm{K}_0\,\Lambda$.

10.1.3 Torsion Sheaves and Vector Bundles Each *indecomposable* coherent sheaf is either a *torsion sheaf* or a *vector bundle*, therefore the splitting

$$\operatorname{coh}\mathbb{X} \;=\; \operatorname{vect}\mathbb{X} \;\vee\; \operatorname{coh}_0\mathbb{X}$$

holds. Here $\operatorname{coh}_0\mathbb{X}$ [resp. $\operatorname{vect}\mathbb{X}$] denotes the full subcategory of all *torsion sheaves* (= *finite length sheaves*) [resp. all *vector bundles*] in $\operatorname{coh}\mathbb{X}$. Moreover,

$$\operatorname{coh}_0\mathbb{X} \;\cong\; \coprod_{x\in\mathbb{X}} \operatorname{mod}_0^{\mathbb{L}(\underline{\mathbf{p}})}\mathcal{O}(\mathbb{X},x)\;;$$

for each $x \in \mathbb{X}$, $\operatorname{mod}_0^{\mathbb{L}(\underline{\mathbf{p}})}\mathcal{O}(\mathbb{X},x)$ is *uniserial* having $p(x)$ simples.

Let $\mathcal{S}_x$ denote the cokernel of

$$0 \to \;\mathcal{O}\; \overset{\cdot x}{\rightarrowtail}\; \mathcal{O}(\vec{c}) \;\twoheadrightarrow\; \mathcal{S}_x \;\to\; 0$$

for any *ordinary point* $x \in \mathbb{X}$, $\mathcal{S}_x$ is a *simple* sheaf (concentrated at x).

Within each *exceptional* tube of $\operatorname{coh}_0\mathbb{X}$, let us distinguish one of its simple sheaves; namely in the i-th exceptional tube (of torsion sheaves concentrated in the point $x_i \in \mathbb{X}$, and having p_i simple sheaves) we distinguish the simple sheaf $\mathcal{S}_{i,0}$ arising as cokernel of the short exact sequence

$$0 \to \mathcal{O}(\vec{c} - \vec{x_i}) \stackrel{\cdot x}{\rightarrowtail} \mathcal{O}(\vec{c}) \twoheadrightarrow \mathcal{S}_{i,0} \to 0$$

Setting $\mathcal{S}_{i,j} := \tau^{-j}\mathcal{S}_{i,0}$ for $j \in \mathbb{Z}_{p_i}$, we are able to encode each *indecomposable* torsion sheaf $\mathcal{F} \in \operatorname{coh}_0 \mathbb{X}$ by the following set of data:

$$\begin{array}{ll} [x, \mathrm{soc}, \mathrm{length}] & \text{if } x = x_1, \ldots, x_t \\ [x, \mathrm{length}] & \text{if } x \in \mathbb{X} \setminus \{x_1, \ldots, x_t\} \end{array} \qquad (10.1)$$

where the length is a natural number and soc $\in \mathbb{Z}_{p_i}$ for $x = x_i$.

The aim of this paper is to present an *encoding* of indecomposable coherent sheaves in coh $\mathbb{X}$ (for $\mathbb{X}$ of *tubular* type) by a *discrete* set of data, which is based on the approach explained above, but refers to some extra deeper knowledge on the category coh $\mathbb{X}$, namely to the homogeneouity of its *tubular structure.* We also describe in the paper how applying this encoding one can effectively determine the class of a given (by encoding data) coherent sheaf in the Grothendieck group $\mathrm{K}_0\,\mathbb{X}$. The presented material, in particular the proposed method, which is phrased in computer accessible data and easily convertible into a collection of the computer algebra routines, form a theoretical basis of the main part of the TUBULAR package (implemented in MapleV.r3 programming language). The package TUBULAR (the idea of Helmut Lenzing) was designed and created by the authors (the main programming work by the second author) for computer dealing with sheaves over weighted projective lines (the first version during the first author's visit at the Universität-Gesamthochschule Paderborn in summer 1993). It actually constitutes an ingredient of the CREP – the project at the Universität Bielefeld for computer algebra treatment of representation theory of finite dimensional algebras.

In Section 2 we demonstrate an *efficient method* of performing actual computations using this datatype. This will involve rather deep analysis of behaviour of telescoping functors (Cf. [6, 7, 3]) and controlling them on the level of Grothendieck group linear transformations.

10.2 Telescoping Functors

10.2.1 The "Sandglass"-Effect in $\mathrm{K}_0\,\mathbb{X}$ The Grothendieck group $\mathrm{K}_0\,\mathbb{X}$ of coh $\mathbb{X}$ [and $\mathcal{D}^b$coh $\mathbb{X}$] is a *free abelian group of finite rank* $n = \sum_{i=1}^{t} p_i - (t-2)$.

Note that by the very definition the simple sheaves $\mathcal{S}_x$ for all ordinary points $x \in \mathbb{X}$, are mapped to *one and the same* class in $\mathrm{K}_0\,\mathbb{X}$. Therefore we will simply denote this class by $[\mathcal{S}]$ in the sequel. Passing from $\operatorname{coh}_0 \mathbb{X}$ to the Grothendieck group $\mathrm{K}_0\,\mathbb{X}$ of coh $\mathbb{X}$ will produce a "sandglass"-effect explained on example of the weight sequence $\mathbf{p} = (2, 4, 4)$ by the following picture.

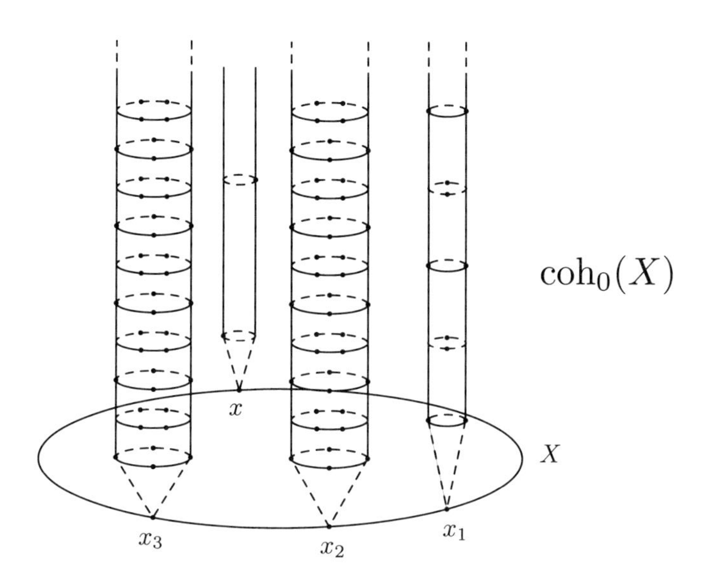

$\mathrm{coh}_0(X)$
x
X
x_3
x_2
x_1

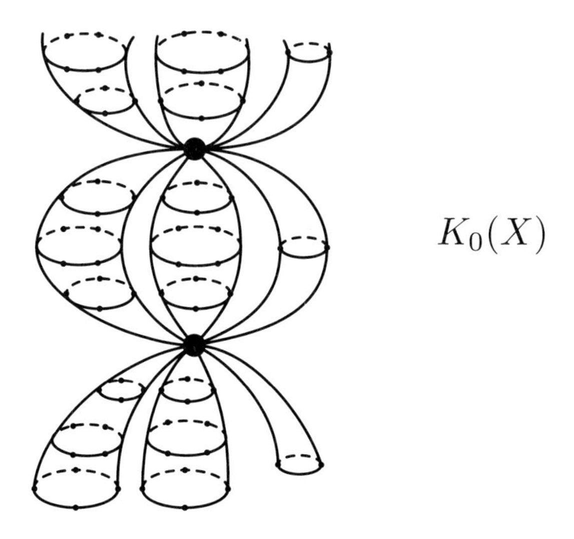

$K_0(X)$

$\mathrm{K}_0\,\mathbb{X}$ is endowed with a non-symmetric *non-degenerate bilinear form* induced by

$$\langle [\mathcal{F}],[\mathcal{G}]\rangle \;=\; \dim_k \mathrm{Hom}(\mathcal{F},\mathcal{G}) - \dim_k \mathrm{Ext}^1(\mathcal{F},\mathcal{G})\ .$$

The *radical* of the corresponding *quadratic form* $\mathrm{q}(x) = \langle x, x\rangle$ is a *direct factor* of $\mathrm{K}_0\,\mathbb{X}$, it has $\mathbb{Z}$-rank 2 for $\mathbb{X}$ being of *tubular* type, where it is generated by the classes

$$\underline{\mathbf{e}} \;=\; \sum_{j\in\mathbb{Z}_{p_t}} \tau^j \mathcal{S}_{t,0} \;=\; [\mathcal{S}] \qquad \text{and} \qquad \underline{\mathbf{u}} \;=\; \sum_{j\in\mathbb{Z}_{p_t}} [\tau^j \mathcal{O}]\ . \tag{10.2}$$

(Here p denotes the least common multiple of weights $(p_1,\ldots,p_t)$; the weights are assumed to be ordered in ascending fashion, then (for $\mathbb{X}$ tubular!) always $p = p_t$.) In the "sandglass"-figure above, the "nodes" are represented by the radical vectors $\mathbb{N}\cdot[\mathcal{S}]$.

The TUBULAR package deals with the following three $\mathbb{Z}$-bases of $\mathrm{K}_0\,\mathbb{X}$:

1. the basis "@$\mathcal{O}$" of indecomposable *projective* Λ-modules:

$$\{\ [\mathcal{O}(x)] \mid \vec{0} \le \vec{x} \le \vec{c}\ \}\ ;$$

2. the basis "@$\mathcal{E}$" of *simple* Λ-modules;

3. the basis "@$\mathcal{S}$":

$$\{\ [\mathcal{O}], [\mathcal{S}_{i,j}], [\mathcal{S}] \mid i = 1,\ldots,t;\ j = 1,\ldots,p_i - 1\}\ .$$

For computational purpose, we will mainly deal with the third basis "@$\mathcal{S}$" in this paper. (This basis is also chosen for all internal computation of the TUBULAR system.)

Furthermore, we consider the linear functions $\mathrm{rk}, \deg : \mathrm{K}_0\,\mathbb{X} \longrightarrow \mathbb{Z}$, and define the *slope* of an arbitrary sheaf $\mathcal{F}$ by

$$\mu : \mathrm{coh}\,\mathbb{X} \longrightarrow \mathbb{Q}\cup\{\infty\}; \qquad\qquad \mu(\mathcal{F}) \;:=\; \frac{\deg(\mathcal{F})}{\mathrm{rk}(\mathcal{F})} \tag{10.3}$$

(setting $\mu(\mathcal{F}) := \infty$ in case $\mathrm{rk}(\mathcal{F}) = 0$). Recall from [2] that $\mathcal{F} \in \mathrm{coh}\,\mathbb{X}$ is *torsion* (i.e. lies in $\mathrm{coh}_0\,\mathbb{X}$) if and only if $\mathrm{rk}(\mathcal{F}) = 0$.

10.2.2 Telescoping Functors and the Tubular Structure of $\mathrm{vect}\,\mathbb{X}$ For each $q \in \mathbb{Q}\cup\{\infty\}$, let

$$\mathcal{C}_q \;:=\; \mathrm{add}\{\ \mathcal{F} \in \mathrm{coh}\,\mathbb{X} \mid \mathcal{F}\ \text{ indecomposable};\ \ \mu(\mathcal{F}) = q\ \}\ .$$

(Cf. to (10.3); note that, by definition of the slope, $\mathcal{C}_\infty = \mathrm{coh}_0\,\mathbb{X}$.)

Theorem 10.2. (*Cf. [2]*) *If* $\mathbb{X}$ *is of* tubular *type, then each* $\mathcal{C}_q$ ($q \in \mathbb{Q}\cup\{\infty\}$) *is an abelian category enjoying the same tubular structure as* $\mathcal{C}_\infty = \mathrm{coh}_0\,\mathbb{X}$.

Note that this classification is similar to the classification of sheaves on elliptic curves [1], see [2] for a link between both theories.

The Theorem gives a chance to describe an arbitrary indecomposable sheaf (not only those lying in $\operatorname{coh}_0 \mathbb{X}$) by a set of data similar to (10.1). This can be accomplished invoking *telescoping functors* as defined by Lenzing and Meltzer in [3].

We recall from [3] and [4] the definition of special right mutations R and S; these are *automorphisms* defined on the level of the *derived category* $\mathcal{D}^b\operatorname{coh}\mathbb{X}$, where S is the right mutation with respect to $\bigoplus_{j\in\mathbb{Z}_p} \tau^j \mathcal{S}_{t,0}$, and R is the right mutation defined with respect to $\bigoplus_{j\in\mathbb{Z}_p} \tau^j \mathcal{O}$. Denoting by $\pi : \mathcal{D}^b\operatorname{coh}\mathbb{X} \longrightarrow \mathrm{K}_0\,\mathbb{X}$ the passage to the Grothendieck group, we have the following commutative diagram:

$$\begin{array}{ccc} \mathcal{D}^b\operatorname{coh}\mathbb{X} & \underset{S}{\overset{R}{\rightrightarrows}} & \mathcal{D}^b\operatorname{coh}\mathbb{X} \\ \downarrow{\scriptstyle\pi} & & \downarrow{\scriptstyle\pi} \\ \mathrm{K}_0\,\mathbb{X} & \underset{\sigma}{\overset{\rho}{\rightrightarrows}} & \mathrm{K}_0\,\mathbb{X} \end{array}$$

ρ, σ are the $\mathbb{Z}$-linear maps induced on $\mathrm{K}_0\,\mathbb{X}$ by R, S; since they preserve the Euler characteristic $\langle -,-\rangle$, they are really *automorphisms on* $\mathrm{K}_0\,\mathbb{X}$.

Remark 10.1. On the level of $\mathcal{D}^b\operatorname{coh}\mathbb{X}$, both right mutations R and S are defined in absolute "analogue" way. However, it turns out that the mutation S simply amounts to the $\mathbb{L}(\mathbf{p})$-shift by $\vec{x}_t$, whereas there is no such "easy" interpretation for the mutation R.

Restricted on $\operatorname{coh}\mathbb{X}$, R is no longer functorial. However, we have the following *isomorphisms* with respect to restriction to categories of the type $\mathcal{C}_q$ defined above:

$$\begin{array}{llll} S: \ \mathcal{C}_q \xrightarrow{\sim} \mathcal{C}_{q+1} & \text{for all} & q \in \mathbb{Q} \cup \{\infty\} \\ R: \ \mathcal{C}_q \xrightarrow{\sim} \mathcal{C}_{\frac{q}{1+q}} & \text{for all} & 0 \le q \le \infty\,. \end{array} \qquad (10.4)$$

The subsequent Theorem 10.3 claims that essentially all indecomposable sheaves can be reached from those lying in $\operatorname{coh}_0 \mathbb{X}$ in unique way (Cf. Proposition 10.1 for uniqueness) by suitable composition of the functors R and S. Such compositions will be called *telescoping functors*; their existence allows to convey the tubular structure of $\operatorname{coh}_0 \mathbb{X} = \mathcal{C}_\infty$ to any $\mathcal{C}_q$, thereby rendering an explicit proof of Theorem 10.2.

Theorem 10.3. (*Cf. [3]*) *For each $q \in \mathbb{Q}^+$ there exists a (unique) sequence of non-negative powers of R and S such that their composition*

$$\Phi_{\infty,q} := S^{a_1} \circ R^{b_1} \circ S^{a_2} \circ R^{b_2} \circ \cdots \circ S^{a_m} \circ R^{b_m} : \quad \mathcal{C}_\infty \xrightarrow{\sim} \mathcal{C}_q$$

yields an isomorphism $\Phi_{\infty,q} : \mathcal{C}_\infty \cong \mathcal{C}_q$.

Remark 10.2. The "restriction" $q \in \mathbb{Q}^+$ can easily be removed (applying S^{-l}, i.e. invoking suitable shifts by $-\vec{x}_t$), as we will see in the next section.

Let us note the following consequences of this Theorem, which describe mainly how telescoping functors may be exploited to "transport" structure from $\mathcal{C}_\infty$ to any $\mathcal{C}_q$:

1. Telescoping functors allow an explicit encoding of indecomposable sheaves within each tubular family $\mathcal{C}_q$ by conveying it from the encoding chosen for the family $\mathcal{C}_\infty$ (see (10.1)).

2. We have the "sandglass"-effect demonstrated for $\mathcal{C}_\infty$, for each $\mathcal{C}_q$. It should be noted at this place that for each $q \in \mathbb{Q}$ the "first" radical vector $\mathbf{w}_q$ (the lowest node in the sandglass-figure) may be computed as $\mathbf{w}_q = r\underline{\mathbf{u}} + d\underline{\mathbf{e}}$ (Cf. (10.2)) where $q = \frac{d}{r}$ with r, d coprime.

3. Passing to the Grothendieck group (and invoking the automorphisms ρ and σ induced by R and S) admits an effective method of "controlling" $\Phi_{\infty,q}$ by some linear automorphism $\phi_{\infty,q} : \mathrm{K}_0\,\mathbb{X} \to \mathrm{K}_0\,\mathbb{X}$.

4. Telescoping functors also yield an effective way of calculating roots and radical vector of the quadratic form q on $\mathrm{K}_0\,\mathbb{X}$.

We are now in the position to introduce a *discrete set of data* which allows to describe indecomposable sheaves in $\mathrm{coh}\,\mathbb{X}$ (comp. (10.1)). To "get rid" of the continuous parameter $x \in \mathbb{X}$ we have to restrict to indecomposable sheaves lying in *exceptional tubes.*

So, given such an indecomposable sheaf $\mathcal{F} \in \mathrm{coh}\,\mathbb{X}$, we encode it by

$$\mathrm{coh}\,\mathbb{X} \ni \mathcal{F} \qquad \equiv \qquad [q, i, s, l]\ , \tag{10.5}$$

where $q \in \mathbb{Q} \cup \{\infty\}$ is given by the *slope* $q = \mu(\mathcal{F})$, i is the *number* of the exceptional tube containing $\mathcal{F}$, s denotes the *socle index* of $\mathcal{F}$ referring to the quasi-filtration of $\mathcal{F}$ within this tube (recall that i and s both refer to an ordering of the exceptional tubes and of the mouth-objects within each exceptional tube, both orderings are *defined* by transfer of the corresponding ordering in $\mathcal{C}_\infty$ invoking telescoping functors), and finally $l \in \mathbb{N}$ amounts to the *quasi-length* of $\mathcal{F}$.

Let us now rephrase one of the central tasks of the TUBULAR system.

Question 10.1. *Given a set of data (10.5) encoding an indecomposable sheaf $\mathcal{F}$ as above, how can we* explicitly *determine its Grothendieck class $\pi(\mathcal{F})$?*

The subsequent diagram visualizes this problem:

$$\begin{array}{ccc}
\mathcal{C}_\infty \ni [\infty, i, s, l] & \xrightarrow{\Phi_{\infty,q}} & [q, i, s, l] \equiv \mathcal{F} \in \mathcal{C}_q \\
\pi\big\downarrow & & \pi\big\downarrow \\
\mathrm{K}_0\,\mathbb{X} \ni \pi([\infty, i, s, l]) & \xrightarrow{\phi_{\infty,q}} & ? \quad \in \mathrm{K}_0\,\mathbb{X}\ .
\end{array}$$

The next section will be devoted to the solution of this task; for developing an efficient routine handling this problem, essentially three kinds of computations will have to be performed:

1. Compute the automorphisms ρ and σ on $\mathrm{K}_0\,\mathbb{X}$.
2. Given $q \in \mathbb{Q}$, determine the decomposition of the linear automorphism $\phi_{\infty,q} : \mathrm{K}_0\,\mathbb{X} \to \mathrm{K}_0\,\mathbb{X}$ ("controlling" the telescoping functor $\Phi_{\infty,q} : \mathcal{C}_\infty \xrightarrow{\sim} \mathcal{C}_q$) into a product of powers of ρ and σ. (Cf. Theorem 10.3.)
3. Efficiently *compute* this product.

10.3 Computation of Telescoping Functors in $\mathrm{K}_0\,\mathbb{X}$

10.3.1 Computation of R and S Recall that the right mutation S with respect to $\bigoplus_{j\in\mathbb{Z}_p} \tau^j \mathcal{S}_{t,0}$ simply amounts to the $\mathbb{L}(\underline{\mathbf{p}})$-shift by the element $\vec{x_t} \in \mathbb{L}(\underline{\mathbf{p}})$;

$$S \;=\; - \otimes \mathcal{O}(\vec{x_t})$$

(as before, we assume the weight sequence $\underline{\mathbf{p}} = (p_1, \dots, p_t)$ ordered such that the corresponding weight $p = p_t$ becomes the largest weight). With respect to the basis "@$\mathcal{S}$" of $\mathrm{K}_0\,\mathbb{X}$ (chosen as "standard basis") the $\mathrm{K}_0\,\mathbb{X}$-automorphism σ induced by this functor becomes an $n \times n$-matrix (invertible over the integers) which for simplicity of notation we will also denote by S in the sequel. (Here $n = \sum_{i=1}^t p_i - (t-2)$ is the $\mathbb{Z}$-rank of $\mathrm{K}_0\,\mathbb{X}$.) Using the interpretation of S as a shift, and bearing in mind that $\mathcal{S}_{t,j}(\vec{x_t}) = \tau^- \mathcal{S}_{t,j} = \mathcal{S}_{t,j+1}$ for $j \in \mathbb{Z}_p$, the matrix S can easily be computed.

The right mutation R with respect to $\bigoplus_{j\in\mathbb{Z}_p} \tau^j \mathcal{O}$, which is an automorhphism on $\mathcal{D}^b\mathrm{coh}\,\mathbb{X}$, is no longer functorial when restricted to $\mathrm{coh}\,\mathbb{X}$. However, when restricted to $\mathcal{F} \in \mathcal{C}_q$ such that $q \in \mathbb{Q}^+ \cup \{\infty\}$, $R\mathcal{F}$ may be evaluated as the middle term of the *universal extension*

$$0 \longrightarrow \bigoplus_{j\in\mathbb{Z}_p} \mathrm{DExt}^1(\mathcal{F}, \tau^j\mathcal{O}) \otimes \tau^j\mathcal{O} \rightarrowtail R\mathcal{F} \twoheadrightarrow \mathcal{F} \longrightarrow 0\,.$$

Therefore, on $\mathrm{K}_0\,\mathbb{X}$ the automorphism ρ induced by R is given by

$$x \quad \mapsto \quad x \;-\; \sum_{j\in\mathbb{Z}_p} \langle x, \mathbf{u}_j\rangle \cdot \mathbf{u}_j \quad \text{for} \quad x \in \mathrm{K}_0\,\mathbb{X}\,,$$

where $\mathbf{u}_j := [\tau^j\mathcal{O}]$. Applying this formula to the standard basis "@$\mathcal{S}$" yields an $n \times n$ matrix which for simplicity we will also denote by R.

10.3.2 Computing the Decomposition of a Telescoping Functor Theorem 10.3 claims that given any $q \in \mathbb{Q}^+$ the *telescoping functor* $\Phi_{\infty,q}$ can uniquely be decomposed as a product $S^{a_1} \circ R^{b_1} \circ S^{a_2} \circ R^{b_2} \circ \cdots \circ S^{a_m} \circ R^{b_m}$ with non-negative integers $a_1, b_1, \dots, a_m, b_m$. To prove uniqueness, we invoke the subsequent Proposition from [3]; investigating the effect of the mutations R and S

on the *slope* of indecomposable sheaves. Namely, R and S induce the following *operations* on *slope*:

Proposition 10.1.

$$R \mapsto \begin{pmatrix} 1 & 0 \\ 1 & 1 \end{pmatrix}; \qquad S \mapsto \begin{pmatrix} 1 & 1 \\ 0 & 1 \end{pmatrix}$$

induces an isomorphism of semi-groups (with units)

$$\varphi : W(\{R,S\},\times) \quad \longrightarrow \quad (SL_2(\mathbb{N}_0),\cdot)\ .$$

(Here, $SL_2(\mathbb{N}_0)$ denotes the semi-group of all 2×2- matrices with non-negative integral entries and determinant 1.)

Moreover,

$$\begin{pmatrix} a & b \\ c & d \end{pmatrix}.q \quad := \quad \frac{aq+b}{cq+d}, \qquad \begin{pmatrix} a & b \\ c & d \end{pmatrix} \in SL_2(\mathbb{N}_0), \quad q \in \mathbb{Q}^+$$

defines an operation of $SL_2(\mathbb{N}_0)$ [and hence, via φ, of $W(\{R,S\},\times)$] on $\mathbb{Q}^+$ such that the composition

$$\varphi : W(\{R,S\},\times) \longrightarrow (SL_2(\mathbb{N}_0),\cdot) \longrightarrow \mathbb{Q}^+; \qquad w \mapsto \varphi(w) \mapsto \varphi(w).1$$

turns $\mathbb{Q}^+$ into a free semi-group *with basis*

$$\left\{ \begin{pmatrix} 1 & 0 \\ 1 & 1 \end{pmatrix}.1 = \frac{1}{2}, \ \begin{pmatrix} 1 & 1 \\ 0 & 1 \end{pmatrix}.1 = 2 \right\}.$$

The actual *construction* of this decomposition basically amounts to a *chain fraction decomposition* of $q \in \mathbb{Q}^+$:

Each (positive) $q \in \mathbb{Q}^+$ admits a decomposition

$$q = S^{a_1} R^{b_1} S^{a_2} R^{b_2} \cdot ....S^{a_m} R^{b_m}.1$$

Namely, if

$$q = x_1 + \frac{1}{x_2 + \frac{1}{\cdots x_n}}$$

then $a_i = x_{2i-1};\ \ b_i = x_{2i}$, for $1 \le i < \frac{n}{2}$ and

$$a_m = \begin{cases} x_{n-1} & \text{if } n \text{ is even} \\ x_n - 1 & \text{if } n \text{ is odd} \end{cases};$$

$$b_m = \begin{cases} x_n - 1 & \text{if } n \text{ is even} \\ 0 & \text{if } n \text{ is odd} \end{cases}.$$

Example 10.1. For

$$q \quad = \quad \frac{7}{4} \quad = \quad 1 + \frac{1}{1+\frac{1}{3}}$$

we obtain $\frac{7}{4} = SRS^2.1$. So in this case $a_1 = 1, b_1 = 1, a_2 = 2, b_2 = 0$.

Example 10.2. For

$$q \quad = \quad \frac{5}{13} \quad = \quad 0 + \cfrac{1}{2 + \cfrac{1}{1 + \cfrac{1}{1 + \frac{1}{2}}}}$$

we obtain $\frac{5}{13} = R^2SRS.1$; so here $a_1 = 0, b_1 = 2, a_2 = 1, b_2 = 1, a_3 = 1, b_3 = 0$.

Let us return to the general situation: If we let $q = \frac{d}{r}$, d, r coprime, then

$$\begin{pmatrix} 1 & 0 \\ 1 & 1 \end{pmatrix} . \frac{d}{r} = \frac{d}{d+r}, \qquad\qquad \begin{pmatrix} 1 & 1 \\ 0 & 1 \end{pmatrix} . \frac{d}{r} = \frac{d+r}{d},$$

so the effect of a *single* mutation R or S on the slope may be described by the figure

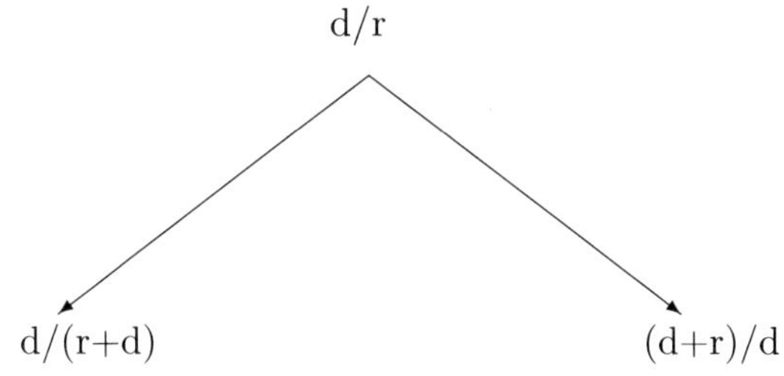

We can visualize the *complete situation* for all $q \in \mathbb{Q}^+$ in the following *binary tree* (where the dotted path corresponds to the Example 10.1 above):

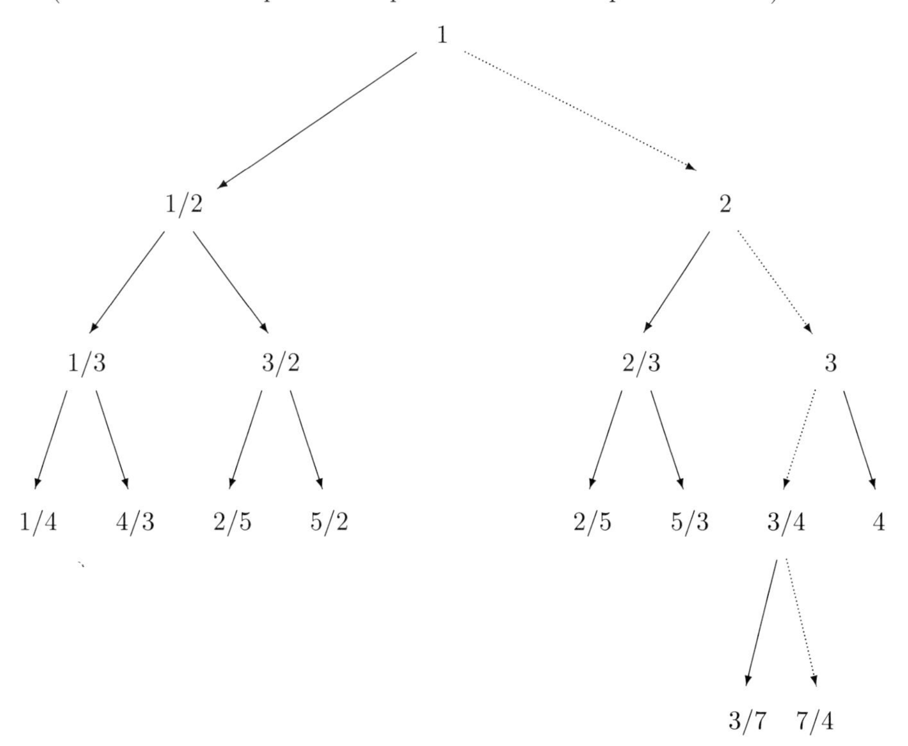

Observe that the root of this binary tree is 1. So, in order to encode $\Phi_{\infty,q} : \mathcal{C}_\infty \longrightarrow \mathcal{C}_q$, we have to invoke *one* application of $R : \mathcal{C}_\infty \longrightarrow \mathcal{C}_1$ ahead of the composition (note that applications are to be read from right to left!).

Finally, to deal with *negative* fractions $q < 0$, we let l be the minimal natural number such that $q + l \in \mathbb{Q}^+$, and apply S^{-l} first.

Therefore, given $q \in \mathbb{Q}$, the complete decomposition of the corresponding telescoping functor $\Phi_{\infty,q} : \mathcal{C}_\infty \longrightarrow \mathcal{C}_q$ reads (using the notations above):

$$\Phi_{\infty,q} \quad := \quad S^{a_1} \circ R^{b_1} \circ S^{a_2} \circ R^{b_2} \circ \cdots \circ S^{a_m} \circ R^{b_m} \circ R \circ S^{-l} . \quad (10.6)$$

10.3.3 Efficient Computation of Matrix Products As the result of the preceding section, given any $q \in \mathbb{Q}$, we may decompose the corresponding telescoping functor $\Phi_{\infty,q} : \mathcal{C}_\infty \longrightarrow \mathcal{C}_q$ into a product of R and S as in (10.6).

On the level of the Grothendieck group $\mathrm{K}_0\,\mathbb{X}$ of $\mathrm{coh}\,\mathbb{X}$, $\Phi_{\infty,q}$ is "controlled" by the linear automorphism $\phi_{\infty,q} : \mathrm{K}_0\,\mathbb{X} \to \mathrm{K}_0\,\mathbb{X}$. The matrix of $\phi_{\infty,q}$ with respect to the standard basis "@$\mathcal{S}$" of $\mathrm{K}_0\,\mathbb{X}$ now clearly turns (10.6) into a corresponding *product* of powers of the matrices R and S discussed before. The final task in actually producing this automorphism is to *compute* this product *efficiently*.

Look, for instance, at the following slope

$$q \quad = \quad \frac{2217945176158455 0335}{2120154320253601 43} ;$$

decomposition will yield the "word"

```
&*(S^104,R ,S ,R ,S ,R^2,S ,R ,S ,R ,S^2,R^81,S ,R ,S ,R^2,S ,R ,
   S ,R^2,S^2,R^3,S^2,R ,S ,R ,S^2,R^82,S ,R ,S ,R^2,S^2,R ,
      S ,R^2,S ,R ,S ,R^2,S^3,R^3,S^2,R ,S ,R ,S^83);
```

the corresponding product amounts to hundreds of multiplications of $n \times n$ integral matrices. Fortunately, we are able to reduce the amount of computation time to an extend hardly conceivable at first sight!

But before, let us first draw some "negative" conclusion from the Proposition 10.1:

Corollary 10.1. *Since*

$$R \mapsto \begin{pmatrix} 1 & 0 \\ 1 & 1 \end{pmatrix}; \qquad S \mapsto \begin{pmatrix} 1 & 1 \\ 0 & 1 \end{pmatrix}$$

generate $W(\{R,S\},\times)$ *as a* free semi-group, *there are* no *relations in* R, S *with* positive *integer exponents.*

There are however relations with *negative* exponents (Cf. [6, 3]).

Now we turn to efficient computation of (10.6), which mainly amounts to the efficient computation of *matrix powers*. Although there is a general possibility of reduction invoking the characteristic polynomial (Cf. Remark 10.4), the situation treated here admits a reduction which is incomparably better:

(A) Computation of powers S^m: As discussed before, the right mutation S amounts to the $\mathbb{L}(\underline{\mathbf{p}})$-shift by $\vec{x}_t$. Now, the shift by $p_t\vec{x}_t = \vec{c}$ is "canonical", this allows for

$$S^m \quad = \quad S^{p_t m'} \cdot S^r, \qquad m = p_t \cdot m' + r, \quad 0 \le r < p_t$$

so that when computing $S^1, \ldots, S^{p_t-1}$ in advance and storing them, computation of $S^{p_t m'} \cdot S^r$ amounts to *one single row-operation* (i.e. *independent* of however large m, m' will become!).

Hence, time complexity with regard to products S^m does *not* depend on m any more!

(B) Computation of powers R^m: We would like to achieve and indeed we do have the same advantage as in (A).

Proposition 10.2 (Conjugation formula). *The functors S^{-1} (induced by the shift by $-\vec{x}_t$) and R are* conjugated. *On the level of* $\mathrm{K}_0\,\mathbb{X}$ *the conjugation functor is "controlled" by the linear transformation $\alpha : \mathrm{K}_0\,\mathbb{X} \to \mathrm{K}_0\,\mathbb{X}$, given by the matrix $A = S^{-1}R$ (with respect to the basis "@S")." Therefore we obtain the equality*

$$R \quad = \quad A \cdot S^{-1} \cdot A^{-1} \;,$$

and as a consequence, we can reduce the evaluation of R^m to that of S^m, namely

$$R^m \quad = \quad A \cdot S^{-m} \cdot A^{-1} \qquad \forall \;\; m \in \mathbb{N} \;.$$

Hence, time complexity with regard to products R^m does *not* depend on m either.

Remark 10.3. The proof of this Proposition invokes sophisticated arguments on the level of functors (for instance, mutations commute with Auslander-Reiten translation, preserve the Euler characteristic, etc.) and should be impossible to obtain by merely looking "closely enough at matrices".

Indeed the conjugation formula reflects the fact that the functors R and S defined on the level of the derived category $\mathcal{D}^b\mathrm{coh}\,\mathbb{X}$ are conjugate themselves. (Cf. [3].)

Remark 10.4. Given a quadratic $n \times n$-matrix B, we can reduce the effort for computing powers B^m of B to that of $B^2, \ldots, B^{n-1}$ (the *characteristic polynomial* evaluated at B equals to 0 by Cayley-Hamilton Theorem).

Bibliography

[1] M. F. Atiyah: *Vector bundles over an elliptic curve*, Proc. London Math. Soc. **7**, (1957), 414–452.

[2] W. Geigle and H. Lenzing: *A class of weighted projective curves arising in representation theory of finite dimensional algebras*, Singularities, Representation of Algebras and Vector Bundles (Lambrecht 1985), Springer Lecture Notes Math. **1273**, (1987), 265–297.

[3] H. Lenzing and H. Meltzer: *Sheaves on a weighted projective line of genus one*, Representations of Algebras, ICRA VI, Ottawa 1992, CMS Conf. Proc. **14**, (1993), 313–337.

[4] H. Meltzer: *Tubular mutations*, Coll. Math. **74**, (1997), 267–274.

[5] C. M. Ringel: *Tame algebras and integral quadratic forms*, Springer Lecture Notes Math. **1099**, (1984).

[6] A. N. Rudakov: *The Markov numbers and exceptional bundles on* $\mathbb{P}^2$, Math. USSR-Izv. **32**, (1989), 99–112.

[7] ______: *Exceptional vector bundles on a quadric*, Math. USSR-Izv. **33**, (1989), 115–138.

Progress in Mathematics, Vol. 173, © 1999 Birkhäuser Verlag Basel/Switzerland

Chapter 11

Some Open Problems in the Theory of Kazhdan-Lusztig polynomials and Coxeter groups

Fokko du Cloux

11.1 Introduction

The central role played in the representation theory of reductive algebraic groups, semisimple Lie algebras and quantum groups by the Hecke algebra of the corresponding Weyl group, as the lieu where the basic combinatorial phenomena take place, has progressively become clear during the past two decades, chiefly through the works of George Lusztig. There has also been great progress in the purely combinatorial theory of Coxeter groups and Hecke algebras; we would like to cite for instance the papers of Kazhdan and Lusztig [17], Deodhar [5, 6, 7], Björner and Wachs [1], Dyer [9, 14], Brink and Howlett [3], and Brenti [2]. In particular, in [17] the celebrated *Kazhdan-Lusztig polynomials* were introduced, which are the "shadows" in the Hecke algebra of the perverse sheaves on the flag manifold; these polynomials will be the main theme in the present paper. They form a fascinating computational challenge, because of the difficulty of performing extended computations about them, because they and their subsequent generalizations provide the key to the understanding of the various representation theories involved, and because there are still many open questions about them, which seem hard to attack in the absence of more example material. For a review of the computational achievements in this area, the reader may consult [4].

We have tried to focus in this paper on those questions for which it seems most likely that they can be resolved computationally, either through a counterexample, or through the insight gained in the course of a computational study. The author would like to thank the referee for several useful remarks.

11.2 The Basic Definitions

11.2.1 Recall that a *Coxeter group* (or more correctly, a Coxeter system) is a group W defined by a set S of generators, where we shall assume that $n = |S|$

is finite and ≥ 1, and relations of the form:

$$(st)^{m_{s,t}} = \varepsilon$$

for some set of exponents $m_{s,t} \in \{1, 2, 3, \dots, \infty\}$ satisfying $m_{s,t} = m_{t,s}$ for all $s, t \in S$, and $m_{s,t} = 1$ if and only if $s = t$; here ε denotes the identity element. In particular, the relation $(ss) = \varepsilon$ for all $s \in S$ tells us that the generators are involutions; we have $m_{s,t} = 2$ if and only if s and t commute in W, and whenever some $m_{s,t}$ equals ∞, the corresponding relation is simply ommitted. The number n of generators of the presentation is called the *rank* of W.

For each $w \in W$, we define the *length* $l(w)$ of w to be the smallest non-negative integer p for which there exist $s_1, \dots, s_p \in S$ such that $w = s_1 \dots s_p$; the word $(s_1, \dots, s_p) \in S^*$ (where S^* is as usual the free monoid generated by S) is then called a *reduced expression* for w. It is easily seen that for all $w \in W$ and $s \in S$, one has $l(ws) = l(w) \pm 1$.

11.2.2 The basic combinatorial property of Coxeter groups, which characterizes them among groups generated by involutions, is the so-called *exchange condition*:

(E) Let $w \in W$ and $s \in S$ be such that $l(ws) = l(w) - 1$. Then for each reduced expression $(s_1, \dots, s_p)$ of w, there is an unique $j \in \{1, \dots, p\}$ such that

$$s_1 \dots s_p s = s_1 \dots \not{s}_j \dots s_p$$

In fact, even more is true. Let us denote by $T = \{wsw^{-1}\}_{w \in W, s \in S}$ the set of all conjugates of all elements of S; the elements of T are called the *reflections* in W. Then we have the following generalized exchange condition:

(GE) Let $w \in W$ and $t \in T$ be such that $l(wt) < l(w)$. Then for each reduced expression $(s_1, \dots, s_p)$ of w, there is an unique $j \in \{1, \dots, p\}$ such that

$$s_1 \dots s_p t = s_1 \dots \not{s}_j \dots s_p$$

Similar conditions hold of course on the left. For the proofs of these properties, as well as for those of the elementary properties of the Bruhat order and Hecke algebras in the sequel, we refer to [16].

11.2.3 We define the *Bruhat order* on W by setting, for each $w \in W$:

$$[\varepsilon, w] = \{s_{i_1} \dots s_{i_q}\}_{1 \leq i_1 < \dots < i_q \leq p}$$

where $(s_1, \dots, s_p)$ is any reduced expression for w. It may be shown that the set $[\varepsilon, w]$ is independent of the choice of the reduced expression. Then it is not hard to see that one may define a partial order relation on W by setting $v \leq w$ if and only if $v \in [\varepsilon, w]$.

It is also easy to see that for any $w \in W$ and $s \in S$, the condition $l(ws) < l(w)$ is equivalent to $ws < w$ in the Bruhat order, and similarly $l(ws) > l(w)$ is equivalent to $ws > w$.

11.3 Hecke algebras and Kazhdan-Lusztig polynomials

11.3.1 Let q be an indeterminate, and let $\mathbb{Z}[q^{1/2}, q^{-1/2}]$ be the ring of Laurent polynomials in $q^{1/2}$. Let $\mathcal{H}$ be the free $\mathbb{Z}[q^{1/2}, q^{-1/2}]$-module generated by W, and let $(e_w)_{w \in W}$ be the canonical basis of $\mathcal{H}$. Then it may be shown that there is a unique $\mathbb{Z}[q^{1/2}, q^{-1/2}]$-algebra structure on $\mathcal{H}$ such that:

$$e_w e_s = \begin{cases} e_{ws} & \text{if } ws > w \\ q e_{ws} + (q-1) e_w & \text{if } ws < w \end{cases}$$

In particular, for each $s \in S$ we have the relation

$$e_s^2 = q e_\varepsilon + (q-1) e_s$$

and since e_ε is the identity in $\mathcal{H}$, this implies that each e_s, and hence each e_w, $w \in W$, is invertible in $\mathcal{H}$.

It may be shown that the above quadratic relations, together with the so-called *braid relations*

$$e_s e_t e_s \ldots = e_t e_s e_t \ldots$$

($m_{s,t}$ factors on each side) for each $s \neq t$ such that $m_{s,t} < \infty$, constitute a presentation of the $\mathbb{Z}[q^{1/2}, q^{-1/2}]$-algebra $\mathcal{H}$. From this it follows that one may define an involution ι of $\mathcal{H}$ as a $\mathbb{Z}$-algebra by setting:

$$\begin{aligned} \iota(q) &= q^{-1} \\ \iota(e_s) &= e_s^{-1} \qquad s \in S \end{aligned} \tag{11.1}$$

Then we get the following fundamental theorem:

Theorem 11.1. *(Kazhdan and Lusztig [17]) There exists a unique basis $(c_w)_{w \in W}$ of $\mathcal{H}$ as a $\mathbb{Z}[q^{1/2}, q^{-1/2}]$-module, such that:*

(i) $\iota(c_w) = c_w$ for each $w \in W$

(ii) $c_w = q_w^{-1/2} \sum_{v \leq w} P_{v,w} e_v$

where we have put $q_w = q^{l(w)}$, and where the $P_{v,w}$ are polynomials in q with integer coefficients, satisfying $\deg(P_{v,w}) \leq (1/2)(l(w) - l(v) - 1)$ *if $v < w$, $P_{w,w} = 1$. (Note that the basis denoted here by (c_w) is the basis (C'_w) in the original paper [17].)*

Although we will refer to [17] or [16] for the full proof of this theorem, we would like to give an outline of the construction. For each $v \leq w$ in W define $\mu_{v,w} \in \mathbb{Z}$ to be the coefficient of $P_{v,w}$ of degree exactly equal to $(1/2)(l(w) - l(v) - 1)$; clearly $\mu_{v,w}$ can be nonzero only if the difference in lengths between v and w is *odd*. It is easy to see that necessarily

$$c_s = q^{-1/2}(1 + e_s)$$

for each $s \in S$. Then we define c_w by induction on $l(w)$ by the formula

$$c_w = c_{ws}c_s - \sum_{\substack{v<ws \\ vs<v}} \mu_{v,ws}c_v$$

where $s \in S$ is any generator such that $ws < s$. It is obvious that $\iota(c_w) = c_w$, and the correction term is exactly what is needed to ensure the degree condition on the coefficients of c_w. Then the uniqueness part of the theorem, which is a bit harder to prove, will ensure that this definition does not depend on the choice of s.

The polynomials $P_{v,w}$ appearing in the theorem are of course the famous *Kazhdan-Lusztig polynomials*. The first important open problem about them is the following:

Question 11.1. *Are the coefficients of $P_{v,w}$ always non-negative?*

This has been proven to be true in several important cases: all finite Coxeter groups, Weyl groups of Kač-Moody algebras, "free" Coxeter groups; it is likely that it will be true in general. A representation-theoretic interpretation of the coefficients of Kazhdan-Lusztig polynomials in general has recently been conjectured [13]; if proved, it would imply their non-negativity. However, on the combinatorial level this fact seems very mysterious. In the case of Weyl groups, it has been proven by identifying the coefficients of $P_{v,w}$ with the dimensions of suitable cohomology spaces. In the other cases, the proof is computational: for the non-crystallographic finite Coxeter groups, all Kazhdan-Lusztig polynomials have been computed by Alvis (unpublished) and independently by the author, verifying positivity in the process; for the free Coxeter groups (this is the case where all the $m_{s,t}$ are equal to ∞; W is then isomorphic to the free product of n copies of $\mathbb{Z}/2$), the Bruhat order is simple enough that the positivity could be checked by hand by Dyer [8].

11.4 Kazhdan-Lusztig polynomials and Bruhat intervals

11.4.1 We have defined in Section 11.2.2 the set T of reflections of W. If $w \in W$ and $(s_1, \dots, s_p)$ is a reduced expression for w, then the generalized exchange condition (GE) implies that the reflections t such that $l(wt) < l(w)$ are exactly those of the form

$$t = s_p s_{p-1} \dots s_{j+1} s_j s_{j+1} \dots s_{p-1} s_p$$

for $j = 1, \dots, p$ (and it is not hard to see that these reflections are all distinct). Therefore this set of reflections is independent of the choice of the reduced expression; we will denote it by $\nu(w)$.

Now let $v \leq w$ in W, and choose a maximal chain $v_0 = v < v_1 < \dots < v_r = w$ from v to w. The for each j in $\{1, \dots, r\}$, we may write $v_{j-1} = v_j t_j$, where $t_j = v_j^{-1} v_{j-1} = v_{j-1}^{-1} v_j$ is in T. We recall the two following theorems from Dyer:

Theorem 11.2. *(Dyer [9])*

(i) *Let $R \subset T$ be a (say finite) set of reflections, $r = |R|$. Then the subgroup $W' \subset W$ generated by R posesses a canonical set S' of Coxeter generators, with $S' \subset T$, $|S'| \leq r$.*

(ii) *Each left or right W'-coset in W contains a unique element of minimal length.*

In fact, the set S' above is defined as follows:

$$S' = \{t \in T \mid \nu(t) \cap W' = \{t\}\}$$

where $\nu(t)$ has been defined in Section 11.4.1. It is shown in [9] that if $r = 2$, then $|S'| = 2$ as well, and that in the general case, $R = S'$ if and only if for each pair $\{t, t'\}$ of distinct reflections in R, the pair $\{t_1, t_2\}$ of Coxeter generators for the group generated by t and t' is just $\{t, t'\}$. If this is not the case, then $l(t_1)+l(t_2) < l(t)+l(t')$ and we replace $\{t, t'\}$ by $\{t_1, t_2\}$; clearly the process has to stop after a finite number of steps. Finally, it may be shown that if t and t' are as above, then they generate a finite subgroup of W if and only if the order of the product tt' divides one of the $m_{s,t}$ in the definition of W; if this is the case, we enumerate this subgroup and find the two reflections of smallest length to get t_1 and t_2. Otherwise, t and t' generate an infinite dihedral subgroup; then one may show that if t and t' are not the canonical generators, one of the two reflections $tt't$ or $t'tt'$ will have smaller length; so again, repeating the process, we will find t_1 and t_2 in a finite number of steps. (in [9] a different description of the algorithm is given in terms of the roots of the canonical geometrical realization of W).

Theorem 11.3. *(Dyer [10]) Let $v \leq w$ and $t_1, \dots, t_r$ be as in Section 11.4.1, and let $W' \subset W$ be the subgroup generated by $t_1, \dots, t_r$. Let z be the element of minimal length in vW'.*

(i) *We have $[v, w] \subset zW'$, and $u \to z^{-1}u$ is an order-preserving isomorphism from $[v, w]$ to $[v', w'] \subset W'$, where $v' = z^{-1}v$, $w' = z^{-1}w$.*

(ii) *Under the above correspondance, we have equality of Kazhdan-Lusztig polynomials $P_{v,w} = P_{v',w'}$ (the latter being computed in W').*

Corollary 11.1. *Fix $r \geq 1$.*

(a) *There are only finitely many isomorphism types of Bruhat intervals of length $\leq r$ for finite Coxeter groups.*

(b) *There are only finitely many polynomials that can arise as Kazhdan-Lusztig polynomials for elements v, w in a finite Coxeter group such that $l(w) - l(v) = r$.*

Question 11.2. *(Dyer, Kazhdan-Lusztig) Does $P_{v,w}$ depend only on the isomorphism type of the interval $[v, w]$ as a poset?*

This question has bee answered in the affirmative by Dyer in his thesis for intervals of length ≤ 4, using the fact that a combinatorial description is available for the coefficient of degree 1 in $P_{v,w}$ in this case (note that when $l(w) - l(v) \leq 4$, $P_{v,w}$ is of degree at most one). But the most convincing result in this area is due to Brenti. Let us say that $[v, w]$ is *adihedral*, if there are no pairs $x < y$ in $[v, w]$, $l(y) - l(x) = 3$, such that the interval $[x, y]$ is isomorphic to the full poset of the Coxeter group of type A_2, shown in Figure 11.1.

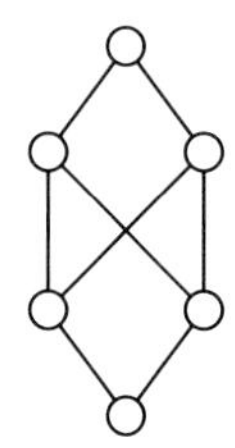

Figure 11.1:

Theorem 11.4. *(Brenti [2]) If $[v, w]$ is adihedral, then $P_{v,w}$ is equal to the Stanley g-polynomial of the dual poset (cf. [18, sect. 3.14]). In particular, $P_{v,w}$ depends only on the isomorphism type of $[v, w]$ as a poset.*

For example, as a corollary of the above theorem we obtain that whenever $[v, w]$ is a Boolean lattice, $P_{v,w}$ is equal to 1.

11.4.2 It would be important to gain a better understanding of the scope of the above result. Although there are certainly a number of isomorphism types of adihedral Bruhat intervals of length 4, say, it would appear that this condition becomes harder and harder to fulfill as the length of the interval grows, at least in the case of finite groups. Also it is easy to show that an interval of the form $[\varepsilon, w]$ is adihedral if and only if it is a Boolean lattice. Hence we would like to pose the following problem:

Question 11.3. *Classify all adihedral Bruhat intervals, say for finite Coxeter groups.*

11.4.3 Another question that arises naturally in this context is the following:

Question 11.4. *Are adihedral Bruhat intervals always lattices? Or even face lattices of convex polyhedra?*

This has been checked computationally by the author for intervals of length 4 in finite Weyl groups; in this case one always finds face lattices of 3-dimensional polyhedra. Admittedly more example material is needed before an opinion may be formed about the above problem.

The referee has pointed out that a more precise formulation of the above problem had been considered by him. Consider W as the reflection group associated to a (possibly infinite or non-crystallographic) root system Φ in a real vector space V. Write $u \to v$, and say that v covers u, if $u, v \in W$, $u \leq v$ and $l(v) = l(u)+1$. For $x \leq y$ in W, there is an associated real vector space $V_{[x,y]}$ constructed in [14]; $V_{[x,y]}$ is spanned by certain elements $\overline{\alpha}_{u,v}$ for $x \leq u \to v \leq y$, and there is a linear map θ from $V_{[x,y]}$ to V such that $\theta(\overline{\alpha}_{u,v}) = \beta_{u,v}$ where $\beta = \beta_{u,v}$ is the unique positive root with $v = s_\beta(u)$ (s_β denotes the reflection in β). Now by [12, prop. 1.4], the $\beta_{x,z}$ with $x \to z \leq y$ are generators of the extreme rays of a pointed polyhedral cone $E'_{x,y}$ in V, and it follows that the $\overline{\alpha}_{x,z}$ with $x \to z \leq y$ are generators of the extreme rays of a pointed polyhedral cone $E_{x,y}$ in $V_{[x,y]}$.

Question 11.5. *If $[x, y]$ is adihedral, is the Bruhat interval $[x, y]$ isomorphic to the face lattice of $E_{x,y}$?*

Note that one could ask the same question with $E'_{x,y}$ replacing $E_{x,y}$, but the modified question cannot have an affirmative answer if $l(y) - l(x) > \dim V$, and so is not very natural.

11.5 Some Problems in Type A

11.5.1 In the final section, we would like to pose some problems that are specific to type A (*i.e.* the case where W is a symmetric group), either because they are false in other types, or because their answer is most interesting in this case. For the first one, recall that as a corollary of a theorem of Dyer's we have remarked that there are only finitely many polynomials that can arise as $P_{v,w}$ when $l(w) - l(v)$ is bounded and v and w are constrained to belong to a finite Coxeter group (it is easy to see that both these constraints are necessary). However, if the constraint on the lengths is lifted it would appear that there are no restrictions apart from the "obvious" ones, and we would like to ask:

Question 11.6. *Is it true that any $P \in \mathbf{N}[q]$ with constant term equal to 1 may arise as a $P_{v,w}$ in type A?* [1]

Notice that if W, W' are two Coxeter groups, $v \leq w$ in W and $v' \leq w'$ in W', then $(v, v') \leq (w, w')$ in the product group $W \times W'$, and $P_{(v,v'),(w,w')} = P_{v,v'} P_{w,w'}$. Since $A_n \times A_{n'}$ is a parabolic subgroup inside $A_{n+n'+1}$, it follows that the set of Kazhdan-Lusztig polynomials for type A is closed under product.

Some simple cases can be treated explicitly. For instance, if we denote for simplicity the generators of S_{n+1} simply by the integers $1, \ldots, n$, we have the following series of polynomials, extending in an obvious fashion, which show that all polynomials of the form $1 + aq^m$, $a \geq 1$, $m \geq 1$, are obtained:

[1] This question has been very recently answered in the affirmative by Patric Polo [19].

$$
\begin{array}{lll}
P_{vw} = 1+q & v = 2 & w = 2132 \\
P_{vw} = 1+2q & v = 23 & w = 213243 \\
P_{vw} = 1+3q & v = 234 & w = 21324354 \\
P_{vw} = 1+4q & v = 2345 & w = 2132435465
\end{array}
$$

$$
\begin{array}{lll}
P_{vw} = 1+q^2 & v = 232 & w = 21321432 \\
P_{vw} = 1+2q^2 & v = 23243 & w = 21321432543 \\
P_{vw} = 1+3q^2 & v = 2324354 & w = 21321432543654 \\
P_{vw} = 1+4q^2 & v = 232435465 & w = 21321432543654765
\end{array}
$$

$$
\begin{array}{lll}
P_{vw} = 1+q^3 & v = 232432 & w = 2132143215432 \\
P_{vw} = 1+2q^3 & v = 232432543 & w = 21321432154326543 \\
P_{vw} = 1+3q^3 & v = 232432543654 & w = 213214321543265437654 \\
P_{vw} = 1+4q^3 & v = 232432543654765 & w = 2132143215432654376548765
\end{array}
$$

and so on. (It is very likely that the intervals above are of minimal length realizing the corresponding polynomials).

11.5.2 Recall that we have denoted by $\mu_{v,w}$ the coefficient of highest possible degree, *viz.* $(1/2)(l(w) - l(v) - 1)$. Then the result of Dyer's recalled in the previous paragraph implies that for each given degree, there are only finitely many polynomials that may arise as $P_{v,w}$ with $\mu_{v,w} \neq 0$, say in type A. So we pose the following problem:

Question 11.7. *Which are the polynomials that can arise as $P_{v,w}$ with $\mu_{v,w} \neq 0$, say in type A? As a first step, make sense of the list below in degree* 3.

Since from Dyer's theorem any P as above with degree m arises already in type A_{2m+1}, it is possible to list all the possibilities for the first values of m. Here are the first few cases:

$$
\begin{array}{lll}
\text{degree } 1: & 1+q & \\
\text{degree } 2: & 1+q^2 & \\
 & 1+q+q^2 & \\
 & 1+2q+q^2 & \\
 & 1+3q+q^2 & \\
 & 1+4q+q^2 & \\
\text{degree } 3: & 1+q^3 & 1+4q+5q^2+q^3 \\
 & 1+4q^2+q^3 & 1+5q+5q^2+q^3 \\
 & 1+q+q^2+q^3 & 1+5q+6q^2+q^3 \\
 & 1+q+2q^2+q^3 & 1+6q+6q^2+q^3 \\
 & 1+2q+q^2+q^3 & 1+6q+7q^2+q^3 \\
 & 1+2q+2q^2+q^3 & 1+6q+8q^2+q^3 \\
 & 1+3q+q^2+q^3 & 1+7q+7q^2+q^3 \\
 & 1+3q+2q^2+q^3 & 1+7q+9q^2+q^3 \\
 & 1+3q+3q^2+q^3 & 1+8q+9q^2+q^3 \\
 & 1+4q+3q^2+q^3 & 1+9q+9q^2+q^3 \\
 & 1+4q+4q^2+q^3 &
\end{array}
$$

11.5.3 Of course, the preceding problem is probably quite difficult. A very special aspect of it is already a longstanding open problem in type A, known as the "0, 1-comjecture" (see for instance [15] for an interesting connection):

Question 11.8. *Is it true that in type A the coefficients $\mu_{v,w}$ are always equal to 0 or 1?*

This has been checked computationally at least up to rank 8. However, it seems plausible to the author that new phenomena might occur even for ranks as high as 12 or 13. It should be noted that this property fails for all other series of finite Coxeter groups (except of course for dihedral groups, for which all Kazhdan-Lusztig polynomials are equal to 1): the first failures occur in types B_5, D_6, E_6, H_3.

Bibliography

[1] A. Björner, M. Wachs: *Bruhat order of Coxeter groups and shellability*, Advances in Math. **43**, (1982), pp. 87–100.

[2] F. Brenti: *A combinatorial formula for Kazhdan-Lusztig polynomials*, Invent. Math. **118**, (1994), pp. 371–394.

[3] B. Brink, R. Howlett: *A finiteness property and an automatic structure for Coxeter groups*, Math. Ann. **296**, (1993), no. 1, pp. 179–190.

[4] F. du Cloux: *The state of the art in the computation of Kazhdan-Lusztig polynomials*, AAECC **7**, (1996), no. 3

[5] V.V. Deodhar: *Some characterizations of Bruhat ordering on a Coxeter group and determination of the relative Möbius function*, Invent. Math. **39**, (1977), pp. 187–198.

[6] V.V. Deodhar: *On some geometric aspects of Bruhat ordering I: A finer decomposition of Bruhat cells*, Invent. Math. **79**, (1985), pp. 499–511.

[7] V.V. Deodhar: *On some geometric aspects of Bruhat ordering II: The parabolic analogue of Kazhdan-Lusztig polynomials*, J. Algebra **111**, (1987), pp. 483–506.

[8] M. Dyer: *On some generalizations of the Kazhdan-Lusztig polynomials for "universal" Coxeter groups*, J. Algebra **116**, (1988), pp. 353–371.

[9] M. Dyer: *Reflection subgroups of Coxeter systems*, J. Algebra **135**, (1990), pp. 57–73.

[10] M. Dyer: *On the "Bruhat graph" of a Coxeter system*, Comp. Math. **78**, (1991), pp. 185-191.

[11] M. Dyer: *The nil Hecke ring and Deodhar's conjecture on Bruhat intervals*, Invent. Math. **111**, (1993), pp. 571–574.

[12] M. Dyer: *Bruhat intervals, polyhedral cones and Kazhdan-Lusztig polynomials*, Math. Z. **215**, (1994), pp. 223–236.

[13] M. Dyer: *Representation theories from Coxeter groups*, in: Representations of groups (N. Allison and G. Cliff, eds.), CMS Conference Proceedings **16**, (1995), pp. 105–139.

[14] M. Dyer: *On coefficients of q in Kazhdan-Lusztig polynomials*,in: Algebraic groups and Lie groups (G. Lehrer, ed.), Cambridge University Press, Cambridge, 1997.

[15] A. Garsia, T. McLarnan: *Relations between Young's natural and Kazhdan-Lusztig representations of* S_n, Adv. in Math. **69**, (1988), no. 1, pp. 32–92.

[16] J.E. Humphreys: *Reflection groups and Coxeter groups*, Cambridge University Press, Cambridge, 1990

[17] D. Kazhdan, G. Lusztig: *Representations of Coxeter groups and Hecke algebras*, Invent. Math. **53**, (1979), pp. 165–184.

[18] R.P. Stanley: *Enumerative Combinatorics*, vol. 1, Cambridge University Press, Cambridge, 1997.

[19] P. Polo: *Construction de polynômes de Kazhdan-Lusztig*, submitted to CRAS.

Progress in Mathematics, Vol. 173, © 1999 Birkhäuser Verlag Basel/Switzerland

Chapter 12

Relative Trace Ideals and Cohen-Macaulay Quotients of Modular Invariant Rings

Peter Fleischmann

Abstract

Let G be a finite group, $\mathbb{F}$ a field whose characteristic p divides the order of G and A^G the invariant ring of a finite-dimensional $\mathbb{F}G$-module V. In analogy to modular representation theory we define for any subgroup $H \leq G$ the (relative) trace-ideal $A_H^G \lhd A^G$ to be the image of the relative trace map $t_H^G : A^H \to A^G$, $f \mapsto \sum_{g \in [G:H]} g(f)$. Moreover, for any family $\mathcal{X}$ of subgroups of G, we define the **relative trace ideals** $A_{\mathcal{X}}^G := \sum_{X \in \mathcal{X}} A_X^G \lhd A^G$ and study their behaviour.
If $\mathcal{X}$ consists of proper subgroups of a Sylow p-group $P \leq G$, then $A_{\mathcal{X}}^G$ is always a proper ideal of A^G; in fact, we show that its height is bounded above by the codimension of the fixed point space V^P. But we also prove that if V is relatively $\mathcal{X}$-projective, then $A_{\mathcal{X}}^G$ still contains all invariants of degree not divisible by p. If V is projective then this result applies in particular to the (absolute) trace ideal $A_{\{e\}}^G$.
We also give a 'geometric analysis' of trace ideals, in particular of the ideal $A_{<P}^G := \sum_{Q<P} A_Q^G o$, and show that $\mathcal{I}^{G,P} := \sqrt{A_{<P}^G}$ is a prime ideal which has the geometric interpretation as 'vanishing ideal' of G-orbits with length coprime to p. It is shown that $A^G/\mathcal{I}^{G,P}$ is always a Cohen-Macaulay algebra, if the action of P is defined over the prime field. This generalizes a well known result of Hochster and Eagon for the case $P = 1$ (see [13]). Moreover we prove that if V is a direct summand of a permutation module (i.e. a 'trivial source module'), then the $A_{\mathcal{X}}^G$ are radical ideals and $A_{<P}^G = \mathcal{I}^{G,P}$. Hence in this case the ideal and the corresponding Cohen-Macaulay quotient can be constructed using relative trace maps.

12.1 Introduction

Let R be an arbitrary commutative ring, V a free R-module of finite rank, and G a finite group, acting linearly on V. Then G acts on $V^* := Hom_R(V, R)$ by the rule $gf := f \circ g^{-1}$. This action can be extended to the symmetric algebra $A = R[V] := S(V^*)$, on which G now acts as graded R-algebra automorphisms. We define $A^G = \oplus_{i \geq 0} A_i^G$ to be the corresponding graded ring of invariants.

Up to now, the relation between module theoretic properties of the RG-modules V or V^* and the ring and ideal structure of A^G does not seem to be well understood. In this paper we investigate these connections by utilizing techniques of modular representation theory to construct certain ideals of A^G. For any subgroup $H \leq G$ we define the (relative) trace-ideal $A^G_H \lhd A^G$ to be the image of the relative trace map $t^G_H : A^H \to A^G$, $f \mapsto \sum_{g \in [G:H]} g(f)$ (see Section 12.3). In classical invariant theory, where R is a field of characteristic zero, these trace maps, in particular the absolute trace map t^G_1, have always been a major tool in constructing invariants. This of course is due to the fact that if the index $|G : H|$ is invertible in R, then t^G_H is surjective. If R is a field of characteristic $p \mid |G : H|$, this is no longer true, as A^G_H cannot contain the constants. Notice that the condition $|G : H| \in R^*$ implies that all V and V^* are relatively H-projective in the sense of modular representation theory. So a natural question is: How far away from being surjective are relative trace maps for subgroups H whose index in G is not a unit in R? We will investigate this question in a broader context, by defining trace ideals $A^G_{\mathcal{X}} := \sum_{H \in \mathcal{X}} A^G_H$, where $\mathcal{X}$ is an arbitrary family of subgroups of G. It turns out that, if $R = \mathbb{F}$ is a field of characteristic $p > 0$ and the elements of $\mathcal{X}$ intersect all Sylow p-subgroups of G properly, then the height of $A^G_{\mathcal{X}}$ is bounded above by the codimension of the fixed point space $\operatorname{codim}_{\mathbb{F}} V^P$ of a Sylow p - group P. Nevertheless we show, that if V (or V^*) is relatively H-projective (or $\mathcal{X}$-projective), then the ideal A^G_H (or $A^G_{\mathcal{X}}$ respectively) is still large enough, as to contain iA^G_i for all $i \in \mathbb{N}_0$. If again R is a field of characteristic p, this implies that t^G_H is surjective in all degrees coprime to p; in particular this applies to projective and regular representations of G. (In the case of regular representations, this result has been obtained independently in [17].)

In the fourth section we investigate in more detail the situation of permutation modules. Generalizing a result of Shank and Wehlau [22] on $A^G_{\{e\}}$, we prove that in this case the $A^G_{\mathcal{X}}$ are always radical ideals. In particular if G is a p-group, the ideal $A^G_< := \sum_{U<P} A^G_U$ is prime and its quotient is a polynomial ring in the 'orbit products' of G on V^* (for the case of cyclic p-groups see also [18] and [20]). For the regular representation of the cyclic group of order p over a field of characteristic p we show that A^G can be generated by elements of a degree less than that given by Goebel's bound for general invariant rings of permutation modules.

In the last section we assume $R = \mathbb{F}$ to be a field of characteristic $p > 0$ and utilize the geometric interpretation of A^G as a ring of regular functions on the orbit variety V/G. For a fixed Sylow p-subgroup P of G we will define the relative trace ideal $A^G_{<P}$ and the prime ideal $\mathcal{I}^{G,P} = \sqrt{A^G_{<P}}$ in A^G. The latter one is the 'vanishing ideal' of G-orbits with length coprime to p. Using a result of Wilkerson [23], it will be shown that the quotient $A^G/\mathcal{I}^{G,P}$ is a Cohen-Macaulay (CM) ring of Krull dimension $\dim_{\mathbb{F}} V^P$, whenever the action of P on V is defined over the prime field of $\mathbb{F}$. Here V^P denotes the subspace of P-fixed points in V. If char $\mathbb{F}$ does not divide $|G|$, i.e. if $P = 1$, then $\mathcal{I}^{G,P} = (0)$ and we recover the well known fact, proved by Eagon and Hochster ([13]), that in this case A^G is Cohen-Macaulay.

All these constructions are viewed as efforts to link the module structure of V to the ring and ideal structure of A^G. In particular we show that if V is a 'trivial source module' (i.e. a direct summand of a permutation module), then all $A^G_{\mathcal{X}}$ are radical ideals and $A^G_{<P} = \mathcal{I}^{G,P}$ is prime (with CM-quotient $A^G/A^G_{<P}$ of dimension $\dim_{\mathbb{F}}\ V^P$).

For all facts about modular representations the reader might refer to [1], [6] or [16].

Througout the paper the groups considered will assumed to be finite.

12.2 Basic Definitions

Definition 12.1. Let R be a commutative ring, A an R-algebra and B an A-bimodule. We let $\mathcal{D}_R(A, B)$ denote the R-module of R-linear derivations from A to B, i.e.

$$\mathcal{D}_R(A, B) := \{\partial \in \mathrm{Hom}_R(A, B) | \partial(aa') = a\partial(a') + \partial(a)a' \ \forall a, a' \in A\}.$$

With ${}_{RG}mod, mod_{RG}$ and ${}_{RG}mod_{RG}$ we denote the categories of RG-(bi)-modules V such that V is also a free R-module of finite rank.

If $Z(A)$ denotes the center of A, then $\mathcal{D}_R(A, B)$ is a $Z(A)$-module with $(a\partial)(f) := a(\partial f)$ for $a \in Z(A)$, $f \in A$ and $\partial \in \mathcal{D}_R(A, B)$. If $B = {}_AA_A$, the regular bimodule, then $\mathcal{D}_R(A) := \mathcal{D}_R(A, A)$ is a $Z(A)$-Lie-algebra. An ideal $I \lhd A$ will be called **derivation-stable** if $\partial(i) \in I$ for all $i \in I$ and $\partial \in \mathcal{D}_R(A)$. For a group G, RG denotes the corresponding group ring.

Let $V \in {}_{RG}mod$. We define

$$T(V) := T_R(V) := \oplus_{i=0}^{\infty} V^{\otimes i}$$

to be the graded tensor algebra over V. The R-linear action of G on V together with the universal property of tensor products induces a representation of G as automorphisms of the (non-commutative) graded ring $T(V)$. For any $v_1 \otimes \cdots \otimes v_n$ and $g \in G$ we have: $g(v_1 \otimes \cdots \otimes v_n) = gv_1 \otimes \cdots \otimes gv_n$

If $A = T(V)$, $d := \ rank_R V$ and B is an $A - A$ bimodule then the map

$$c : \mathrm{Hom}_R(V, B) \cong B^d \to \mathcal{D}_R(A, B)$$

$$\alpha \mapsto d\alpha : (v_{i_1} \otimes \cdots \otimes v_{i_k} \mapsto \sum_j v_{i_1} \otimes \cdots \alpha(v_{i_j}) \ldots \otimes v_{i_k})$$

is an isomorphism of R-modules. In fact the functors $\mathrm{Hom}_R(V, \ )$, $(\ )^d$ and $\mathcal{D}_R(A, \ )$ from A-bimodules to R-modules are all isomorphic and exact. If $B = T(V)$, then each $\alpha \in \mathrm{Hom}_R(V, V) \subseteq \mathrm{Hom}_R(V, B)$ has the unique extension $d(\alpha) \in \mathcal{D}_R(A)$, thus we have the canonical embedding $d : \mathcal{E} := \mathrm{End}_R(V) \hookrightarrow \mathcal{D}_R(A)$; in particular $d(Id_V)(a) = deg(a) \cdot a$ for homogeneous $a \in T(V)$. Let $\mathcal{I}$ be a derivation-stable homogeneous ideal of A and $c_{\mathcal{I}} : A \to A/\mathcal{I}$ the canonical epimorphism. Then the exactness of $\mathcal{D}_R(A, \ )$ implies that for any

$\partial \in \mathcal{D}_R(A, A/\mathcal{I})$ there is $\tilde{\partial} \in \mathcal{D}_R(A)$ such that $\partial = c_\mathcal{I} \circ \tilde{\partial}$. Hence $\partial(\mathcal{I}) = 0$ and $\mathcal{D}_R(A, A/\mathcal{I}) = \mathcal{D}_R(A/\mathcal{I}, A/\mathcal{I})$. Again we get isomorphisms

$$c : \mathrm{Hom}_R(V, A/\mathcal{I}) \to \mathcal{D}_R(A/\mathcal{I}, A/\mathcal{I}), \alpha \mapsto d\alpha.$$

If moreover $\mathcal{I} \cap V = 0$ then $V \subseteq A/\mathcal{I}$ and again each $\alpha \in \mathrm{End}_R(V)$ can be uniquely extended to an element $d\alpha \in \mathcal{D}_R(A/\mathcal{I})$ (i.e. satisfying $d\alpha(v) = \alpha(v)$ for $v \in V \subseteq (A/\mathcal{I})_1$) such that the element $\vartheta := d(Id_V)$ satisfies $\vartheta(f) = deg(f)f$ for each homogeneous $f \in A/\mathcal{I}$.

Now let $\mathcal{I}$ be the homogeneous ideal of $T(V)$, generated by the tensors $v \otimes w - w \otimes v$ for $v, w \in V$ and $S_R(V) = S(V)$ the symmetric algebra $S(V) := T(V)/\mathcal{I}$. Then

$$S(V) := \oplus_{i=0}^{\infty} S_i(V)$$

is also a graded R-algebra with $S_i(V) := < v_1 \otimes \cdots \otimes v_i + \mathcal{I} >_R$ and $v_i \in V$. We also define $\Lambda_R(V) = \Lambda(V)$ to be the exterior algebra $\Lambda(V) := T(V)/\mathcal{J}$ where $\mathcal{J}$ is the homogeneous ideal generated by the tensors $v \otimes v$ for $v \in V$ and denote $v_1 \wedge \cdots \wedge v_i := v_1 \otimes \cdots \otimes v_i + \mathcal{J}$. Notice that $v_1 \wedge v_2 = -v_2 \wedge v_1$.

Again

$$\Lambda(V) := \oplus_{i=0}^{\infty} \Lambda_i(V)$$

is a graded R-algebra with $\Lambda_i(V) := < v_1 \wedge \cdots \wedge v_i >_R$ and $v_i \in V$. In particular the ideals $\mathcal{I}$ and $\mathcal{J}$ are G-stable and hence we also have a natural action of G on $S(V)$ and $\Lambda(V)$ as automorphisms of graded R-algebras. Choosing an R-basis $\mathcal{B} := \{v_1, \ldots, v_d\}$ of V we can identify $S(V)$ with the polynomial ring $R[v_1, \ldots, v_d]$ in the indeterminates $v_1, \ldots, v_d$.

For $w_i := \sum_{j_i} w_{ij_i} v_{j_i}$ and $i = 1, \ldots, k$ we have

$$w_1 \wedge \cdots \wedge w_k = \sum_{j_1, \ldots, j_k} w_{1j_1} \cdot \cdots \cdot w_{kj_k} v_{j_1} \wedge \cdots \wedge v_{j_k}.$$

If $k > d$ then in each sum the v_{j_r} can not be pairwise different, so $w_1 \wedge \cdots \wedge w_k = 0$. We see that each $\Lambda_i(V)$ has an R basis

$$\{v_{j_1} \wedge \cdots \wedge v_{j_i} \mid j_1 < j_2 < \cdots < j_i, \ 1 \le j_k \le d\}$$

of cardinality $\binom{d}{i}$. In particular $\Lambda_i(V) = 0$ for $i > d$.

Let $v, w \in V$ then for any $\delta \in \mathcal{D}_R(A)$ we have $\delta(v \otimes w - w \otimes v) = \delta v \otimes w + v \otimes \delta w - \delta w \otimes v - w \otimes \delta v \in \mathcal{I}$ and $\delta(v \otimes v) = \delta v \otimes v + v \otimes \delta v = (\delta v + v) \otimes (\delta v + v) - \delta v \otimes \delta v - v \otimes v \in \mathcal{J}$. Hence $\mathcal{I}$ and $\mathcal{J}$ are derivation-stable.

If $A = \oplus_{i=0}^{\infty} S_i(V)$ then $\mathrm{Hom}_R(V, A) \cong A^d$ with $d = rank(V)$, hence $\mathcal{D}_R(A)$ is a free A-module of rank d. Now a straightforward induction, using the defining property of derivations (product rule), shows that c_I assigns to each vector $\alpha := (\alpha_1, \ldots, \alpha_n) \in A^n$ a unique derivation $d\alpha \in \mathcal{D}_R(A)$ given by the formula

$$d\alpha(f) = \sum_i \partial_{v_i}(f)\alpha_i \tag{12.1}$$

(Here ∂_{v_i} is a shorthand for the formal partial differential operator $\frac{\partial}{\partial_{v_i}}$.)

12.3 The Trace Map

We collect some further notation which will be used in the sequel: Let A be an arbitrary graded R-G-algebra i.e. a graded R-algebra on which G acts as graded algebra isomorphisms. For $W \in {}_{RG}mod$ and any subgroup $H \leq G$ we define $W^H := \{w \in W | hw = w \; \forall h \in H\}$ to be the R-submodule of H-fixed points in W. In particular A^H is the **algebra of H-invariants** in A. We have the following homomorphism of R-modules:

$$t_H^G : W^H \to W^G, \; w \mapsto \sum_{g \in [G:H]} gw,$$

where $[G : H]$ denotes an arbitrary cross section of (left) H-cosets in G such that $G = \uplus_{g \in [G:H]} gH$. This is called the **(relative) trace homomorphism**, its image will be denoted with W_H^G. If the index $|G : H|$ is invertible in R, then t_H^G is an epimorphism; if $W = A = R[V] = S(V^*)$ then the trace homomorphism preserves degrees. For any $a, a' \in A^G$ and $b \in A^H$ we have $t_H^G(aba') = at_H^G(b)a'$, showing that the image $A_H^G := t_H^G(A^H)$ is an ideal in A^G. If $A = R[V]$ then A_U^H is a homogeneous ideal. If $\mathcal{X}$ is a set of subgroups of H, then we define $W_{\mathcal{X}}^H := \sum_{U \in \mathcal{X}} W_U^H$ and $W_<^H := \sum_{U<H} W_U^H$. Notice that again $A_{\mathcal{X}}^H$ and $A_<^H$ are ideals in A^H and homogenous if $A = R[V]$.

Let $U, H \leq G$ be subgroups and suppose we are given a cross section $[U : G : H]$ of U, H double-cosets with $G = \uplus_{g \in [U:G:H]} UgH$. For each $g \in [U : G : H]$ the subset UgH is a disjoint union of H-cosets of the form ugH. Although UgH is not a subgroup in general, we can define $[UgH : H]$ to be a cross section in U of these H-cosets, i.e. $UgH = \uplus_{t \in [UgH:H]} tgH$. Then $[UgH : H]$ is also a cross section $[U : U \cap {}^gH]$ of $U \cap {}^gH$ cosets in U, i.e. $U = \uplus_{u \in [UgH:H]} u(U \cap {}^gH)$. Here, as in the following ${}^gH$ denotes gHg^{-1}. For each $a \in A^U$ and $b \in A^H$ we have the following 'Mackey formulae':

$$t_H^G(b) = \sum_{g \in [U:G:H]} t_{U \cap {}^gH}^U(gb) \in A_{\mathcal{X}}^U \tag{12.2}$$

$$t_U^G(a) \cdot t_H^G(b) = \sum_{g \in [U:G:H]} t_{U \cap {}^gH}^G(agb) \in A_{\mathcal{X}}^G \tag{12.3}$$

where $\mathcal{X} := \{U \cap {}^xH \mid x \in G\}$ (e.g. see [3], [16]). The group G acts on $\mathcal{E}(V) := \mathrm{End}_R(V)$ by the rule $\alpha^g(v) = g^{-1}\alpha g(v)$ for $\alpha \in \mathcal{E}(V)$, $g \in G$ and $v \in V$. Notice that $\mathcal{E}(V)^G = \mathrm{End}_{RG}(V)$.

Now fix a set $\mathcal{X}$ of subgroups of G. In representation theory the module $V \in RG - mod$ is called **relatively $\mathcal{X}$-projective** if and only if $\mathcal{E}(V)^G = \mathcal{E}(V)_{\mathcal{X}}^G$ or in other words, if the identity Id_V is contained in $\mathcal{E}(V)_{\mathcal{X}}^G$. Notice that V is relatively $\mathcal{X}$-projective if and only if so is V^*. If $\mathcal{X} = \{H\}$ then the $\mathcal{X}$-projective module V is called relatively H-projective (e.g. if the index $[G : H]$ is invertible in R). It is easy to see that $t_H^G(a) = t_{{}^gH}^G({}^ga)$ for any $g \in G$ and $a \in A^H$; hence $A_H^G = A_{{}^gH}^G$ and a module V is relatively H projective iff it is so for any

G-conjugate of H. Notice that V is projective in ${}_{RG}mod$ if and only if it is relatively 1-projective, i.e. with respect to the trivial subgroup ([2] 3.6.4).

Here we propose to extend the above machinery to graded R-G-algebras, where the endomorphism ring $\mathcal{E}(V)$ is replaced by the R-Lie-algebra of derivations and the identity element by the element $\vartheta = d(Id)$. So let A be an R-G-algebra and $U \le H \le G$ be subgroups. It is clear that G acts on $\mathcal{D}_R(A)$ by conjugation. Again we can define $\mathcal{D}_R(A)^H$ to be the Lie-subalgebra of H-fixed points and $\mathcal{D}_R(A)^H_U$ to be the image of the trace homomorphism

$$t_U^H : \mathcal{D}_R(A)^U \to \mathcal{D}_R(A)^H, \ \delta \mapsto \sum_{h \in [H:U]} h\delta h^{-1}.$$

As above we define: $\mathcal{D}_R(A)^H_< := \sum_{Y<H} \mathcal{D}_R(A)^H_Y$. Now $\mathcal{D}_R(A)^H_U$ and $\mathcal{D}_R(A)^H_<$ are **Lie-algebra ideals** in $\mathcal{D}_R(A)^H$. For any two subsets $D \subseteq \mathcal{D}_R(A)$ and $X \subseteq A$ we define $D \cdot X := \{\partial(b) \mid b \in X \text{ and } \partial \text{ in } D\}$. If $\delta = \sum_{g\in[G:H]} g\partial g^{-1}$ with $\partial \in \mathcal{D}_R(A)^H$, then we have for any $a \in A^H$, $h\partial(a) = \partial(ha) = \partial(a) \in A^H$. If moreover $a \in A^G \le A^H$, then

$$\delta(a) = \sum_{g\in[G:H]} g\partial g^{-1}(a) = \sum_{g\in[G:H]} g\partial(a) \in A^G_H$$

hence $\mathcal{D}_R(A)^G_H \cdot A^G \subseteq A^G_H$. From this we immediately get $\mathcal{D}_R(A)^G_{\mathcal{X}} \cdot A^G \subseteq A^G_{\mathcal{X}}$. We are now ready to prove:

Proposition 12.1. *Let G be a group, $\mathcal{X}$ a set of subgroups and $V \in {}_{RG}mod$. For a homogeneous derivation-stable and G-stable ideal $\mathcal{I}$ of $T(V)$ with $V \cap \mathcal{I} = 0$ we set $A := T(V)/\mathcal{I}$. Then the following are equivalent:*

(i): *V is relatively $\mathcal{X}$-projective;*

(ii): *$\vartheta = d(Id_V) \in \mathcal{D}_R(A)^G_{\mathcal{X}}$;*

(iii): *Each homogeneous component A_i with i invertible in R is relatively $\mathcal{X}$-projective.*

If one of (i),(ii) *or* (iii) *holds, then we have the inclusions:*

$$\oplus_{i\in\mathbb{N}} i \cdot (A_i)^G \subseteq \mathcal{D}_R(A)^G_{\mathcal{X}} \cdot A^G \subseteq A^G_{\mathcal{X}}.$$

In particular if $R = \mathbb{F}$ is a field of characteristic $p > 0$ and V is relatively $\mathcal{X}$-projective, then $A^G_{\mathcal{X}}$ contains all invariants of degree coprime to p.

Proof. The injection $d : \mathcal{E}(V) \to \mathcal{D}_R(A)$, $\alpha \mapsto d(\alpha)$ is an RG-module homomorphism, so $t^G_H(d(\alpha)) = d(t^G_H(\alpha))$. Moreover $Id_V \in \mathcal{E}(V)^G_{\mathcal{X}}$ implies that $Id_V = \sum_{H\in\mathcal{X}} t^G_H(\alpha_H)$ for some $\alpha_H \in \mathcal{E}(V)^H$, hence $\vartheta = d(Id_V) = \sum_{H\in\mathcal{X}} t^G_H(d(\alpha_H)) \in \mathcal{D}_R(A)^G_{\mathcal{X}}$. This proves '(i) $\Rightarrow$ (ii)'.

(ii) $\Rightarrow$ (iii): Suppose that $i \in R^*$ and $\vartheta \in \mathcal{D}_R(A)^G_{\mathcal{X}}$. Let π_i denote the projection of A onto A_i; then $\pi_i \in \mathrm{Hom}_{RG}(A, A_i)$ and for any $\partial \in \mathcal{D}_R(A)^H$

we have $\pi_i \circ \partial_{|A_i} \in \mathcal{E}(A_i)^H$. By assumption $\vartheta = d(Id_V) = \sum_{H\in\mathcal{X}} t_H^G(\partial_H)$ with $\partial_H \in \mathcal{D}_R(A)^H$. So we get

$$Id_{|A_i} = \pi_i(i^{-1}\vartheta_{|A_i}) = \sum_{H\in\mathcal{X}} \pi_i(t_H^G(i^{-1}\partial_H)) = \sum_{H\in\mathcal{X}} t_H^G(i^{-1}\pi_i \circ \partial_H) \in \mathcal{E}(A_i)_{\mathcal{X}}^G.$$

Hence A_i is relatively $\mathcal{X}$-projective. The implication (iii) $\Rightarrow$ (i) is obvious, since $V \cong A_1$. If $\vartheta \in \mathcal{D}_R(A)_{\mathcal{X}}^G$ then for any $a \in (A_i)^G$ we get:

$$ia = \vartheta a \in \mathcal{D}_R(A)_{\mathcal{X}}^G \cdot A^G \subseteq A_{\mathcal{X}}^G.$$

□

Corollary 12.1. *If R is a field of characteristic p, dividing the group order $|G|$, V a projective module (e.g. the regular representation of G) and $A = R[V] = S(V^*)$, then the usual (absolute) trace map*

$$t_{<1>}^G : A \to A^G, \ a \mapsto \sum_{g\in G} g(a)$$

is surjective in all degrees $i \neq 0 \mod p$.

Remark 12.1. For the case of the regular representation of G the result above also appears in [17].

Corollary 12.2. *Let R be a commutative ring and $V \in {}_{RG}mod$ relatively H-projective for the subgroup $H \leq G$. Then for any $i \in \mathbb{N}$ which is invertible in R, the homogeneous components $S_i(V^*)$ and $\Lambda_i(V^*)$ of the symmetric or exterior tensors $S(V^*)$ respectively $\Lambda_i(V^*)$ are also relatively H-projective.*

Remark 12.2. Suppose that R a field of characteristic p which divides $|G|$ and $A = R[V]$ with $V \in {}_{RG}mod$.

a) Relative projectivity of V with respect to H will in general not imply relative projectivity of $S_{p\cdot i}(V^*)$ or $\Lambda_{p\cdot i}(V^*)$. If R is a field of characterisitic $p \mid |G|$ then each projective RG-module W has dimension divisible by p (because its restriction $W_{|P}$ to a Sylow p-subgroup P is a free RP-module; see [1]). If V has dimension $d = p^k\ell$ with $\gcd(\ell, p) = 1$, then $S_{p^k}(V^*)$ has dimension $\binom{d+p^k-1}{p^k}$ which is $\neq 0 \mod p$; hence $S_{p^k}(V^*)$ is not be projective. Similarly $\Lambda_{p^k}(V^*)$ has dimension $\binom{d}{p^k} \neq 0 \mod p$ and $\Lambda(V^*)_{p^k}$ is not projective either.

b) If P is a Sylow p-subgroup of G, then V is relatively P-projective. If moreover V is indecomposable, then there always exists a subgroup $Q \leq P$, unique up to conjugacy in G, which is minimal such that V is relatively Q-projective. In modular representation theory, such a subgroup is called a **vertex** of V (e.g. see [1] pg. 67 ff). In this language, 12.1 says that the relative trace ideal A_Q^G of a vertex Q of V contains all invariants of degree coprime to p. This also holds for $A_{\mathcal{X}}^G$, whenever $\mathcal{X}$ contains a vertex of V.

c) Due to a theorem of D.Higman ([2] 3.6.4) V is relatively H-projective if and only if V is a direct summand of an induced module $W^{\uparrow G}$, where $W \in {}_{RH}mod$ (e.g. $W = V_{|H}$). In this case A_H^G contains all invariants of degree coprime to p.

12.4 Invariant Rings of Permutation Modules

Now let G be a finite group, Ω a finite G-set and R^Ω the set of functions from Ω to R. We consider the permutation module $V_\Omega := R^\Omega \cong \oplus_{\omega\in\Omega} R \cdot e_\omega$, where $e_\omega(\omega') = 1$ if and only if $\omega = \omega'$ and zero otherwise. This module is selfdual, so $V_\Omega^* \cong V_\Omega$ and we can set $A = A(R) := S_R(V_\Omega) = \oplus_{f\in\mathbb{N}_0^\Omega} R\pi(f)$, where for each $f \in \mathbb{N}_0^\Omega$ we define $\pi(f) := \prod_{\omega\in\Omega} e_\omega^{f(\omega)}$. Notice that for $\alpha, \beta \in \mathbb{N}_0^\Omega$ we have $\pi(\alpha)\cdot\pi(\beta) = \pi(\alpha+\beta)$ where $(\alpha+\beta)(\omega) = \alpha(\omega)+\beta(\omega)$, hence we can view $S(V_\Omega)$ as the semigroup-algebra $R[\mathbb{N}_0^\Omega]$.

The group Σ_Ω and its subgroup G act on A by permuting the monomials $\pi(f)$ according to the rule:

$$\sigma(\pi(f)) = \prod_{\omega\in\Omega} e_{\sigma(\omega)}^{f(\omega)} = \prod_{\omega\in\Omega} e_\omega^{f\circ\sigma^{-1}(\omega)} = \pi(f\circ\sigma^{-1}).$$

Hence the RG-module $S(V_\Omega)$ is isomorphic to the permutation module $V_{\mathbb{N}_0^\Omega}$ and in each degree i the module $S(V_\Omega)_i$ is isomorphic to the permutation module $V_{(\mathbb{N}_0^\Omega)_i}$, where $(\mathbb{N}_0^\Omega)_i := \{f \in \mathbb{N}_0^\Omega \mid \sum_\omega f(\omega) = i\}$ with the natural action of Σ_Ω and G. In particular the R-module A^G has a basis consisting of the orbit sums $\pi(f)^+ := \sum_{g\in[G:G_f]} g\pi(f) = t_{G_f}^G(\pi(f))$ for $f \in \mathbb{N}_0^\Omega$.

Now the Mackey formula of Section 12.3 describes the multiplication of orbit sums:

Lemma 12.1 ([9]). *For $\alpha,\ \beta \in \mathbb{N}_0^\Omega$ we have in $A(R)^G$:*

$$\pi(\alpha)^+ \cdot \pi(\beta)^+ = \sum_{g\in[G_\alpha:G:G_\beta]} \frac{|G_{\alpha+g(\beta)}|}{|G_\alpha \cap {}^gG_\beta|}\pi(\alpha+g\beta)^+.$$

In particular if R has characteristic $p > 0$ and G is a p-group we get

$$\pi(\alpha)^+ \cdot \pi(\beta)^+ = \sum_{g\in[G_\alpha:G:G_\beta],\ G_{\alpha+g(\beta)}=G_\alpha\cap\, {}^gG_\beta} \pi(\alpha+g\beta)^+.$$

The following is an extended version of a result of [10], where it has been used to obtain degree bounds for the generators of modular invariant rings of trivial source modules:

Theorem 12.1. *Let $\mathcal{X}$ be a set of subgroups of G, $A_\mathcal{X} := \sum_{H\in\mathcal{X}} t_H^G(A^H) \triangleleft A^G$ and $A_<^G := A_{\{H\ \mid\ H<G\}}^G$. Then:*

(i): $A_\mathcal{X}^G = \oplus_{\alpha\in\mathbb{N}_0^\Omega} Rg_\mathcal{X}(\alpha)\cdot\pi(\alpha)^+$, *where* $g_\mathcal{X}(\alpha) :=\ gcd(\frac{|G_\alpha|}{|H_\alpha|} \mid H \in \mathcal{X})\cdot 1_R$.

(ii): *For $T \subseteq \Omega$ let $\chi_T : \Omega \to \mathbb{N}_0$ be the characteristic function of T (i.e. 1 on T and 0 on $\Omega \backslash T$). Then $S := \langle \pi(\alpha) \mid \alpha \in \mathbb{N}_0^\Omega,\ G_\alpha = G \rangle_R \cong R[\pi(\chi_{\omega^G}) \mid \omega^G \in \Omega/G]$ with algebraically independent elements $\pi(\chi_{\omega^G})$. In other words, S is a polynomial subalgebra of A^G in $|\Omega/G|$ many variables $\pi(\chi_{\omega^G})$.*

(iii): *$A^G/A^G_< \cong S/(A^G_< \cap S)$ with $A^G_< \cap S = gcd(\frac{|G|}{|H|} \mid H < G) \cdot S$.*

Proof.

(i): Let $H \in \mathcal{X}$ and $a \in A^H$, then $a = \sum_\alpha c_\alpha t^H_{H_\alpha}(\pi(\alpha))$. Hence $t^G_H(a) = \sum_\alpha c_\alpha t^G_{H_\alpha}(\pi(\alpha)) = \sum_\alpha c_\alpha t^G_{G_\alpha}(t^{G_\alpha}_{H_\alpha}(\pi(\alpha))) = \sum_\alpha c_\alpha \frac{|G_\alpha|}{|H_\alpha|} t^G_{G_\alpha}(\pi(\alpha))$. This is contained in the set on the right hand side of (i). Let $g_{\mathcal{X}} = \sum_{H \in \mathcal{X}} \frac{|G_\alpha|}{|H_\alpha|} c_H$ with $c_H \in \mathbb{Z}$. Then

$$g_{\mathcal{X}} \pi(\alpha)^+ = \sum_{H \in \mathcal{X}} \frac{|G_\alpha|}{|H_\alpha|} c_H t^G_{G_\alpha}(\pi(\alpha)) = \sum_{H \in \mathcal{X}} c_H t^G_{G_\alpha} t^{G_\alpha}_{H_\alpha}(\pi(\alpha))$$
$$= \sum_{H \in \mathcal{X}} c_H t^G_H t^H_{H_\alpha}(\pi(\alpha)) \in A^G_{\mathcal{X}}.$$

(ii): If $\alpha \in \mathbb{N}_0^\Omega$ with $G_\alpha = G$, then α is constant on G-orbits in Ω. In this case α induces a function $\hat{\alpha} \in \mathbb{N}_0^{\Omega/G}$, $\omega^G \mapsto \alpha(\omega)$ and we get

$$\pi(\alpha) = \prod_{\omega^G \in \Omega/G} (\pi(\chi_{\omega^G})^{\hat{\alpha}(\omega^G)}.$$

Since different G-orbits are disjoint, the set $\{\pi(\chi_{\omega^G}) \mid \omega \in \Omega/G\}$ is algebraically independent.

(iii): Obviously $A^G = A^G_< + S$ with

$$A^G_< \cap S = \oplus_{\alpha \in \mathbb{N}_0^\Omega,\ G_\alpha = G} R g_<(\alpha) \cdot \pi(\alpha)^+ = gcd(\frac{|G|}{|H|} \mid H < G) \cdot S.$$

□

The next result shows, that in case of permutation modules over fields of positive characteristic, the $A^G_{\mathcal{X}}$ are always radical ideals. This generalizes Theorem 6.1 of [22]:

Proposition 12.2. *Let $\mathbb{F}$ be a field of characteristic $p > 0$, $G \leq \Sigma_\Omega$ a permutation group, $\mathcal{X}$ a set of subgroups of G and $S_{\mathcal{X}} := \oplus_{\alpha \in \mathbb{N}_0^\Omega,\ p \mid |G_\alpha : H_\alpha|,\ \forall H \in \mathcal{X}} \mathbb{F}\pi(\alpha)^+$. Then $A^G = A^G_{\mathcal{X}} \oplus S_{\mathcal{X}}$ and*

$$\sqrt{A^G_{\mathcal{X}}} = A^G_{\mathcal{X}} = \oplus_{\alpha \in \mathbb{N}_0^\Omega,\ p \text{ does not divide } |G_\alpha : H_\alpha|, \text{ for some } H \in \mathcal{X}} \mathbb{F}\pi(\alpha)^+.$$

Proof. Obviously $g_{\mathcal{X}}(\alpha) \neq 0$ if and only if p does not divide $|G_\alpha : H_\alpha|$ for some $H \in \mathcal{X}$ so we get $A^G = A^G_{\mathcal{X}} \oplus S_{\mathcal{X}}$ and the second equation for $A^G_{\mathcal{X}}$. Thus we only have to prove that $A^G_{\mathcal{X}}$ is a radical ideal. If this is false, we can assume without loss, that there is $0 \neq s \in S_{\mathcal{X}} \cap \sqrt{A^G_{\mathcal{X}}}$ with $s = c_1\pi(\alpha_1)^+ + \cdots + c_k\pi(\alpha_k)^+ \in S_{\mathcal{X}}$. Let $p^m\alpha_i$ denote the p^m-fold sum of α_i in the semigroup $\mathbb{N}_0^\Omega$. Then

$$(\pi(\alpha_i)^+)^{p^m} = (t^G_{G_{\alpha_i}}(\pi(\alpha_i)))^{p^m} = t^G_{G_{\alpha_i}}(\pi(p^m\alpha_i)) = t^G_{G_{p^m\alpha_i}}(\pi(p^m\alpha_i)) = \pi(p^m\alpha_i)^+.$$

Hence each s^{p^m} involves $\pi(p^m\alpha_i)$ with coefficient $c_i^{p^m}$. Since $G_{p^m\alpha_i} = G_{\alpha_i}$, we conclude $s^{p^m} \notin A^G_{\mathcal{X}}$ for all $m \in \mathbb{N}$, a contradiction. □

From Theorem 12.1 it is easy to see, that if $R = \mathbb{F}$ is a field of characteristic $p > 0$ and P is a p-group, then $A^G_< \cap S = 0$ and we get the following result:

Theorem 12.2. *If P is a p-group, $R = \mathbb{F}$ a field of characteristic $p > 0$ and $V = \mathbb{F}^\Omega$ is a permutation module, then the relative trace ideal $A^P_<$ of A^P is a prime ideal of height $|\Omega| - |\Omega/P|$; moreover A^P is generated by $A^P_<$ together with the subalgebra S of orbit products $\pi(\omega^G)$. In particular the quotient $A^P/A^P_< \cong S$ is a polynomial ring in $|\Omega/P|$ variables. This holds for any projective $\mathbb{F}P$-module V, in particular for the regular representation $V \cong \mathbb{F}P$.*

Remark 12.3. If $V = V_\Omega$ is a permutation module of the p-group P, then

$$V = \oplus_{\omega^P \in \Omega/P} V_{\omega^P},$$

where the $V_{\omega^P} \cong R_{P_\omega}^{\uparrow P}$ are induced from the trivial module of the stabilizer P_ω, so they are P_ω-projective. The group P acts on Ω without fixed points (i.e. orbits of size 1) if and only if V_Ω is $\mathcal{X}$-projective where $\mathcal{X}$ is the set of all proper subgroups of P. In this case Theorem 12.2 illustrates the statement of Proposition 12.1 that $A^P_<$ contains all invariants of degree coprime to p.

12.5 The Regular Representation of $\mathbb{Z}_p$

Now we consider more closely the groups of prime order. So let p be a prime and $P = \mathbb{Z}_p = < g >$. In this situation $A^P_<$ coincides with the absolute trace ideal $A^G_{\{e\}}$ of A^G. Hence Theorem 12.2 says that if R is a field of characteristic p, then for any permutation representation of $\mathbb{Z}_p$ on a set Ω, the trace ideal is prime with quotient being a polynomial algebra in as many variables as there are P orbits on Ω. We want to look at the regular representation $V = RP$ in more detail.

We can choose $g = (1\ 2\ \dots\ p) \in \Sigma_p$ and view V as the natural permutation module $R^{\underline{p}}$ where $\underline{p} := \{1, 2, \dots, p\}$; hence $A^G/A^G_< \cong R[N(e_1)]$ where $N(e_1)$ is the orbit product $e_1e_2\dots e_p = e_1 g(e_1) g^2(e_1) \dots g^{p-1}(e_1)$. Due to a fundamental result of M. Göbel on invariant rings of permutation modules, the invariant ring A^P is generated by orbit sums of monomials whose degree is less than or equal to $\binom{p}{2}$. In our special case we can do better:

Proposition 12.3. *Let R be an arbitrary commutative ring and $A = S(V)$ with $V = RP$ and $P \cong \mathbb{Z}_p$. Then A^G is generated by $N(e_1)$ together with elements of the trace ideal $A^P_{\{e\}}$ of degree less than or equal to $max\{3, \binom{p-1}{2} + 1\}$.*

Proof. For $p \in \{2,3\}$ the result can be verified easily. So we can assume that $p \geq 5$. As before we will identify the monomial $e_1^{f(1)} \dots e_p^{f(p)}$ with the exponent function $f \in \mathbb{N}_0^{\underline{p}}$. A homogeneous invariant $a \in A^P$ of degree i will be called **indecomposable** if it cannot be written as a sum of products of invariants of strictly lower degree; if it is not indecomposable it is called decomposable. Two homogeneous invariants of the same degree will be called **equivalent** if their difference is decomposable.

Following Göbel [12] we call an orbit sum $f^+ := \sum_{g \in P_f} g(f)$ 'special' if $f(\underline{p})$ is an interval containing 0. So $f = (0,1,2,2,3)$ is special while $h = (0,1,3,4,4)$ and $h' = (1,2,3,4,5)$ are not. In [12] it is proved that A^P is generated by special orbit sums, thus giving the degree bound $0+1+\cdots+p-1$. So we have to investigate the special orbit sums for indecomposability. Now the following lemma yields our result: □

Lemma 12.2. *Let $p \geq 5$ and $f^+ \in A^P_<$ be an indecomposable and special orbit sum. Then $(\#f^{-1}(\{0\}) \geq 2)$ or $(\#f^{-1}(\{0\}) = 1$ and $\#f^{-1}(\{1\}) \geq 2)$ or $(\#f^{-1}(\{0\}) = \#f^{-1}(\{1\}) = 1$ and $\#f^{-1}(\{2\}) = p-2)$. In particular f^+ is of degree less or equal to $\leq 0+1+1+2+3+\cdots+p-2$.*

Proof. As f is special, $\#f^{-1}(\{0\})$ and $\#f^{-1}(\{1\}) \geq 1$ and we can assume that $\#f^{-1}(\{0\}) = \#f^{-1}(\{1\}) = 1$. Let $f(i_1) = 1$ and $f(i_0) = 0$ with $i_0 \equiv i_1 + c$ and $p > c \in \underline{p}$. (By cycling we can assume that $1 \leq i_1 < i_0 \leq p$). Then we get

$$f^+ = (f_1-1, f_2-1, \dots, 1, \dots, 0, \dots, f_p-1)^+ \cdot (1,1,\dots,1,0,1,\dots,1,0,1,\dots,1)^+ -$$

$$\sum_{g^i \neq 1} [(f_1-1, f_2-1, \dots, 1, \dots, 0, \dots, f_p-1) \cdot g^i((1,1,\dots,0,1,\dots,1,0,1,\dots,1))]^+.$$

All terms that have to be subtracted are decomposable, except for $i = c$, when $g^c((1,1,\dots,0,1,\dots,1,0,1,\dots,1))$ has its zero entries at $i_1 + c = i_0$ and $i_0 + c \neq i_1 \pmod p$. If $f(i_0+c) > 2$, then

$$[(f_1-1, f_2-1, \dots, 1, \dots, 0, \dots, f_p-1) \cdot g^c((1,1,\dots,0,1,\dots,1,0,1,\dots,1))]^+$$

has no entry 1, so is decomposable again; we conclude $f(i_0+c) = 2$ and $-f^+$ is equivalent to

$$f'^+ := (f_1, f_2, \dots, 2, \dots, 0, \dots, 1, \dots, f_p)^+$$

with $f'(i_1) = 2$, $f'(i_0) = 0$, $f'(i_0+c) = 1$ and $f'(j) = f(j)$ for all the other j's.

Now consider

$$\begin{aligned} f'^{+} &= (f_1-1, f_2-1, \dots, 1, \dots, 0, \dots, 0, \dots, f_p-1)^{+} \\ &\quad \cdot (1,1,\dots,1,\dots,0,\dots,1,\dots,1)^{+} \\ &\quad - \sum_{g^i \neq 1} [(f_1-1, f_2-1, \dots, 1, \dots, 0, \dots, 0, \dots, f_p-1) \\ &\quad \cdot g^i((1,1,\dots,1,\dots,0,\dots,1,\dots,1))]^{+}. \end{aligned}$$

A similar argument as before shows that $-f'^{+}$ is equivalent to

$$f''^{+} := (f_1, f_2, \dots, 2, \dots, 1, \dots, 0, \dots, f_p)^{+}$$

with $f''(i_1) = 2$, $f''(i_0) = 1$, $f''(i_0 + c) = 0$ and $f''(j) = f(j)$ for all the other j's. Now iterating this process shows that $f(i_1 + \ell \cdot c) = 2$ for all $\ell \in \underline{p} \backslash \{1, p\}$ (modulo p). Hence $f(i) = 2$ for all $i \neq i_1, i_0$. □

Remark 12.4. If $p = 5$ the bound in 12.3 gives 7, which in this case can be shown to be sharp by computer calculations (this has been verified by Gregor Kemper using a MAGMA implementation of his algorithms, see also [14]).

12.6 A canonical Trace Ideal and Cohen-Macaulay Quotients

In this section we will discuss a special feature of modular invariant theory over a field $\mathbb{F}$ of positive characteristic $0 < p \mid |G|$. As before let V be a finitely generated $\mathbb{F}G$-module for the finite group G, $A := A(V) := \mathbb{F}[V]$ its symmetric algebra and A^G the invariant ring. Let $\overline{\mathbb{F}}$ denote the algebraic closure of $\mathbb{F}$ and $\hat{A} := A \otimes_{\mathbb{F}} \overline{\mathbb{F}}$. Then clearly $\hat{V} := V \otimes_{\mathbb{F}} \overline{\mathbb{F}}$ can be viewed as an algebraic $\mathbb{F}$-variety and the maximal ideal spectrum $max - spec\ \hat{A}$ is in bijection to the points in $\hat{V}$.

Now let $U, H \leq X \leq G$ be subgroups; notice that A is integral and finite over A^H and A as well as A^H are integrally closed in their quotient fields $Q(A)$ and $Q(A)^H$ respectively. Moreover $Q(A) : Q(A)^H$ is a Galois extension with Galois group H. The dual $i^* := i_H^*$ of the inclusion map $i_H : A^H \hookrightarrow A$ is the map $spec\ A \to spec\ A^H$, $m \mapsto m \cap A^H$. In this situation the classical 'lying over', 'going up', 'going down' and transitivity theorems of commutative algebra apply (see [3] Theorem 1.4.4. on pg. 7). Notice that $A^G \otimes_{\mathbb{F}} \overline{\mathbb{F}} = (A \otimes_{\mathbb{F}} \overline{\mathbb{F}})^G = \hat{A}^G$ and the above statements also hold for $\hat{A}$ instead of A. Hence $max - spec\ \hat{A}^H$ can be identified with the set $\hat{V}/H$ of H-orbits on $\hat{V}$ and $\hat{i}_H^*$, the dual of $\hat{i}_H := i_H \otimes_{\mathbb{F}} \overline{\mathbb{F}}$, can be identified with the orbit map $v \mapsto v^H$. Thus a strict geometric quotient $\hat{V}/H$ exists and the invariant ring $\hat{A}^H$ can be interpreted as the ring of regular functions $\mathcal{O}(\hat{V}/H)$ of the quotient variety (e.g. see [11] Theorem 5.52 pg. 187 or [3] pg. 8). Notice that all H-orbits are Zariski-closed, since H is finite. The orbit map $\hat{i}_H^*$, being the comorphism of a finite embedding, is surjective and closed, i.e. sends closed irreducible subvarieties of $\hat{V}$ to closed irreducible subvarieties of $\hat{V}/H$.

The U-fixed points in $\hat{V}$ form a linear subspace, hence an irreducible closed $\mathbb{F}$-subvariety $\hat{V}^U$ of $\hat{V}$. The image $\hat{i}_H^*(\hat{V}^U)$ is therefore a closed and irreducible subvariety of $\hat{V}/H$; it coincides with the set of orbits v^H such that $U \leq_H G_v$. Here and in the following, $U \leq_H X$ means that an H-conjugate ${}^hU = hUh^{-1}$ of U is contained in the subgroup $X \leq G$.

In particular if $H = P$ is a Sylow p-subgroup of G, then

$$\hat{i}_G^*(\hat{V}^P) := \{v^G \mid p \not| \, |v^G|\} = \{v^G \mid P \leq_G G_v\} \subseteq \hat{V}/G.$$

For any $v^H \in \hat{i}_H^*(\hat{V}^U)$, the fibre $(\hat{i}_H^*)_{|\hat{V}^U}^{-1}(v^H)$ consists of all $v' \in \hat{V}^U$ that are H-conjugate to v, hence it is a finite set and we conclude that the dimension of the variety $\hat{i}_H^*(\hat{V}^U) \subseteq \hat{V}/H$ is equal to the dimension of the $\overline{\mathbb{F}}$ - vector space $\hat{V}^U$ or of the $\mathbb{F}$-vector space V^U. Notice that A and A^H still induce regular functions on V and V/H respectively, but the maps $A \to \mathcal{O}(V)$ and $A^H \to \mathcal{O}(V/H)$ are no longer injective and neither $\mathcal{O}(V)$ nor $\mathcal{O}(V/H)$ need to be integral domains.

Now let $\theta^* : \hat{Y} \hookrightarrow \hat{V}$ be the inclusion of a closed G-stable $\mathbb{F}$-subvariety with $\mathcal{O}(\hat{Y}) = \hat{B} = B \otimes_{\mathbb{F}} \overline{\mathbb{F}} = \theta(\hat{A}) \cong \hat{A}/\hat{\mathcal{I}}$ and $\hat{\mathcal{I}} := \ker\theta$. Again there is a strict quotient $\hat{Y}/G$ and $\hat{i}^* \circ \theta^* : \hat{Y} \to \hat{V}/G$ factors through the quotient map $j^* : \hat{Y} \to \hat{Y}/G$, which is the dual of the inclusion $\mathcal{O}(\hat{Y})^G \hookrightarrow \mathcal{O}(\hat{Y})$. Then $\hat{Z} := \hat{i}^* \circ \theta^*(\hat{Y})$ is a closed $\mathbb{F}$-subvariety of $\hat{V}/G$ whose corresponding ring of regular functions $\mathcal{O}(\hat{Z}) =: \hat{C}$ is a certain quotient of $\hat{A}^G$. We summarize the situation in the following commutative diagrams:

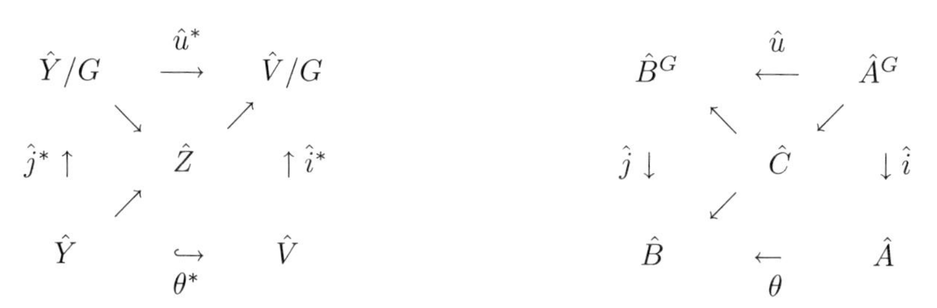

Since the morphism $\omega^* = i^* \circ \theta^* : \hat{Y} \to \hat{Z}$ is surjective, its comorphism $\omega : \hat{C} \to \hat{B}$ is an injection and the kernel of $\phi : \hat{A}^G \to \hat{C}$ coincides with the kernel of $\hat{u}$, so $\hat{C} \cong \hat{A}^G/(\ker\, \theta\hat{i}) = \hat{A}^G/(\ker\, \theta \cap \hat{A}^G)$. Although the morphism $\gamma^* : \hat{Y}/G \to \hat{Z}$ is bijective by construction, it is in general **not** an isomorphism. Indeed, the comorphism $\gamma : \hat{C} \hookrightarrow \hat{B}^G$ makes $\hat{B}^G$ a purely inseparable, integral extension. This result can be found (without proof) in [21] pg. 51. For convenience we include a short proof here:

Lemma 12.3. *The injection $\gamma : A^G/(\ker\ \theta \cap A^G) \cong C \hookrightarrow B^G$ is a purely inseparable, integral ring extension.*

Proof. Let $b \in B^G$ and $a \in A$ with $\theta(a) = b$. Fix a Sylow p-group P of G and define $m := \prod_{h \in P} h(a) \in A^P$. Then $\theta(m) = \prod_{h \in P}(h\theta(a)) = b^{|P|}$. Hence $\frac{1}{|G:P|}\theta(t_P^G(m)) = \frac{1}{|G:P|}t_P^G(\theta(m)) = b^{|P|} \in \theta i_G(A^G) = \gamma(C)$. □

Remark 12.5. Notice that we have not yet used the linear structure of $\hat{V}$, so all the statements above remain valid, if we start with an arbitrary affine $\mathbb{F}$-variety $\hat{X}$ instead of the linear variety $\hat{V}$.

We identify A and A^G with $A \otimes 1 \subseteq \hat{A}$ and $A^G \otimes 1 \subseteq \hat{A}^G$ respectively and define

$$\hat{\mathcal{I}}^{H,U} := \mathcal{I}(\hat{i}_H^*(\hat{V}^U)) = \{f \in \hat{A}^H \mid f(v^H) = 0,\ \forall\ v^H \in \hat{i}_H^*(\hat{V}^U)\}$$

i.e. the (prime) ideal in $\hat{A}^H$ corresponding to the irreducible and closed subset $\hat{i}_H^*(\hat{V}^U) \subseteq \hat{V}/H$; hence $\hat{A}^H/\hat{\mathcal{I}}^{H,U} \cong \mathcal{O}(\hat{i}_H^*(\hat{V}^U))$. We also define $\mathcal{I}^{H,U}$ to be the prime ideal $\hat{\mathcal{I}}^{H,U} \cap A \lhd A$. In particular we have $\mathcal{I}^{H,1} = 0$.

Lemma 12.4. *For subgroups $U, H \leq X \leq G$ we have $\mathcal{I}^{X,U} = \mathcal{I}^{H,U} \cap A^X$ and $\mathbb{F}[V^H]$ is a polynomial ring which is a finite integral and purely inseperable extension of $A^H/\mathcal{I}^{H,H}$.*

Proof. Notice that $\mathcal{I}^{X,U} = A^X \cap \mathcal{I}^{1,U} = A^X \cap A^H \cap \mathcal{I}^{1,U} = A^X \cap \mathcal{I}^{H,U}$. Now let $\theta^* : \hat{Y} := \hat{V}^H \hookrightarrow \hat{V}$ be the canonical injection. Clearly $\hat{Y} = \hat{Y}/H$ and we have $\hat{B} = \hat{B}^H \cong \hat{A}/\hat{\mathcal{I}}^{1,H}$ in our notation above. Since $\hat{Y}$ is a linear $\mathbb{F}$-variety, both algebras $B = \mathbb{F}[V^H]$ and $\hat{B} \cong B \otimes_{\mathbb{F}} \overline{\mathbb{F}}$ are polynomial subalgebras of A and $\hat{A}$ respectively. We also have $C \cong A^H/(A^H \cap \mathcal{I}^{1,H}) = A^H/\mathcal{I}^{H,H}$. From Lemma 12.3 we know that $B = B^H$ is a finite, integral and purely inseparable extension of C. $\square$

The following example shows that $A^H/\mathcal{I}^{H,H}$ will in general be a proper subring of $\mathbb{F}[V^H]$.

Example 12.1. We use the notation of Section 12.3. Let $V = \mathbb{F}^{\Omega} \cong V^*$ be a permutation module and P a p-group. Then $V^P = \langle v^+ \mid v \in V\rangle_{\mathbb{F}}$ is the span of the orbit sums in V. If $\delta_\omega \in V^*$ denotes the element dual to e_ω, then $\mathcal{I}^{1,P} = (\delta_\omega - \delta_{g(\omega)} \mid g \in P)$. Hence $\mathbb{F}[V^P] = \mathbb{F}[V]/\mathcal{I}^{1,P} \cong \mathbb{F}[\delta_{[\omega]} \mid [\omega] \in \Omega/P\}$. On the other hand $\mathcal{I}^{P,P} = \mathbb{F}[V]^P_<$ (see Proposition 12.5), hence

$$\mathbb{F}[V]^P/\mathcal{I}^{P,P} \cong \mathbb{F}[\pi(\omega^P) \mid \omega^P \in \Omega/P]$$

and its image in $\mathbb{F}[V^P]$ coincides with the isomorphic but proper subalgebra $\mathbb{F}[\delta_{[\omega]}^{|\omega^P|} \mid [\omega] \in \Omega/P\}$.

In the example above, $\mathbb{F}[V]^P/\mathcal{I}^{P,P}$ and $\mathbb{F}[V^P]$ are still abstractly isomorphic polynomial rings. As the next result shows, this is always true for finite groups whose action on V is defined over the prime field. This follows from results of Wilkerson on inseparable extensions of algebras over the Steenrod algebra [23]. At this point I like to thank David Wehlau and Jim Shank for pointing out a gap in an earlier version of Lemma 12.4 and also for putting my attention to the paper [23].

Let $\mathbb{F}_p$ denote the prime field of characteristic p, W be a finite-dimensional $\mathbb{F}_p$-space and $\mathcal{A}_p$ the *mod*(p)-Steenrod algebra. The following is part of Theorem II in [23]:

Theorem 12.3. *(Wilkerson) Let R^* be a sub-$\mathcal{A}_p$-algebra of $\mathbb{F}_p[W]$ such that the graded field of quotients $F\mathbb{F}_p[W]$ is algebraic over FR^*. Define a filtration*

$U_i(R^*)$ of W by the rule that $v \in U_i(R^*)$ if and only if v^{p^i} is separable over FR^*. Let $D^*(R^*)$ be the 'diagonal' subalgebra of $F\mathbb{F}_p[W]$ generated by $U_i(R^*)^{p^i}$.

Then $D^*(R^*)$ is separable over R^* and $D^*(R^*)$ is isomorphic to

$$\otimes_i(\mathbb{F}_p[U_i(R^*)/U_{i-1}(R^*)])^{p^i}$$

as $\mathcal{A}_p$-algebras.

Proposition 12.4. *Let G be a finite group, and $V \in \mathbb{F}G - mod$ such that the G-action is defined over the prime field. Then $\mathbb{F}[V]^G/\mathcal{I}^{G,G}$ and $\mathbb{F}[V^G]$ are abstracly isomorphic polynomial rings.*

Proof. By flat base change we can assume that $\mathbb{F} = \mathbb{F}_p$ is the prime field. It is well known that the polynomial ring $\mathbb{F}[V]$, as well as its subring $\mathbb{F}[V]^G$ are unstable $\mathcal{A}_p$-modules (e.g. see [19]). The ideal $\mathcal{I}^{1,G}$ is generated by the elements in $(V^G)^\perp$, so it is $\mathcal{A}_p$-stable as well as $\mathcal{I}^{G,G} = \mathcal{I}^{1,G} \cap A^G$. Hence $\mathbb{F}[V]^G/\mathcal{I}^{G,G} =: R^*$ is a sub-$\mathcal{A}_p$-algebra of $\mathbb{F}[V^G]$ and, by Lemma 12.4, the extension is purely inseparable. Using the notation of Theorem 12.3, we see that $R^* = D^*(R^*)$ is a polynomial ring. Since its transcendence degree is the same as that of $\mathbb{F}[V^G]$, both rings are isomorphic. $\square$

Remark 12.6. In [23] only the case $\mathbb{F} = \mathbb{F}_p$ is considered. Replacing the *mod(p)*-Steenrod algebra by the Steenrod algebra $\mathcal{A}_q$ with $q = p^s$ as in [19] Chapter 11, we expect Theorem 12.3 and Proposition 12.4 to hold as well for $\mathbb{F}_q$ instead of $\mathbb{F}_p$ but leave the corresponding details to a subsequent paper.

For any ideal $\mathcal{J} \lhd A^G$ let $\mathcal{V}(\mathcal{J}) := \{v^G \in V/G \mid f(v) = 0, \ \forall f \in \mathcal{J}\}$. In [22] it has been shown that $\mathcal{V}(A^G_{\{e\}})) = \{v^G \mid p \text{ divides } |G_v|\}$ and that $\sqrt{A^G_{\{e\}}} = \mathcal{P}_G \cap A^G$, where $\mathcal{P}_G$ is the ideal in A, generated by $(g-1)V^*$ for all elements $g \in G$ of order p. (Part of this result is also contained in [8]). We are, in some sense, looking at the other extreme, namely the subvariety of orbits whose stabilisers contain a Sylow p-group of G. Obviously we have

$$i^*(V^P) = \{v^G \mid |P| \le |G_v|\} = \{v^G \in V/G \mid p \text{ does not divide } |v^G|\},$$

where P denotes a fixed Sylow p-group of G. In order to make the corresponding ideal $\mathcal{I}^{G,P}$ computable, we like to describe it in terms of relative traces. To do this, we first go the other way round and give a geometrical description of relative trace ideals. Notice that, by Hilbert's Nullstellensatz, we have for any ideals $\mathcal{I}, \mathcal{J} \lhd A = A^{\{e\}}$, $\sqrt{\mathcal{I}} = \sqrt{\mathcal{J}}$ if and only if $\mathcal{V}(\mathcal{I}) = \mathcal{V}(\mathcal{J})$. Using the orbit map i_G^* one can easily see that the same is true for ideals in A^G. We need the following

Lemma 12.5. (i): *For $U \le G$, $f \in (V^*)^U$ and $v \in V^U$ we have*

$$t_U^G(f)(v) = f(t_U^G(v)).$$

(ii): *For every $v \in V$ and every Sylow p-group Q of the stabilizer G_v, there is an element $f \in A^Q$ with $t_Q^G(f)(v) \neq 0$.*

Proof. (i): We have $t_U^G(f)(v) = \sum_{g\in[G:U]}(gf)(v)$, where $[G:U]$ denotes an arbitrary system of representatives for the right cosets of U in G (since $f \in A^U$ the value $t_U^G(f)$ is independent of the choice of coset representatives). Since in this situation we have to consider the G-action on V and on V^* simultaneously we take some care in choosing a system which is appropriate for our needs: it is well known that for any subgroup $U \le G$ there always exists a system R of coset representatives for U in G which represents at the same time left- and right U-cosets, i.e. $G = \uplus_{x\in \mathrm{R}} xU = \uplus_{x\in\mathrm{R}} Ux = \uplus_{x\in\mathrm{R}} x^{-1}U$. This is a combinatorical consequence of Ph. Hall's 'marriage theorem'.

Now we get $t_U^G(f)(v) = \sum_{x\in\mathrm{R}}(xf)(v) = \sum_{x\in\mathrm{R}}(x^{-1}f)(v)$ (because of the special property of R) $= \sum_{x\in\mathrm{R}} f(xv) = f(t_U^G(v))$.

(ii): This can be seen in a similar way as in the proof of Theorem 2.1. of [22]: let $A = \mathbb{F}[X_1 \ldots, X_n]$, and for $V \ni w \ne v$ define $i(w) := \min\{i \mid X_i(w-v) \ne 0\}$ and $\tilde{f} := \prod_{w\in v^G,\ w\ne v}(X_{i(w)} - X_{i(w)}(w))$. Then $\tilde{f}(w) = 0$ for all $w \in v^G\setminus\{v\}$ and $\tilde{f}(v) \ne 0$. Now let $f := \prod_{q\in Q} q(\tilde{f}) \in A^Q$; then

$$f(v) = \prod_{q\in Q} \tilde{f}(q^{-1}v) = (\tilde{f}(v))^{|Q|} \ne 0$$

and for every $w = gv \ne v$: $f(w) = \tilde{f}(w)\prod_{q\in Q\setminus\{1\}} \tilde{f}(q^{-1}w) = 0$. Hence by a suitable choice of coset representatives as in (i) one gets

$$t_Q^G(f)(v) = |G_v : Q| \sum_{g\in[G:G_v]} f(gv) = |G_v : Q| \cdot f(v) \ne 0.$$

$\square$

Theorem 12.4. *Let $\mathbb{F}$ be a field of characteristic $p > 0$ and $A := \mathbb{F}[V]$ for $V \in \mathbb{F}G - mod$. Let $\mathcal{X}$ be a set of subgroups of G, $P \in Syl_p(G)$, the set of Sylow p-subgroups of G and $\mathcal{X}_P := \{U \le P \cap {}^gX \mid X \in \mathcal{X},\ g \in G\}$. Then we have for $A_{\mathcal{X}}^G := \sum_{X\in\mathcal{X}} t_X^G(A^X)$:*

(i): $$A_{\mathcal{X}}^G = A_{\mathcal{X}_P}^G.$$

(ii): $\mathcal{V}(A_{\mathcal{X}}^G) = \{v^G \mid P_v \in Syl_p(G_v),\ P_v \cap {}^gX < P_v,\ \forall X \in \mathcal{X},\ g \in G\} = \{v^G \mid p \mid |G_v : ({}^gX)_v|,\ \forall X \in \mathcal{X}, \forall g \in G\} = i_G^*(\cup_{Q\le P,\ \text{minimal with } Q\notin\mathcal{X}_P} V^Q)$.

(iii): $$\sqrt{A_{\mathcal{X}}^G} = \cap_{Q\le P,\ \text{minimal with } Q\notin\mathcal{X}_P} \mathcal{I}^{G,Q}$$
$$= (\cap_{Q\le P,\ \text{minimal with } Q\notin\mathcal{X}_P} ((g-1)V^* \mid g \in Q)A) \cap A^G.$$

Proof.

(i): Let $X \in \mathcal{X}$ and $a \in A^X$, then the first Mackey formula of Section 12.3 yields: $t_X^G(a) = \sum_{g\in[P:G:X]} t_{P\cap {}^gX}^P(ga) \in A^P$. Hence

$$|G : P| t_X^G(a) = t_P^G(t_X^G(a)) = \sum_{g\in[P:G:X]} t_{P\cap {}^gX}^G(gb) \in A_{\mathcal{X}_P}^G.$$

This shows '$\subseteq$'. The other inclusion follows immediately from transitivity and conjugacy relations of trace maps.

(ii): The second equality is clear. So we only have to show the first and last ones. Let RHS denote the right hand side of the first equation. To show '$\subseteq$' we assume $v \in V$ such that $v^G \notin$ RHS. By Sylow's theorem, all Sylow p-groups of G are conjugate, so any G-orbit v^G can be represented by a suitable $v \in V$, such that $Q := P_v$ is a Sylow p-group of G_v. Then there is $X \in \mathcal{X}$ and $g \in G$ with $Q = Q \cap {}^gX$. From Lemma 12.5 (ii) there is $f \in A^Q$ with $0 \neq t_Q^G(f)(v) = t_{^gX}^G(t_Q^{^gX}(f))(v)$. But for $\phi := t_Q^{^gX}(f) \in A^{^gX}$ we have ${}^{g^{-1}}\phi = t_{^{g^{-1}}Q}^X({}^{g^{-1}}f) \in A^X$ and $t_X^G({}^{g^{-1}}\phi) = t_{^gX}^G(\phi) = t_Q^G(f) \in A_{\mathcal{X}}^G$ and $v \notin \mathcal{V}(A_{\mathcal{X}}^G)$.

To prove '$\supseteq$' let $v^G \in$ RHS, $X \in \mathcal{X}$ and $f \in A_X^G$. Again we can assume that $Q := P_v \in Syl_p(G_v)$ and that $f = t_X^G(b)$ with $b \in A^X$. Then from the first Mackey formula we get $f = \sum_{g \in [Q:G:X]} t_{Q\cap\ {}^gX}^Q(gb)$ and

$$t_{Q\cap\ {}^gX}^Q(gb)(v) = \frac{|Q|}{|Q \cap\ {}^gX|}(gb)(v) = 0.$$

Hence $v \in \mathcal{V}(A_{\mathcal{X}}^G)$.

To prove the third equality notice that $\mathcal{X}_P$ is closed under taking subgroups, so

$$\begin{aligned}\mathcal{V}(A_{\mathcal{X}}^G) &= \cup_{Q \leq P,\ Q \notin \mathcal{X}_P}\{v^G \mid Q \in Syl_p(G_v)\} = \cup_{Q \leq P,\ Q \notin \mathcal{X}_P} i_G^*(V^Q) \\ &= \cup_{Q \leq P,\ \text{minimal with } Q \notin \mathcal{X}_P} i_G^*(V^Q).\end{aligned}$$

(iii): The first equality follows from (ii) and Hilbert's Nullstellensatz. To prove the second statement, notice that V^Q is the socle $soc(V_{|Q})$ of V restricted to Q and it is well known that $(soc(V_{|Q}))^\perp = Rad(\mathbb{F}[Q]) \cdot V_{|Q}^*$, where Rad denotes the Jacobson radical. In case of p-groups in characteristic $p > 0$, the Jacobson radical coincides with the augmentation ideal $\mathbb{F}[Q](h-1 \mid h \in Q)$, hence $(soc(V_{|Q}))^\perp = \langle (h-1)V^* \mid h \in Q \rangle_{\mathbb{F}}$. So we conclude that $\hat{\mathcal{I}}(V^Q) = \hat{\mathcal{I}}^{1,Q} = \hat{A}((h-1)V^* \mid h \in Q)$. Now the rest follows by intersection with A and A^G.

□

The following consequences include a characterization of the ideal $\mathcal{I}^{G,P}$ in terms of relative traces:

Proposition 12.5. *Let P be a fixed Sylow p-group of G. Then the following hold:*

(i): *If $\mathcal{X} = \{H\}$ with a normal subgroup $H \triangleleft G$, then $\mathcal{X}_P = \{U \leq H \cap P\}$ and $\mathcal{V}(A_H^G) = i_G^*(\cup_{x \in P\setminus(H\cap P),\ x^p \in H\cap P} V^x)$ and*

$$\sqrt{A_H^G} = (\cap_{x \in P\setminus(H\cap P),\ x^p \in H\cap P}\ (x-1)V^*A)\ \cap\ A^G.$$

In particular we obtain Feshbach's theorem (see [8]):

$$\sqrt{A_1^G} = (\cap_{x\in P\setminus\{e\},\ x^p=1}\ (x-1)V^*A)\ \cap\ A^G.$$

(ii): *For* $\mathcal{X} := \{Q \mid Q < P\}$ *and* $A_{\mathcal{X}}^G = A_{<P}^G$ *we have:*

$$\mathcal{V}(A_{<P}^G) = \mathcal{V}(\mathcal{I}^{G,P}) = \{v^G \in V/G \mid p \text{ does not divide } |v^G|\} = i_G^*(V^P).$$

$$\sqrt{A_{<P}^G} = \mathcal{I}^{G,P} = A^G \cap (hf - f \mid f \in A,\ h \in P)A.$$

In particular height $A_{<P}^G = \text{codim } i^*(\hat{V}^P) = \text{codim}_{\mathbb{F}}\ V^P$.

(iii): *For every family* $\mathcal{X}$ *with* $\mathcal{X}_P$ *consisting of proper subgroups of* P *we have*

$$\text{height } A_{\mathcal{X}}^G \leq \text{codim}_{\mathbb{F}}\ V^P.$$

Proof. Most of these statements are immediate consequences of Theorem 12.4.

In (i) let $Q = \langle q_1, \ldots, q_s\rangle$ be a subgroup of P minimal subject to the condition not lying in $\mathcal{X}_P$, i.e. not being a subgroup of $H \cap P$; suppose moreover that the q_i form a minimal generating set. Then some $q_i \notin H \cap P$, hence $Q = \langle q_i\rangle$ and $q_i^p \in H \cap P$.

(ii) is clear by Theorem 12.4 and the fact that height $A_{<P}^G = \text{height } \sqrt{A_{<P}^G} = \text{height } \mathcal{I}^{G,P} = \text{codim } \mathcal{V}(\mathcal{I}^{G,P}) = \text{codim } i_G^*(V^P) = \text{codim}_{\mathbb{F}}(V^P)$.

(iii) follows from the fact $A_{\mathcal{X}}^G = A_{\mathcal{X}_P}^G \leq A_{<P}^G$. □

If V is projective relative to a proper subgroup of a Sylow p-group of G, and hence with respect to some $Q < P$, we know from Proposition 12.1, that $A_{<P}^G$ is 'large' in the sense that it contains all invariants of degree not divisible by p.

If G has a cyclic Sylow p-group P, then all trace ideals can be described in an easy way: notice that in this case, the lattice of subgroups of P is linearly ordered, so, for any family $\mathcal{X}$ there is a unique subgroup $U \leq P$ such that $\mathcal{X}_P = \{X \leq U\}$. If $U < P$ then there is also a unique subgroup $Q \leq P$ containing U as a maximal subgroup. So we get:

Corollary 12.3. *Let* $P \in Syl_p(G)$ *be cyclic, then every trace ideal* $A_{\mathcal{X}}^G$ *is either* A^G *(if* $P \in \mathcal{X}_P$*), or of the form* $A_{<Q}^G$ *for a subgroup* $Q \leq P$*. Moreover we have*

$$\mathcal{V}(A_{<Q}^G) = i_G^*(V^Q)$$

and $\sqrt{A_{<Q}^G}$ *is the prime ideal* $\mathcal{I}^{G,Q} = ((V^Q)^\perp A) \cap A^G$ *of height* $\text{codim}_{\mathbb{F}} V^Q$.

In the following we are going to construct certain Cohen-Macaulay quotients of modular invariant rings. Recall that a (positively) graded $\mathbb{F}$-algebra R is Cohen-Macaulay (CM) if and only if it is free as a module over the subalgebra generated by a homogeneous system of parameters (h.s.o.p.) of R. In that case R is free over any such h.s.o.p.

Definition 12.2. (see [4] pg.270) Let $S \leq R$ be a subring; a **Reynolds operator** for the pair (R, S) is a split epimorphism of S modules $\pi: \ {}_SR \to S$, i.e. with a right inverse $\rho: \ S \hookrightarrow R$. For example t_P^G is a Reynolds operator for the pair (A^P, A^G).

Due to a theorem of Eagon and Hochster ([13]) it is well known that A^G is always a Cohen-Macaulay ring if char $\mathbb{F}$ is coprime to $|G|$. Using the Reynolds operator t_P^G this has been generalized by Campbell, Hughes and Pollack, who proved that if char $\mathbb{F} = p > 0$, the invariant ring A^G is still Cohen-Macaulay provided A^P is ([5]). The next proposition will be a further generalization thereof.

Crucial for all these results is the following theorem of Hochster-Eagon (see [4] 6.4.5):

Theorem 12.5. *Suppose S is a subring of R such that R is integral over S and π is a Reynolds operator for (R, S). Then, if R is CM, so is S.*

First notice that if $\mathcal{I} \triangleleft A^H$ is an ideal then, as $A^G \leq A^H$ we have $t_H^G(\mathcal{I})A^G = t_H^G(\mathcal{I}A^G) \subseteq t_H^G(\mathcal{I})$; hence $t_H^G(\mathcal{I})$ is an ideal of A^G.

Proposition 12.6. *Let $\mathbb{F}$ be a field of characteristic p and P a Sylow p-subgroup of G. Suppose that $\mathcal{I} \triangleleft A^P$ is an ideal satisfying*

$$t^P_{P \cap {}^gP}(g\mathcal{I}) \subseteq \mathcal{I} \quad (*)$$

for all $g \in G$. Then $t_P^G(\mathcal{I}) = \mathcal{I} \cap A^G \triangleleft A^G$ and t_P^G induces a Reynolds operator for the pair $(A^P/\mathcal{I}, A^G/(\mathcal{I} \cap A^G))$. In particular if $A^P/\mathcal{I}$ is Cohen-Macaulay so is $A^G/(\mathcal{I} \cap A^G)$.

Proof. Let $a \in \mathcal{I} \cap A^G$, then $|G : P|a = t_P^G(a)$, hence $a \in t_P^G(\mathcal{I})$. Let $b \in t_P^G(\mathcal{I})$; then, by the Mackey-formula, $b = t_P^G(i) = \sum_{g \in [P:G:P]} t^P_{P \cap {}^gP}(gi) \subseteq \mathcal{I} \cap A^G$. Now we have a surjection $t_P^G : \ A^P/\mathcal{I} \to A^G/t_P^G(\mathcal{I}) = A^G/(\mathcal{I} \cap A^G)$ and a canonical embedding $\rho : A^G/\mathcal{I} \cap A^G \hookrightarrow A^P/\mathcal{I}$; both are homomorphisms of $A^G/(\mathcal{I} \cap A^G)$-modules with $t_P^G \circ \rho = |G : P| \cdot Id$ invertible. Hence t_P^G induces a Reynolds operator for the pair $(A^P/\mathcal{I}, (A^G/\mathcal{I} \cap A^G))$. Clearly $A^P/\mathcal{I}$ is integral over $A^G/(\mathcal{I} \cap A^G)$ and the claim follows from Theorem 12.5. □

Remark 12.7.

(i): Notice that for $\mathcal{I} = 0$ we recover the above mentioned result of Campbell, Hughes and Pollack.

(ii): The condition $(*)$ above is satisfied whenever $g\mathcal{I} = \mathcal{I}$ for all g in the normalizer $N_G(P)$ and $A^P_< \leq \mathcal{I}$.

(iii): Let $\mathcal{Y}$ be a family of subgroups of $P \in Syl_p(G)$ such that $Y \in \mathcal{Y} \Rightarrow P \cap {}^gY \in \mathcal{Y}$ for every $g \in G$, then condition $(*)$ of Proposition 12.6 is satisfied for $\mathcal{I} = A^P_{\mathcal{Y}}$. Indeed, for $Y \in \mathcal{Y}$, $a \in A^Y$ we have

$$\begin{aligned} t^P_{P\cap\ {}^gP}(gt^P_Y(a)) &= t^P_{P\cap\ {}^gP}(t^{{}^gP}_{{}^gY}(ga)) \\ &= \sum_{h\in[P\cap\ {}^gP:{}^gP:{}^gY]} t^P_{P\cap\ {}^gP}(t^{P\cap\ {}^gP}_{P\cap\ {}^gP\cap\ {}^{hg}Y}(hga)) \\ &= \sum_{h\in[P\cap\ {}^gP:{}^gP:{}^gY]} t^P_{P\cap\ {}^gP\cap\ {}^{hg}Y}(hga) \\ &= \sum_{h\in[P\cap\ {}^gP:{}^gP:{}^gY]} t^P_{P\cap\ {}^{hg}Y}(t^{P\cap\ {}^{hg}P}_{P\cap\ {}^gP\cap\ {}^{hg}Y}(hga)) \in A^P_{\mathcal{Y}}. \end{aligned}$$

Now we can prove the following theorem on Cohen-Macaulay quotients of modular invariant rings, which generalizes the theorem of Hochster and Eagon on the CM-property of non-modular invariant rings:

Theorem 12.6. *Let* char $\mathbb{F} = p$ *and* P *a Sylow* p*-subgroup of* G*. Suppose that the action of* P *on* V *is defined over the prime field. Then* $\mathcal{I}^{G,P} = t^G_P(\mathcal{I}^{P,P}) = \mathcal{I}^{P,P} \cap A^G$ *and* $A^G/\mathcal{I}^{G,P}$ *is a Cohen-Macaulay ring.*

Proof. From Proposition 12.5 we see that $A^P_< \le \mathcal{I}^{P,P}$. Clearly for any $g \in N_G(P)$ we have $g\mathcal{I}^P_P = \mathcal{I}^P_P$, so $\mathcal{I}^{P,P}$ satisifies the assumption of Proposition 12.6. By Proposition 12.4 $A^P/\mathcal{I}^P_P$ is CM, so the result follows from 12.6. Notice that if $P = 1$ we get $\mathcal{I}^P_P = \mathcal{I}(V^*) = (0) \triangleleft A$ and A^G is CM. □

We consider the ideals $A^G_{<P}$ and $\mathcal{I}^{G,P}$ as links between the $\mathbb{F}G$-module structure of V and the ideal structure of $\mathbb{F}[V]^G$ and therefore propose to study conditions on V, implying the equality

$$A^G_{<P} = \mathcal{I}^{G,P} \ (*),$$

which, by Proposition 12.5, is equivalent to the primality of $A^G_{<P}$. From the point of view of **constructive** modular invariant theory the analysis of $\mathbb{F}[V]^G$ might benefit in this case from the fact that $A^G_{<P}$ is constructible using relative trace maps and the quotient can be handled by methods, developed for Cohen-Macaulay rings.

By the results above, to prove equality $(*)$ it suffices to consider the case where G is a p-group: in fact $A^P_< = \mathcal{I}^{P,P}$ implies that $\mathcal{I}^{G,P} = t^G_P(\mathcal{I}^{P,P}) = t^G_P(A^P_<) = A^G_{<P}$. But the following example shows, that for p-groups in general, $A^G_{<P}$ can be stricly smaller than $\sqrt{A^G_{<P}}$:

Example 12.2. (Kathrin Kuhnigk [15]) Let $p > 2$, $\mathbb{F}$ of characteristic p and $G = \langle g \rangle \cong \mathbb{Z}_p$, acting on $\mathbb{F}[X,Y]$ by $g^i(X) = X + iY$ and $g^i(Y) = Y$ for every i. Then, since $t^G_1(X^p) = t^G_1(X)^p$, it is clear that A^G_1 is generated by $t^G_1(X^i)$ for $i = 1, \ldots, p-1$. We get

$$t^G_1(X^i) = \sum_{\ell=0}^{p-1}(X+\ell Y)^i = \sum_{\ell=0}^{p-1}\sum_{j=0}^{i}\binom{i}{j}X^{i-j}Y^j\ell^j = \sum_{j=0}^{i}\binom{i}{j}X^{i-j}Y^j\ (\sum_{\ell=0}^{p-1}\ell^j).$$

Whenever $0 < j < p-1$, the subset $\{\ell^j \mid 0 < \ell < p\}$ of $\mathbb{F}_p^*$ is a nontrivial multiplicative subgroup S. If S^+ is the sum of its elements, then $(s-1)S^+ = 0$ for $1 \neq s \in S$, hence $S^+ = 0 = \sum_{\ell=0}^{p-1} \ell^j$. This last equation also holds for $j = 0$, while for $j = p-1$ we get $\sum_{\ell=0}^{p-1} \ell^j = -1 \in \mathbb{F}$. Hence $t_1^G(X^i) = -Y^{p-1}$ for $i = p-1$ and zero otherwise. We conclude that

$$A_{<P}^G = (Y^{p-1}) < (Y) = \sqrt{A_{<P}^G},$$

We have already seen in Proposition 12.2 that all relative trace ideals $A_{\mathcal{X}}^G$ are radical ideals if V is a permutation representation. In the following theorem we will prove that this is still true for direct summands of permutation modules, in particular for projective modules.

Theorem 12.7. *If $V \in {}_{\mathbb{F}G}mod$ is a direct summand of a permutation module, then all (relative) trace ideals $A_{\mathcal{X}}^G$ are radical ideals. In particular if $P \in Syl_p(G)$, then*

$$A_{<P}^G = \mathcal{I}^{G,P}.$$

is a prime ideal of height codim V^P *and the quotient $A^G/A_{<P}^G$ is a Cohen-Macaulay ring of Krull dimension* dim V^P.

If moreover P is cyclic, then each trace ideal $A_{\mathcal{X}}^G$ with $P \notin \mathcal{X}_P$ is a prime ideal of the form $\mathcal{I}^{G,Q}$ with height $\mathrm{codim}_{\mathbb{F}}\ V^Q$ *and $Q \leq P$ minimal subject to ${}^gX \cap P < Q$ for all $X \in \mathcal{X}$ and $g \in G$.*

Proof. From Theorem 12.4 and the Remark 12.7 (iii) we see that

$$A_{\mathcal{X}}^G = A_{\mathcal{X}_P}^G = t_P^G(A_{\mathcal{X}_P}^P) = A_{\mathcal{X}_P}^P \cap A^G.$$

So, together with 'lying over' and 'going up and down' for prime ideals and integral extensions, we can restrict to the Sylow p-group P, and hence assume right away that G is a p-group. For p-groups in characteristic p any transitive permutation module is indecomposable and so, by the Krull-Schmidt theorem, any direct summand of a permutation module is again a permutation module (see [7] Chapter IX, 3.4.). Hence we can assume that V is a permutation module and in this case we know the results already from Proposition 12.2 and Theorem 12.2.

The last statement is an immediate consequence of Corollary 12.3. □

Remark 12.8.

(i): Notice that the proof of Theorem 12.7 is independent of Proposition 12.4 and hence of the result 12.3 in [23].

(ii): The last statement of Theorem 12.7 generalizes a result of Shank and Wehlau [22] Corollary 6.2.

(iii): Let V be indecomposable with vertex $Q \leq P$, then V is a direct summand of $W^{\uparrow G}$ where W can be chosen to be an indecomposable summand of

$V_{|Q}$ and then is unique up to isomorphism and conjugation. In modular representation theory such an $\mathbb{F}Q$-module W is called a **source** of V. It is well known (and proved by 'Mackey decomposition') that V is a direct summand of a permutation module if and only if the trivial module $\mathbb{F}_Q$ is a source of V (see e.g. [16], chapter II, 12.5). Thus we have shown that $A^G_{<P} = \mathcal{I}^{G,P}$ for 'trivial source modules'. In particular if $Q < P$ (which is equivalent to the fact that the trivial $\mathbb{F}P$-module is not a direct summand of V_P), then A^G is generated by $A^G_{<P}$ together with preimages of generators for the CM-algebra $A^G/A^G_{<P}$; these latter ones can all be chosen to be of degree divisible by p.

Acknowledgements:

I thank Jacques Alev (Reims), David Green (Wuppertal) and Gregor Kemper (Heidelberg) for stimulating and fruitful discussions during the process of writing this paper. I also thank David Wehlau and Jim Shank (Kingston, Ontario) for their very valuable and constructive criticism.

Bibliography

[1] J. Alperin: *Local Representation Theory*, Cambridge Univ. Press 1986.

[2] D. Benson: *Representations and Cohomology I*, Cambridge Univ. Press 1991.

[3] D. Benson: *Polynomial Invariants of Finite Groups*, Cambridge Univ. Press 1993.

[4] W. Bruns, J. Herzog: *Cohen-Macaulay Rings*, Cambridge Univ. Press 1993.

[5] H.E.A. Campbell, I. Hughes, R.D. Pollack: *Rings of Invariants and p-Sylow Subgroups*, Canad. Math. Bull. **34**, (1), 42–47, (1991).

[6] C.W. Curtis, I. Reiner: *Methods of Representation theory I,II*, Wiley Interscience 1981, 1987.

[7] W. Feit: *The Representation Theory of Finite Groups*, North-Holland (1982).

[8] M. Feshbach: *p-subgroups of compact Lie groups and torsion of infinite height in $H^*(BG;\mathbb{F}_p)$*, Mich. Math.J. **29**, 299–306, (1982).

[9] P. Fleischmann: *A new degree bound for Vector Invariants of Symmetric Groups*, to appear in Transactions of the AMS
`http://www.exp-math.uni.essen.de/~peter`

[10] P. Fleischmann, W. Lempken: *On generators of modular invariant rings of finite groups*, Bull. London Math. Soc. **29**, (585–591), (1997).
`http://www.exp-math.uni.essen.de/~peter`

[11] J. Fogarty: *Invariant theory*, Benjamin (1969).

[12] M. Göbel: *Computing bases for rings of permutation- invariant polynomials*, J. Symb. Comp. **19**, 285–291, (1995).

[13] M. Hochster, J.A. Eagon: *Cohen-Macaulay rings, invariant theory and the generic perfection of determinantal loci*, Amer. J. Math. **93**, 1020–1058, (1971).

[14] G. Kemper: *Calculating invariant rings of finite groups over arbitrary fields*, J. of Symbolic Computation, (1995).

[15] K. Kuhnigk: *Der Transfer in der modularen Invariantentheorie*, Diplomarbeit, Universität Göttingen, (1998).

[16] P. Landrock: *Finite Group Algebras and their Modules*, LMS Lecture Note Series, Cambridge University Press (1984).

[17] M. Neusel: *Mme Bertin's $\mathbb{Z}/4$ revisited*, preprint (1997).

[18] M. Neusel: *The transfer in the invariant theory of modular representations*, preprint, (1997).

[19] L. Smith: *Polynomial Invariants of Finite Groups*, A.K. Peters Ltd., (1995).

[20] L. Smith: *Modular vector invariants of cyclic permutation representations*, preprint (1997).

[21] J.P. Serre: *Algebraic groups and Class Fields*, Springer, (1975).

[22] R.J. Shank, D.L. Wehlau: *The transfer in modular invariant theory*, preprint, `http://www.mast.queensu.ca/~shank/transfer.dvi` (1997).

[23] C. Wilkerson: *Rings of invariants and inseparable forms of algebras over the Steenrod algebra*, preprint `http://hopf.math.purdue.edu/pub/hopf.html` (1979).

Progress in Mathematics, Vol. 173, © 1999 Birkhäuser Verlag Basel/Switzerland

Chapter 13

On Sims' Presentation for Lyons' Simple Group

Holger W. Gollan and George Havas

Abstract

This paper gives a verification that the corrected Sims presentation is indeed a presentation of Lyons simple group. We prove that Lyons original characterization of his new simple group works inside a permutation group of degree 8 835 156.

13.1 Introduction

In his original paper [7], Lyons describes a new finite simple group Ly with the following two properties: Ly has an involution t such that its centralizer $Cen_{Ly}(t)$ is isomorphic to the double cover $2A_{11}$ of the alternating group A_{11}, and additionally $t \notin \mathbf{Z}^*(Ly)$. Sims [11] outlined a presentation for Ly, together with a proof for existence and uniqueness of such a group, but the details have never been published. Nevertheless, others have used this presentation to verify matrix representations for Ly (see e.g. Meyer, Neutsch and Parker [8], and Jansen and Wilson [6]), although the original presentation was not fully correct [6]. Cooperman, Finkelstein, Tselman and York [1] have used probabilistic algorithms and the representation of $Ly \leq GL(111, 5)$ by Meyer, Neutsch and Parker to compute a permutation group G of degree 9 606 125 that should be isomorphic to Lyons simple group Ly. The first author [2] has then used Sims' ideas of bases and strong generating sets [10], together with some lengthy calculations and additional arguments, to prove that G is indeed a group in which the two hypotheses of Lyons' original paper hold.

In this paper, we first recall how to use Sims corrected presentation to produce a permutation group P on 8 835 156 points, where the stabilizer of any point is isomorphic to $G_2(5)$. In particular, we get $|P| = |Ly|$. To check Lyons' original characterization, we have to find the centralizer of an involution in P. To do this, we use the concept of standard generators, as introduced by Wilson [12], as an intermediate step. A recipe to go from Sims' generators to standard generators is given by Jansen and Wilson [6], and we use some results from [2] to find generators for the centralizer as words in the standard generators. To do this, we do some calculations inside the permutation group G of degree 9 606 125, computed by Cooperman, Finkelstein, Tselman and York, using the concept of bases and strong generating sets. During the computations we use some results from [2] to describe our elements, but the actual proof is independent of [2], and can be reproduced with the details given here.

13.2 The Permutation Group of Degree 8 835 156

We start with the corrected Sims presentation as given in [4]. This is actually a presentation on the five generators a, b, c, d, z. The four elements a, b, c, d, together with their relations, generate a group isomorphic to $G_2(5)$ [6, 4]. To get a permutation representation on the cosets of this group we use the Adaptive Coset Enumerator (ACE), developed by the second author. As stated in [4], this run needs roughly half a gigabyte of memory with the parameters chosen. For more details on this enumeration the reader is referred to [4] and the remarks therein. The next lemma states the results of this coset enumeration.

Lemma 13.1. *Sims corrected presentation defines a finite group S with $|S| = |Ly|$. The action of S on the cosets of the subgroup $G_2(5) = \langle a, b, c, d\rangle$ gives a faithful permutation representation for S of degree* 8 835 156.

In the following we use the name P for this permutation group, and we want to show that $P \simeq Ly$. To do this we want to find an involution $t \in P$ such that $Cen_P(t) \simeq 2A_{11}$ and $t \notin \mathbf{Z}^*(P)$. To find generators for the centralizer, we go via so-called standard generators. They have been introduced by Wilson [12] to give a unified treatment in the computational theory of sporadic simple groups. Standard generators are sets of elements that are relatively easy to find, with only a small number of defining relations. In the case of Lyons simple group we have the following from [12].

Proposition 13.1. *Standard generators for Lyons simple group Ly are given by a pair (g_1, g_2) of elements of Ly such that*

$$\begin{aligned} o(g_1) &= 2, \\ g_2 &\in 5A, \\ o(g_1g_2) &= 14, \\ o((g_1g_2)^3g_2) &= 67. \end{aligned}$$

(Note that there are two conjugacy classes of elements of order 5 in Ly, but we have $(25A)^5 = 5A$, and $25A$ is unique of its order.)

As mentioned in the introduction, Jansen and Wilson [6] give a way to find standard generators when given Sims' generators. For completeness we state this in the following proposition.

Proposition 13.2. *With Sims' original generators a, b, c, d, z, and the additional element*

$$k_0 = (bzdz)^2dz,$$

the pair of elements

$$\begin{aligned} k_1 &= (ck_0)^{-8}a^4(ck_0)^8, \\ k_2 &= (ck_0ck_0^2)^{-17}c(ck_0ck_0^2)^{17}. \end{aligned}$$

is a pair of standard generators, if $S \simeq Ly$.

To get to the centralizer of an involution, we now need a way to find generators for such a centralizer as words in the standard generators. This is done in the next section in the permutation representation of Ly on 9 606 125 points.

13.3 The Permutation Group of Degree 9 606 125

In [2] it is shown that the permutation group G of degree 9 606 125, as computed by Cooperman, Finkelstein, Tselman and York, is isomorphic to Lyons simple group. This is done by checking Lyons' characterization from his original paper. To do this, two elements c_3, c_4 are computed that are shown to generate the full centralizer of an involution in G. Unfortunately, standard generators are not introduced in [2], but the first author has made additional calculations in an extended version ([3]) to make compatible the different sets of generators for Ly that are available in the literature. The first step is to go from the two generators of [1], an involution x and an element w of order 67, to standard generators. This is done by the following proposition.

Proposition 13.3. *Given the two generators x and w from [1], the pair of elements*

$$\begin{aligned} s_1 &= x, \\ s_2 &= ((xw)^5)^{xw^{12}xw^{33}} \end{aligned}$$

is a pair of standard generators for $G = \langle x, w \rangle \simeq Ly$.

The next step in [3] is to find a strong generating set for the group generated by s_1 and s_2. This enables us to use Schreier vectors when searching for words for the centralizing elements c_3 and c_4. From [3] we get the following proposition.

Proposition 13.4. *Define*

$$\begin{aligned} s_3 &= (s_2s_1)^2s_2^3s_1(s_2^2s_1)^3s_2^3s_1(s_2^4s_1)^3(s_2^2s_1s_2^4s_1)^2s_2^3, \\ s_4 &= s_3^3(s_1s_3^2)^2s_1(s_3^5s_1s_3^3)^{-1}. \end{aligned}$$

Then the set $\{s_1, s_2, s_3, s_4\}$ is a strong generating set for G with stabilizer chain

$$\{1\} \leq T = \langle s_1, s_4 \rangle \leq H = \langle s_1, s_3 \rangle \leq G.$$

This is enough preparation to write the elements c_3 and c_4 as words in the strong generators s_1, s_2, s_3, s_4, hence as words in the standard generators s_1 and s_2. Having computed the Schreier vectors for the different groups in the stabilizer chain above, we get the following result

Lemma 13.2. *Inside G we have*

$$\begin{aligned} c_3 &= s_1 s_4 s_1 s_4^6 (s_1 s_4)^3 s_4^4 s_1 s_4^3 s_1 s_4 s_3^3 s_1 s_3^2 s_1 s_3 s_1 s_3^3 (s_1 s_3)^2 \cdot \\ &\quad \cdot s_2^2 s_1 (s_2 s_1)^3 s_2^4 s_1 s_2^3 s_1 s_2^2 (s_1 s_2 s_1 s_2^4)^2, \\ c_4 &= s_4^3 s_1 s_4^2 s_1 s_4^3 (s_1 s_4)^4 s_4 s_1 s_4^4 s_3^3 s_1 s_3 s_1 s_3^5 s_1 s_3^3 (s_1 s_3)^3. \end{aligned}$$

This gives a way to find the centralizer of an involution, actually s_1, inside Lyons simple group, when given a pair (s_1, s_2) of standard generators for Ly. The next section deals with the question whether this also works in the permutation group P arising from Sims' group S as described in Section 13.2.

13.4 The final Verification

The standard generators for P, given in Proposition 13.2, together with the words given in Proposition 13.4 and Lemma 13.2, now give us two elements in P, say d_3 and d_4 that should generate the centralizer of the involution g_1, if $P \simeq Ly$. To prove that $P \simeq Ly$, we have to show the following: first we have to prove that $C = \langle d_3, d_4 \rangle$ is the full centralizer of g_1 in P; then we have to show that $C \simeq 2A_{11}$; and finally we need that $g_1 \notin \mathbf{Z}^*(P)$. We begin with the middle part, since this is easy by the calculations done in [2]. There is a classical presentation for $2A_{11}$, already known to Schur [9], given by the following theorem.

Theorem 13.1. *The following are defining relations for the double cover $2A_{11}$ of the alternating group A_{11} as a finitely presented group on the ten generators $b_1, b_2, b_3, b_4, b_5, b_6, b_7, b_8, b_9, j$:*

$$\begin{aligned} b_1^3 &= j, \\ b_\alpha^2 &= j \qquad \text{for } \alpha = 2, \dots, 9, \\ (b_1 b_2)^3 &= j, \\ (b_1 b_\lambda)^2 &= j \qquad \text{for } \lambda = 3, \dots, 9, \\ (b_\beta b_{\beta+1})^3 &= j \qquad \text{for } \beta = 2, \dots, 8, \\ b_\gamma b_\delta &= j b_\delta b_\gamma \qquad \text{for } \gamma = 2, \dots, 7 \text{ and } \delta = \gamma + 2, \dots, 9. \end{aligned}$$

To find words in our two elements d_3 and d_4 for the above relations, we once again quote [2], where another application of Schreier vectors gives such words in the case of the permutation group G of degree 9 606 125. Using the same words inside P, we get the following theorem.

Theorem 13.2. *Define the elements*

$$\begin{aligned} u_0 &= d_3, \\ u_1 &= d_4, \\ u_2 &= u_0^7 u_1 u_0^3 (u_1 u_0)^2 (u_0 u_1)^2 u_0^{-2} u_1 u_0^{-4} (u_1 u_0^{-1})^4 u_0^{-2}, \\ u_3 &= (u_0^2 u_1)^2 u_0 (u_0^2 u_1)^2 u_0 u_1 u_0^{-1} u_1 u_0^{-3} u_1 u_0^{-2} (u_1 u_0^{-1})^3, \\ u_4 &= (u_3 u_2)^2 (u_2 u_3)^2 u_3^2 u_2 u_3^{-1} u_2^{-2} u_3^{-3} u_2^{-1} \end{aligned}$$

as intermediate results. Then the following elements

$$
\begin{aligned}
b_1 &= u_4^4 u_2 u_3^3 u_2^4 u_3 (u_1 u_0)^3 (u_0 u_1)^3, \\
b_2 &= u_4^6 (u_2 u_3)^2 u_3 u_2^5 u_3 (u_0 u_1)^2 u_0^5 (u_1 u_0)^2 u_0^3 u_1, \\
b_3 &= u_4^4 u_3^2 u_2^2 u_3^4 (u_0 u_1)^3 u_0^2 u_1 u_0^5 u_1 u_0 u_1, \\
b_4 &= u_4^4 u_2 u_3 u_2^2 u_3^4 u_2^4 u_1 u_0^6 (u_1 u_0)^3 u_0^5, \\
b_5 &= u_3^2 u_2^2 u_3^3 u_1 u_0^3 (u_1 u_0)^2 u_0^2 u_1, \\
b_6 &= u_4^5 u_2^2 u_3 u_2^4 u_3^2 u_0 u_1 u_0^4 (u_1 u_0^2)^2 u_1 u_0, \\
b_7 &= u_4^2 u_2^3 u_3^3 u_2^2 u_3^2 u_0^6 (u_1 u_0)^2 u_0^3 u_1, \\
b_8 &= u_4^6 u_2^2 u_3^3 u_2 (u_0^2 u_1)^4 u_0^2, \\
b_9 &= u_4^6 u_3^2 u_2^3 u_3 u_2 u_0^6 u_1 u_0^7 u_1 u_0 u_1
\end{aligned}
$$

generate a subgroup D of C that is isomorphic to $2A_{11}$, which is confirmed by checking the relations in Theorem 13.1.

The next step is to prove that $C = D \simeq 2A_{11}$ and that C is the full centralizer of g_1 in P. To do this, we first note that $C \leq Cen_P(g_1)$ by checking the generators d_3 and d_4 of C. To compute the order of $Cen_P(g_1)$, we use the formula for induced characters from page 64 of Isaacs' book on character theory [5]. It is easy to compute that g_1 has 2772 fixpoints in our permutation representation. Now choose any stabilizer $U \simeq G_2(5)$ with $g_1 \in U$. Then U has only one class of involutions, and the abovementioned formula gives

$$\varphi^P(g_1) = |\, Cen_P(g_1) \,| \cdot \frac{\varphi(g_1)}{|\, Cen_U(g_1) \,|}$$

where φ is the trivial character of U. Therefore we get the following lemma.

Lemma 13.3.

$$
\begin{aligned}
|\, Cen_P(g_1) \,| &= \varphi^P(g_1) \cdot |\, Cen_H(g_1) \,| \\
&= 2772 \times 14\,400 = 39\,916\,800 = |2A_{11}|
\end{aligned}
$$

Since $2A_{11} \simeq D \leq C \leq Cen_P(g_1)$, this gives

$$Cen_P(g_1) = C \simeq 2A_{11}$$

To end our proof that Lyons' characterization works for P, we have to show that $g_1 \notin \mathbf{Z}^*(P)$. But we have $g_1 \in U \simeq G_2(5)$, and $\mathbf{Z}^*(G_2(5)) = \{1\}$, hence $g_1 \notin \mathbf{Z}^*(G_2(5))$. This gives the following lemma immediately.

Lemma 13.4.

$$g_1 \notin \mathbf{Z}^*(P)$$

Now we have checked everything from Lyons' characterization, hence conclude with the following final result.

Theorem 13.3. *Sims corrected presentation describes Lyons simple group Ly.*

Acknowledgements

The second author was partially supported by the Australian Research Council.

Bibliography

[1] G. Cooperman, L. Finkelstein, M. Tselman and B. York: *Constructing Permutation Representations for Matrix Groups*, J. Symbolic Comput. **24**, (1997) 1–18.

[2] H.W. Gollan: *A new existence proof for Ly, the sporadic simple group of R. Lyons*, Preprint **30**, Institut für Experimentelle Mathematik, Universität GH Essen, 1995.

[3] H.W. Gollan: *A Contribution to the Revision Project of the Sporadic Groups: Lyons' simple group Ly*, Vorlesungen aus dem FB Mathematik, volume **26**, Universität GH Essen, 1997.

[4] G. Havas and C.C. Sims: *A presentation for the Lyons simple group*, These proceedings, chapter 14.

[5] I.M. Isaacs: *Character Theory of Finite Groups*,. Academic Press, New York (1976).

[6] C. Jansen and R.A. Wilson: *The minimal faithful 3-modular representation for the Lyons group*, Commun. Algebra **24**, (1996) 873–879.

[7] R. Lyons: *Evidence for a new finite simple group*, J. Algebra **20**, (1972) 540–569.

[8] W. Meyer, W. Neutsch and R. Parker: *The minimal 5-representation of Lyons' sporadic group*, Math. Ann. **272**, (1985) 29–39.

[9] I. Schur: *Über die Darstellung der symmetrischen und der alternierenden Gruppe durch gebrochene lineare Substitutionen*, Journal Reine Angew. Math. **139**, (1911) 155–250.

[10] C.C. Sims: *Computational methods in the study of permutation groups*, in: Computational problems in abstract algebra (Pergamon Press, 1970) 169–183.

[11] C.C. Sims: *The existence and uniqueness of Lyons' group*, in: Finite Groups '72 (North Holland, 1973) 138–141.

[12] R.A. Wilson: *Standard generators for sporadic simple groups*, J. Algebra **184**, (1996) 505–515.

Progress in Mathematics, Vol. 173, © 1999 Birkhäuser Verlag Basel/Switzerland

Chapter 14

A Presentation for the Lyons Simple Group

George Havas and Charles C. Sims

Abstract

We give a presentation of the Lyons simple group together with information on a complete computational proof that the presentation is correct. This fills a longstanding gap in the literature on the sporadic simple groups. This presentation is a basis for various matrix and permutation representations of the group.

14.1 Introduction

Lyons [9, 10] gave a detailed description of a new simple group. Sims [13] outlined a presentation for the group and a proof of its correctness. The proof itself is in a long manuscript which was never published. That manuscript in fact does much more than just provide a presentation; it gives a proof of existence and uniqueness for the group. The original proof relied upon a large number of special purpose computer programs.

With the dramatic improvement of computer technology in both hardware and software we now give a complete presentation for the Lyons group and a proof of correctness which can be verified readily enough. This proof follows the initial lines of that in Sims' manuscript, with minor modifications. It requires a computer with adequate random access memory and also computational group theory facilities such as provided by GAP [12] or MAGMA [1]. A major procedural requirement is the ability to undertake large coset enumerations. All computations described in this paper can be carried out using GAP and/or MAGMA.

The lemmas in this paper are based on only a small part of the original (mainly hand-written) manuscript, which comprises four chapters and ten appendices. The four chapters run to 124 pages and the appendices are another 62 pages. The original manuscript gives the following theorem.

Theorem 14.1. *Up to isomorphism there is exactly one finite simple group G possessing an involution whose centralizer in G is isomorphic to the 2-fold covering group of A_{11}.*

The proof in the original manuscript goes this way. The group G is constructed as a permutation group of degree 8835156 in which the stabilizer of a point is $G_2(5)$. The proof uses in effect the presentation given in this paper. The existence and uniqueness proofs for G reduce essentially to the verification of a large number of facts about $G_2(5)$, much of which is done by computer. Algorithms and programs are explicitly provided. Chapter two of the manuscript presents the proof as a sequence of 58 traditional lemmas, traditional in the sense that their statements do not refer to computers or programs. However the proofs of these lemmas do rely heavily on computer results.

14.2 A Presentation for the Lyons Group

In this section we give an amended version of the presentation which appears in Sims' manuscript. (It has previously been noted by Jansen and Wilson [8] that the presentation in the original manuscript is not fully correct.)

The presentation given by Sims is on five generators: a, b, c, d, z, where $\langle a, b, c, d\rangle \simeq G_2(5)$. A number of redundant generators e, a_i, c_j, u and x_k, all in $\langle a, b, c, d\rangle$, are used as in the original construction.

Let G be the group generated by the 39 elements $\{a$, b, c, d, z, e, a_1, a_2, a_3, a_4, c_1, c_2, c_3, c_4, c_5, u, x_{23}, x_{24}, x_{25}, x_{29}, x_{47}, x_{96}, x_{129}, x_{162}, x_{163}, x_{164}, x_{165}, x_{166}, x_{167}, x_{168}, x_{169}, x_{364}, x_{365}, x_{366}, x_{367}, x_{368}, x_{369}, x_{370}, $x_{371}\}$ satisfying the following 86 relators and relations:

$$a^8, \tag{14.1}$$
$$b^5, \tag{14.2}$$
$$(ab)^4, \tag{14.3}$$
$$(a^2, b), \tag{14.4}$$
$$(a, b)^3, \tag{14.5}$$
$$c_1 = c, \tag{14.6}$$
$$c_2 = c^a, \tag{14.7}$$
$$c_3 = c^b, \tag{14.8}$$
$$c_4 = c^{b^{-1}}, \tag{14.9}$$
$$c_5 = (c_1, c_2), \tag{14.10}$$
$$c_1^5, \tag{14.11}$$
$$c_5^5, \tag{14.12}$$
$$(c_1, c_5), \tag{14.13}$$
$$(c_2, c_5), \tag{14.14}$$
$$(c_3, c_5), \tag{14.15}$$
$$(c_4, c_5), \tag{14.16}$$
$$(c_1, c_3) = c_5^{-2}, \tag{14.17}$$
$$(c_1, c_4) = c_5^2, \tag{14.18}$$
$$(c_2, c_3) = c_5^{-1}, \tag{14.19}$$

$$(c_2, c_4), \tag{14.20}$$
$$(c_3, c_4) = c_5^{-1}, \tag{14.21}$$
$$c_2^a = c_1^{-2}, \tag{14.22}$$
$$c_3^a = c_3^{-2} c_1 c_4, \tag{14.23}$$
$$c_4^a = c_1 c_2^{-1} c_3^{-1} c_4 c_1 c_4, \tag{14.24}$$
$$c_5^a = c_5^2, \tag{14.25}$$
$$c_2^b = c_2 c_3 c_4^{-1} c_1^{-1} c_4^{-1}, \tag{14.26}$$
$$c_3^b = c_1^2 c_4 c_3^{-2}, \tag{14.27}$$
$$c_5^b = c_5, \tag{14.28}$$
$$e = b^{-2}(a, c_1) a^{-1} c_1 a d c_1 b a b^{-1} d c_1 d, \tag{14.29}$$
$$e^2, \tag{14.30}$$
$$c_1^e = b c_3 b c_1^{-1}, \tag{14.31}$$
$$c_3^e = b^2 a c_2 c_1^{-1} c_4^{-1} a^{-1}, \tag{14.32}$$
$$c_4^e = b^2 c_3^{-1} c_4, \tag{14.33}$$
$$c_5^e = c_1 c_3^{-1} c_1 c_4^{-1}, \tag{14.34}$$
$$e a^2 e = a b^2 a b^{-1} a c_1 c_5 c_4^{-1}, \tag{14.35}$$
$$b^e = c_1 c_3^{-1} c_4^{-2}, \tag{14.36}$$
$$e a b^2 a b^{-1} a e = a^2 c_1 c_5 c_4^{-1}, \tag{14.37}$$
$$e c_2 e c_2^{-1} e c_2 c_4^{-1} a c_3 a^3, \tag{14.38}$$
$$d = a b^2 a^{-1} b^{-1} c_1^{-1} c_4^{-1} c_5^{-1} a e a e a c_2^{-1} b e, \tag{14.39}$$
$$d^2 = a^4, \tag{14.40}$$
$$(b, d), \tag{14.41}$$
$$a^d = a^3, \tag{14.42}$$
$$(d c_5)^5, \tag{14.43}$$
$$d c_5 d c_5^{-1} d c_5 a^{-2}, \tag{14.44}$$
$$a_1 = a, \tag{14.45}$$
$$a_2 = a^{b^{-1}}, \tag{14.46}$$
$$a_3 = a^{b^{-2}}, \tag{14.47}$$
$$a_4 = a^{b^2}, \tag{14.48}$$
$$u = a^2, \tag{14.49}$$
$$x_{23} = d, \tag{14.50}$$
$$x_{24} = d^{-1}, \tag{14.51}$$
$$x_{25} = b^2 a^{-1} b^{-1} a^{-1} b^2 c^{-1} a c a^{-1} c^{-1} d c^{-1} b^2 a^{-1} c b, \tag{14.52}$$
$$x_{29} = a b a^{-1} b^2 a^{-1} b^2 c^{-1} a c b c b^{-1} d c b^{-2} d c b c, \tag{14.53}$$
$$x_{47} = c^{-1} d^{-1} a^{-1} c^{-1} a^{-1} b c^{-1} d^{-1}, \tag{14.54}$$
$$x_{96} = b d, \tag{14.55}$$

$$x_{129} = d^{-1}b^{-1}, \tag{14.56}$$
$$x_{162} = a^{-1}b^{-1}a^{-1}bc^{-1}b^{-1}ac^{-1}ba, \tag{14.57}$$
$$x_{163} = b^{-2}aca^{-1}bcb, \tag{14.58}$$
$$x_{164} = a^{-1}b^{-1}a^{-1}caba, \tag{14.59}$$
$$x_{165} = b^{-1}a^{-1}bcb^{-1}ab, \tag{14.60}$$
$$x_{166} = b^2a^{-1}b^{-1}cbacb^{-2}c^{-1}, \tag{14.61}$$
$$x_{167} = a^{-3}, \tag{14.62}$$
$$x_{168} = ba^{-1}b^{-1}a^{-2}, \tag{14.63}$$
$$x_{169} = ab^2ab^{-2}a^{-1}, \tag{14.64}$$
$$x_{364} = a^{-1}db^{-1}c^{-1}a^{-1}bc^{-1}d^{-1}bc^{-1}b^{-1}a^{-1}c^{-1}ab^{-1}abab^{-1}, \tag{14.65}$$
$$x_{365} = ab^{-2}a^{-1}cbc^{-1}ba^{-1}c^{-2}dcaca^{-1}cb^{-1}d^{-1}, \tag{14.66}$$
$$x_{366} = bab^{-1}a^{-1}b^{-2}a^{-1}b^{-1}cb^2cbdcb^{-2}dcbc, \tag{14.67}$$
$$x_{367} = c^{-1}b^{-1}c^{-1}d^{-1}b^2c^{-1}d^{-1}b^{-1}c^{-1}b^{-2}c^{-1}bab^2aba^{-1}b^{-1}, \tag{14.68}$$
$$x_{368} = a^{-1}c^{-1}d^{-1}ac^{-1}b^{-1}ac^{-1}d^{-1}a^{-1}c^{-1}acb^{-2}c^{-1}b^{-1}a^{-1}b^{-2}, \tag{14.69}$$
$$x_{369} = a^{-1}c^{-1}d^{-1}ba^{-1}b^{-1}c^{-1}d^{-1}ba^{-1}b^{-1}c^{-1}b^{-1}ab^{-2}ab^{-1}, \tag{14.70}$$
$$x_{370} = d^{-1}acbca^{-1}c^{-1}d^{-1}ab^{-1}a^{-1}ca^{-1}b^{-2}a^{-1}ba^{-1}, \tag{14.71}$$
$$x_{371} = c^{-1}a^{-1}cbc^{-1}d^{-1}ca^{-1}cacb^{-1}a^{-1}b^{-1}a^{-1}b, \tag{14.72}$$
$$c_1^z = x_{162}, \tag{14.73}$$
$$c_2^z = x_{163}, \tag{14.74}$$
$$c_3^z = x_{164}, \tag{14.75}$$
$$c_4^z = x_{165}, \tag{14.76}$$
$$c_5^z = x_{166}, \tag{14.77}$$
$$a_1^z = x_{167}, \tag{14.78}$$
$$a_2^z = x_{168}, \tag{14.79}$$
$$a_3^z = x_{169}, \tag{14.80}$$
$$z^2 = u, \tag{14.81}$$
$$(a_4z)^3u, \tag{14.82}$$
$$x_{47}zx_{364}zx_{368}z, \tag{14.83}$$
$$z^{-1}x_{24}z^{-1}x_{369}zx_{23}zx_{29}, \tag{14.84}$$
$$zx_{366}zx_{365}z^{-1}x_{367}z^{-1}x_{370}, \tag{14.85}$$
$$zx_{96}zx_{25}z^{-1}x_{129}z^{-1}x_{371}. \tag{14.86}$$

Theorem 14.2. *G is isomorphic to the Lyons simple group.*

The proof is given in the following section where we identify relations from this presentation by their numbers. It should be noted that it is this presentation (in effect) for the Lyons group which is used as part of proofs in [11, 8, 15]. Thus proofs that various matrix representations of the group are correct rely upon the fact that their generators satisfy all the relations in a presentation for the group.

Furthermore, the final coset enumeration described in this paper gives a minimal degree permutation representation. An alternative independent existence proof for the Lyons group is given in [5].

14.3 A computational Proof

We give steps in the proof that the presentation is correct as a sequence of lemmas. Each lemma is readily proved computationally. (The lemmas are derived from those given by Sims, but are varied in the light of current computational capabilities.)

Consider the following four 7×7 matrices over $GF(5)$ (given in [13]).

$$a = \begin{pmatrix} 4 & 0 & 0 & 0 & 0 & 0 & 0 \\ 0 & 0 & 0 & 0 & 3 & 0 & 0 \\ 0 & 0 & 0 & 2 & 0 & 0 & 0 \\ 0 & 0 & 1 & 0 & 0 & 0 & 0 \\ 0 & 1 & 0 & 0 & 0 & 0 & 0 \\ 0 & 0 & 0 & 0 & 0 & 0 & 2 \\ 0 & 0 & 0 & 0 & 0 & 3 & 0 \end{pmatrix} \qquad b = \begin{pmatrix} 4 & 0 & 0 & 0 & 0 & 0 & 4 \\ 0 & 2 & 0 & 0 & 2 & 0 & 0 \\ 0 & 0 & 0 & 3 & 0 & 0 & 0 \\ 0 & 0 & 3 & 2 & 0 & 0 & 0 \\ 0 & 2 & 0 & 0 & 0 & 0 & 0 \\ 0 & 0 & 0 & 0 & 0 & 0 & 1 \\ 3 & 0 & 0 & 0 & 0 & 1 & 4 \end{pmatrix}$$

$$c = \begin{pmatrix} 1 & 1 & 0 & 0 & 2 & 0 & 0 \\ 0 & 1 & 0 & 0 & 0 & 0 & 0 \\ 3 & 1 & 1 & 0 & 3 & 2 & 4 \\ 1 & 3 & 0 & 1 & 4 & 3 & 4 \\ 0 & 0 & 0 & 0 & 1 & 0 & 0 \\ 0 & 1 & 0 & 0 & 1 & 1 & 0 \\ 0 & 3 & 0 & 0 & 2 & 0 & 1 \end{pmatrix} \qquad d = \begin{pmatrix} 1 & 0 & 0 & 0 & 0 & 0 & 0 \\ 0 & 0 & 0 & 3 & 0 & 0 & 0 \\ 0 & 0 & 0 & 0 & 2 & 0 & 0 \\ 0 & 3 & 0 & 0 & 0 & 0 & 0 \\ 0 & 0 & 2 & 0 & 0 & 0 & 0 \\ 0 & 0 & 0 & 0 & 0 & 1 & 0 \\ 0 & 0 & 0 & 0 & 0 & 0 & 1 \end{pmatrix}$$

Lemma 14.1. *The matrices a, b, c, d are elements of $G_2(5)$ and satisfy relations (14.1–14.44). (These are all the relations in the presentation for G which involve only $a, b, c, d, e, c_1, c_2, c_3, c_4, c_5$, wherein $c_1, c_2, c_3, c_4, c_5, e$ are defined.)*

The family of groups $G_2(q)$ was originally described by Dickson [3, 4]. That the matrices a, b, c and d are elements of $G_2(5)$ follows from Dickson's characterization of $G_2(5)$ as the group of 7×7 matrices with determinant 1 which preserve a particular quadratic form and map a specific subspace of alternating bilinear forms into itself. The group $G_2(5)$ has order 5859000000. That matrices a, b, c and d satisfy the specified relations for G can be verified by straightforward computation. We call the group presented by these generators and relations H. Further, by an easy computation with MAGMA we obtain the following result.

Lemma 14.2. *The matrices a, b, c, d generate $G_2(5)$.*

We now identify H by studying subgroups. In H let $M = \langle a, b, c \rangle$, $T = \langle a, b \rangle$, and $P = \langle c_1, c_2, c_3, c_4, c_5 \rangle$.

Lemma 14.3. *T is a homomorphic image of $GL(2, 5)$.*

This can be verified by straightforward computation. Coset enumeration shows that the group $\langle a, b \mid a^8, b^5, (ab)^4, (a^2, b), (a, b)^3 \rangle$ defined by relations (14.1–14.5) of the presentation for G has order $480 = |GL(2,5)|$. In $GL(2,5)$ define $a' = \left(\begin{smallmatrix} 0 & 3 \\ 1 & 0 \end{smallmatrix}\right)$, $b' = \left(\begin{smallmatrix} 2 & 2 \\ 2 & 0 \end{smallmatrix}\right)$. Then relations (14.1–14.5) are satisfied if a and b are replaced by a' and b'. Since a', b' and (a', b') have orders 8, 5 and 3, respectively, it follows that $GL(2,5) = \langle a', b' \rangle$ and so relations (14.1–14.5) give a presentation for $GL(2,5)$. (This presentation omits a redundant relator which appeared in earlier presentations.)

Lemma 14.4. *The order of P divides 5^5, T normalizes P, and $M = PT$. The order of M divides* 1500000.

That the order of P divides 5^5 follows from its generators satisfying relations (14.10–14.21) which, together with the orders of c_2, c_3 and c_4 being 5, define a group of order 5^5 (easily verified by coset enumeration, for example). That T normalizes P follows from relations (14.7–14.9) and (14.22–14.28), which show the actions of a and b on the generators of P. It follows that $|M|$ divides $5^5|GL(2,5)|$.

Lemma 14.5. *The index of M in H is* 3906. *The order of H divides* 585900000. *H is isomorphic to $G_2(5)$.*

The coset enumeration for $|H : M|$ is straightforward. That $H \simeq G_2(5)$ follows. Since the matrices a, b, c, d satisfy the defining relations for H and generate $G_2(5)$, that group is a quotient group of H. But the order of H is at most the order of $G_2(5)$. Therefore they are isomorphic. Also, each of the subgroups T, P and M are isomorphic to the above specified preimages, with orders 480, 5^5 and 1500000, respectively.

The final step in the proof comes from the following lemma.

Lemma 14.6. *The index of H in G is* 8835156. *G has order* 51765179004000000.

That the permutation representation coming from this enumeration is a Lyons group is shown in [6]. (Here we choose to refer to this more accessible description of a construction, rather than follow the method used in the original manuscript, which accounts for much of its length.) The coset enumeration involved in this step is the only one which is challenging in the context of todays computing facilities. We give some details in the next section.

14.4 The Large Coset Enumeration

The enumeration to determine $|G : H|$ requires the definition of some nine million cosets, at least. Thus this step requires some care. Recent descriptions of relevant aspects of coset enumeration are given in [7, 14, 2]. Of particular importance is the fact that in broad terms the space required (memory locations) for a standard coset enumeration to complete is at least the maximum number of cosets defined times twice the number of group generators.

The presentation for G is given on 39 generators, most of which are redundant. A regular coset enumeration using this presentation would need about three gigabytes of random access memory, at least, which is more than available on all but the biggest supercomputers. Furthermore, it would take a very long time.

This can be readily alleviated by eliminating most of the redundant generators. Thus, using Tietze transformations, it is easy to obtain a six generator presentation on $\{a, b, c, d, e, z\}$. This is straightforward, even by hand, using relators (14.6–14.10) and (14.45–14.72), and can be done using built-in capabilities of either GAP or MAGMA. It yields a 53 relator presentation.

Using this presentation reduces the memory requirements by a factor of better than six. This brings the computation within the range of computers with as little as half a gigabyte of random access memory.

Coset enumerations using such six generator presentations have now been performed a number of times using two independent coset enumeration programs. With Felsch-type methods the maximum number of cosets defined is equal to the index and the total is less than 13 million, depending somewhat on specific details of presentation used and enumeration strategy. The first time it was done (in July 1996), using a descendant of the program described in [7] and available in MAGMA, the enumeration completed in about 15 cpu hours on a CONVEX computer, indeed using about half a gigabyte of random access memory.

As foreshadowed in [2] this enumeration has successfully been done in parallel. In this case it used a version of the enumerator which is in GAP and due to Martin Schönert. It has been done in a four processor environment on a CONVEX and in a 16 processor environment on an SGI Power Challenge. In each case, approximately linear speed-up with processors was realized.

Input files to reproduce this coset enumeration in GAP, MAGMA, or with stand-alone programs are available from the first author.

Of course it is simple to derive presentations for the Lyons group on fewer generators. In our presentation, relation (14.29) allows us to express e in terms of a, b, c and d, and relation (14.39) allows us to express d in terms of a, b, c and e. However, how easy it is to find a presentation suitable for coset enumeration on fewer than six generators is unclear.

Thus, it is trivial to use Tietze transformations to eliminate generator e from the six generator presentation. However this replaces 23 occurrences of the generator e by a product 17 symbols long. It is possible that coset enumeration in such a presentation would be much more difficult than compensated for by the saving due to fewer generators. This is evidenced by coset enumeration behaviour in $G_2(5)$. For index 3906 enumerations of $\langle a, b, c \rangle$ in $G_2(5)$ (that is, corresponding to the enumeration $H|M$), the maximum number of cosets required explodes from 3906 for five generator presentations on $\{a, b, c, d, e\}$ to the extent that enumerations do not complete given a maximum of five million cosets for corresponding four generator presentations on $\{a, b, c, d\}$.

It is likewise easy to eliminate d from the six generator presentation for G. This replaces occurrences of the generator d by a product 18 symbols long. In one sense this is more promising, since the index 3906 enumerations in corresponding four generator presentations for $H|M$ require the same maximum and similar total cosets as for the five generator presentations. However, there are many more occurrences of d in the presentation for G in view of the way many of the x_i are defined involving d. At the time of writing, enumerations in five generator presentations for G have not been attempted.

Acknowledgements

The first author was partially supported by the Australian Research Council.

Bibliography

[1] W. Bosma, J. Cannon and C. Playoust: *The Magma algebra system I: The user language*, J. Symbolic Comput. **24**, (1997) 235–265. See also `http://www.maths.usyd.edu.au:8000/comp/magma/Overview.html`

[2] G. Cooperman and G. Havas: *Practical parallel coset enumeration*, in: Workshop on High Performance Computing and Gigabit Local Area Networks, Lecture Notes in Control and Information Systems **226**, (1997) 15–27.

[3] L.E. Dickson: *Theory of linear groups in an arbitrary field*, Trans. Amer. Math. Soc. **2**, (1901) 363–394.

[4] L.E. Dickson: *A new system of simple groups*, Math. Ann. **60**, (1905) 137–150.

[5] H.W. Gollan: *A new existence proof for Ly, the sporadic simple group of R. Lyons*, Preprint 30, Institut für Experimentelle Mathematik, Universität GH Essen, 1995.

[6] H.W. Gollan and G. Havas: *On Sims' presentation for Lyons' simple group*, These proceedings, chapter 13.

[7] G. Havas: *Coset enumeration strategies*, in: ISSAC'91, Proc. 1991 Internat. Sympos. Symbolic and Algebraic Computation (ACM Press, New York, 1991) 191–199.

[8] C. Jansen and R.A. Wilson: *The minimal faithful 3-modular representation for the Lyons group*, Commun. Algebra **24**, (1996) 873–879.

[9] R. Lyons: *Evidence for a new finite simple group*, J. Algebra **20**, (1972) 540-569.

[10] R. Lyons: *Evidence for a new finite simple group*, (Errata) J. Algebra **34**, (1975) 188–189.

[11] W. Meyer, W. Neutsch and R. Parker: *The minimal 5-representation of Lyons' sporadic group*, Math. Ann. **272**, (1985) 29–39.

[12] M. Schönert et al.: GAP – *Groups, Algorithms and Programming*, (Lehrstuhl D für Mathematik, Rheinisch-Westfälische Technische Hochschule, Aachen, Germany, fifth edition, 1995). See also `http://www-groups.dcs.st-and.ac.uk/~gap`

[13] C.C. Sims: *The existence and uniqueness of Lyons' group*, in: Finite Groups '72 (North Holland, 1973) 138–141.

[14] C.C. Sims: *Computation with finitely presented groups*, Cambridge University Press (1994).

[15] R.A. Wilson: *A representation for the Lyons group in* $GL_{2480}(4)$, *and a new uniqueness proof*, Arch. Math. **70**, (1998) 11–15.

Progress in Mathematics, Vol. 173, © 1999 Birkhäuser Verlag Basel/Switzerland

Chapter 15

Reduction of Weakly Definite Unit Forms

Hans-Joachim von Höhne

15.1 Introduction

In the following we consider integral quadratic forms $q : \mathbb{Z}^n \longrightarrow \mathbb{Z}$ of the form $q(x) = \sum_i x_i^2 + \sum_{i<j} q_{ij} x_i x_j$, called *unit forms*. Such unit forms appear naturally (as Tits– or Euler forms) in various areas of representation theory in connection with the characterization of the representation type and the examination of the Jordan-Hölder multiplicities of the indecomposable representations (see for instance [11, 1, 10]). In this context the main problems concerning a given unit form q are the following. 1) Decide whether or not q is *weakly non-negative* (resp. *weakly positive*), that is, q satisfies $q(x) \geq 0$ (resp. $q(x) > 0$) for all $0 \neq x \in \mathbb{N}^n$. 2) Describe the set $\mathcal{O}_q = \{0 \neq x \in \mathbb{N}^n \mid q(x) = 0\}$ of *positive zero roots* if q is weakly non-negative and the set $\mathcal{R}_q = \{x \in \mathbb{N}^n \mid q(x) = 1\}$ of *positive roots* if q is weakly positive. There are several concepts to approach these problems; a survey is given by the author in [10].

In the present note we emphasize the use of *reduction* of unit forms. This technique was introduced in [3] as an adaptation of the 'edge reduction' formulated originally by Kleiner and Roiter for DGC's and bocses and has been proved to be a powerful tool to decide weak positivity and to describe the positive roots in this case. In [5] it was shown that weak non-negativity behaves also well under reduction, however the algorithm presented there (using exhaustive reductions) is not very useful in practice. Here now we give an algorithm for weak non-negativity which may be considered as a generalization of the algorithm presented in [3] and works very efficiently in practice. Moreover, in the weakly non-negative case, the algorithm produces the set of all critical vectors of the considered unit form; this set plays a decisive role for the description of all zero roots (see Section 15.4).

15.2 Proper Reductions

We first recall some basic notions and results from [3]. Let $q : \mathbb{Z}^n \longrightarrow \mathbb{Z}$ be a unit form and $a < b$ be indices such that $q_{ab} = -1$.

We define a new unit form $q' : \mathbb{Z}^{n+1} \longrightarrow \mathbb{Z}$ by the formula

$$q'(y) = q\rho(y) + y_a y_b$$

where $\rho : \mathbb{Z}^{n+1} \longrightarrow \mathbb{Z}^n$ denotes the linear map which maps the canonical basis vector $e_i \in \mathbb{Z}^{n+1}$ onto $e_i \in \mathbb{Z}^n$ if $1 \le i \le n$ and onto $e_a + e_b$ if $i = n+1$. In other words, the integral symmetric matrix associated with q' is obtained from that of q by writing the sum of the ath and bth column (row) into the $n+1$st column (row), setting the $n+1$st diagonal entry $= 2$ and changing the entries at the places (a, b) and (b, a) from -1 to 0. We say that q' is obtained from q by *proper reduction* with respect to a and b.

The unit form q can be recovered from q' using the non-linear map $\sigma : \mathbb{Z}^n \longrightarrow \mathbb{Z}^{n+1}$ defined as follows: $\sigma(x)_i = x_i$ for $i \notin \{a, b, n+1\}$ and

$$(\sigma(x)_a, \sigma(x)_b, \sigma(x)_{n+1}) = \begin{cases} (0, x_b - x_a, x_a) & \text{if } x_a \le x_b\,, \\ (x_a - x_b, 0, x_b) & \text{otherwise.} \end{cases}$$

Indeed, we have $\rho\sigma = \text{id}$ and therefore $q'\sigma = q$.

The following result describes the main relation between q and q'.

Proposition 15.1. *[3, 5] q is weakly non-negative (resp. weakly positive) if and only if q' is weakly non-negative (resp. weakly positive). If this is the case, the maps σ and ρ induce inverse bijections between $\mathcal{O}_q$ and $\mathcal{O}_{q'}$ (resp. $\mathcal{R}_q$ and $\mathcal{R}_{q'}$).*

Proof. of the weakly non-negative part. The equivalence of the weak non-negativity of q and q' and the inclusion $\sigma(\mathcal{O}_q) \subseteq \mathcal{O}_{q'}$ hold since σ and ρ map positive vectors into positive vectors and $q = q'\sigma$. Remains to show that $\rho(\mathcal{O}_{q'}) \subseteq \mathcal{O}_q$ and $\sigma\rho\,|_{\mathcal{O}_{q'}} = \text{id}_{\mathcal{O}_{q'}}$ if q is weakly non-negative. Indeed, the inequality $0 = q'(y) = q\rho(y) + y_a y_b \ge y_a y_b \ge 0$ for $y \in \mathcal{O}_{q'}$ yields $y_a y_b = 0$ and therefore $\rho(y) \in \mathcal{O}_q$ and $\sigma\rho(y) = y$. □

Remark 15.1.

1) In [3] the above result gives rise to a simple algorithm for weak positivity, roughly described as follows. Given a unit form q on $\mathbb{Z}^n$ one produces a sequence of unit forms $q^0, q^1, q^2, \ldots$ by forming iterated proper reductions of $q^0 = q$. If the process stops at some step $m \le 6^n - n$ with $q^m_{ab} \ge 0$ for all $a < b$ then q^m and therefore q is weakly positive and $\mathcal{R}_q = \rho^m(\{e_1, \ldots, e_{n+m}\})$; otherwise q is not weakly positive. The bound 6^n comes from the well-known result of Ovsienko [7] saying that $x_i \le 6$ for all $x \in \mathcal{R}_q$ and $1 \le i \le n$ if q is weakly positive.

2) In contrast, the same procedure of taking arbitrary iterated proper reductions does not work in order to decide weak non-negativity. The reason for that is twofold: first, the set $\mathcal{O}_q$ does not offer an a priori bound as $\mathcal{R}_q$ does in the weakly positive case and secondly, there may occur self-reproductions. The first problem can be avoided if we consider instead of $\mathcal{O}_q$ the set of critical vectors of q; this set is a finite subset of $\mathcal{O}_q$ and contains essential information about $\mathcal{O}_q$ and weak definiteness properties of q (see Section 15.4). The problem with prevalent selfreproductions can be reduced by using a further kind of reduction, which we describe now.

15.3 Restricted Reductions

Given a unit form $q : \mathbb{Z}^n \longrightarrow \mathbb{Z}$ and indices $a < b$ such that $q_{ab} < 0$ (arbitrary) we define a further unit form $q'' : \mathbb{Z}^n \longrightarrow \mathbb{Z}$ by the formula

$$q''(y) = q(y) - q_{ab} y_a y_b$$

and say that q'' is obtained from q by *restricted reduction* with respect to a and b. Note that, for $q_{ab} = -1$, the unit form q'' is the restriction to the first n variables of the proper reduction of q with respect to a and b.

Remark 15.2.

1) If q is weakly non-negative then so is q'', but the converse does not hold in general.

2) We associate with q'' the maps $\rho = \sigma = \mathrm{id}_{\mathbb{Z}^n}$. These maps satisfy similar relations as the maps ρ and σ associated with proper reductions: $\rho\sigma = \mathrm{id}_{\mathbb{Z}^n}$, $q''(y) = q\rho(y) - q_{ab} y_a y_b$ for all $y \in \mathbb{Z}^n$ and, in particular, $q''(y) \geq q\rho(y)$ for all $y \in \mathbb{N}^n$. In contrast, the equality $q(x) = q''\sigma(x)$ does hold only if $x_a = 0$ or $x_b = 0$.

15.4 Critical Vectors

Let $q : \mathbb{Z}^n \longrightarrow \mathbb{Z}$ be a unit form. A non-zero vector $x \in \mathbb{N}^n$ is called a *critical vector* of q if either it has the form $x = e_a + e_b$ for some indices $a < b$ with $q_{ab} < -2$ or else the restriction of q to the coordinate subspace $\bigoplus_{x_i > 0} \mathbb{Z} e_i$ is non-negative and its radical is generated by x. We denote by $\mathcal{C}_q$ the set of all critical vectors of q. The following lemma summarizes the main properties of this set.

Lemma 15.1.

a) [2] For all $x \in \mathcal{C}_q$, we have $x_i \leq 6$ for $1 \leq i \leq n$. In particular, $\mathcal{C}_q$ is finite.

b) [6] q is weakly positive if and only if $\mathcal{C}_q = \emptyset$.

c) [8] q is weakly non-negative if and only if $Bx \in \mathbb{N}^n$ for all $x \in \mathcal{C}_q$, here B denotes the integral symmetric matrix of q.

d) [9] If q is weakly non-negative then we have $\mathcal{C}_q \subseteq \mathcal{O}_q \subseteq \sum_{x \in \mathcal{C}_q} \mathbb{Q}^+ x$, here $\mathbb{Q}^+$ denotes the set of non-negative rational numbers.

e) [5] If q' is obtained from q by proper reduction then we have $\sigma(\mathcal{C}_q) \subseteq \mathcal{C}_{q'}$.

15.5 Algorithm for Weak Definiteness

15.5.1 Let $q : \mathbb{Z}^n \longrightarrow \mathbb{Z}$ be a fixed unit form. In order to decide whether or not q is weakly definite we will construct a sequence of triples (q^m, z^m, L^m), $m = 0, 1, 2, \ldots$ where

$q^m : \mathbb{Z}^{r_m} \longrightarrow \mathbb{Z}$ is a unit form in $r_m \geq n$ variables,

$z^m : \{1, \ldots, r_m\} \longrightarrow \mathbb{N}^n$ is a sequence of r_m vectors $\in \mathbb{N}^n$, and

L^m is a subset of $\mathbb{N}^n$.

The intention of these data is the following. The unit forms $q^1, q^2, \ldots$ will be obtained by iterated (proper or restricted) reductions of q, the sequence z^m then describes the image under ρ^m of the canonical base vectors of the domain of q^m, and in the set L^m we save certain vectors of the form $\rho^{m-1}(e_a + e_b)$ which are candidates for being critical vectors of q but may be got lost if the mth performed reduction (with respect to a, b) is restricted.

A further essential ingredient of our algorithm is the choice of a finite subset $S \subseteq \mathbb{N}^n$ satisfying $\mathcal{C}_q \subseteq S$ and being closed with respect to predecessors, that is, we have $x \in S$ whenever $x \in \mathbb{N}^n$, $y \in S$ and $x < y$. This set will be relevant to the decision of what kind of reduction is performed in each step. Later we will discuss different choices of S. A good compromise here is to take $S := \{x \in \mathbb{N}^n \mid x_i \leq 6 \text{ for all } i\}$.

15.5.2 Now we describe the algorithm in detail. We start with $q^0 = q$, $z^0(i) = e_i$ for $1 \leq i \leq n$ and $L^0 = \emptyset$. Assume now that $m \geq 0$ and (q^m, z^m, L^m) is already constructed. First we check the following

Stopping condition: $q^m_{ab} \geq 0$ for all $a < b$.

If this condition is not satisfied we choose indices $a < b$ such that $q^m_{ab} < 0$ and set $x := z^m(a) + z^m(b)$. We denote by B the integral symmetric matrix of q and check then the following

Failing condition: $q^m_{ab} < -2$, or $q^m_{ab} = -2$ and $Bx \notin \mathbb{N}^n$.

If this condition is not satisfied we consider two different scenarios.

(*) $q^m_{ab} = -1$ and $x \in S$.

(**) $q^m_{ab} = -2$, $Bx \in \mathbb{N}^n$ and $x \in S$.

Then we set

$$q^{m+1} = \begin{cases} \text{prop. red. of } q^m \text{ w.r.t. } a, b & \text{if (*) holds,} \\ \text{restr. red. of } q^m \text{ w.r.t. } a, b & \text{otherwise,} \end{cases}$$

$$z^{m+1} = \begin{cases} \text{extension of } z^m \text{ by } x & \text{if (*) holds,} \\ z^m & \text{otherwise,} \end{cases}$$

$$L^{m+1} = \begin{cases} L^m \cup \{x\} & \text{if (**) holds,} \\ L^m & \text{otherwise.} \end{cases}$$

Theorem 15.1. *Let $q : \mathbb{Z}^n \longrightarrow \mathbb{Z}$ be a unit form. There is an $m \geq 0$ such that the procedure described in 15.5.2 stops or fails at step m. If it fails then q is not weakly non-negative. If it stops then q is weakly non-negative and we have $\mathcal{C}_q \subseteq L^m \subseteq \mathcal{O}_q \cap S$. Moreover, q is weakly positive if and only if the procedure stops with $L^m = \emptyset$; in this case, we have $\mathcal{R}_q = \{z^m(1), \ldots, z^m(n+m)\}$.*

Proof. We start with some useful observations.

i) All vectors belonging to some z^m or L^m are contained in S.

ii) In every step $m \geq 0$ we have $z^m(i) = \rho^m(e_i)$ for all $1 \leq i \leq r_m$, here $\rho^m : \mathbb{Z}^{r_m} \longrightarrow \mathbb{Z}^n$ denotes the mth 'power' of the maps ρ. (This follows easily by induction on m using the factorization $\rho^m = \rho^{m-1}\rho$.)

iii) If the process does not stop at m, the indices $a < b$ are such that $q^m_{ab} < 0$ and $x = z^m(a) + z^m(b)$, then we have $q(x) \leq 2 + q^m_{ab}$. (Indeed, using ii) we have $x = \rho^m(e_a) + \rho^m(e_b) = \rho^m(e_a + e_b)$ and therefore $q(x) = q\rho^m(e_a + e_b) \leq q^m(e_a + e_b) = 2 + q^m_{ab}$.)

Now we prove the assertions of the theorem. First we show that the process terminates. Assume that this is not the case. Then the sequence $q^0, q^1, q^2, \ldots$ involves infinitely many proper reduction steps and therefore the sequence $z^0, z^1, z^2, \ldots$ gives rise to an infinite sequence $z(1), \ldots, z(n), z(n+1), \ldots$ of non-zero elements of the finite set S. So this sequence meets at least one vector infinitely many often and we can choose such a vector y of minimal length (≥ 2). On the other hand, by construction, every $z(r)$, $r > n$, has the form $z(r) = z(a) + z(b)$ for a pair of numbers $a < b$ smaller than r and different r give rise to different pairs of numbers. Applying this in particular to all r with $z(r) = y$ we get a contradiction to the minimal choice of y.

Next we show that q is not weakly non-negative if the process fails at some m. Indeed, if $q^m_{ab} < -2$ then iii) yields $q(x) \leq 2 + q^m_{ab} < 0$. On the other hand, if $q^m_{ab} = -2$ we get $q(x) \leq 0$, and if moreover $Bx \notin \mathbb{N}^n$ we have $e^t_i Bx < 0$ for some i and therefore $q(e_i + 2x) = 1 + 2e^t_i Bx + 4q(x) < 0$.

Now suppose that the process stops at some m. Then clearly we have $\mathcal{C}_{q^m} = \emptyset$ and part b) of Lemma 15.2 below implies that $\mathcal{C}_q \subseteq L^m$. By the construction of L^m this means that every $y \in \mathcal{C}_q$ satisfies the condition (**) of Section 15.5.2, in particular, $By \in \mathbb{N}^n$. So q is weakly non-negative by Lemma 15.1 c). On the other hand we have $L^m \subseteq \mathcal{O}_q \cap S$ by part a) of Lemma 15.2 and i).

Finally, q is weakly positive if and only if it is weakly non-negative with $\mathcal{O}_q = \emptyset$ or, equivalently, $\mathcal{C}_q = \emptyset$. This shows the claimed characterization of weak positivity. It is not hard to see that in this case all performed reductions are proper. The description of $\mathcal{R}_q$ follows therefore from ii) and Proposition 15.1. □

Lemma 15.2. *In every step $m \geq 0$ of the procedure described in 15.5.2, the following assertions hold.*

a) $L^m \subseteq \mathcal{O}_q$.

b) Every vector $y \in \mathcal{C}_q$ satisfies $\sigma^m(y) \in \mathcal{C}_{q^m}$ or $y \in L^m$.

Proof.

a) By construction (condition (**)), every element $x \in L^m$ has the form $x = z^k(a) + z^k(b)$ for some $0 \leq k < m$ and $1 \leq a < b \leq r_k$ with $q^k_{ab} = -2$, moreover, x satisfies $Bx \in \mathbb{N}^n$. Now, using the observation iii) in the proof of Theorem 15.1 we get $q(x) \leq 2 + q^k_{ab} = 0$. On the other hand $Bx \in \mathbb{N}^n$ implies $q(x) \geq 0$ and therefore $x \in \mathcal{O}_q$.

b) We procede by induction on m. The case $m = 0$ is trivial. So assume that the assertion holds for $m \geq 0$, let q^{m+1} be obtained from q^m by (proper or restricted) reduction with respect to indices $a < b$ and set $x = z^m(a) + z^m(b)$. Consider a fixed vector $y \in \mathcal{C}_q$. We have to show that $\sigma^{m+1}(y) \in \mathcal{C}_{q^{m+1}}$ or $y \in L^{m+1}$. Since $L^m \subseteq L^{m+1}$ we can assume that $\sigma^m(y) \in \mathcal{C}_{q^m}$. If the performed reduction is proper then Lemma 15.1 e) yields that $\sigma^{m+1}(y) \in \mathcal{C}_{q^{m+1}}$. So we can assume further on that the reduction is restricted. If $\sigma^m(y)_a = 0$ or $\sigma^m(y)_b = 0$ then clearly we have again $\sigma^{m+1}(y) = \sigma^m(y) \in \mathcal{C}_{q^{m+1}}$. So we can assume that $\sigma^m(y)_a \neq 0 \neq \sigma^m(y)_b$. Then we have $e_a + e_b \leq \sigma^m(y)$ and therefore $x = z^m(a) + z^m(b) = \rho^m(e_a + e_b) \leq \rho^m\sigma^m(y) = y$. This implies that $x \in S$ since $y \in \mathcal{C}_q \subseteq S$. Now, the condition '$x \in S$ but the reduction is not proper' means that $q^m_{ab} < -1$. On the other hand we have $q^m_{ab} \geq -2$, therefore $q^m_{ab} = -2$ and then $Bx \in \mathbb{N}^n$ since the process does not fail at m (by assumption). So the condition $(**)$ of 15.5.2 holds and we have $x \in L^{m+1}$. Since moreover we have $L^{m+1} \subseteq \mathcal{O}_q$ (by a)) and $x \leq y \in \mathcal{C}_q$ we get $x = y$ and therefore $y \in L^{m+1}$.

This finishes the proof of the lemma and therefore of Theorem 15.1. □

15.5.3 Now we give some final remarks how the running time of the algorithm in practice can be improved. Clearly the algorithm becomes faster if the number of involved proper reduction steps can be reduced.

The first way to do this is to include in the scenarios $(*)$ and $(**)$ the additional condition that x is L^m-*free*, that means, the support $\operatorname{supp} x = \{i \mid x_i > 0\}$ of x does not contain $\operatorname{supp} y$ for any $y \in L^m$.

The second way is to choose the set S more restrictive. For instance, consider the following subsets of S.

$$S_2 = \{x \in S \mid |\operatorname{supp} x| \leq 9 \text{ and } |x| \leq 30\},$$
$$S_2' = \{x \in S \mid |\operatorname{supp} x| \geq 10\,,\ |x| \leq 2n - 4 \text{ and } x_i \leq 2 \text{ for all } i\},$$
$$S_1 = S_2 \cup S_2',$$

here $|\operatorname{supp} x|$ denotes the number of elements of $\operatorname{supp} x$ and $|x| = \sum_i x_i$ the length of x. In [2] it was shown that $\mathcal{C}_q \subseteq S_1$, and from [4] we know that the criterion of Lemma 15.1 c) can be improved replacing the set $\mathcal{C}_q$ by the set $\mathcal{C}_q \cap S_2$. This has the following consequences. Running the algorithm with S_1 instead of S we can decide weak non-negativity and produce $\mathcal{C}_q$ but have to dispense with a complete description of $\mathcal{R}_q$ in the weakly positive case. Moreover, running the algorithm with S_2 instead of S we can still decide weak non-negativity but in general we do not get the complete set $\mathcal{C}_q$ in that case.

The algorithm running with S_1 was implemented on computer and works quite efficiently. It is included in the CREP program package in form of a PASCAL program.

Bibliography

[1] Gabriel, P., Roiter, A.V.: *Algebra VIII, Representations of finite-dimensional algebras*, Encyclop. Math. Sciences **73**, (Springer, Berlin, 1992).

[2] von Höhne, H.-J.: *On weakly positive unit forms*, Comment. Math. Helv. **63**, (1988), 312–336.

[3] von Höhne, H.-J.: *Edge reduction for unit forms*, Arch. Math. **65**, (1995), 300–302.

[4] von Höhne, H.-J.: *On weakly non-negative unit forms*, Proc. Lond. Math. Soc. **73**, (1996), 47–67.

[5] von Höhne, H.-J., de la Peña, J.A.: *Edge reduction for weakly non-negative quadratic forms*, Archiv Math. **70**, (1998), 270–277.

[6] Ovsienko, S.A.: *Weakly positive integral quadratic forms*, in: Schurian matrix problems and quadratic forms. Ukr. Akad. Nauk (1978), 3–17 (in russian).

[7] Ovsienko, S.A.: *Boundedness of roots of weakly positive integral quadratic forms*, in: Representations and quadratic forms. Ukr. Akad. Nauk (1979), 106–123 (in russian).

[8] de la Peña, J.A.: *Quadratic forms and the representation type of an algebra*, Sonderforschungsbereich 343, Ergänzungsreihe **90-003**, (Bielefeld, 1990).

[9] de la Peña, J.A.: *On the corank of the Tits form of a tame algebra*, J. Pure Appl. Alg. **117**, (1996), 89–105.

[10] Workshop "Representation Theory" November 95: *Quadratic Forms in the Representation Theory of Finite-Dimensional Algebras*, Sonderforschungsbereich 343, Ergänzungsreihe **96-001**, (Bielefeld, 1996).

[11] Ringel, C.M.: *Tame algebras and integral quadratic forms*, Lecture Notes in Mathematics **1099**, (Springer, Berlin, 1984).

Progress in Mathematics, Vol. 173, © 1999 Birkhäuser Verlag Basel/Switzerland

Chapter 16

Decision Problems in Finitely Presented Groups

Derek F. Holt

16.1 Introduction

Let $G = \langle X \mid R \rangle$ be a group defined by a finite presentation, let $A = X \cup X^{-1}$ and let A^* denote the set of words in A, including the empty word ε. For $v, w \in A^*$, $v = w$ will mean that they are equal as words, and $v =_G w$ will mean that they map onto the same element of G. We shall also use $\overline{w}$ to denote the image of the word w in G. In this article, we shall be concerned with the following three decision problems.

The word problem: for $w \in A^*$ is $\overline{w} = 1$?

The conjugacy problem: for $v, w \in A^*$ does there exist $g \in A^*$ with $vg =_G gw$?

The generalized word problem: given $w \in A^*$ and a finite set of words in A^* generating a subgroup H of G, is $\overline{w} \in H$?

As with almost all such decision problems, they are theoretically undecidable in general; in particular, the celebrated Novikov-Boone theorem asserts that the word problem is undecidable. (For an accessible proof of this result, see Chapter 13 of [11].) However, we shall be concerned here with examples and situations in which one or more of these problems is decidable, and more importantly in efficient computational solutions.

Notice that we have phrased the problems in such a way that they make sense in the more general context of monoids, although we do not know whether the conjugacy problem is of any interest in this more general setting.

16.2 The Word Problem

Probably the first efficient method devised for solving the word problem was Dehn's algorithm for small cancellation groups, which is described in Section 4, Chapter V of [10]. After enlarging the set R of defining relators, any element equal to the identity in such a group can be reduced in length by a single substitution, using a defining relator.

More generally, a *rewriting system* for a group (or monoid) G is a set S of rewrite-rules $v \to w$ with $v, w \in A^*$, such that $v =_G w$ for all rewrite-rules, and the set of all such equations forms a set of defining relations for G as a monoid. (So, in particular, it contains rules of the form $xx^{-1} \to \varepsilon$ for generators $x \in A$.) There is a fairly extensive literature on rewriting systems, so we shall give only brief details here. For example, there is a detailed theoretical and practical treatment of their applications to group theory in [13].

The idea is that for any word $w \in A^*$, we attempt to reduce the word by substituting the right hand sides of the rules in S for the left hand sides. Usually we ensure that this is a genuine reduction by imposing a suitable well-ordering $\prec$ of A^*, which must be preserved under left and right multiplication by words in A^*, and arrange the rewrite rules such that $w \prec v$. (For example, in a *shortlex* ordering, shorter words precede longer ones, and a lexicographical ordering is used to distinguish between words of equal length.)

Words that contain no occurrences of left-hand-sides of rules in S are called S-irreducible. S is called *complete* if every word $w \in A^*$ can be reduced, by a finite number of reductions, to an S-irreducible word that is uniquely determined by w. This turns out to be equivalent to the property that $v =_G w$ if and only if v and w reduce to the same irreducible word, and so we can use a complete rewriting system to solve the word problem in G.

Example 16.1. A complete rewriting system S for the free abelian group $\langle a, b \mid ab = ba \rangle$ of rank 2 (using a shortlex ordering on A^*) consists of the 8 reduction rules

$$aa^{-1} \to 1,\ a^{-1}a \to 1,\ bb^{-1} \to 1,\ b^{-1}b \to 1,$$
$$ba \to ab,\ ba^{-1} \to a^{-1}b,\ b^{-1}a \to ab^{-1},\ b^{-1}a^{-1} \to a^{-1}b^{-1}.$$

In general, we can try to find a complete presentation for G by starting with rules corresponding to the defining relators of G, and applying the *Knuth-Bendix completion procedure* which adjoins new rules to S, and halts when S is complete. However, in most cases when G is infinite, there is no finite complete system, and so the procedure will never halt.

Finite complete systems have been found for various particular types of groups, including polycyclic-by-finite groups, two-dimensional surface groups, and many Coxeter groups. However, there is no really satisfactory underlying theory, and the existence of such systems often depends on judicious or fortuitous choice of generators X for G and the ordering $\prec$ on A^*.

Although a finite complete rewriting system usually seems to lead to a reasonably efficient computational solution of the word problem, this is not always the case, as the following monoid example illustrates.

Example 16.2. $A = \{e, c, o, z\}$, S consists of the three rules

$$ze \to ce,\ zc \to oz,\ oc \to cz.$$

Then S is complete (using a shortlex ordering on A^*), and the word $z^n e$ reduces to $c^n e$ for all $n > 0$, but it requires $3.2^n - 2n - 3$ reduction steps.

An alternative method of solving the word problem efficiently in a group G is to try to find an *automatic structure* for G. These are defined in terms of finite state automata, which we shall abbreviate to fsa in both singular and plural. See for example [7] for the general theory of fsa.

Definition 16.1. The group $G = \langle A \mid R \rangle$ is said to be *automatic* (with respect to A), if there exist finite state automata W, and M_a for each $a \in A$, such that:

(i) W has input alphabet A, and accepts a unique word mapping onto each element of G.

(ii) Each M_a has input alphabet $A' \times A'$, and it accepts the padded pair (v^+, w^+) for $v, w \in A^*$ if and only if $v, w \in L(W)$ and $va =_G w$.

Here $A' = A \cup \{\$\}$, where \$ is used as a padding-symbol. For $w, x \in A^*$, the associated *padded pair* $(w^+, x^+) \in (A' \times A')^*$ is obtained by adjoining symbols \$ to the shorter of w and x to make them have equal length. The language of the fsa W is denoted by $L(W)$.

The fsa W is called the *word-acceptor* and the M_a the *multiplier* automata. The complete collection $\{W, M_a\}$ is known as an *automatic structure* for G.

The definition of the class of *automatic groups* evolved in the mid-1980's following the appearance of a paper by Jim Cannon on hyperbolic groups [1]. Bill Thurston noticed that some of the properties that were proved about these groups could be reformulated in terms of finite state automata. The class of word-hyperbolic groups, which is a generalization due to Gromov of the class considered by Cannon, is one of the most important family of examples. A finitely generated group G is called word-hyperbolic if, for any geodesic triangle in the Cayley graph of G, any point on any side of the triangle lies within a bounded distance of the union of the other two sides. This property turns out to be independent of the choice of finite generating set for G.

The best general reference for the theory of automatic groups is the book [2]. In particular, it turns out that the automaticity of G is independent of the choice of generating set X. All finite groups are easily seen to be automatic; in fact the class of automatic groups is invariant under finite variations, such as sub- and super-groups of finite index. It is also closed under direct and free products, and includes, for example, all word-hyperbolic groups, braid groups, Coxeter groups and Artin groups of finite type. All automatic groups have finite presentations.

Given an automatic structure, the word problem can be solved in time $O(n^2)$, where n is the length of the input word u, by using the multiplier automata to reduce u to a G-equivalent word in $L(W)$ (which can be thought of as the normal form for u).

A collection of programs has been written at Warwick for computing automatic structures. Currently, they only work for shortlex structures, in which $L(W)$ consists of the least words under the shortlex ordering that map onto each group element. Many, but not all, of the known classes of automatic groups are known to possess shortlex structures. In particular, word-hyperbolic groups are shortlex automatic with respect to any set of group generators and any ordering of the generators and their inverses.

The programs are described in some detail in [3] and [5], and more generally in [2].

The latest version is part of a package called kbmag and is available by anonymous `ftp` from `ftp.maths.warwick.ac.uk` in the directory `people/dfh/kbmag2`.

16.3 The Conjugacy Problem

One method of solving the conjugacy problem in a group G is to try to find a *bi-automatic structure* for G.

Definition 16.2. The group $G = \langle A \,|\, R \rangle$ is said to be *bi-automatic*, if there exist finite state automata W, M_a and ${}_aM$ for each $a \in A$, such that:

(i) W has input alphabet A, and accepts a unique word mapping onto each element of G.

(ii) Each M_a has input alphabet $A' \times A'$, and it accepts the padded pair (v^+, w^+) for $v, w \in A^*$ if and only if $v, w \in L(W)$ and $va =_G w$.

(iii) Each ${}_aM$ has input alphabet $A' \times A'$, and it accepts the padded pair (v^+, w^+) for $v, w \in A^*$ if and only if $v, w \in L(W)$ and $av =_G w$.

The bi-automatic structure for G consists of the collection $\{W, M_a, {}_aM\}$ of fsa in the definition. Not every automatic structure is part of a bi-automatic structure, but no examples are known of groups that are automatic but not bi-automatic. In particular, any shortlex automatic structure of a word-hyperbolic group is part of a bi-automatic structure.

Given a bi-automatic structure and words $v, w \in A^*$, we can construct the composite multipliers M_v and ${}_wM$, which recognize multiplication on the right by v and on the left by w, respectively. Then

$$L(M_v) \cap L({}_wM) = \{\, (g^+, h^+) \in (A' \times A')^* \mid g, h \in L(W),\ gv =_G h,\ wg =_G h \,\}$$

so v and w are conjugate in G if and only if $L(M_v) \cap L({}_wM)$ is nonempty. Unfortunately, the algorithm for constructing M_v has complexity exponential in the length of v, so this method is only practical for short words v, w.

An $O(n \log n)$ algorithm for the conjugacy problem for word-hyperbolic groups has been devised by Epstein, and we expect it to be efficient, but it requires a prior knowledge of the constant of hyperbolicity of G. Currently no practical general method is known for computing this constant, so it is difficult at this time to assess Epstein's conjugacy algorithm properly.

16.4 The Generalized Word Problem

A method of solving the generalized word problem efficiently for a subgroup H of a group G is to try to find an *automatic coset system* for G with respect to H.

Definition 16.3. The group $G = \langle A \mid R \rangle$ is said to be *coset automatic* with respect to the finitely generated subgroup H, if there exist finite state automata W, and M_a for each $a \in A$, such that:

(i) W has input alphabet A, and accepts a unique word in each coset Hg of H in G.

(ii) Each M_a has input alphabet $A' \times A'$, and it accepts the padded pair (v^+, w^+) for $v, w \in A^*$ if and only if $v, w \in L(W)$ and $Hva = Hw$.

The automata $W, M_a (a \in A)$ constitute an *automatic coset system* for G with respect to H. Given an automatic coset system, the generalized word problem can be solved in time $O(n^2)$, where n is the length of the input word u, by using the multiplier automata to reduce u to a word in $L(W) \cap Hu$. (In practice, the representative of H in $L(W)$ is virtually always the empty word ε, although the definition does not require this.)

The kbmag package mentioned above is capable of computing shortlex automatic coset systems under certain additional technical conditions which we shall not explain in detail here, but which do appear to hold in many examples, including all quasiconvex subgroups of word-hyperbolic groups.

Suppose that we have an automatic coset system $\{W, M_a\}$ for G with respect to a subgroup H, and let

$$Y = \{\, \overline{va}\,\overline{x}^{-1}) \mid a \in A,\ (v^+, x^+) \in L(M_a) \,\}.$$

So Y is a set of Schreier generators of H with respect to G.

We now make the additional assumption that Y is finite. (In fact, this is one of the assumptions required for the implementation in kbmag to succeed.) Then H can be shown to be finitely presented, and there is a straightforward algorithm, which is implemented in kbmag, for computing a presentation of H with Y as generating set.

The idea is to regard the elements of Y as initial states of the coset multiplier automata M_a. Then we build the composites M_r for the defining relators r of G, and the defining relators for H are the initial states of the M_r. See [8] or [6] for further details.

Example 16.3. Let G be the Heineken group

$$\langle\, x, y, z \mid [x,[x,y]] = z,\ [y,[y,z]] = x,\ [z,[z,x]] = y \,\rangle$$

and

$$H = \langle\, [x,y],\ [y,z],\ [z,x] \,\rangle.$$

Then kbmag calculated that H is free of rank 3 (and hence free on the given generating set). In fact some related calculations enabled us to prove that G itself is word-hyperbolic.

There is an alternative approach to solving the generalized word problem, which starts with an automatic structure $\{W, M_a\}$ for the full group G. Let $L = L(W)$. The subgroup H is called L-rational if $L \cap H$ is a regular language (i.e. the language of an fsa). Such subgroups were studied in [4], where it is proved that L-rational is equivalent to L-quasiconvex. This means that any prefix of a word in $L \cap H$ lies within a bounded distance of H in the Cayley graph of G. Such subgroups are always finitely generated.

An algorithm for constructing an fsa W_H with language $L \cap H$, which takes as input an automatic structure for G and a set of generators for an L-rational subgroup H of G, is described in [9]. A practical version is described in [8], and an implementation is available in kbmag.

The fsa W_H can be used together with the automatic structure of G to determine whether a given word in A^* lies in H. First use the the multiplier automata to reduce the word to one in L, and then use W_H to test whether it lies in H.

As a further application, if we can find W_{H_1} and W_{H_2} for two subgroups H_1 and H_2, then we can compute the intersection of their languages, and hence derive a (finite) generating set for $H_1 \cap H_2$.

Bibliography

[1] J.W. Cannon: *The combinatorial structure of cocompact discrete hyperbolic groups*, Geom. Dedicata **16**, 123–148 (1984).

[2] David B. A. Epstein J.W. Cannon, D. F. Holt, S. Levy, M. S. Patterson and W. Thurston: *Word Processing in Groups*, Jones and Bartlett, 1992.

[3] D. B. A. Epstein, D. F. Holt and S. E. Rees: *The use of Knuth-Bendix methods to solve the word problem in automatic groups*, J. Symbolic Computation **12**, (1991), 397–414.

[4] S. M. Gersten, H. B. Short: *Rational Subgroups Of Biautomatic Groups*, Ann. Math. **134**, (1991), 125–158.

[5] Derek F. Holt: *The Warwick automatic groups software*, in: Geometrical and Computational Perspectives on Infinite Groups DIMACS Series in Discrete Mathematics and Theoretical Computer Science **25**, ed. Gilbert Baumslag et. al., 1995, 69–82.

[6] Derek F. Holt and Darren F. Hurt: *Computing automatic coset systems and subgroup presentations*, to appear in J. Symbolic Computation.

[7] John E. Hopcroft and Jeffrey D. Ullman: *Introduction to automata theory, languages and computation*, Addison-Wesley, 1979.

[8] Darren F. Hurt: *The Use of Knuth-Bendix Methods and Automatic Coset Systems for Solving the Generalized Word Problem and Finding Subgroup Presentations*, PhD Thesis, University of Warwick, 1997.

[9] Ilya Kapovich: *Detecting quasiconvexity: algorithmic aspects*, in: Geometric and Computational Perspectives on Infinite Groups, DIMACS Series in Discrete Mathematics and Theoretical Computer Science **25**, ed. Gilbert Baumslag et. al., 1995, 91–99.

[10] R.C. Lyndon and P.E. Shupp: *Combinatorial Group Theory*, Springer-Verlag, Berlin-Heidelberg-New York, 1977.

[11] Joseph J. Rotman: *An Introduction to the Theory of Groups*, Third Edition, Allyn and Bacon, Inc., 1984.

[12] I. D. Redfern: *Automatic Coset Systems*, PhD Thesis, University of Warwick, 1993.

[13] C.C. Sims: *Computation with Finitely Presented Groups*, Cambridge University Press, 1994.

Progress in Mathematics, Vol. 173, © 1999 Birkhäuser Verlag Basel/Switzerland

Chapter 17

Some Algorithms in Invariant Theory of Finite Groups[1]

Gregor Kemper[2] and Allan Steel

Abstract

We present algorithms which calculate the invariant ring $K[V]^G$ of a finite group G. Our focus of interest lies on the modular case, i.e., the case where $|G|$ is divided by the characteristic of K. We give easy algorithms to compute several interesting properties of the invariant ring, such as the Cohen-Macaulay property, depth, the β-number and syzygies.

17.1 Introduction

This paper presents various algorithms for invariant theory of finite groups, which were implemented in the computer algebra system Magma ([4] or [6]) during a visit of the first author to Sydney. We focus on those algorithms which are new or which have never been written up before, and only sketch those that can already be found in the literature. Due to improvements of existing algorithms and a better usage of computational resources by the Magma system, this recent implementation generally produces much better timings than the Invar package which was implemented by the first author in Maple (see [8]). For general reading on invariant theory, we refer the reader to the books by Sturmfels [11], Benson [3], and Smith [10].

As in the Invar package, the primary goal is the computation of the invariant ring of a given finite matrix group over a base field of arbitrary characteristic. Of particular interest is the **modular case**, i.e., the case where the characteristic of the base field K divides the group order, since in that case the structure of invariant rings is still not very well understood. We give easy algorithms to calculate properties of modular invariant rings, such as the Cohen-Macaulay property, depth, free resolutions, the Hilbert series, and the complete intersection property. Our approach to calculating the invariant ring is divided into two

[1]This research was supported in part with the assistance of grants from the Australian Research Council.

[2]The first author thanks John Cannon and the Magma group for their hospitality during his stay in Sydney.

major steps: we first construct a system of **primary invariants**, i.e., homogeneous invariants $f_1, \ldots, f_n$ which are algebraically independent, such that the invariant ring is a finitely generated module over $A = K[f_1, \ldots, f_n]$. In the next step we calculate **secondary invariants**, which is just another term for generators of the invariant ring as an A-module.

Section 17.2 is concerned with the problem of how to produce invariants (of some given degree) most effectively. In the next section we come to the methods of finding primary and secondary invariants, where for the secondary invariants we offer completely different algorithms for each of the modular and the non-modular cases. In Section 17.5 we discuss how properties of the invariant ring can easily be calculated.

Let us fix some notation. Throughout this article, K will be a field (which in the implementation is assumed to be either an algebraic number field or a finite field) and $G \leq \mathrm{GL}_n(K)$ is a finite matrix group acting from the right on an n-dimensional vector space $V \cong K^n$ with basis $x_1, \ldots, x_n$. Thus G also acts on the symmetric algebra $K[V] = S(V)$, which is the polynomial ring in the variables $x_1, \ldots, x_n$. The invariant ring $\{f \in K[V] \mid f^\sigma = f \;\forall \sigma \in G\}$ is denoted by $K[V]^G$. Since the action of G preserves the natural grading on $K[V]$, this is a graded algebra over K.

All timings in the examples were obtained on a 200 MHz Sun Ultrasparc 2, running Solaris 5.5.1.

17.2 Calculating Homogeneous Invariants

The most basic task is to calculate homogeneous subspaces of invariants, i.e., vector spaces $K[V]^G_d$ consisting of all homogeneous invariants of degree d. All subsequent algorithms depend on effective methods for this. There are two basic approaches. The first one consists of the application of the **Reynolds operator**

$$\pi_G\colon K[V] \to K[V]^G,\; f \mapsto \frac{1}{|G|} \sum_{\sigma \in G} f^\sigma$$

on all (or a sufficient number of) monomials of degree d. This method is only available in the non-modular case. If G is a permutation group, one can take sums over orbits of monomials. Since G acts on the set of monomials in this case, the desired basis is given by all these sums, irrespective of the characteristic of K.

The second method, which we call the **linear algebra method**, exploits the exact sequence

$$\begin{array}{ccccccc} 0 & \longrightarrow & K[V]^G & \longrightarrow & K[V] & \longrightarrow & \bigoplus_{\sigma \in S(G)} K[V] \\ & & f & \mapsto & (f^\sigma - f)_{\sigma \in S(G)} & & \end{array},$$

where $S(G)$ is a generating set for G. This sequence restricts to the homogeneous components. The map whose kernel is $K[V]^G_d$ is explicitly given, and hence $K[V]^G_d$ can be calculated by solving a homogeneous system of linear equations in $k = \dim(K[V]_d) = \binom{n+d-1}{n-1}$ unknowns. This method is available for any base field K.

In the non-modular case, both methods are at hand, so we need to assess the computational cost of them. The rank of the linear system involved in the linear algebra method is $\dim(K[V]_d) - \dim(K[V]_d^G)$. The following proposition gives a reasonable estimate for this rank.

Proposition 17.1. *Let* $a_d = \dim(K[V]_d^G)$ *and* $b_d = \dim(K[V]_d)$. *Then*

$$\lim_{N\to\infty} \frac{\sum_{d=0}^N a_d}{\sum_{d=0}^N b_d} = \frac{1}{|G|}.$$

Proof. We choose a maximal homogeneous subset S of $B := K[V]$ which is linearly independent over $A := K[V]^G$. By Galois theory, S has $|G|$ elements. Let M be the free A-module generated by S. Every homogeneous element of B can be written as a linear combination of elements of S with fractions from A as coefficients, whose denominators are all homogeneous. Since B is finitely generated over A, it suffices to take one homogeneous denominator a. In other words,

$$M \subset B \subset a^{-1}M.$$

Let $f(t) = \sum_{s\in S} t^{\deg(s)}$ and $e = \deg(a)$, then for the Hilbert series of A and B it follows

$$f(t)\cdot H(A,t) \le H(B,t) \le t^{-e} f(t)\cdot H(A,t) \quad \text{(coefficient-wise)}.$$

Writing $f(t) = \sum_{i=0}^m c_i t^i$, we obtain for $N \in \mathbb{N}$

$$\sum_{d=0}^{N} \sum_{i=0}^{\min(m,d)} c_i a_{d-i} \le \sum_{d=0}^{N} b_d \le \sum_{d=0}^{N} \sum_{i=0}^{\min(m,d+e)} c_i a_{d+e-i}. \tag{17.1}$$

The left-hand side can be estimated as follows:

$$\begin{aligned} \sum_{d=0}^{N} \sum_{i=0}^{\min(m,d)} c_i a_{d-i} &= \sum_{i=0}^{m} \sum_{d=i}^{N} c_i a_{d-i} = \sum_{i=0}^{m} c_i \left(\sum_{d=0}^{N-i} a_d\right) \\ &\ge \left(\sum_{i=0}^{m} c_i\right)\left(\sum_{d=0}^{N-m} a_d\right) = |G|\cdot \sum_{d=0}^{N-m} a_d. \end{aligned}$$

Similarly, the right-hand side of Inequality (17.1) is bounded from above by $|G|\cdot\sum_{d=0}^{N+e} a_d$. Extending (17.1) by these estimates and dividing through the middle term yields

$$|G|\cdot \frac{\sum_{d=0}^{N-m} a_d}{\sum_{d=0}^{N} b_d} \le 1 \le |G|\cdot \frac{\sum_{d=0}^{N+e} a_d}{\sum_{d=0}^{N} b_d}.$$

Now the difference between the right-hand side and the left-hand side of this inequality converges to 0 as $N\to\infty$ since

$$\frac{\sum_{d=N-m+1}^{N+e} a_d}{\sum_{d=0}^{N} b_d} \le (e+m)\cdot \frac{b_{N+e}}{\sum_{d=0}^{N} b_d} = (e+m)\cdot \frac{\binom{N+e+n-1}{n-1}}{\binom{N+n}{n}} \to 0.$$

Hence the limits of both sides are equal, and the result follows. $\square$

The cost of the linear algebra method can now be estimated as follows: We have a linear system with $k = \binom{n+d-1}{n-1}$ unknowns and $s \cdot k$ equations (where s is the number of generators by which G is given), which has rank $k - k/|G|$. The cost of solving this by Gaussian echelonization and then back-substitution is

$$(s - 1/3)k^3 - (s - 1/2)k^2 + 5/6 \cdot k + O\left(\frac{k^2(k+s)}{|G|}\right)$$

arithmetic operations in the base field K, i.e., additions and multiplications equally distributed. In the case of small finite fields, Magma uses packed representations of matrices which speed up the solution of the linear system considerably.

To put up the linear system, we have to apply the group generators to all monomials of degree d. For each generator and each monomial, this means forming a product of d linear forms, which requires

$$2n \cdot \sum_{i=0}^{d-1} \binom{n+i-1}{n-1} = 2n \cdot \binom{n+d-1}{n} = 2dk$$

field operations. This is not the best way to evaluate all products, but it will yield a sufficiently good upper bound to see that the solution of the linear system is dominant. The total cost of the linear algebra method is thus estimated by

$$(s - 1/3)k^3 + \left((2d-1)s + 1/2\right)k^2 + 5/6 \cdot k,$$

where, as before, $k = \binom{n+d-1}{d}$.

To assess the cost of applying the Reynolds operator to a monomial $t = x_1^{e_1} \cdots x_n^{e_n}$, we must first look at the stabilizer $G_t \leq G$ of t. Taking the sets $M_e = \{i \mid e_i = e\} \subset \{1, \ldots, n\}$ for $e = 1, \ldots, d$, we see by unique factorization that a $\sigma \in G$ lies in G_t if and only if for all e and for all $i \in M_e$ the ith row of σ only has one non-zero entry α_i which occurs at a column j with $j \in M_e$, and furthermore $\prod_{e=1}^{d} \prod_{i \in M_e} \alpha_i^e = 1$. This gives a very quick procedure to decide whether any given $\sigma \in G$ stabilizes t. On the other hand, we see that we cannot in general expect to have non-trivial stabilizers. Hence the cost of applying the Reynolds operator to m monomials can be estimated by

$$2mdk \cdot |G|.$$

The number m of applications of the Reynolds operator which are required depends on two factors: The first one is whether we want to calculate all invariants of degree d or maybe only one or a few, and the second one is luck. Whether or not the application of the Reynolds operator to just a few monomials will yield linearly independent invariants depends on the choice of these monomials but also on the choice of a basis of V. We shall see in Section 17.4.1 how the chances can be optimized in the special context of calculating secondary invariants. Certainly m lies between 1 and k. Making our estimates slightly coarser, we see that the break-even point for k lies somewhere between

$$\sqrt{\frac{2d \cdot |G|}{s}} \quad \text{and} \quad \frac{2d \cdot |G|}{s}$$

(where as above s is the number of generators by which G is given), depending on the value assumed for m. The Reynolds operator will perform better if $k = \binom{n+d-1}{n-1}$ exceeds this point. In order to take the speedup of the linear algebra method arising from packed representations (see above) into account, we assign a multiplicative constant depending only on K to the above break-even point.

In the case that G is a permutation group, the invariants are calculated in any case by using sums over orbits of monomials. Thus the above decision only comes into play when G is not a permutation group and $\text{char}(K) \nmid |G|$. A set of monomials can be submitted to the algorithm with the effect that if the Reynolds operator is used, then it is applied to these monomials first. This will become important in Section 17.4.1.

17.3 Constructing Primary Invariants

The first major step in the calculation of an invariant ring is the construction of a system of primary invariants $f_1, \ldots, f_n$. These are not uniquely determined by the group, and a good choice of primary invariants turns out to be crucial for the effectiveness of the calculation of secondary invariants. In fact, if $d_1, \ldots, d_n$ are the degrees of the f_i, then there are $d_1 \cdots d_n/|G|$ secondary invariants in the Cohen-Macaulay case. In all cases, this is a lower bound. Furthermore, the maximum degree of a secondary invariant in the Cohen-Macaulay case is $d_1 + \cdots + d_n + a$, where a is the degree of the Hilbert series $H(K[V]^G, t)$ as a rational function in t (i.e., the degree of the numerator minus the degree of the denominator). Hence it is important that the $d_1, \ldots, d_n$ are chosen as small as possible. The Magma implementation has a new algorithm given in [9] which is guaranteed to yield an optimal system of primary invariants $f_1, \ldots, f_n$. This means that the product of the degrees of the f_i will be minimal, and among the systems of primary invariants having minimal degree product, $f_1, \ldots, f_n$ will have a minimal degree sum. We shall not repeat the entire algorithm from [9], but rather give an overview and discuss some important aspects.

Let us for the moment call a degree vector $(d_1, \ldots, d_n) \in \mathbb{N}^n$ **primary** if there exists a system $f_1, \ldots, f_n$ of primary invariants with $\deg(f_i) = d_i$. There are some very strong constraints on a primary degree vector. First, the Hilbert series of $K[V]^G$ must have the form

$$H(K[V]^G, t) = \frac{f(t)}{(1 - t^{d_1}) \cdots (1 - t^{d_n})} \tag{17.2}$$

with $f(t)$ a polynomial having integral coefficients (see Section 17.5.3). In particular, the product $d_1 \cdots d_n$ must be divisible by $|G|$, since the coefficient of $(1-t)^{-n}$ in the Laurent expansion of $H(K[V]^G, t)$ about $t = 1$ is $1/|G|$ (see [10, Theorem 5.5.3]). If the invariant ring is Cohen-Macaulay, the coefficients of $f(t)$ must be non-negative and are in fact the number of secondary invariants of the corresponding degrees. The constraint given by Equation (17.2) is applicable whenever the Hilbert series is known. In the non-modular case and in the case of a permutation group G, this can be calculated by Molien's formula. In the other cases, a further constraint is used, which follows from the fact that any system $f_1, \dots, f_n$ of homogeneous invariants is a system of primary invariants if and only if

$$\dim\left(K[V]/(f_1, \dots, f_n)\right) = 0, \tag{17.3}$$

where the dimension is the Krull-dimension. From this it follows by Krull's principal ideal theorem that $\dim\left(K[V]/(f_1, \dots, f_i)\right) = n - i$ for all i.

Proposition 17.2 ([9]). *If $(d_1, \dots, d_n)$ is a primary degree vector with $d_1 \leq d_2 \leq \dots \leq d_n$, then the inequality*

$$d_i \geq \min\{d \mid \dim\left(K[V]/(\textstyle\sum_{j=1}^{d} K[V]_j^G)\right) \leq n - i\}$$

holds for all i.

This gives a lower bound for the d_i which can quite easily be evaluated since the calculation of Krull dimensions of ideals is available in Magma.

Example 17.1. Take $G = W_3(F_4)$, the 3-modular reduction of the Weyl group of type F_4. Historically, this is the first (modular) reflection group whose invariant ring is not a polynomial ring ([12]).

Setting $d_i = \min\{d \mid \dim\left(K[V]/(\sum_{j=1}^{d} K[V]_j^G)\right) \leq n - i\}$, one obtains $(d_1, d_2, d_3, d_4) = (2, 4, 18, 24)$. It turns out that the first random try of invariants of these degrees already yields a system of primary invariants. Since these primary invariants are optimal, this already shows that $K[V]^G$ is not a polynomial ring. The total running time for this example (which is dominated by the time for calculating all invariants up to degree 24) is about 5 minutes.

If a degree vector satisfies the constraint given by Equation (17.2) or by Proposition 17.2, it is a primary degree vector in most cases. Moreover, being a system of primary invariants is a Zariski-open condition on a tuple $(f_1, \dots, f_n) \in K[V]_{d_1}^G \times \dots \times K[V]_{d_n}^G$ (see [9]), hence a random choice $f_1, \dots, f_n$ of invariants of the degrees d_i will usually yield a system of primary invariants. Having such $f_1, \dots, f_n$, it is easy to check Equation (17.3) to see whether they provide primary invariants. It is hence a good strategy to get the best degree vector satisfying the above constraints which are applicable for the particular group, calculate random invariants of these degrees and test the condition (17.3). In most cases, this will yield an optimal system of primary invariants and only requires one Gröbner basis calculation (for the final dimension test), and a quite limited number of Gröbner basis calculations if Proposition 17.2 is used. However, the problem is that there are degree vectors which satisfy the constraints but are not primary. Furthermore, it might happen that a vast number of unlucky invariants of some degrees $d_1, \dots, d_n$ are tried without success even if $(d_1, \dots, d_n)$ is a primary degree vector. If K is a finite field, this can be excluded by simply looping over all homogeneous invariants of some degrees. Hence to obtain an algorithm which is guaranteed to yield an optimal system of primary invariants, we need a criterion which decides whether a degree vector is primary or not in the case that K is infinite. This is provided by the following proposition, which is the key to the general algorithm.

Proposition 17.3 ([9]). *Let $A = \oplus_{d=0}^{\infty} A_d$ be a graded commutative algebra over an infinite field $K = A_0$ and let $n \in \mathbb{N}_0$ and $d_1, \ldots, d_k \in \mathbb{N}$. Then the following conditions are equivalent:*

1. *There exist homogeneous $f_1, \ldots, f_k \in A$ with $\deg(f_i) = d_i$ such that*
$$\dim\left(A/(f_1, \ldots, f_k)\right) \leq n - k.$$
2. *For each subset $M \subset \{1, \ldots, k\}$ we have*
$$\dim\left(A \Big/ \big(\sum_{i \in M} A_{d_i}\big)\right) \leq n - |M|.$$

If K is a finite field, then the implication "(1) ⇒ (2)" still holds.

Observe that the conditions in (2) can be checked algorithmically. In the proof of the implication "(2) ⇒ (1)" the existence of a homogeneous element $f_1 \in A_{d_1}$ such that $A/(f_1)$ satisfies the conditions in (2) for the degree vector $(d_2, \ldots, d_k)$ is shown. This f_1 can be found by a loop over elements of A_{d_1}. Then f_1 can be extended to a system $f_1, \ldots, f_k$ with $\dim\left(A/(f_1, \ldots, f_k)\right) \leq n-k$. We thus arrive at the following method: Loop over all degree vectors $(d_1, \ldots, d_n)$, ordered by rising products and sums, and check the conditions (2) from Proposition 17.3. When a degree vector is found which satisfies (2), recursively construct primary invariants $f_1, \ldots, f_n$. If K is a finite field and on some recursion level no f_i has been found even after looping through the complete space A_{d_i}, then proceed to the next degree vector.

Clearly this approach always produces an optimal system of primary invariants, but it has the drawback that it requires $2^{n+1}-1$ Gröbner basis calculations for the dimension test even if the first degree vector that is tested is actually a primary degree vector. The "random approach" described above is in a sense complementary to it and only requires one Gröbner basis calculation if successful. The actual algorithm implemented in Magma brings these two approaches together. It has an outer loop over degree vectors which satisfy the applicable constraints. For each such degree vector, a random choice of f_i is tested first. If that is unsuccessful, then increasingly more of the conditions from Proposition 17.3(2) are brought in. More precisely, if at some stage an f_1 is chosen which fails to be extendible to a system $f_1, \ldots, f_k$, then the conditions for this k are brought in, which in the case of infinite K guarantees that the next f_1 will be extendible further. This algorithm combines the virtue of always producing an optimal system of primary invariants with probabilistically good running times. Indeed, we obtained quite good timings in comparison with other algorithms for the construction of primary invariants (see [9]).

17.4 Calculating Secondary Invariants

In this section we assume that primary invariants $f_1, \ldots, f_n$ for G have been chosen and we set $A = K[f_1, \ldots, f_n]$. The next task is to calculate secondary invariants, i.e., generators for $K[V]^G$ as a module over A. We have completely different algorithms for each of the modular and the non-modular cases.

17.4.1 The non-modular Case If $\operatorname{char}(K) \nmid |G|$, we can easily calculate the Hilbert series $H(K[V]^G, t)$ by Molien's formula. Comparing this to Equation (17.2) gives us complete information about the number and degrees of secondary invariants which are needed.

In particular, their number is $\deg(f_1) \cdots \deg(f_n)/|G|$. We then find secondary invariants by the following consideration: for any homogeneous invariants $g_1, \ldots, g_m$ it is equivalent that they generate $R := K[V]^G$ as a module over A and that they generate the vector space $R/(f_1, \ldots, f_n)_R$, where the index R means that we are taking an ideal in R. This is seen by an easy induction on degrees (Proposition 17.4 on page 279). But due to the Reynolds operator, the natural map

$$R/(f_1, \ldots, f_n)_R \to K[V]/(f_1, \ldots, f_n)$$

(with the right ideal taken in $K[V]$) is injective. If $m = \deg(f_1) \cdots \deg(f_n)/|G|$, then $g_1, \ldots, g_m$ are a complete set of secondary invariants if and only if they are linearly independent modulo $(f_1, \ldots, f_n)$. Note that we can reduce the g_i modulo $(f_1, \ldots, f_n)$ by using a Gröbner basis, which has already been calculated for doing the dimension test involved in constructing the primary invariants.

A further optimization ensues from the fact that the Reynolds operator π_G is a homomorphism of modules over $K[V]^G$ and in particular of modules over A. Hence all secondary invariants can be obtained by applying π_G to a basis of $K[V]$ over A. But such a basis is obtained by taking a basis of $K[V]/(f_1, \ldots, f_n)$, which can be chosen to consist of monomials. Restricting the set of monomials in such a way, we can substantially increase the chances of finding suitable invariants by just a few applications of π_G.

The algorithm, which is listed as Algorithm 17.1 on page 275, does not only calculate a minimal system of secondary invariants, but it also produces a subset of *irreducible* secondary invariants such that each secondary invariant is a power product of the irreducible ones. Here a secondary invariant is called **irreducible** if it cannot be written as a polynomial expression in the primary invariants and the other secondary invariants, and 1 is considered as the empty power product. Hence the subset M of irreducible secondary invariants produced by Algorithm 17.1 is a minimal system of generators of $K[V]^G$ as an algebra over A. The calculation of a subset of irreducible secondary invariants has several important benefits: first, we gain more insight into the structure of the invariant ring. Second, fewer "fresh" invariants, especially of high degrees, have to be calculated. Third, the calculation of syzygies will be considerably simpler (see Section 17.5.5).

Input: Primary invariants $f_1, \dots, f_n$.

Output: Secondary invariants $g_1, \dots, g_m$, and a subset M of irreducible secondary invariants such that each g_i is a power product of the elements of M.

Begin

Calculate a Gröbner basis B of $(f_1, \dots, f_n)$ w.r.t. any term order, and monomials $m_1, \dots, m_r$ forming a basis of $K[V]/(f_1, \dots, f_n)$.

Obtain numbers $k_0, \dots, k_e \in \mathbb{N}_0$ from $(1-t^{d_1}) \cdots (1-t^{d_n}) \cdot H(K[V]^G, t) = \sum_{i=0}^{e} k_i t^i$, where $d_i = \deg(f_i)$.

Set $M := \emptyset$ and $m := 1$.

For $d = 0, \dots, e$ **do**

Set $k := 0$ and $h := 0$ (*h will be a linear polynomial in indeterminates $t_1, \dots, t_{k_d}$ with coefficients in $K[V]$*).

For all power products g of elements of M having degree d **do**

If $k = k_d$ **then break**.

Calculate the normal form g_{red} of g w.r.t. B.

If the linear system $g_{red} = h(t_1, \dots, t_{k_d})$ is not solvable for the t_i-variables **then**

set $g_m := g$, $m := m + 1$, $k := k + 1$, and $h := h + g_{red} \cdot t_k$.

end for

For $i = 1, 2, \dots$ **do**

If $k = k_d$ **then break**.

Calculate the ith linearly independent invariant g of degree d, using the monomials $m_1, \dots, m_r$ if the Reynolds operator is applied (see Section 17.2).

Calculate the normal form g_{red} of g w.r.t. B.

If the linear system $g_{red} = h(t_1, \dots, t_{k_d})$ is not solvable for the t_i-variables **then**

set $g_m := g$, $m := m + 1$, $k := k + 1$, $h := h + g_{red} \cdot t_k$, and $M := M \cup \{g\}$.

end for

end for

end.

Algorithm 17.1: Calculate secondary invariants in the non-modular case

Example 17.2.

1. A three-dimensional representation of $G = A_5$ is given by the generators
$$\begin{pmatrix} 1 & 0 & -\alpha \\ 0 & 0 & -1 \\ 0 & 1 & -\alpha \end{pmatrix}, \begin{pmatrix} -1 & -1 & \alpha \\ -\alpha & 0 & \alpha \\ -\alpha & 0 & 1 \end{pmatrix},$$
where $\alpha^2 - \alpha - 1 = 0$ and $K = \mathbb{Q}(\alpha)$. The Hilbert series is calculated by Molien's formula to be
$$H(K[V]^G, t) = \frac{1 + t^{15}}{(1-t^2)(1-t^6)(1-t^{10})}.$$
Trying the first invariants of degree 2, 6 and 10 yields primary invariants at once. The first of the basis monomials $m_1, \ldots, m_r$ obtained in the first step of Algorithm 17.1 having degree 15 is $x_2^2 x_3^{13}$, and applying the Reynolds operator to it yields the missing secondary invariant of degree 15. The entire calculation takes less than half a second.

2. We consider the group $G \leq \mathrm{GL}_4(\mathbb{C})$ of order 36 generated by the matrices
$$\begin{pmatrix} \zeta & 0 & 0 & 0 \\ 0 & -\zeta & 0 & 0 \\ 0 & 0 & 1 & 0 \\ 0 & 0 & 0 & 1 \end{pmatrix} \quad \text{and} \quad \begin{pmatrix} 1 & 0 & 0 & 0 \\ 0 & 1 & 0 & 0 \\ 0 & 0 & \zeta & 0 \\ 0 & 0 & 0 & -\zeta \end{pmatrix},$$
where $\zeta = e^{2\pi i/3}$. The Hilbert series is
$$H(K[V]^G, t) = \frac{1 + 2t^3 + 3t^6 + 2t^9 + t^{12}}{(1-t^3)^2(1-t^6)^2}.$$
Again in less than 2 seconds, Magma finds primary invariants $x_1^3, x_3^3, x_2^6, x_4^6$ and secondary invariants $1, h_1 = x_1 x_2^2, h_2 = x_3 x_4^2, h_1^2, h_1 h_2, h_2^2, h_1^2 h_2, h_1 h_2^2, h_1^2 h_2^2$. The irreducible secondary invariants are h_1 and h_2. Observe that we only had to calculate invariants up to degree 6.

17.4.2 The modular Case In [8], the first author gave an algorithm for calculating secondary invariants in the modular case. Here we present a variant of this algorithm.

We first choose a subgroup $H \lneqq G$ (for example, $H = \{1\}$) and calculate secondary invariants $h_1, \ldots, h_r$ of H, where we take the primary invariants $f_1, \ldots, f_n$ of G as primary invariants for H also. This is done either by recursion or by the non-modular Algorithm 17.1. If $K[V]^H$ is not Cohen-Macaulay, it is useful to calculate the A-linear relations between the h_i also by the method described in Section 17.5.2. In other words, we calculate the kernel S of the map $A^r \to K[V]^H$ given by the h_i, where A^r is a free module over A with free generators to which we assign the degrees of the h_i. The map
$$K[V]^H \longrightarrow \bigoplus_{\sigma \in S(G/H)} K[V], \ f \mapsto (f^\sigma - f)_{\sigma \in S(G/H)}$$

has the kernel $K[V]^G$, where $S(G/H)$ is a subset of G which together with H generates G. We obtain the following commutative diagram of graded A-modules with exact rows and columns:

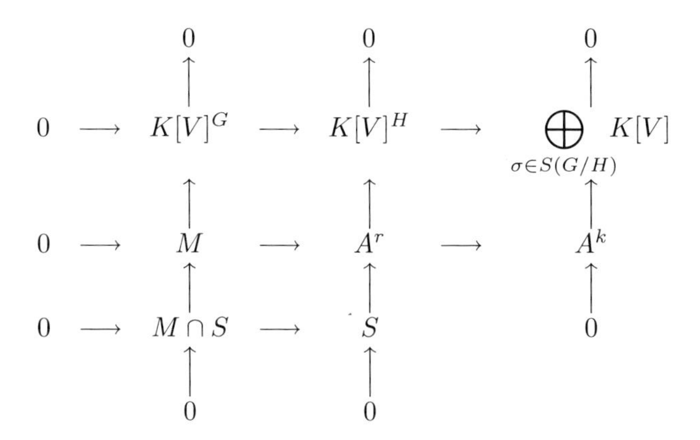

Here $A^k \xrightarrow{\sim} \bigoplus_{\sigma \in S(G/H)} K[V]$ is given by the fact that $K[V]$ is a free module of rank $\prod_{i=1}^n \deg(f_i)$ over A, hence k is $|S(G/H)|$ times this rank. The map $A^r \to A^k$ is defined by the commutativity, and M is its kernel. Observe that all maps in the above diagram are degree-preserving. Now by calculating generators for M, one obtains generators for $K[V]^G$, i.e., secondary invariants. But M is the kernel of a linear map between two free modules over the polynomial algebra A, so its generators can be calculated by the syzygy function of Magma.

In fact the effort of putting up the map $A^r \to A^k$ turns out to be comparable to or even greater than the effort of the actual syzygy calculation. To obtain this map, we have to find the representations of all $h_i^\sigma - h_i$ ($i = 1, \ldots, r$, $\sigma \in S(G/H)$) as elements of $\bigoplus_{j=1}^l A \cdot m_j$, where the m_j are (free) generators of $K[V]$ as an A-module, which are pre-calculated. This is done by equating $h_i^\sigma - h_i$ to a general element of the homogeneous K-subspace of $\bigoplus_{j=1}^l A \cdot m_j$ of degree $d = \deg(h_i)$ with unknown coefficients. Comparing coefficients then leads to an inhomogeneous system of linear equations over K for these unknown coefficients, and solving it yields the desired representation.

In this situation it is quite common that calculations in the same degree, say d, occur very often. A significant speedup arises from the fact that all the related inhomogeneous systems can be solved with only one nullspace computation. The general element of the homogeneous K-subspace of degree d is formed only once and only a slight extension of the usual nullspace algorithm allows the determination of all of the desired representations simultaneously. Since only one construction and echelonization of the space is needed to determine all the representations it is *much* more efficient to use this method than to build and solve separate systems for each of the polynomials having degree d.

Example 17.3.

1. We continue Example 17.1, where the primary invariants of the 3-modular reduction of the Weyl group of type F_4 have already been calculated. To calculate secondary invariants, we choose a Sylow subgroup of order 128 as the subgroup H, hence $r = 2 \cdot 4 \cdot 18 \cdot 24/128 = 27$. It takes more than two hours to set up the map $A^r \to A^k$ and then only about two minutes to calculate its kernel. The result is secondary invariants of degrees 0, 10 and 20.

2. The 6-dimensional indecomposable representation of the cyclic group Z_8 of order 8 over $K = \mathbb{F}_2$ is generated by the matrix

$$\begin{pmatrix} 1 & 0 & 0 & 0 & 0 & 0 \\ 1 & 1 & 0 & 0 & 0 & 0 \\ 0 & 1 & 1 & 0 & 0 & 0 \\ 0 & 0 & 1 & 1 & 0 & 0 \\ 0 & 0 & 0 & 1 & 1 & 0 \\ 0 & 0 & 0 & 0 & 1 & 1 \end{pmatrix}.$$

   It takes 5.3 seconds to find (optimal) primary invariants of degrees 1, 2, 2, 2, 4, 8. Secondary invariants are then calculated in about 8 minutes. The resulting number of secondary invariants is 43 and their maximal degree is 13. To the best knowledge of the authors, invariant rings of indecomposable representations of cyclic groups of dimensions exceeding 5 have never been calculated before.

Since in the above algorithm the number r of secondary invariants for $K[V]^H$ is bounded from below by $\deg(f_1) \cdots \deg(f_n)/|H|$ with equality in the Cohen-Macaulay case, it is important that the subgroup H is chosen as large as possible. This will lead to a minimal number of linear systems to be solved for putting up the map $A^r \to A^k$, and it will minimize the effort of calculating the kernel of this map. By default, the Magma implementation chooses H as a p'-Sylow subgroup of maximal order with $p' \neq \operatorname{char}(K)$. Other subgroups can be submitted by the user. Another, complementary, approach is to calculate secondary invariants for a p-Sylow subgroup P of G first, where $p = \operatorname{char}(K)$, and then to obtain secondary invariants for $K[V]^G$ by using the relative Reynolds operator $\pi_{G/P}\colon K[V]^P \to K[V]^G$ (see [5] or [10, p. 28]). This approach is in development. One could also calculate $K[V]^P$ by applying the algorithm recursively, where each step consists of an extension of the current subgroup by an index p. There is some experimenting involved as to what approach should be best, and the answer will probably depend very much on the special situation.

17.5 Properties of the Invariant Ring

In this section we discuss how some important properties of the invariant ring can be computed after primary and secondary invariants have been found. The properties dealt with in 17.5.1–17.5.3 are only relevant in the modular case. We use the fact that the algorithm from Section 17.4.2 yields the invariant ring $K[V]^G$ as the quotient $M/(M \cap S)$ of submodules of A^r.

17.5.1 Minimal Secondary Invariants and the Cohen-Macaulay Property We first make a general, well-known remark on generating systems of homogeneous modules.

Proposition 17.4. *Let $A = \sum_{d=0}^{\infty} A_d$ be a commutative graded algebra over a field $K = A_0$ and $M = \sum_{d=N}^{\infty} M_d$ a graded A-module with $N \in \mathbb{Z}$. Then it is equivalent for a subset $S \subset M$ of homogeneous elements that S generates M as an A-module and that S generates M/A_+M as a vector space over K. Here A_+M is the submodule of M generated by the elements $a \cdot g$ with $a \in A$ homogeneous of positive degree and $g \in M$.*

In particular, if M is finitely generated, it is equivalent for a homogeneous generating set to be minimal in the sense that no generator can be omitted and to have minimal cardinality.

Proof. Clearly if S generates M, it also generates M/A_+M as a K-vector space.

Now suppose that S generates M/A_+M and let $g \in M$ be homogeneous of some degree d. Then by assumption

$$g = \sum_{i=1}^{m} \alpha_i g_i + \sum_{j=1}^{r} a_j h_j$$

with $g_1, \dots, g_m \in S$, $\alpha_i \in K$, $a_j \in A_+$ and $h_j \in M$. By multiplying out homogeneous parts and omitting those summands which are not of degree d, we can assume that the a_j and h_j are homogeneous with $\deg(a_j h_j) = d$. Hence $\deg(h_j) < d$ and h_j lies in the submodule spanned by S by induction on d, which works since $\{N, N+1, N+2, \dots\}$ is a well-ordered set. Hence g lies in the module spanned by S.

The assertion about minimality of generating systems now follows from the corresponding property of vector spaces. □

In the non-modular case, Algorithm 17.1 already yields a minimal system of secondary invariants, so we now turn to the modular case and use the notation from Section 17.4.2.

If $K[V]^H$ is Cohen-Macaulay, then $K[V]^G \cong M$, and a minimal system of generators for M yields a minimal system of secondary invariants of $K[V]^G$. Obtaining a minimal generating system for M from a possibly redundant one amounts to a series of membership tests of submodules, which are done by linear algebra. In general, however, $K[V]^G \cong M/(M \cap S) \cong (M+S)/S$. If B_1 generates S, then a subset $B_2 \subset M$ generates $M/(M \cap S)$ if and only if $B_1 \cup B_2$ generates $M+S$. Hence a minimal system of secondary invariants for $K[V]^G$ can be found by minimally completing B_1 to a system of generators for $M+S$. By default, this minimization is performed by the Magma implementation.

If m is the cardinality of a minimal system of secondary invariants, then $K[V]^G$ is Cohen-Macaulay if and only if

$$m = \frac{\deg(f_1)\cdots\deg(f_n)}{|G|} \tag{17.4}$$

(see, for example, [8, Proposition 12]). Hence checking the Cohen-Macaulay property is also an easy exercise once the primary and secondary invariants have been calculated.

17.5.2 Module-Syzygies and Depth If $K[V]^G$ is not Cohen-Macaulay, then our minimal generating set of $K[V]^G$ over A will contain relations. Let us call these relations **module-syzygies** in order to avoid confusion with the syzygies treated in Section 17.5.5. We use the notation of Section 17.4.2 and complete a generating system of S to a generating system of $M+S$ in order to obtain a minimal system of secondary invariants (see above). This yields an epimorphism $F' \oplus F \to M + S \subset A^r$, where F and F' are free A-modules with F corresponding to the generators of S and F' corresponding to its completion. The kernel N of this epimorphism can be calculated by the standard syzygy function of Magma. We obtain the following commutative diagram of A-modules with exact rows and columns:

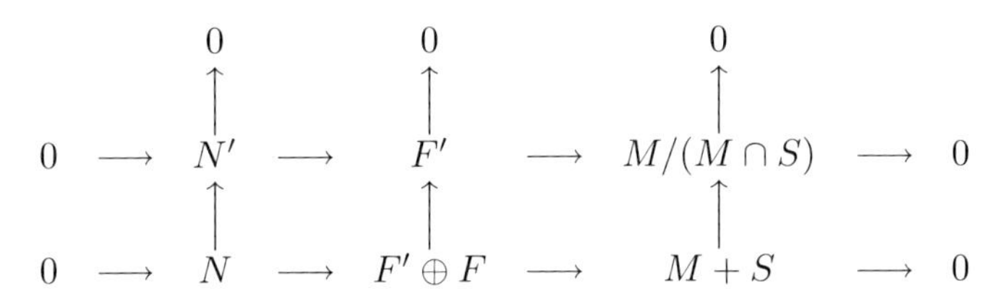

$$\begin{array}{ccccccccc} & & 0 & & 0 & & 0 & & \\ & & \uparrow & & \uparrow & & \uparrow & & \\ 0 & \longrightarrow & N' & \longrightarrow & F' & \longrightarrow & M/(M\cap S) & \longrightarrow & 0 \\ & & \uparrow & & \uparrow & & \uparrow & & \\ 0 & \longrightarrow & N & \longrightarrow & F'\oplus F & \longrightarrow & M+S & \longrightarrow & 0 \end{array}$$

Here $F' \oplus F \to F'$ is the first projection, and N' is the image of N under it. We have to show the exactness at F'. By the commutativity, any element from N' is mapped to 0 under $F' \to M/(M \cap S)$. Conversely, suppose that $v_2 \in F'$ is mapped to 0. Then it is mapped to an element $m \in M \cap S$ under $F' \to M$. Hence there exists $v_1 \in F$ which is mapped to m under $F \to S$, and $v_2 - v_1$ lies in N. Now $v_2 \in N'$ by construction.

As $K[V]^G \cong M/(M \cap S)$, N' consists exactly of the module-syzygies by the above diagram. So we see that module-syzygies can be calculated quite easily.

Carrying the calculations further in this way and calculating syzygies of N' and so on, we obtain a minimal free resolution

$$0 \longrightarrow F_r \longrightarrow \cdots \longrightarrow F_1 \longrightarrow F_0 \longrightarrow K[V]^G \longrightarrow 0 \tag{17.5}$$

of $K[V]^G$ (as a graded A-module). Each step consists of a simple application of the standard syzygy function and a minimization of generators. If $F_r \neq 0$, then by the Auslander-Buchsbaum formula we obtain

$$\operatorname{depth}(K[V]^G) = n - r$$

(see [8]). All these calculations are relatively easy once the secondary invariants have been found.

Example 17.4. In Example 17.3(2), it is clear by the criterion given in Equation (17.4) that the invariant ring of the 6-dimensional indecomposable representation of Z_8 over $\mathbb{F}_2$ cannot be Cohen-Macaulay. Hence the depth is at most 5. It takes 4.5 seconds to calculate a minimal free resolution of the invariant ring. This turns out to have the length $r = 3$, hence the depth is in fact $6 - 3 = 3$, in accordance with the famous result by Ellingsrud and Skjelbred [7], which says that the invariant ring of an indecomposable representation of a cyclic group always has depth 3 (provided that the dimension of the representation is at least 3).

17.5.3 The Hilbert Series There are two methods to obtain the Hilbert series of the invariant ring.

1. If a free resolution (see Equation 17.5) has been calculated, then

$$H(K[V]^G, t) = \sum_{i=0}^{r} (-1)^i H(F_i, t).$$

   Note that all F_i occurring in (17.5) are graded free modules, where the grading is such that the homomorphisms are degree-preserving. If $d_1, \ldots, d_r$ are the degrees of the free generators of an F_i, then

$$H(F_i, t) = \frac{t^{d_1} + \cdots + t^{d_r}}{(1 - t^{\deg(f_1)}) \cdots (1 - t^{\deg(f_n)})}.$$

2. Since $H(K[V]^G, t) = H(M, t) - H(M \cap S, t)$ (again using the notation from Section 17.4.2), The Hilbert series can be obtained by calculating Gröbner bases for M and for $M \cap S$ and then computing the Hilbert series of these modules by an algorithm given by Bayer and Stillman [2]. This algorithm is purely combinatorial and only uses the leading monomials of the Gröbner bases. Note that in many cases $S = 0$ due to the choice of the subgroup H.

If a free resolution has been calculated, the method (1) amounts to a mere bookkeeping task. But according to our experience, the second method never takes any noticeable amount of time compared to the time required for the calculation of primary and secondary invariants. Hence the implementation uses method (2) in all cases.

Example 17.5. We consider the cyclic group G of order 7 generated by the matrix

$$\begin{pmatrix} 1 & 0 & 0 & 0 & 0 \\ 1 & 1 & 0 & 0 & 0 \\ 0 & 1 & 1 & 0 & 0 \\ 0 & 0 & 1 & 1 & 0 \\ 0 & 0 & 0 & 1 & 1 \end{pmatrix} \in \mathrm{GL}_5(\mathbb{F}_7).$$

The invariant ring is calculated in about 4 minutes. The degrees of the primary invariants are 1, 2, 2, 6, and 7. The subsequent calculation of the Hilbert series takes 4 seconds. The result is

$$H(K[V]^G,t) = \frac{t^{13}+2t^{10}+2t^9+4t^8+3t^7+3t^6+4t^5+2t^4+2t^3+1}{(1-t)(1-t^2)^2(1-t^6)(1-t^7)}.$$

This coincides with a result of Almkvist and Fossum [1], which gives Hilbert series of all invariant rings of indecomposable representations of cyclic groups and in this case reads

$$H(K[V]^G,t) = \frac{1}{7}\sum_{\zeta\in\mu_7}\prod_{j=0}^{4}\frac{1}{1-\zeta^{4-2j}t},$$

where the sum runs over the set μ_7 of the 7th complex roots of unity.

17.5.4 Fundamental Invariants and β Given homogeneous invariants $f_1,\ldots,f_m \in K[V]^G$ it is equivalent that they generate $K[V]^G$ as a K-algebra and that they generate the ideal $K[V]^G_+ \trianglelefteq K[V]^G$ spanned by the invariants of positive degree. This observation, which is proved as Proposition 17.4, is in fact the starting point of Hilbert's Proof of the Finiteness Theorem. Now by Proposition 17.4 the latter condition is in turn equivalent to saying that $f_1,\ldots,f_m$ generate the quotient module $K[V]^G_+/(K[V]^G_+)^2$ as a vector space over K. So again we see that a system of generators of $K[V]^G$ as an algebra over K which is minimal in the sense that no element can be omitted also has minimal cardinality, and the degrees of its members are uniquely determined. The maximum of these degrees is often denoted by $\beta(V,G)$, and members of a minimal generating system are called **fundamental invariants**.

By taking the primary and the secondary invariants together, we have a system of generators for $K[V]^G$ as a K-algebra. It is in fact sufficient to take the irreducible secondary invariants if they have been calculated (see Section 17.4.1). This system can be minimized by taking the elements f in turn and for each one testing if f is contained in the algebra generated by the current system of fundamental invariants minus f. The test again comes down to a linear system over K, obtained by equating f to a general element of the K-subspace of degree-d elements of the algebra generated by the current fundamental invariants minus f. If it is solvable, then f can be omitted from the system of fundamental invariants. Again this method is optimized by collecting all relevant calculations of one degree into a single system of equations (see page 277).

Example 17.6. In Example 17.3(2), the minimal secondary invariants of the 6-dimensional indecomposable representation over $\mathbb{F}_2$ of the cyclic group Z_8 were calculated. When extracting fundamental invariants from these and the primary invariants, 9 secondary invariants, including those of degrees 12 and 13, can be omitted. Hence $\beta(V,G) = 11$ and the minimal number of algebra generators of the invariant ring is 40. The extraction of fundamental invariants takes about 4 minutes.

17.5.5 Syzygies Sections 17.3 and 17.4 were devoted to finding generators of $K[V]^G$ as an algebra over K. In the non-modular case, these are given by the primary invariants and the irreducible secondary invariants $h_1, \dots, h_r$, where all secondary invariants $g_1, \dots, g_m$ are power products of the h_i. We write $g_i = p_i(h_1, \dots, h_r)$, where p_i are power products of indeterminates $t_1, \dots, t_r$. In the modular case, let $h_1, \dots, h_r$ be all secondary invariants which have degree > 0. We are now interested in algebraic relations between these generators, i.e., we are are looking for the kernel J of the homomorphism

$$A[t_1, \dots, t_r] \to K[V]^G, \ t_i \mapsto h_i$$

of graded K-algebras, where we set $\deg(t_i) = \deg(h_i)$. To calculate these relations is one of the basic tasks of invariant theory, since they define the quotient variety $V/G = \operatorname{Spec}(K[V]^G)$. Since the primary invariants are algebraically independent, there is no necessity to replace them by indeterminates in this paper, though they are of course represented symbolically in the Magma implementation. There is a standard Gröbner basis method using tag-variables for calculating this kernel. But Gröbner bases can be avoided since the following proposition reduces the calculation of syzygies to a pure linear algebra problem.

Proposition 17.5. *In the above situation, let $S \subset J$ be a set containing*

1. *generators for the module-syzygies, i.e., for the A-module $J \cap \oplus_{i=1}^m A \cdot p_i$,*
2. *for each pair g_i and h_k such that $p_i \cdot t_k$ is none of the p_j, a relation of the form $p_i t_k - f_{i,k}$ with $f_{i,k} \in \oplus_{j=1}^m A \cdot p_j$.*

Then J is spanned by S.

Proof. Let J' be the ideal generated by S. For a power product p of the t_j, let p_i be a maximal subproduct of p which is one of the p_j and write $D(p)$ for the length (i.e., number of factors) of p/p_i. We prove by induction on $D(p)$ that p is congruent to an element of $\oplus_{j=1}^m A \cdot p_j$ modulo J'. If $D(p) = 0$, we are done. Otherwise, $p = p_i \cdot t_k \cdot q$ for some k. By the assumption, $p_i \cdot t_k$ is congruent to an element of $\oplus_{j=1}^m A \cdot p_j$. But for all j we have $D(p_j \cdot q) \leq \operatorname{length}(q) < D(p)$, hence by induction all $p_j \cdot q$ lie in $\oplus_{j=1}^m A \cdot p_j$ modulo J', and so does p.

Now let $f \in A[t_1, \dots, t_r]$ be a polynomial which maps to 0 under $A[t_1, \dots, t_r] \to K[V]^G$. By the above, $f \equiv g \mod J'$ with $g \in \oplus_{j=1}^m A \cdot p_j$, hence g maps to 0, too. By the first assumption, g lies in J' and thus also $f \in J'$. $\square$

The calculation of module-syzygies, which only exist in the modular case, was discussed in Section 17.5.2. Finding the representations of $g_i h_k$ is done by the following linear algebra method: We form a general degree-d element of $\sum_{j=1}^m A g_j$ (with unknown coefficients) and equate it to $g_i h_k$, where $d = \deg(g_i h_k)$. This is an inhomogeneous system of linear equations over K. Solving it will yield the desired representation. As above, a considerable speedup is achieved by doing all computations of one degree in a single linear system.

We can obtain a *minimal* generating set for J and at the same time avoid a considerable amount of computation by using the following strategy: First order the products $g_i h_k$ by rising degrees. Then for each product $g_i h_k$ of degree d, consider a general degree-d element of the ideal $J' \subset J$ generated by the relations obtained so far. It is again a linear algebra problem to decide if there exists a specialization of the unknown coefficients involved which yields the desired representation for $g_i h_k$. If such a specialization exists, then we do not have to calculate the representation, and no new generator has to be added to J.

It is of considerable interest to obtain a minimal generating system for J, since $K[V]^G$ is said to be a complete intersection if such a system has exactly r elements.

Example 17.7. The group $G = W_3(F_4)$ studied in examples 17.1 and 17.3(1) has secondary invariants 1 and g_{10} of degree 10, and a further secondary invariant of degree 20, for which g_{10}^2 can be taken. Since Equation (17.4) is fulfilled, they are linearly independent over A, and we only need one relation for g_{10}^3, and $K[V]^G$ is a complete intersection. The relation is found in 13 seconds, but it is too messy to be printed here.

On the other hand, consider the cyclic group $G \leq \mathrm{GL}_2(\mathbb{C})$ generated by $\left(\begin{smallmatrix} \zeta & 0 \\ 0 & \zeta \end{smallmatrix}\right)$ with $\zeta = e^{2\pi i/3}$. We find primary invariants $f_1 = x_1^3$ and $f_2 = x_2^3$, and secondary invariants 1, $h_1 = x_1 x_2^2$ and $h_2 = x_1^2 x_2$, from which h_1 and h_2 are irreducible. A minimal system of relations is

$$h_1^2 = f_2 h_2, \quad h_2^2 = f_1 h_1, \quad h_1 h_2 = f_1 f_2,$$

hence $K[V]^G$ is not a complete intersection.

Bibliography

[1] Gert Almkvist: *Invariants of* $\mathbf{Z}/p\mathbf{Z}$ *in Charcteristic* p, in: Invariant Theory (Proc. of the 1982 Montecatini Conference), Lecture Notes in Math. **996**, Springer-Verlag, Heidelberg, Berlin 1983.

[2] Dave Bayer, Mike Stillman: *Computation of Hilbert Functions*, J. Symbolic Computation **14**, (1992), 31–50.

[3] David J. Benson: *Polynomial Invariants of Finite Groups*, Lond. Math. Soc. Lecture Note Ser. **190**, Cambridge Univ. Press, Cambridge 1993.

[4] Wieb Bosma, John J. Cannon, Catherine Playoust: *The Magma Algebra System I: The User Language*, J. Symbolic Computation **24**, (1997).

[5] H. E. A. Campbell, I. Hughes, R. D. Pollack: *Rings of Invariants and p-Sylow Subgroups*, Canad. Math. Bull. **34(1)**, (1991), 42–47.

[6] John J. Cannon, Catherine Playoust: *Magma: A new computer algebra system*, Euromath Bulletin **2(1)**, (1996), 113–144.

[7] Geir Ellingsrud, Tor Skjelbred: *Profondeur d'anneaux d'invariants en caractéristique p*, Compos. Math. **41**, (1980), 233–244.

[8] Gregor Kemper: *Calculating Invariant Rings of Finite Groups over Arbitrary Fields*, J. Symbolic Computation **21**, (1996), 351–366.

[9] Gregor Kemper: *Calculating Optimal Homogeneous Systems of Parameters*, J. Symbolic Computation, to appear.

[10] Larry Smith: *Polynomial Invariants of Finite Groups*, A. K. Peters, Wellesley, Mass. 1995.

[11] Bernd Sturmfels: *Algorithms in Invariant Theory*, Springer-Verlag, Wien, New York 1993.

[12] H. Toda: *Cohomology mod 3 of the Classifying Space BF_4 of the Exceptional Group F_4*, J. Math. Kyoto Univ. **13**, (1972), 97–115.

Progress in Mathematics, Vol. 173, © 1999 Birkhäuser Verlag Basel/Switzerland

Chapter 18

Coxeter Transformations associated with Finite Dimensional Algebras

Helmut Lenzing

Abstract

Let Λ be a finite dimensional algebra of finite global dimension. On the K-theoretic level the Auslander-Reiten translation on the bounded derived category of Λ-modules induces the Coxeter transformation which is an important invariant, preserved under derived equivalence, and displaying a lot of interesting homological information. The paper deals with the ramifications of this concept, stresses its homological nature and the computational aspects with a focus on hereditary and canonical algebras.

18.1 Introduction

For a finite dimensional k-algebra Λ of finite global dimension the Coxeter transformation is the automorphism Φ_Λ of the Grothendieck group $\mathrm{K}_0(\Lambda)$ which is induced by the Auslander-Reiten translation $\tau_{\mathcal{D}}$ of the bounded derived category $\mathcal{D}$ of Λ-modules. The Coxeter transformation preserves the homological bilinear form (Euler form) on $\mathrm{K}_0(\Lambda)$. Its characteristic polynomial, the Coxeter polynomial χ_Λ, further controls the growth behaviour of Φ_Λ, hence of $\tau_{\mathcal{D}}$. The Coxeter transformation (and associated polynomials) form important invariants under derived equivalence but also provide natural links between the representation theory of finite dimensional algebras and other theories, notably the theory of Lie algebras (classical and Kac-Moody) [4], the theory of C^*-algebras [15], the spectral theory of graphs [6] and also to number theory and computing [24, 5]. The Coxeter-data are moreover easy to calculate from the Cartan matrix of Λ, whose entries are the dimensions of the Hom-spaces between the indecomposable projective Λ-modules. On the other hand, except for particular classes of algebras, fairly little is known on the general structure of the Coxeter transformation and its characteristic polynomial.

Throughout we work over an algebraically closed field k. The algebras, we consider, are associative with a unit element, and moreover finite dimensional over k if not stated otherwise. For a finite dimensional k-algebra Λ we consider usually finite dimensional right modules and denote by $\mathrm{mod}\,(\Lambda)$ the category of such modules. The ordinary *k-duality* $\mathrm{Hom}_k\,(-,k)$ will be denoted by D.

The *stable module category* $\underline{\mathrm{mod}}(\Lambda)$ modulo projectives (resp. $\overline{\mathrm{mod}}\,(\Lambda)$ modulo injectives) has the same objects as $\mathrm{mod}\,(\Lambda)$, its morphism spaces are given as the factor spaces $\underline{\mathrm{Hom}}(X,Y)$ (resp. $\overline{\mathrm{Hom}}\,(X,Y)$) of $\mathrm{Hom}_\Lambda\,(X,Y)$ of Λ-linear maps modulo morphisms factorizing through a projective (resp. an injective) Λ-module.

Important for the finite dimensional representation theory is the concept of an *almost-split sequence* $0 \to A \to B \to C \to 0$ in $\mathrm{mod}\,(\Lambda)$ and the associated concept of the *Auslander-Reiten translation(*s) τ_Λ and τ_Λ^{-1} relating the end terms $C = \tau_\Lambda^- A$ and $A = \tau_\Lambda C$ of almost-split sequences. Actually, τ_Λ is a functor from $\underline{\mathrm{mod}}\,(\Lambda)$ to $\overline{\mathrm{mod}}\,(\Lambda)$ with inverse $\tau_\Lambda^{-1} : \overline{\mathrm{mod}}\,(\Lambda) \to \underline{\mathrm{mod}}\,(\Lambda)$. The two functors are uniquely determined by the validity of *Auslander-Reiten duality*

$$\underline{\mathrm{Hom}}_\Lambda\left(\tau_\Lambda^{-1}Y, X\right) = \mathrm{DExt}^1_\Lambda\,(X,Y) = \overline{\mathrm{Hom}}_\Lambda\,(Y, \tau_\Lambda X)$$

to be interpreted as isomorphisms which are functorial in $X \in \underline{\mathrm{mod}}\,(\Lambda)$ and $Y \in \overline{\mathrm{mod}}\,(\Lambda)$. As general references to the subject we refer to [1], [11] and [36].

By Jordan-Hölder's Theorem, the *Grothendieck group* $\mathrm{K}_0\,(\Lambda)$ of $\mathrm{mod}\,(\Lambda)$ modulo short exact sequences is the free abelian group on the classes $[S_1]$, $[S_2]$, ..., $[S_n]$ of simple Λ-modules. The class $[M]$ of a Λ-module equals $[M] = \sum_{i=1}^n [M : S_i]\,[S_i]$, where $[M : S_i]$ denotes the multiplicity of the simple module S_i in a composition series of M. In order to achieve a sensible K-theory, we assume for the rest of the paper that Λ has finite global dimension, in which case the classes $[P_1]$, $[P_2]$, ... , $[P_n]$ of indecomposable projective Λ-modules form another natural basis of $\mathrm{K}_0\,(\Lambda)$. A similar assertion holds for the classes $[Q_1]$, $[Q_2]$, ... , $[Q_n]$ of indecomposable injective Λ-modules. We arrange the numbering in such a way that P_i (resp. Q_i) is the projective cover (resp. injective envelope) of S_i. The (usually non-symmetric) *Euler form* $\langle -,- \rangle$ on $\mathrm{K}_0\,(\Lambda)$ is the homological bilinear form given on classes of Λ-modules by the formula

$$\langle [X],[Y] \rangle = \sum_{i=0}^{\infty} (-1)^i \dim_k \mathrm{Ext}^i_\Lambda(X,Y).$$

We always consider $\mathrm{K}_0\,(\Lambda)$ to be equipped with this additional structure, resulting in a *bilinear lattice* in the sense of [25]. Accordingly, isomorphisms between Grothendieck groups are assumed to preserve the respective Euler forms. By Schur's Lemma we get the *dual basis formula* $\langle [P_i]\,, [S_j] \rangle = \delta_{ij}$, relating the classes of indecomposable projectives and simples, respectively, as dual bases with respect to the Euler form. In particular, the Euler form is non-degenerate and has determinant ± 1. With respect to the basis of classes of indecomposable projective modules the Euler form is given by the *Cartan matrix* $C_\Lambda = (c_{ij})$, where $c_{ij} = \langle [P_i]\,, [P_j] \rangle = \dim_k \mathrm{Hom}_\Lambda\,(P_i, P_j)$.

Without going into technical details we review some basic facts on the *bounded derived category* $\mathrm{D}^b\,(\mathrm{mod}\,(\Lambda))$ of Λ-modules; see [16] and [17] for further details. The main function of the derived category is to form the natural environment for the homological algebra for $\mathrm{mod}\,\Lambda$. This can be seen, in particular, from the fact that $\mathcal{D} = \mathrm{D}^b\,(\mathrm{mod}\,(\Lambda))$ contains for each integer $n \in \mathbb{Z}$ a copy $(\mathrm{mod}\,\Lambda)[n]$ of $\mathrm{mod}\,(\Lambda)$, with objects written $X[n]$, such that

$$\mathrm{Hom}_{\mathcal{D}}(X[n],Y[m]) = \mathrm{Ext}_{\Lambda}^{m-n}(X,Y) \quad \text{for all } X,Y \in \mathrm{mod}\,(\Lambda)\,.$$

The derived category $\mathcal{D}$ carries the structure of a *triangulated category*, that is $\mathcal{D}$ is equipped with a translation functor $T : \mathcal{D} \to \mathcal{D}$ such that $T\,(X\,[n]) = X\,[n+1]$ for each Λ-module X; moreover $\mathcal{D}$ disposes of a distinguished system of triangles $X \to Y \to Z \to X\,[1]$. Finally, a triangle $X \xrightarrow{u} Y \xrightarrow{v} Z \xrightarrow{\mu} X\,[1]$ in $\mathcal{D}$ has all the terms X, Y and Z in $\mathrm{mod}\,(\Lambda)$ if and only if $\mu : 0 \to X \xrightarrow{u} Y \xrightarrow{v} Z \to 0$ is an exact sequence of Λ-modules. This latter fact implies that the inclusion $\mathrm{mod}\,(\Lambda) \hookrightarrow \mathcal{D}$, $X \mapsto X\,[0]$, induces an isomorphism between the Grothendieck group $\mathrm{K}_0\,(\Lambda)$ and the Grothendieck group $\mathrm{K}_0\,(\mathcal{D})$ of the triangulated category $\mathcal{D}$ modulo distinguished triangles. Moreover, this isomorphism maps the Euler form of $\mathrm{K}_0\,(\Lambda)$ to the Euler form on $\mathrm{K}_0\,(\mathcal{D})$ given on classes of objects from $\mathcal{D}$ by the expression $\langle [X]\,,[Y]\rangle = \sum_{n\in\mathbb{Z}} (-1)^n \dim_k \mathrm{Hom}_{\mathcal{D}}\,(X, Y\,[n])$. We are thus going to identify $\mathrm{K}_0\,(\Lambda)$ and $\mathrm{K}_0\,(\mathcal{D})$ with their structure of Euler forms from now on.

A distinguished triangle $A \to B \to C \to A\,[1]$ is called an *Auslander-Reiten triangle* if its "end terms" A and C are indecomposable, and further each non-isomorphism $A \to X$ into an indecomposable object X of $\mathcal{D}$ extends to B or, equivalently, each non-isomorphism $Y \to C$ from an indecomposable object Y of $\mathcal{D}$ lifts to B. The Auslander-Reiten translation $\tau_{\mathcal{D}}$ of $\mathcal{D}$ carries C to A.

If Λ has finite global dimension, the bounded derived category $\mathrm{D}^b(\mathrm{mod}\,\Lambda)$ of Λ-modules has Auslander-Reiten triangles [19]. Moreover, the corresponding Auslander Reiten-translation $\tau_{\mathcal{D}} : \mathrm{D}^b(\mathrm{mod}\,\Lambda) \to \mathrm{D}^b(\mathrm{mod}\,\Lambda)$ is an equivalence of categories preserving the triangulated structure.

Auslander-Reiten duality, in the context of the derived category, takes the form $\mathrm{DHom}_{\mathcal{D}}\,(X, Y\,[1]) = \mathrm{Hom}_{\mathcal{D}}\,(Y, \tau_{\mathcal{D}} X)$ for all $X, Y \in \mathcal{D}$. Note that this adjunction formula determines the functor $\tau_{\mathcal{D}}$. We are now in a position to introduce the object of our study properly.

The Coxeter transformation is an automorphism Φ_Λ of $\mathrm{K}_0\,(\Lambda)$, called the *Coxeter transformation* for Λ, which is uniquely determined by any of the following properties (cf. [19, 36])

(i) $\langle x, \Phi_\Lambda y\rangle = -\langle y, x\rangle$ for all $x, y \in \mathrm{K}_0\,(\Lambda)$;

(ii) $\Phi_\Lambda\,([P_i]) = -\,[Q_i]$ for each $i = 1, \ldots, n$;

(iii) $\Phi_\Lambda\,([X]) = [\tau_{\mathcal{D}}\,(X)]$ for each $X \in \mathcal{D}^b\,(\Lambda)$.

Moreover, if C_Λ is the Cartan matrix then, with respect to the basis of classes of indecomposable projectives, the Coxeter transformation is given by the matrix $-C_\Lambda^{-1} C_\Lambda^t$, called the *Coxeter matrix* which we denote by the same symbol Φ_Λ.

The Coxeter transformation Φ_Λ thus measures how far the Euler form deviates from symmetry, but certainly its main role is to describe the action of the Auslander-Reiten translation $\tau_{\mathcal{D}}$ on the K-theoretic level. The link between Φ_Λ and the Auslander-Reiten translation τ_Λ for $\mathrm{mod}\,(\Lambda)$ is less direct. We refer to [36, p. 74 ff.] for a detailed discussion, and just quote from there:

Proposition 18.1. *Let M be a Λ-module, then the following holds:*

(i) *If* $\mathrm{pd}_\Lambda M \leq 1$ *and* $\mathrm{Hom}\,(M,\Lambda) = 0$ *then* $[\tau_\Lambda M] = \Phi_\Lambda\,[M]$.

(ii) *If* $\mathrm{pd}_\Lambda M \leq 2$ *and* $\mathrm{id}_\Lambda \leq 2$ *then for some injective Λ-module Q_M we get* $[\tau_\Lambda M] = \Phi_\Lambda[M] + [Q_M]$.

This relationship between τ_Λ and Φ_Λ is very good for hereditary algebras since in this case $\tau_\mathcal{D}$ and τ_Λ agree on indecomposable modules which are non-projective. The connection is still reasonably good for a *quasi-tilted algebra* where in addition to $\mathrm{gl.\,dim}\,\Lambda \leq 2$ each indecomposable module is requested to have projective or injective dimension at most one. This applies in particular for the canonical algebras, see [29] for further details. The relationship between the two concepts gets very weak, however, if the global dimension increases.

Two algebras Λ and Λ' are called *derived-equivalent* if their bounded derived categories $\mathrm{D}^b\,(\mathrm{mod}\,(\Lambda))$ and $\mathrm{D}^b\,(\mathrm{mod}\,(\Lambda'))$ are equivalent as triangulated categories (triangle-equivalent for short). Since the K-theory $(\mathrm{K}_0\,(\Lambda)\,, \langle -,-\rangle)$ is an invariant of the derived category, derived-equivalent algebras Λ, Λ' have the same K-theory, equivalent Coxeter transformations, and in particular the same Coxeter polynomial. From the computational point it is further important that the K-theory $(\mathrm{K}_0(\Lambda), \langle -,-\rangle)$ is easy to calculate in terms of the Cartan-matrix.

We recall that a Λ-module E is called *exceptional* if $\mathrm{End}_\Lambda\,(E) = k$ and $\mathrm{Ext}^i_\Lambda\,(E,E) = 0$ for all $i \geq 1$. This concept is very important for all K-theoretic studies of representations since for many classes of algebras, in particular algebras derived equivalent to a hereditary or a canonical algebra, each exceptional module E is determined by its class $[E]$, which is a *root* of $\mathrm{K}_0\,(\Lambda)$ (or the quadratic form $q_\Lambda(x) = \langle x, x\rangle$), i.e. satisfies $\langle [E]\,, [E]\rangle = 1$.

For instance, each indecomposable representation of a finite quiver of Dynkin type is exceptional; similarly each indecomposable preprojective or preinjective representation of any finite quiver (without oriented cycles) is exceptional. If the algebra Λ has a presentation $\Lambda = k[\vec{\Delta}]/I$, where $\vec{\Delta}$ is a finite quiver without oriented cycles, and I is an admissable ideal in the path algebra $k[\vec{\Delta}]$, then each indecomposable projective Λ-module is exceptional.

18.2 Bilinear Lattices

Though our main interest is in representations of finite dimensional algebras, it is convenient to develop the theory in the more general setting of bilinear lattices. A *bilinear lattice* is a finitely generated (free) abelian group, equipped with a (usually) non-symmetric bilinear form such that the canonical map $V \to \mathrm{Hom}_\mathbb{Z}\,(V,\mathbb{Z})$, $x \mapsto \langle x, -\rangle$ is an isomorphism. Of course, our main example of a bilinear lattice is $\mathrm{K}_0\,(\Lambda)$ equipped with the Euler form, where Λ is of finite global dimension.

It is not difficult to see that a bilinear lattice V is determined by the following three data, again underlining the role of the Coxeter transformation.

1. the quadratic form $q_V(x) = \langle x, x\rangle$, equivalently the symmetric bilinear form $(x|y) = \langle x, y\rangle + \langle y, x\rangle$;
2. the Coxeter transformation $\tau : V \to V$ defined by $\langle y, x\rangle = -\langle x, \tau y\rangle$;
3. the skew-symmetric restriction of $\langle -, -\rangle$ to the *radical* $\mathrm{rad}(V)$ which is defined as the fixed point set of τ, equivalently as the set of all x with $(x|-) = 0$.

In particular, if $\mathrm{rad}(V) = 0$ we have $\langle x, y\rangle = ((1-\tau)^{-1}x|y)$.

Note that here, and later, we denote the Coxeter transformation by τ in order to stress its similarity to the Auslander-Reiten translation.

An element $a \in V$ is called a *root* if $\langle a, a\rangle = 1$. The characteristic polynomial of the Coxeter transformation is called the *Coxeter polynomial* of V and is denoted χ_V. If $V = \mathbb{Z}^n$, and $\langle x, y\rangle = x^t C y$ for some $n \times n$-matrix C, then τ is given as the mapping $x \mapsto \Phi x$ obtained determined by the matrix $\Phi = -C^{-1}C^t$. In particular, the Coxeter polynomial $\chi_V = \alpha_0 + \alpha_1 T + \cdots + \alpha_{n-1}T^{n-1} + \alpha_n T^n$ is self-reciprocal, i.e. $a_p = a_{n-p}$ holds for $p = 0, \ldots, n$.

An isomorphism $\alpha : V \to W$ of bilinear lattices is a bijective $\mathbb{Z}$-linear mapping, preserving the bilinear forms and hence commuting with the respective Coxeter transformations. $\mathrm{Aut}(V)$ denotes the group of all automorphisms of V.

To include the K-theory of an algebra over a not algebraically closed field, the previous definitions (bilinear lattice, root, etc.) need to be modified. For this, also for missing proofs in Section 18.4 we refer to [25].

18.3 Coxeter Polynomials and Growth

18.3.1 Perpendicular Calculus From now on V denotes a bilinear lattice. We provide a K-theoretic analogue of the formation of the perpendicular category with respect to an exceptional object E (compare [13]). For a root a of V we define the left (resp. right) *perpendicular lattice* ${}^{\perp}a$ (resp. $a^{\perp}$) as the subgroup $\{x \in V | \langle x, a\rangle = 0\}$ (resp. $\{x \in V | \langle a, x\rangle = 0\}$). Note that $a^{\perp} = {}^{\perp}(\tau a)$, allowing to restrict to right perpendicular lattices. The terminology "perpendicular lattice" is justified by the proposition below.

Proposition 18.2. *Let a be a root of V. The mappings*

$$\begin{aligned} \ell_a &: V \to V, \quad x \mapsto x - \langle a, x\rangle a \\ r_a &: V \to V, \quad x \mapsto x + \langle a, x\rangle \tau a \end{aligned}$$

are projections from V onto $a^{\perp} = {}^{\perp}\tau a$ with kernel $\ker \ell_a = \mathbb{Z}.a$, resp. $\ker r_a = \mathbb{Z}.\tau a$, satisfying for $x \in V$ and $y \in a^{\perp}$ the conditions

(i) $\langle x, y\rangle = \langle \ell_a x, y\rangle$.

(ii) $\langle y, x\rangle = \langle y, r_a x\rangle$.

Moreover, $a^{\perp}$ is a bilinear sublattice of V. Further $\ell_a x$ and $r_a x$ are uniquely determined by (i), resp. (ii).

Proof. Straightforward. □

Corollary 18.1. *For $x, y \in a^{\perp}$ we get $\langle y, x\rangle = -\langle x, (r_a\tau)y\rangle = -\langle(\ell_a\tau^{-1})x, y\rangle$.*
In particular, $r_a\tau$ is the Coxeter transformation for $a^{\perp}$; its inverse is given as $\ell_a\tau^{-1}$.

18.3.2 Hilbert-Poincaré Series For each $s \in V$ we introduce the *Hilbert-Poincaré series* of (V, s) as $P_{(V,s)} = \sum_{p=0}^{\infty}\langle s, \tau^p s\rangle T^p$. Note that the series gets invertible if s is a root. The Hilbert-Poincaré series is easy to calculate.

Lemma 18.1. *Let $\chi_V = \sum_{j=0}^{n}\alpha_j T^j$ be the Coxeter polynomial of V, and let s be a member of the bilinear lattice V. Then*

$$P_{(V,s)} = \frac{\sum_{p=0}^{n-1}\left(\sum_{j=0}^{p}\alpha_j\langle s, \tau^{p-j}s\rangle\right)T^p}{\chi_V}. \tag{18.1}$$

Proof. Since $\chi_V P_{(V,s)} = \sum_{p=0}^{\infty}\langle s, \left(\sum_{j=0}^{\min(n,p)}\alpha_j\tau^{p-j}\right)s\rangle T^p$, the Theorem of Cayley-Hamilton implies the claim in view of the self-reciprocity of χ_V. □

If s is a root, the next proposition identifies the polynomial in the nominator of (18.1) in terms of the Coxeter polynomials for V and $a^{\perp}$. We note that the arguments follow [29], where a special case is treated.

Proposition 18.3. *Let a be a root in V. Then the Coxeter polynomials χ for V (resp. χ_0 for $V_0 = a^{\perp}$) and the Hilbert-Poincaré series of (V, a) are related by the formula*

$$\frac{\chi_0}{\chi} = \frac{1 - P_{(V,a)}}{T}, \quad \textit{i.e. } P_{(V,a)} = \frac{\chi - T\chi_0}{\chi}. \tag{18.2}$$

Moreover, $\gamma = \limsup_{n\to\infty}\sqrt[n]{|\langle s, \tau^n s\rangle|}$ is in the range $1 \leq \gamma \leq \rho$, where ρ is the spectral radius of the Coxeter transformation for V.

Proof. On $V[[T]] = \mathbb{Z}[[T]] \otimes V$ the linear operator $1 - \tau^{-1}T$ is invertible with inverse $\sum_{n=0}^{\infty}\tau^{-n}T^n$, and easy calculation yields

$$\left(1 - \ell_a\tau^{-1}T\right)\left(1 - \tau^{-1}T\right)^{-1}(v) = v + \left(\sum_{n=1}^{\infty}\langle a, \tau^{-n}v\rangle T^n\right)a,$$

for each $v \in V$. Hence composition $h = \left(1 - \ell_a\tau^{-1}T\right)\left(1 - \tau^{-1}T\right)^{-1}$ satisfies

$$\begin{aligned} h(v) &\in v + \mathbb{Z}[[T]]\,a \text{ for each } v \in V_0, \\ h(a) &= \left(\sum_{n=0}^{\infty}\langle a, \tau^{-n}a\rangle T^n\right)a \end{aligned}$$

so that $\det(h) = \sum_{n=0}^{\infty}\langle a, \tau^{-n}a\rangle T^n$.

Note that $\ell_a \tau^{-1}$ equals τ_0^{-1} on V_0 and $\ell_a(a) = 0$, hence $\det\left(1 - \tau_0^{-1}T\right) = \det\left(1 - \ell_a \tau^{-1} T\right)$ in view of the decomposition $V = V_0 \oplus \mathbb{Z}\tau a$. The above expression for $\det(h)$ thus yields

$$\frac{\det\left(1 - \tau_0^{-1}T\right)}{\det\left(1 - \tau^{-1}T\right)} = \sum_{n=0}^{\infty} \langle a, \tau^{-n} a \rangle T^{-n} = \frac{1 - P_{(V,a)}}{T}$$

which implies the first assertion since $-\tau^{-1}$ and $-\tau_0^{-1}$ have determinant one.

Since χ and χ_0 are self-reciprocal and the degree of χ_0 is less then the degree of χ, the rational function χ_0/χ has a pole in the range $1/\rho \leq |z| \leq 1$, but no pole for $|z| < \rho$, implying that the radius of convergence r of $P_{(V,a)}$ satisfies $1/\rho \leq r \leq 1$. □

18.3.3 One-Point Extensions Formation of the one-point sink extension $\Lambda[M] := \left(\begin{smallmatrix} \Lambda & 0 \\ M & k \end{smallmatrix}\right)$, where M is a right Λ-module, is a process inverse to the formation of the right perpendicular category, and like the former process is a convenient tool for all type of induction arguments. This process is also easy to control on the K-theretic level:

We start with a pair (V, s), consisting of a bilinear lattice V and an element s of V. The *one-point sink extension* $V[s]$ of V by s has the underlying group $V^* = V \oplus \mathbb{Z}s^*$ with $\mathbb{Z}s^* \cong \mathbb{Z}$; its bilinear form extends the bilinear form of V and is otherwise determined by the properties

1. $\langle s^*, s^* \rangle_{V^*} = 1$,
2. $\langle s^*, v \rangle_{V^*} = 0$ and $\langle v, s^* \rangle_{V^*} = \langle v, s \rangle_V$ for all $v \in V$.

For each s in V the extension vertex s^* is a root of $V[s]$ and the perpendicular lattice $(s^*)^{\perp}$ equals V.

Theorem 18.1. *Let a be a root of a bilinear lattice V and $s = r_a(a) = a + \tau a$. If V_0 is the bilinear lattice right perpendicular to a, then there is a natural isomorphism $V \cong V_0[s]$, where a corresponds to s^*.*

Moreover, the Coxeter polynomials $\chi = \chi_V$, $\chi_0 = \chi_{V_0}$ and the Hilbert-Poincaré series $P = P_{(V,a)}$ and $P_0 = P_{(V_0,s)}$ are related by the reciprocity law

$$\begin{aligned} \frac{\chi}{\chi_0} &= 1 - \langle s, s \rangle + T + P_0 \\ &= \frac{T}{1 - P}. \end{aligned}$$

Proof. For the first assertion note that $V = V_0 \oplus \mathbb{Z}a$ and $V_0[s] = V_0 \oplus \mathbb{Z}s^*$. The morphism $V \to V_0[s]$ which is the identity on V_0 and sends a to s^* has the requested properties. Next we derive a formula relating P and P_0. Let τ (resp. τ_0) denote the Coxeter transformations for V (resp. V_0), then $\tau_0 x = \tau x - \langle x, a \rangle \tau a$ for each $x \in V_0$. By induction this yields $\tau_0^n s = \tau^n s + \langle \tau_0^n s, s \rangle a - \sum_{j=0}^{n} \langle \tau_0^j s, s \rangle \tau^{n-j} a$

for each $n \geq 0$, hence

$$\langle s, \tau_0^n s\rangle + \langle s, \tau_0^{n+1} s\rangle\langle s, s\rangle - \sum_{j=0}^{n}\langle s, \tau_0^{j+1} s\rangle \left(\langle a, \tau^{n-j} a\rangle + \langle a, \tau^{n-j-1} a\rangle\right) = \langle s, \tau^n s\rangle.$$

Since $\langle s, \tau^n s\rangle = 2\langle a, \tau^n a\rangle + \langle a, \tau^{n-1} a\rangle + \langle a, \tau^{n+1} a\rangle$, passage to power series yields

$$P_0 + \frac{P_0 - \langle s, s\rangle}{T}(1 - P - TP) = 2P + \langle s, s\rangle - 1 + TP + \frac{P-1}{T}.$$

Multiplication by T and collecting terms finally leads to

$$(1+T)(1-P)P_0 = -(1+T)\left[(1-P) - \langle s, s\rangle(1-P) - TP\right]$$

which together with Proposition 18.3 implies the result. □

Corollary 18.2. *Let V be a bilinear lattice and $V[s]$ be the one-point sink extension with respect to $s \in V$. The Coxeter polynomial of $V[s]$ is given by*

$$\chi_{V[s]} = \chi_V\left(1 - \langle s, s\rangle + T + P_{(V,s)}\right). \tag{18.3}$$

Proof. The assertion is just a reformulation of Theorem 18.1. □

18.4 The K-Theory of Canonical Algebras

The K-theory of canonical algebras is well understood (cf. [25], where the more general situation of arbitrary base fields is covered) and will be summarized in this section. Since the bilinear lattice $K_0(\Lambda)$ is an invariant under derived equivalence, this section deals with the K-theory of *derived canonical* algebras, i.e. with algebras which are derived-equivalent to a canonical algebra. We note that this class contains the tame hereditary algebras, the tame concealed algebras, the tubular algebras and moreover all algebras with a separating tubular family of standard stable (resp. semiregular) tubes [30, 31].

18.4.1 Attaching Tubes Any rank one direct summand $\mathbb{Z}.w$ of $\mathrm{rad}(V)$ is called a 1-*tube* of V. For $p \geq 2$ a p-*tube* T in V is a subgroup having a basis consisting of a τ-orbit

$$s, \tau s, \ldots, \tau^{p-1} s, \tau^p s = s \tag{18.4}$$

of finite period p, where s is a root of V, and additionally

$$\langle \tau^i s, \tau^j s\rangle = \begin{cases} \langle s, s\rangle \\ -\langle s, s\rangle \\ 0 \quad \text{else.} \end{cases} \tag{18.5}$$

The subgroup $\mathbb{Z}.z$, $z = \sum_{j\in\mathbb{Z}_p} \tau^j s$, agrees with $\mathrm{rad}(V) \cap T$, and is called the *axis of* T. Hence, z is uniquely determined by T up to sign, the same holds for the *length function* $\ell : T \to \mathbb{Z}$, given by $\sum_{j\in\mathbb{Z}_p} \tau^j v = \ell(v) z$.

Lemma 18.2. *The basis (18.4) of a p-tube T can be recovered from T up to sign as the τ-orbit of any root of T of length one.*

Let V be a bilinear lattice with non-zero radical. We are going to change a 1-tube $\mathbb{Z}.w$ of V into a p-tube ($p \geq 2$). Let T be the free $\mathbb{Z}$-module on $\mathbb{Z}_p$ with basis $s_1, \ldots, s_p$, we turn T into a bilinear group by requesting

$$\langle s_i, s_j \rangle = \begin{cases} 1 & \text{if } j = i \text{ in } \mathbb{Z}_p \\ -1 & \text{if } j = i+1 \text{ in } \mathbb{Z}_p \\ 0 & \text{else,} \end{cases} \tag{18.6}$$

and $\tau s_j = s_{j+1}$ for $j \in \mathbb{Z}_p$. Note that $\sum_{j \in \mathbb{Z}_p} s_j$ belongs to the kernel of T, so T itself is not a bilinear lattice.

Next we define a bilinear form on $V \oplus T$, extending the forms on V and T by setting

$$\langle u_1, u_2 \rangle = \langle u_1, u_2 \rangle_T \quad \text{for all } u_1, u_2 \in T \tag{18.7}$$

$$\langle v_1, v_2 \rangle = \langle v_1, v_2 \rangle_V \quad \text{for all } v_1, v_2 \in V \tag{18.8}$$

$$-\langle s_{j-1}, v \rangle = \langle v, s_j \rangle = \begin{cases} \langle v, w \rangle_V & \text{if } j = 1 \text{ in } \mathbb{Z}_p,\ v \in V \\ 0 & \text{if } j \neq 1 \text{ in } \mathbb{Z}_p,\ v \in V \end{cases} \tag{18.9}$$

It follows that each s_i, $i = 1, \ldots, p$, is a root in $V \oplus T$, and that $(\sum_{j \in \mathbb{Z}_p} s_j) - w$ generates the kernel of $V \oplus T$. There results a bilinear lattice

$$\bar{V} = (V \oplus T) / \mathbb{Z}(\sum_{j \in \mathbb{Z}_p} s_j - w),$$

said to arise from V by *attaching a p-tube at $\mathbb{Z}.w$.*

Proposition 18.4.

(i) *$\bar{V}$ has rank* $\operatorname{rk} \bar{V} = \operatorname{rk} V + p - 1$.

(ii) *The Coxeter transformation $\bar{\tau}$ on $\bar{V}$ is given by*

$$\begin{aligned} \bar{\tau}(s_j) &= s_{j+1} \\ \bar{\tau}(v) &= \tau(v) - \langle v, w \rangle s_1 + \langle v, w \rangle w \quad \text{for } v \in V. \end{aligned}$$

(iii) *T and V become subgroups of $\bar{V}$, and $s_1, \ldots, s_p$ becomes a system roots of $\bar{V}$.*

Corollary 18.3. *The Coxeter polynomials $\chi_{\bar{V}}$ for $\bar{V}$ and χ_V for V are related by*

$$\chi_{\bar{V}}(X) = \chi_V(X) \frac{X^p - 1}{X - 1}.$$

Proof. With the map $\mu : T \to V$, $v \mapsto \langle v, w\rangle - \langle v, w\rangle s_1$, we get a commutative diagram with exact rows

$$\begin{array}{ccccccccc} 0 & \to & \mathbb{Z} & \to & V \oplus T & \to & \bar{V} & \to & 0 \\ & & \| & & \beta \downarrow & & \downarrow \bar{\tau} & & \\ 0 & \to & \mathbb{Z} & \to & V \oplus T & \to & \bar{V} & \to & 0 \end{array}$$

where $\beta = \begin{pmatrix} \tau_V & 0 \\ \mu & \tau_T \end{pmatrix}$. Now pass to characteristic polynomials. □

Lemma 18.3. *Assume that V_2 is a bilinear lattice of rank 2 with non-zero radical and having a root. Then $V_2 = \mathbb{Z}a \oplus \mathbb{Z}w$, where a is a root and w generates the radical of V. The bilinear form is given with regard to the basis a, w by the matrix $\begin{pmatrix} 1 & 1 \\ -1 & 0 \end{pmatrix}$, accordingly $\tau(a) = a - 2w$, $\tau(w) = w$. Moreover $v \in V$ is a root if and only if $v = \pm(a + nw)$, $n \in \mathbb{Z}$.*

Definition 18.1. A bilinear lattice V arising from V_2, by an attachment of tubes $T_1, \dots, T_t$ with identical axes for a 1-tube $\mathbb{Z}.w$ of V is called a *canonical bilinear lattice* with of type $p_1, \dots, p_t$, where p_i denotes the rank of T_i.

The next statement provides a normal form for canonical bilinear lattices.

Proposition 18.5. *If V is a canonical bilinear lattice of type $p_1, \dots, p_t$, then*

$$a,\ \tau^j s_i\ (1 \le i \le t,\ 0 \le j \le p_i - 2),\ w, \tag{18.10}$$

is a basis of V, where

$$\langle a, a\rangle = 1, \quad \langle a, w\rangle = 1, \quad \langle a, s_i\rangle = 1, \quad \langle s_i, s_i\rangle = 1, \quad \langle w, w\rangle = 0,$$

and $\langle a, \tau^j s_i\rangle = 0$ for $0 < j \le p_i - 2$.

The next statement (cf. [29]) implies that the type of a canonical bilinear lattice is actually an invariant, and shows that the spectrum of its Coxeter polynomial consists of roots of unity.

Proposition 18.6. *Let V be a canonical bilinear lattice of type $p_1, \dots, p_t$, then the Coxeter polynomial of V is given by*

$$\chi_V(X) = (X - 1)^2 \prod_{i=1}^{t} \frac{X^{p_i} - 1}{X - 1}.$$

In particular, $\operatorname{rad}(V)$ *has rank one or two.*

Proof. Immediate from Corollary 18.3. □

Proposition 18.7. *A bilinear lattice is canonical if and only if it arises as the Grothendieck group* $\mathrm{K}_0(\Lambda)$ *of a canonical algebra Λ (or an algebra Σ derived-equivalent to a canonical algebra).*

Proof. Assume that Λ is a canonical algebra. Then $\operatorname{mod}(\Lambda)$ has a separating tubular family $(\mathcal{T}_x)$ of standard stable tubes, indexed by $\mathbb{P}_1(k)$ [36, 12]. Let $p(x)$ denote the rank of $\mathcal{T}_x$, i.e. the number of indecomposables in the mouth of the tube $\mathcal{T}_x$, let S be a quasi-simple module from a tube with $p(x) = 1$, and put $w = [S]$. Since $\mathrm{K}_0(\Lambda)$ as a $\mathbb{Z}$-module has finite rank, only for finitely many points x, say for $x_1, \ldots, x_t$, we can have $p(x) > 1$. Next we pick an indecomposable module L of rank $\langle [L], w\rangle = 1$, and for each $\mathcal{T}_{x_i}$ a quasi-simple S_i with $\operatorname{Hom}(L, S_i) \neq 0$. Putting $a = [L]$, $s_i = [S_i]$, $p_i = p(x_i)$ does the job in view of Proposition 18.5. □

To dispose of an explicit form for the Coxeter polynomial is also interesting in view of a recent result of Happel [20], stating (for an algebra Λ of finite global dimension) that the alternating sum of the dimensions of the Hochschild cohomology groups $\mathrm{H}^i(\Lambda))$ equals the trace of the Coxeter transformation. In particular, the above formula for χ_V yields that a derived canonical algebra has $\mathrm{H}^2(\Lambda) \neq 0$ if $t \geq 4$.

18.4.2 Riemann-Roch Formula and Genus V denotes a canonical lattice, and we keep the previous notation. We define the rank function as $\mathrm{rk} = \langle -, w\rangle$ and introduce an average on the bilinear form setting

$$\langle\langle x, y\rangle\rangle = \sum_{j=0}^{p-1} \langle \tau^j x, y\rangle = \sum_{j=0}^{p-1} \langle x, \tau^{-j} y\rangle.$$

Fixing a root a of rank one from $V_2 \subset V$, we define the *degree function* through

$$\deg x = \langle\langle a, x\rangle\rangle - \mathrm{rk}\,(x)\ \langle\langle a, a\rangle\rangle.$$

Note that

$$\deg w = 0, \quad \deg w = p, \quad \deg \tau^j s_i = \frac{p}{p_i},$$

in particular, deg is τ-stable on $V_0 = \ker[V \stackrel{\mathrm{rk}}{\to} \mathbb{Z}]$.

Lemma 18.4.

(i) $\langle\langle a, \tau x\rangle\rangle - \langle\langle a, x\rangle\rangle = \delta[V]\mathrm{rk}\,(x)$ *for each* $x \in V$.

(ii) $2\langle\langle a, a\rangle\rangle - p\,\delta[V]$.

Theorem 18.2 (Riemann-Roch). *For all x, y from a canonical bilinear lattice V we get*

$$\frac{1}{p}\langle\langle x, y\rangle\rangle = (1 - g[V])\,\mathrm{rk}\,x\,\mathrm{rk}\,y + \frac{1}{p}\begin{vmatrix} \mathrm{rk}\,x & \mathrm{rk}\,y \\ \deg x & \deg y \end{vmatrix},$$

where $g[V]$ is given by $g[V] = 1 + \frac{1}{2}\delta[V] = 1 + \frac{p}{2}\left((t-2) - \sum_{i=1}^{t} \frac{1}{p_i}\right)$.

Proof. In view of the definition of the degree and of Lemma 18.4, the formula is satisfied for $x = a$. For x, y from the set formed by w and all $\tau^j s_i$ both sides evaluate to zero. Observing $\langle\langle y, x\rangle\rangle = -\langle\langle x, \tau y\rangle\rangle$ finishes the proof. □

Accordingly $g[V]$ is called the *genus* of V or its corresponding quadratic form. For $g[V] < 1$ (according to $\delta[V] < 0$) resp. $g[V] = 1$ (according to $\delta[V] = 0$) we call V *domestic* (resp. *tubular*).

If x is in V and $\deg x \neq 0$ or $\mathrm{rk}\, x \neq 0$, we call $\mu(x) = \deg x/\mathrm{rk}\, x$ the *slope* of x.

Proposition 18.8. *Let V be a canonical bilinear lattice.*

(*i*) *If $\delta[V] < 0$, then the radical of V has rank one, and q_V is positive semi-definite.*

(*ii*) *If $\delta[V] = 0$, then the radical of V has rank two, and q_V is positive semi-definite.*

(*iii*) *If $\delta[V] > 0$, then the radical of V has rank one, and q_V is indefinite.*

Proof. Assume first that $\delta[V] \neq 0$, and $x \in \mathrm{rad}(V)$ has non-zero rank, so that the slope $\mu(x) = \deg x/\mathrm{rk}\, x$ is defined. Then $\mu(\tau x) = \mu(x) + \delta[V]$, contradiction. Hence $\mathrm{rad}(V)$ belongs to the subgroup V_0 of elements of rank zero. Since $\mathrm{rad}(V) \cap V_0 = \mathbb{Z}.w$, $\mathrm{rad}(V)$ has rank one in cases (*i*) and (*iii*).

If $\delta[V] = 0$, then $\tau^p = 1$, hence $u = \sum_{j=0}^{p-1} \tau^j a$ is a non-zero τ-stable element of rank p. In view of Proposition 18.6 $\mathrm{rad}(V)$ thus has rank two.

Concerning the definiteness of q_V, we only deal with (*iii*), i.e. assume $\delta[V] > 0$. In view of the Riemann-Roch formula, the element $u = \sum_{j=0}^{p-1} \tau^j a$ satisfies $q_V(u) < 0$. □

Theorem 18.3. *The subgroup G of* $\mathrm{Aut}(V)$ *generated by the shift automorphisms $\sigma_0, \dots, \sigma_t$ is isomorphic to the abelian group on symbols $\sigma_0, \dots, \sigma_t$ with relations*

$$\sigma_1^{p_1} = \sigma_2^{p_2} = \cdots = \sigma_t^{p_t}.$$

Each root v of rank one has the form $\sigma(a)$ for a uniquely defined σ from G.

18.4.3 Coxeter Transformation and Shifts The Coxeter transformation for a bilinear canonical lattice V is generated by shifts corresponding to tubes: Let T_i denote the p_i-tube with axis $\mathbb{Z}.w$ and basis $\tau^j s_i$, $j \in \mathbb{Z}_{p_i}$ $(1 \le i \le t)$, then the map $\sigma_i : V \to V$, $x \mapsto x - \sum_{j \in \mathbb{Z}_{p_i}} \langle \tau^j s_i, x\rangle \tau^j s_i$ is an automorphism of V, called the *shift automorphism* or mutation corresponding to the tube T_i. We further put $\sigma_0(x) = x - \langle w, x\rangle w$.

Proposition 18.9. *Let a denote a rank one root of $V_2 \subseteq V$, then*

$$\tau a = a - \sum_{i=1}^{t} s_i + (t-2)w.$$

Proof. If $t = 0$ we have $V = V_2$, where the formula holds (Lemma 18.3). The general case follows by induction on t from Proposition 18.4. □

Corollary 18.4. *Let $p = \operatorname{lcm}(p_1, \dots, p_t)$, and let*

$$\delta[V] = p\left((t-2) - \sum_{i=1}^{t} \frac{1}{p_i}\right).$$

Then for each $x \in V$ we get

$$\tau^p x = x + \delta[V]\operatorname{rk}(x)w.$$

In particular τ^p acts as the identity on V_0, the subgroup of elements of rank zero.

Proof. The above expression for τa yields $\tau^p a = a - \sum_{i=1}^{t} e_i(s_i + \tau s_i + \dots + \tau^{p-1}s_i) + p(t-2)w$, which implies the claim for $x = a$. For the remaining elements from the basis (Equation 18.10) the claim is obvious. □

Theorem 18.4. *The Coxeter transformation is a composition*

$$\tau = \sigma_0^{t-2} \prod_{i=1}^{t} \sigma_i^{-1}$$

of shift autormorphisms. Moreover, $\tau^p = \sigma_0^{\delta[V]}$.

Proof. Invoking Proposition 18.9 the formula for τ is easily checked on the basis (Equation 18.10). □

18.4.4 Algorithm for Canonical K-Theory We briefly outline an algorithm which decides whether a bilinear lattice V is isomorphic to $\mathrm{K}_0(\Lambda)$ for a canonical algebra, equivalently to $\mathrm{K}_0(\mathbb{X})$ for a weighted projective line $\mathbb{X}$, and if so determines the weight type of Λ (resp. $\mathbb{X}$), and a standard basis for V like in Proposition 18.5. We refer to the "tubular package" of Dowbor and Hübner in [10] for further algorithms related to canonical algebras.

First we check whether the Coxeter polynomial χ_V has canonical type

$$(T-1)^2 \prod_{i=1}^{t} \frac{T^{p_i-1}-1}{T-1}.$$

For this we determine the cyclotomic factors Φ_m of χ_V and use recursive substitution $\Phi_m = V_m / \prod_{d|m, 1<d<m} \Phi_d$, where $V_m = (T^m - 1)/(T-1)$. This works since the Euler function satisfies $m \le 4\varphi(m)^2$ for each m.

If χ_V has canonical type, we store the weight data $(p_1, \dots, p_t)$ and further determine the radical of V, which for canonical type needs to have rank one or two, and in this case fix a rank one summand $\mathbb{Z}w$ of the radical. By means of w we form the subgroup V_0 consisting of all $x \in V$ with $\langle x, w \rangle = \dot{0}$, such that

$V = \mathbb{Z}b \oplus V_0$ and $\langle b, w\rangle = 1$. In addition to the rank function $\langle -, w\rangle$ we also form an analogue of the degree function, putting $d(x) = \sum_{i=0}^{p-1} \langle \tau^j b, x\rangle$, where $p = \operatorname{lcm}(p_1, \dots, p_t)$. [Note that for a canonical lattice d will give the correct value of the degree on the subgroup V_0.]

Next we determine whether the induced bilinear lattice $V_0/\mathbb{Z}w$ has a positive definite quadratic form, and if so, whether $V_0/\mathbb{Z}w$ is isomorphic to the product $\prod_{i=1}^{t} \mathrm{K}_0(A_i)$, where A_i denotes the path algebra of the linear quiver of type $\mathbb{A}_{p_i-1}$. This is done by calculating all the roots s of $V_0/\mathbb{Z}w$ of positive degree, and then collecting those which are simple in the sense that $\pi(s)d(s) = p$, where $\pi(s)$ denotes the τ-period of s, which needs to be a divisor of p. Lifting these roots back to V_0, they need to decompose into t τ-orbits $(\tau^j s_i)$, wich generate tubes $T_1, \dots, T_t$ of ranks $\dot{p}_1, \dots, p_t$ in a suitable numbering.

In the last step we check whether one of the expressions $a + \sum_{i,j} \kappa_{ij} \tau^j s_i$ with $\kappa_{ij} \in \pm\{0, 1, \dots, p_i - 1\}$ is a root. If yes, up to a renumbering in the τ-orbits $(\tau^j s_i)$, the system $a, \tau^j s_i, w$ is a standard basis for a canonical lattice; if non of the above expressions is a root, then V is not of canonical type.

18.5 The K-Theory of Hereditary Algebras

This section deals with the spectral properties of hereditary algebras. That these properties are important for the representation theory becomes clear from the following theorem. We refer to [34] and the literature quoted there for further information on Coxeter transformations for hereditary algebras.

Theorem 18.5. *Let Λ be a finite dimensional connected hereditary algebra. Let ρ_Λ denote the spectral radius of the Coxeter transformation, i.e. the maximal modulus of a root of χ_Λ. Then the following properties hold:*

1. *Λ is representation-finite if and only if the quadratic form q_Λ positive, that is $(x \neq 0 \Rightarrow q_\Lambda(x) > 0)$, equivalently $\rho_\Lambda = 1$ and $\chi_\Lambda(1) \neq 0$.*

2. *Λ is tame but not representation-finite if and only if q_Λ is non-negative, that is $(x \neq 0 \Rightarrow q_\Lambda(x) \geq 0)$, equivalently $\rho_\Lambda = 1$ and $\chi_\Lambda(1) = 0$.*

3. *Λ is wild if and only if q_Λ indefinite, that is there exists x with $q_\Lambda(x) < 0$, equivalently $\rho_\Lambda > 1$.*

18.5.1 Representation-Finite and Tame Quivers For a representation-finite or tame (connected) quiver the K-theory is well understood since the path algebra of an extended Dynkin quiver is tilting-equivalent, in particular derived-equivalent, to a canonical algebra, and further the representation-finite case arises from the tame case by perpendicular calculus. The next two tables display the Coxeter polynomials for the classes of quivers. Here Φ_n denotes the n-th cyclotomic polynomial.

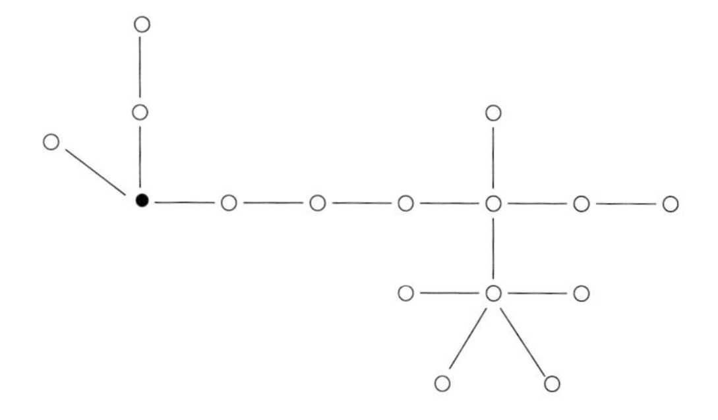

Figure 18.1: The tree $[2,3,5*[2,3,2*[2,2,2,2]]]$

Dynkin diagram	Coxeter polynomial
$\mathbb{A}_n$	$\frac{T^{n+1}-1}{T-1}$
$\mathbb{D}_n$, $n \geq 4$	$\frac{(T^{2n-2}-1)(T+1)}{T^{n-1}-1}$
$\mathbb{E}_6$	$\Phi_3\,\Phi_{12}$
$\mathbb{E}_7$	$\Phi_2\,\Phi_{18}$
$\mathbb{E}_8$	Φ_{30}

Extended Dynkin diagram	
$\overline{\mathbb{A}}_{pq}$	$(T-1)^2\,\frac{T^p-1}{T-1}\,\frac{T^q-1}{T-1}$
$\overline{\mathbb{D}}_n$	$(T-1)^2\left(\frac{T^2-1}{T-1}\right)^2\,\frac{T^{n-2}-1}{T-1}$
$\overline{\mathbb{E}}_n$ $n=6,7,8$	$(T-1)^2\,\frac{T^2-1}{T-1}\,\frac{T^3-1}{T-1}\,\frac{T^{n-3}-1}{T-1}$

18.5.2 Boldt's Algorithm for Trees For trees – unlike for quivers with cycles – we have

1. an efficient notation of recursive character, for instance

```
tree:=[2,3,5*[2,3,2*[2,2,2,2]]]
```

   denotes the tree with base point in figure 18.1

2. a fast algorithm (A. Boldt [3]) implemented in MAPLE following the recursion:

```
coxpoly(tree,T);
```

$$T^{16}+T^{15}-9\,T^{14}-30\,T^{13}-44\,T^{12}-35\,T^{11}-9T^{10}+16\,T^9+26\,T^8+16\,T^7-9\,T^6-35\,T^5-44\,T^4-30\,T^3-9\,T^2+T+1$$

This algorithm is based on a formula for the Coxeter polynomial for the union Δ of two trees Δ_1, Δ_2 with **one** common vertex p:

$$\chi_\Delta = \chi_{\Delta_1} \chi_{\Delta_2'} + \chi_{\Delta_1'} \chi_{\Delta_2} - (T+1)\chi_{\Delta_1'} \chi_{\Delta_2'},$$

where $\Delta_i' = \Delta_i \setminus \{p\}$.

18.5.3 Spectral Properties, Wild Case Let $\vec{\Delta}$ be a connected quiver, which in this section is mostly assumed to be wild. The most important spectral information is Ringel's result [37] stating that the spectral radius is always a simple root of the Coxeter polynomial. This generalizes previous results [8, 35, 38].

If $\vec{\Delta}$ is a tree, its K-theory only depends on the underlying graph Δ, not on the orientation, since any two orientations are related by a sequence of tiltings, which preserve the K-theory. Note that, in general, the K-theory does depend on the orientation. For instance, if $\vec{\Delta}$ denotes a non-oriented cycle with p arrows going clockwise and q going counter-clockwise, its Coxeter polynomial is

$$(T-1)^2 \frac{T^p - 1}{T-1} \frac{T^q - 1}{T-1},$$

so evidently depends on the orientation.

For each tree $\vec{\Delta}$ we may thus assume that it is given a bipartite quiver, i.e. each vertex is a sink or a source. For bipartite quivers in general the spectral information is quite good, since for such quivers the spectrum is always contained in the union of the unit circle $\mathbb{S}^1$ and the real line $\mathbb{R}$, see [35]. We refer to [35] for a detailed discussion of the influence of spectral properties on representation theory of $\operatorname{mod} k[\vec{\Delta}]$.

As the next example shows, for wild quivers in general the above situation does no longer hold. We note that a similar example was independently found by S. Lache.

Example 18.1. The Coxeter polynomial of the quiver

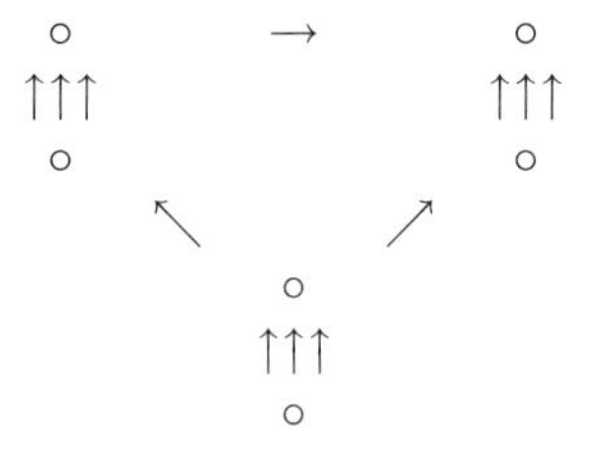

is given by

$$1 - 24T + 158T^2 - 363T^2 + 158T^4 - 24T^5 + T^6.$$

None of its four non-real roots lies on the unit circle.

It is further easy to see that 6 is the minimal number of vertices such that the above effect occurs.

18.6 Examples, Problems and Questions

The following basic *recognition for Grothendieck groups* is largely unsolved: *Given a class $\mathcal{K}$ of (connected) algebras of finite global dimension and a bilinear lattice V, decide either by a structure theorem or by an algorithm whether V is isomorphic to* $\mathrm{K}_0(A)$ *for some algebra A in $\mathcal{K}$.* Clearly, the problem does not change if we assume the class $\mathcal{K}$ to be closed under derived equivalence.

For instance, if $\mathcal{K}$ is the *class of all algebras* of finite global dimension it is not known whether the Cartan matrix has determinant one (see [39, 41]).

As we have seen, the recognition problem has a positive answer if $\mathcal{K}$ denotes the class of all canonical algebras or of all derived canonical algebras. Notice however, an algebra Σ whose Grothendieck group is a canonical bilinear lattice does not need to be derived canonical: A simple example is provided by the algebra Σ, given by the quiver

$$\begin{array}{ccc} \circ & \stackrel{x}{\longrightarrow} & \circ \\ x\uparrow & & \downarrow x \\ \circ & \longrightarrow & \circ \end{array}$$

and subject to all three relations $x^2 = 0$.

Also the *recognition problem for Coxeter polynomials* is largely open: *Given a class $\mathcal{K}$ of (connected) algebras of finite global dimension and a self-reciprocal polynomial $f \in \mathbb{Z}[X]$, decide either by a structure theorem or by an algorithm whether f is the Coxeter polynomial for some algebra A in $\mathcal{K}$.*

18.6.1 Search for Derived Equivalence Let Λ_n be the representation-finite algebra given by the linear quiver with n vertices

$$\circ \stackrel{x}{\longrightarrow} \circ \stackrel{x}{\longrightarrow} \circ \stackrel{x}{\longrightarrow} \cdots \stackrel{x}{\longrightarrow} \circ \stackrel{x}{\longrightarrow} \circ \stackrel{x}{\longrightarrow} \circ$$

bound by the relations $x^3 = 0$. It is easily checked that the characteristic polynomial χ_n of Λ_n is a product of cyclotomic factors. Moreover Λ_n is derived hereditary of Dynkin type for $n = 1, \dots, 8$, where the Dynkin types $\mathbb{A}_1, \mathbb{A}_2, \mathbb{A}_3$, $\mathbb{D}_4, \mathbb{D}_5$ and $\mathbb{E}_6, \mathbb{E}_7, \mathbb{E}_8$ occur.

Exactly for $n = 9, 10, 11$ the algebra Λ_n is derived-equivalent to a canonical algebra; more precisely Λ_9, Λ_{10} and Λ_{11} are of canonical type $(2,3,5)$; $(2,3,6)$ and $(2,3,7)$, respectively. We recall that for Λ canonical, the derived category $\mathrm{D}^b(\mathrm{mod}\,\Lambda)$ has the shape $\bigvee_{n\in\mathbb{Z}} \mathcal{H}[n]$, where $\mathcal{H}$ is the hereditary k-category of coherent sheaves on a weighted projective line. Let L denote the structure sheaf, and let S be the unique simple sheaf of τ-period 5 with $\mathrm{Hom}\,(L, S) \neq \dot{0}$, where $\tau = \tau_{\mathcal{H}}$. Then

$$L \to S \to \tau L[1] \to \tau S[1] \to \tau^2 L[2] \to \tau^2 S[2] \to \tau^3 L[3] \to \tau^3 S[3] \to \tau^4 L[4]$$

is a tilting complex for $\mathcal{H}$ with endomorphism ring Λ_9. It is interesting to note that the realization of Λ_9 as a tilting complex takes 5 copies of $\mathcal{H}$. A similar method yields realizations of Λ_{10} and Λ_{11} as tilting complexes on a weighted projective line of weight type $(2,3,6)$ and $(2,3,7)$, respectively.

18.6.2 Algebra of Poset Type The algebra Λ of the poset (fully commutative quiver)

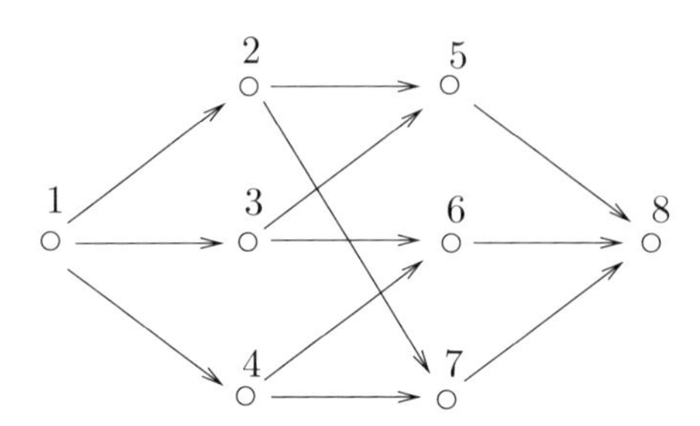

is derived-equivalent to canonical (3,3,3). To see this we first calculate the Coxeter polynomial $(T-1)^2\left(\frac{T^3-1}{T-1}\right)^3$ which is of canonical tubular type $(3,3,3)$. It is then not difficult to show that $\mathrm{K}_0(\Lambda)$ is of canonical type. Invoking the `tubular`-subpackage of CREP, written by T. Hübner and P. Dowbor [9, 10], it is straightforward to determine a realization of Λ by a tilting complex over the weighted projective line of type $(3,3,3)$, as follows. Let $\mathcal{O}$ denote the structure sheaf, and S the simple sheaf given by the exact sequence $0 \to \mathcal{O}(-\vec{x}_3) \to \mathcal{O} \to S \to 0$. Then $P_1 = \tau^{-1}S[-1]$, $P_2 = \mathcal{O}$, $P_3 = \mathcal{O}(\vec{x}_1-\vec{x}_2)$, $P_4 = \mathcal{O}(2\vec{x}_1-2\vec{x}_2)$, $P_5 = \mathcal{O}(\vec{x}_1)$, $P_6 = \mathcal{O}(2\vec{x}_1 - \vec{x}_2)$, $P_7 = \mathcal{O}(\vec{x}_2)$ and $P_8 = S$ yields a tilting complex with endomorphism ring Λ. Actually this realization shows that Λ is quasi-tilted of type $(3,3,3)$, cf. [31].

Returning to the question above, the situation is much less clear if $\mathcal{K}$ is the class of hereditary algebras. If $\mathcal{K}$ is the class of tame hereditary algebras, the problem amounts to characterize the theory of canonical bilinear lattices of genus < 1 and is thus solved. If $\mathcal{K}$ is the class of representation-finite hereditary algebras, the solution then follows by perpendicular calculus or by a direct argument.

However, the problem is largely open for the class of wild hereditary case, characterized by an indefinite quadratic form. This in spite of substantial work on the Coxeter transformation, its spectrum [8], [29], [32], [35], [37], [38], [40], [42] and the Hochschild homology of hereditary algebras [18, 20].

For instance it is not known for which wild hereditary algebras Σ the radical of $\mathrm{K}_0(\Sigma)$ is non-zero. Actually, even for wild tree algebras the rank of the radical can become as large as one wants:

Example 18.2. Consider the family of snowflake trees $S_n(\Gamma)$ having n leaves $(n \geq 2)$, all agreeing with some extended Dynkin diagram Γ. Here is the snowflake $S_6(\overline{\mathbb{D}}_4)$:

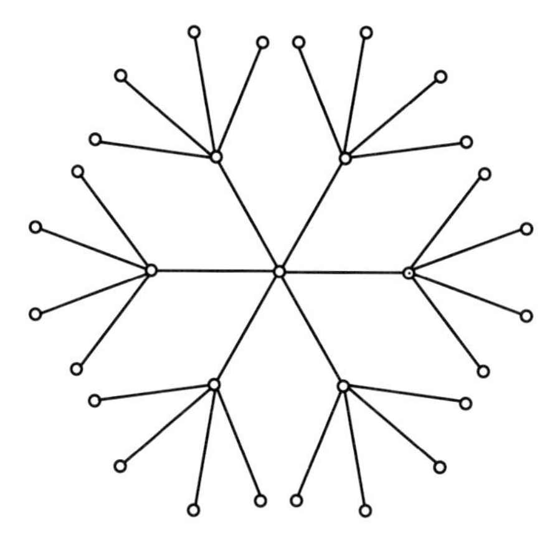

Note that $S_n = S_n(\overline{\mathbb{D}_4})$ has $5n+1$ points. As may be derived from [3], the Coxeter polynomial of S_n is given as

$$(T-1)^{2(n-1)}\left(T^2-2(n-2)T+1\right)(T+1)^{3n+1},$$

accordingly S_n has spectral radius $\sqrt{2(n-2)}$. Moreover the radical of the quadratic form for S_n has rank $n-1$.

We close this report with the famous problem of D.H. Lehmer [24], a problem open since 1933. We recall that the *Mahler measure* of a (self-reciprocal) polynomial is the product of the moduli of all its roots outside the unit circle. Lehmer's problem asks whether there exists an integer, self-reciprocal, normalized polynomial having a smaller Mahler measure > 1 than the Coxeter polynomial

$$\chi_{[2,3,7]} = T^{10} + T^9 - T^7 - T^6 - T^5 - T^4 + T + 1$$

of the wild hereditary tree $[2, 3, 7]$, whose Mahler measure agrees with the spectral radius $\rho_{[2,3,7]} \approx 1.1762808\ldots$ of its Coxeter transformation. Note in this context that computer calculations [5] have confirmed the conjecture for such polynomials up to degree 20. Further C.C. Xi has shown [40] that $\rho_{[2,3,7]}$ is a lower bound for the spectral radii, hence for the Mahler measures, of all wild hereditary algebras (arbitrary base fields allowed).

It would be interesting to confirm Lehmer's conjecture for the subclass of Coxeter polynomials of algebras or bilinear lattices with spectral radius > 1.

Bibliography

[1] M. Auslander, I. Reiten, S. O. Smalø: *Representation Theory of Artin Algebras*, Cambridge Studies in Advanced Mathematics. **36**, Cambridge University Press, Cambridge (1995).

[2] D. Baer, W. Geigle, H. Lenzing: *The preprojective algebra of a tame hereditary Artin algebra*, Comm. Algebra, **15**, 425-457 (1987).

[3] A. Boldt: *Methods to determine Coxeter polynomials*, J. Linear Algebra Appl., **230**, 151–164 (1995).

[4] N. Bourbaki: *Groupes et Algèbres de Lie*, chapitres 4,5 et 6. Hermann, Paris (1968).

[5] D.W. Boyd: *Reciprocal polynomials having small measure II*, Math. Comput., **53**, 355–357 (1989).

[6] D. Cvetković, M. Doob, H. Sachs: *Spectra of graphs*, Academic Press, 1980.

[7] V. Dlab, C. M. Ringel: *Indecomposable Representations of Graphs and Algebras*, Mem. Am. Math. Soc. 173 (1976).

[8] V. Dlab, C. M. Ringel: *Eigenvalues of Coxeter transformations and the Gelfand-Kirillov dimension of the preprojective algebras*, Proc. Am. Math. Soc., **83**, 228-232 (1981).

[9] P. Dowbor, T. Hübner: *A Computer Algebra approach to sheaves over weighted projective lines*, these Proceedings.

[10] P. Dräxler, R. Nörenberg (eds.): *CREP Manual Part II*, Ergänzungsreihe 97-009 des SFB 343, Universität Bielefeld (1997).

[11] P. Gabriel, A.V. Roiter: *Algebra VIII: Representations of Finite-dimensional Algebras*, Encyclopaedia of Mathematical Sciences **73**, Springer-Verlag, Berlin (1992).

[12] W. Geigle, H. Lenzing: *A class of weighted projective curves arising in representation theory of finite dimensional algebras*, in: Singularities, representations of algebras, and vector bundles, Lect. Notes Math., **1273**, 265–297 (1987).

[13] W. Geigle, H. Lenzing: *Perpendicular categories with applications to representations and sheaves*, J. Algebra, **144**, 273-343 (1991).

[14] I. M. Gelfand, V. A. Ponomarev: *Model algebras and representations of graphs*, Funct. Anal. Appl., **13**, 157–166 (1979).

[15] F.M. Goodman, P. de la Harpe, V.F.R. Jones: *Coxeter Graphs and Towers of Algebras*, Springer-Verlag New York-Berlin-Heidelberg, 1989.

[16] P.P. Grivel: *Catégories derivées et foncteurs derivés*, in: Algebraic D-modules, Perspectives in Math., 2, Academic Press 1987.

[17] D. Happel: *Triangulated Categories in the Representation Theory of finite dimensional Algebras*, LMS Lecture Note Series 119, Cambridge 1988.

[18] D. Happel: *Hochschild cohomology of finite-dimensional algebras*, in: Séminaire d'algebre P. Dubreil et M.-P. Malliavin. Lect. Notes Math.,; **1404**, 108-126 (1989).

[19] D. Happel: *Auslander-Reiten triangles in derived categories of finite dimensional algebras*, Proc. Amer. Math. Soc., **112**, 641–648 (1991).

[20] D. Happel: *The trace of the Coxeter matrix and Hochschild cohomology*, Linear Algebra Appl., **258**, 169–177 (1997).

[21] D. Happel, I. Reiten, S. Smalø: *Tilting in Abelian Categories and Quasitilted Algebras*, Mem. Am. Math. Soc. **575**, (1996).

[22] D. Happel, C. M. Ringel: *Tilted Algebras*, Trans. Am. Math. Soc., **274**, 399–443 (1982).

[23] O. Kerner: *Tilting wild algebras*, J. London Math. Soc., **39**, 29–47 (1989).

[24] D.H. Lehmer: *Factorization of certain cyclotomic functions*, Ann. Math., **34**, 461–479 (1933).

[25] H. Lenzing: *A K-theoretic study of canonical algebras*, CMS Conf. Proc., **18**, 433-454 (1996).

[26] H. Lenzing: *Hereditary noetherian categories with a tilting complex*, Proc. Am. Math. Soc., **125**, 1893-1901 (1997).

[27] H. Lenzing, H. Meltzer: *Sheaves on a weighted projective line of genus one, and representations of a tubular algebra*, Representations of algebras, Sixth International Conference, Ottawa 1992. CMS Conf. Proc., **14**, 313–337 (1993).

[28] H. Lenzing, H. Meltzer: *Tilting sheaves and concealed-canonical algebras, Representations of algebras*, Seventh International Conference, Cocoyoc (Mexico) 1994. CMS Conf. Proc., **18**, 455–473 (1996).

[29] H. Lenzing, J. A. de la Peña: *Wild canonical algebras*, Math. Z., **224**, 403-425 (1997).

[30] H. Lenzing, J. A. de la Peña: *Concealed-canonical algebras and algebras with a separating tubular family*, Proc. London Math. Soc. (to appear).

[31] H. Lenzing and A. Skowroński: *Quasi-tilted algebras of canonical type*, Colloq. Math., **71**, 161–181 (1996).

[32] F. Lukas: *Elementare Moduln über wilden erblichen Algebren*, Dissertation Düsseldorf 1993.

[33] H. Meltzer: *Tubular mutations*, Colloq. Math., **74**, 267–274 (1997).

[34] J. A. de la Peña: *Coxeter transformations and the representation theory of algebras*, in: V. Dlab and L.L. Scott (eds.), Finite Dimensional Algebras and Related Topics, 222-253, Kluwer 1994.

[35] J. A. de la Pena, M. Takane: *Spectral properties of Coxeter transformations and applications*, Arch. Math., **55**, 120–134 (1990).

[36] C. M. Ringel: *Tame Algebras and Integral Quadratic Forms*, Lect. Notes Math. **1099**, Springer, Berlin-Heidelberg-New York (1984).

[37] C. M. Ringel: *The spectral radius of the Coxeter transformations for a generalized Cartan matrix*, Math. Ann., **300**, 331–339 (1994).

[38] M. Takane: *On the Coxeter transformation of a wild algebra*, Arch. Math., **63**, 128–135 (1994).

[39] G. Wilson: *The Cartan map on categories of graded modules*, J. Algebra, **85**, 390–398 (1983).

[40] Ch. Xi,: *On wild hereditary algebras with small growth numbers*, Commun. Algebra, **18**, 3413–3422 (1990).

[41] D. Zacharia: *On the Cartan matrix of an algebra of global dimension two*, J. Algebra, **82**, 353–357 (1983).

[42] Y. Zhang: *Eigenvalues of Coxeter transformations and the structure of regular components of an Auslander-Reiten quiver*, Comm. Algebra, **17**, 2347–2362 (1989).

Progress in Mathematics, Vol. 173, © 1999 Birkhäuser Verlag Basel/Switzerland

Chapter 19

Condensation of induced Representations and an Application: the 2-Modular Decomposition Numbers of Co_2

Jürgen Müller and Jens Rosenboom

Abstract

We present an algorithm to condense induced modules for a finite group over a finite field. It is built on existing tools of computational group and representation theory, the MeatAxe for matrix algebras over finite fields and the Schreier-Sims methods for permutation groups, and has been implemented in GAP and as an extension of the MeatAxe. As an application we construct and analyze an induced module for the sporadic simple Conway group Co_2. The result of this analysis is used to complete the 2-modular Brauer character table of Co_2.

19.1 Introduction

In recent years, condensation has become one of the most valuable tools in computational representation theory of finite groups. Especially, the determination of many of the Brauer character tables of sporadic simple groups known today, see [9], would not have been possible without this powerful technique. As it turns out that many of the interesting modules for finite groups are too large to be constructed or analyzed by means of a computer directly, one tries to 'condense' these modules to smaller ones which still reflect enough of the original structure.

The functorial formalism to describe the theoretical correctness of the condensation method, which is introduced in Section 19.2, dates back at least to [6]. This formalism gives us a detailed description of which structural information is retained and which is lost under condensation. Under certain circumstances condensation turns out to induce an equivalence between the module categories of an arbitrary given algebra and of a Morita equivalent basic algebra. This explains why condensation has become a standard tool in the representation theory of finite-dimensional algebras.

But besides its theoretical value, condensation has become a valuable computational tool, as the formal recipe can be translated into a series of explicit computational steps which for suitable types of modules for finite groups can efficiently be performed by help of a computer. This has been implemented e. g. for permutation modules [21, 18, 16, 4] and for tensor product modules over finite fields [22, 14, 17]. In this note we present an algorithm to condense arbitrary induced modules for a finite group over a finite field, thereby enlarging the set of types of modules which can be handled computationally using the condensation technique. In Section 19.3 we describe how the functorial description of the condensation method translates into group theoretical computational steps, in Section 19.4 we then show how these steps can be performed explicitly.

Our algorithm is built on existing tools of computational group and representation theory and we assume the reader to be familiar with the MeatAxe, see [15], and the Schreier-Sims method to handle permutation groups. The permutation group part of our algorithm, see 19.4.1, runs in GAP [19]. For the matrix group part, see page 314, we have two implementations, one using the MeatAxe implementation [18], the other using a new finite field arithmetic [11] implemented in GAP. The uncondensation program has been implemented as extensions of the MeatAxe implementations [18] and [7]. Of course, all these programs are available from the authors on request.

In Section 19.5, we apply our algorithm to an induced module for the sporadic simple Conway group Co_2. A careful analysis of the resulting condensed module and an application of the uncondensation technique allows us to complete the determination of the 2-modular Brauer character table of this group begun in [20]. The table is given at the end of this note. This even completes the determination of all the Brauer character tables of Co_2 for all primes dividing its order.

19.2 Condensation

Let k be a field, A be a finite-dimensional k-algebra, $e \in A$ be an idempotent and mod–A be the category of finitely generated unital right A-modules. Then the condensation functor with respect to e is given as

$$C_e := ? \otimes_A Ae : \text{mod-}A \longrightarrow \text{mod-}eAe\,.$$

Under this functor, $M \in \text{mod-}A$ is mapped to $M \otimes_A Ae \in \text{mod-}eAe$. The latter can be identified with the subset $Me \subseteq M$, let $\iota : M \otimes_A Ae \to Me \subseteq M$ denote the corresponding injection. Using this identification, a homomorphism $\alpha \in \mathrm{Hom}_A(M,N)$ is mapped to $\alpha|_{Me} \in \mathrm{Hom}_A(Me,Ne)$. We have $C_e \cong \mathrm{Hom}_A(eA,?)$ as functors, hence C_e is an exact functor.

Let $e = \sum_S \sum_{j=1}^{n_S} e_S^{(j)} \in A$ be an orthogonal decomposition into primitive idempotents, where S runs through the isomorphism types of simple A-modules. For each summand we have $e_S^{(j)} A / \mathrm{Rad}(e_S^{(j)} A) \cong S$ and $n_S \in \mathbb{N}_0$. By $n_S = 0$ we indicate that this type of idempotent does not occur in the above decomposition. We have $\dim_k(Se) = n_S \cdot \dim_k(\mathrm{End}_A(S))$. By the Morita Theorems, see [5, Section 3.D.], C_e is an equivalence of categories if and only if $n_S > 0$ for all S; in this case e is called faithful.

19.2.1 Duality Let $\Delta : \text{mod--}A \to A\text{--mod}$ be the contravariant duality functor which maps $M \in \text{mod--}A$ to $M^* := \text{Hom}_A(M, k)$ and $\alpha \in \text{Hom}_A(M, N)$ to $(\lambda \mapsto \alpha \cdot \lambda) \in \text{Hom}_A(N^*, M^*)$. Furthermore, let σ be a k-algebra anti-automorphism of A. This defines a contravariant functor $\Sigma : \text{mod--}A \to A\text{--mod}$ which acts as the identity on homomorphism spaces and maps $M \in \text{mod--}A$ to $M^\sigma \in A\text{--mod}$, where the A-action on M^σ is given by $a \cdot m := m \cdot (a^\sigma)$ for $m \in M$, $a \in A$,

If additionally $e^\sigma = e$ holds, then this analogously defines a functor $\text{mod--}eAe \to eAe\text{--mod}$, again called Σ. It is easily seen that $\Sigma \cdot \Delta = \Delta \cdot \Sigma$ holds as equality of functors, both on mod–A and on mod–eAe. This defines duality auto-equivalences of mod–A and of mod–eAe. A module M is called self-dual, if $M \cong M^{\Sigma\Delta}$ holds. Note that the action of the duality auto-equivalence induces an anti-automorphism of the submodule lattice of a self-dual module. Finally, it is also easily seen that $(\Sigma \cdot \Delta) \cdot C_e \cong C_e \cdot (\Sigma \cdot \Delta)$ are naturally equivalent functors. Hence C_e maps a self-dual module to a self-dual module.

19.2.2 Uncondensation The uncondensation functor with respect to e is given as

$$? \otimes_{eAe} eA : \text{mod--}eAe \longrightarrow \text{mod--}A.$$

Under this functor, $\tilde{M} \in \text{mod--}eAe$ is mapped to $\tilde{M} \otimes_{eAe} eA \in \text{mod--}A$ and $\tilde{\alpha} \in \text{Hom}_{eAe}(\tilde{M}, \tilde{N})$ is mapped to $\tilde{\alpha} \otimes \text{id} \in \text{Hom}_A(\tilde{M} \otimes_{eAe} eA, \tilde{N} \otimes_{eAe} eA)$. If e is faithful, then this is the two-sided inverse functor to the condensation functor.

Uncondensation is applied in the following way. Let $N \in \text{mod--}A$ and $\tilde{N} := N \otimes_A Ae \in \text{mod--}eAe$ be its condensed module. The embedding $\iota : \tilde{N} \to N$ induces the A-homomorphism $\iota^A : \tilde{N} \otimes_{eAe} eA \to N : n \otimes ea \mapsto n \cdot a$ for $n \in \tilde{N}$, $a \in A$. Now let $\tilde{M} \leq \tilde{N}$ be an eAe-submodule, given by an embedding α. Then the image of the A-homomorphism $\alpha^A := (\alpha \otimes \text{id}) \cdot \iota^A : \tilde{M} \otimes_{eAe} eA \to N$ is called the uncondensed module of $\tilde{M}$.

Note that in the applications ι is always known explicitly. Hence given a submodule $\tilde{M} \leq \tilde{N}$, which usually is found during the structural analysis of $\tilde{N}$, the image $\text{Im}(\alpha \cdot \iota) \subseteq N$ can effectively be computed. Then $\text{Im}(\alpha^A) \leq N$ equals the smallest A-submodule of N containing $\text{Im}(\alpha \cdot \iota)$, hence can be found by standard techniques. This is a valuable tool to prove the existence of certain submodules of N, especially in face of the problem of generation we will discuss now.

19.2.3 Generation Problem Let A be generated as a k-algebra by the subset $\mathcal{A} \subseteq A$. We then let $\mathcal{C} := \langle eae; a \in \mathcal{A} \rangle_{k-\text{algebra}} \leq eAe$. Now the condensation subalgebra $\mathcal{C}$ is not necessarily equal to eAe, but may be a strict subalgebra of eAe. To prove equality in practice, only relatively weak criteria are known. They are stated and successfully used e. g. in [12], but unfortunately not practicable in the example we are envisaging. Hence, only the action of $\mathcal{C}$ on condensed modules can be computed explicitly, and so we are faced with the task to analyze condensed modules with respect to their structure as $\mathcal{C}$-modules and then to draw conclusions about their eAe-module structure from this analysis.

19.2.4 Fixed Point Condensation We will apply condensation in the following situation. Let G be a finite group and $A := k[G]$ be its group algebra over k. Let $K \leq G$ be a subgroup such that $|K| \neq 0 \in k$; in the sequel K is called the condensation subgroup. Then

$$e = e_K := \frac{1}{|K|} \cdot \sum_{g \in K} g \in k[K] \subseteq k[G]$$

is the centrally primitive idempotent of $k[K]$ belonging to the trivial K-module. We have $e \cdot k[G] \cong (1_K)^G$, where $(1_K)^G$ is the permutation module afforded by the permutation action of G on the cosets of K. Hence by the adjointness of tensor product and homomorphism functors, see [5, Theorem 2.19.], we have $Me \cong \mathrm{Hom}_G((1_K)^G, M) \cong \mathrm{Hom}_K(1_K, M_K) \cong \mathrm{Fix}_M(K)$ as vector spaces, where M is a $k[G]$-module, M_K denotes its restriction to $k[K]$ and $\mathrm{Fix}_M(K) \subseteq M$ consists of the elements of M being fixed by K. Hence this method is called fixed point condensation. As mod–$k[K]$ is a semisimple category, the dimension of the condensed module of M can be computed as the character theoretic scalar product of the trivial K-character and the restriction to K of the Brauer character of M.

Note that by letting $\sigma : k[G] \to k[G] : g \mapsto g^{-1}$ for $g \in G$, the duality $\Sigma \cdot \Delta = \Delta \cdot \Sigma$, see 'Duality' page 311, specializes to the operation of taking contragradient modules.

19.3 Condensation of Induced Modules, theoretically ...

Let $H \leq G$ be a subgroup and $V \in$ mod–$k[H]$. Let $\{g_i; i \in I\}$ for a suitable index set I be a set of representatives for the H-K double cosets in G and for each $i \in I$ let $\{k_{ij}; j \in I_i\}$ for a suitable index set I_i be a set of representatives for the $H^{g_i} \cap K$ right cosets in K. By the Mackey Decomposition Theorem we have $V^G = V \otimes_{k[H]} k[G] = \oplus_{i \in I} \oplus_{j \in I_i} V \otimes g_i k_{ij}$ as a vector space and hence $(V^G)_K \cong \oplus_{i \in I}((V^{g_i})_{H^{g_i} \cap K})^K$ as $k[K]$-modules, where $V^{g_i} \in$ mod–$k[H^{g_i}]$ is defined by $v \cdot h^{g_i} := v \cdot h$ for $v \in V$, $h \in H$. This means we have

$$\begin{aligned} V^G e &\cong \mathrm{Hom}_K(1_K, (V^G)_K) \\ &\cong \bigoplus_{i \in I} \mathrm{Hom}_K(1_K, ((V^{g_i})_{H^{g_i} \cap K})^K) \\ &\overset{\star}{\cong} \bigoplus_{i \in I} \mathrm{Hom}_{H^{g_i} \cap K}(1_{H^{g_i} \cap K}, (V^{g_i})_{H^{g_i} \cap K}) \\ &\cong \bigoplus_{i \in I} \mathrm{Hom}_{H \cap {}^{g_i}K}(1_{H \cap {}^{g_i}K}, V_{H \cap {}^{g_i}K}) \\ &\cong \bigoplus_{i \in I} \mathrm{Fix}_V(H \cap {}^{g_i}K), \end{aligned}$$

where the isomorphism '$\star$' maps $\varphi \in \mathrm{Hom}_{H^{g_i} \cap K}(1_{H^{g_i} \cap K}, (V^{g_i})_{H^{g_i} \cap K})$ by [1, Lemma III.8.6.] to

$$\left(\lambda \mapsto \sum_{j \in I_i} \varphi(\lambda k_{ij}^{-1}) k_{ij} = \varphi(\lambda) \cdot \sum_{j \in I_i} k_{ij}\right) \in \mathrm{Hom}_K(1_K, ((V^{g_i})_{H^{g_i} \cap K})^K).$$

Hence we explicitly obtain in terms of the Mackey decomposition of V^G given above:

$$V^G e = \bigoplus_{i \in I} \left(\mathrm{Fix}_V(H \cap {}^{g_i}K) \otimes g_i \sum_{j \in I_i} k_{ij}\right) = \bigoplus_{i \in I} \left(\mathrm{Im}_V(e_i) \otimes g_i \sum_{j \in I_i} k_{ij}\right),$$

where $e_i := e_{H \cap {}^{g_i}K}$ denotes the idempotent belonging to $H \cap {}^{g_i}K \leq G$.

19.3.1 The action of $ege \in ek[G]e$ for some $g \in G$ on $V^G e$ is derived from the action of e and that of g on V^G in terms of its Mackey decomposition. For $v \in V$, we have $(v \otimes g_i k_{ij}) \cdot g = vh' \otimes g_{i'} k_{i'j'}$, where the indices $i' \in I$, $j' \in I_{i'}$ and $h' \in H$ are uniquely determined by $g_i k_{ij} g = h' g_{i'} k_{i'j'}$. For $v \in \mathrm{Fix}_V(H \cap {}^{g_i}K)$ we hence have

$$(v \otimes g_i \sum_{j \in I_i} k_{ij}) \cdot eg = \sum_{j \in I_i} (v \otimes g_i k_{ij}) \cdot g = \sum_{j \in I_i} vh' \otimes g_{i'} k_{i'j'},$$

where the indices i', j' and $h' \in H$ of course depend on i and j.

The action of e on V^G is described as follows. For each $i \in I$ we have the decomposition

$$e = \frac{|H^{g_i} \cap K|}{|K|} \cdot e_i \cdot \sum_{j \in I_i} k_{ij} \in k[K],$$

and so for $v \in V$ this gives

$$(v \otimes g_i k_{ij}) \cdot e = (v \otimes g_i) \cdot e = \frac{|H^{g_i} \cap K|}{|K|} \cdot ve_i \otimes g_i \sum_{j \in I_i} k_{ij}.$$

19.4 ... and practically

19.4.1 Collecting coset Information Let $\mathcal{G}$ be a generating set for G given as set of permutations acting on the right cosets of H in G, i. e. we assume the permutation module $(1_H)^G$ to be given explicitly. Let us further assume K also to be given as a set $\mathcal{K}$ of permutation generators acting on the right cosets of H in G. In this paragraph we describe the part of the computations which are entirely in the context of permutation groups.

First we compute a Schreier chain for G with respect to this permutation representation. As in our application we already know the group order $|G|$, this can be done using a randomized Schreier-Sims algorithm. Such an algorithm is already available in GAP. But since we have to sift many elements of G through the Schreier chain, we have to make this easily accessible and more flexible and so we use our own GAP implementation of a randomized Schreier-Sims algorithm instead.

Now the H-K double cosets in G are in bijection with the K-orbits on $(1_H)^G$, hence suitable $\{g_i; i \in I\}$ and their decomposition into $\mathcal{G}$ are found by the Schreier-Sims algorithm. Since $H = \mathrm{Stab}_G(H \cdot 1)$ this also gives us a generating set $\mathcal{H}$ and a Schreier chain for H. Furthermore, we have $H^{g_i} \cap K = \mathrm{Stab}_K(H \cdot g_i)$, hence suitable $\{k_{ij}; j \in I_i\}$, a generating set $\mathcal{K}_i$ and a Schreier chain for $H^{g_i} \cap K$ are also found by a run of the Schreier-Sims algorithm. Conjugating gives a generating set $\mathcal{H}_i := {}^{g_i}\mathcal{K}_i$ for $H \cap {}^{g_i}K$, and sifting the elements of $\mathcal{H}_i$ through the Schreier chain for H, gives us decompositions of these elements into $\mathcal{H}$.

19.4.2 Precondensation Now let k be a finite field and $V \in \text{mod-}k[H]$ be given in terms of the generating set $\mathcal{H}$ for H as a set of representing k-matrices. In this paragraph we describe those computations using V which as a precondensation step can be done once and for all elements of G to be condensed. The necessary matrix computations are done using the MeatAxe [18] or using the new finite field arithmetic [11] implemented in GAP.

By the formulae developed in 19.3, a basis of V^Ge is given in terms of bases for the subspaces $\mathrm{Fix}_V(H \cap {}^{g_i}K) = \mathrm{Im}_V(e_i) \leq V$. Note that we have already computed the decompositions of the elements of $\mathcal{H}_i$ into $\mathcal{H}$, hence we can find representing matrices for the elements of $\mathcal{H}_i$, too. Now, to find a basis for $\mathrm{Fix}_V(H \cap {}^{g_i}K)$, we could compute the intersection of the fixed spaces for the action of the elements of $\mathcal{H}_i$ on V, which would be a standard application of the MeatAxe. But since we do need the action of e_i on V anyway, we instead proceed as follows. To compute a representing matrix for e_i, we use the Schreier chain for $H^{g_i} \cap K$, which gives us a factorization of e_i, analogous to the factorization of e given at the end of 19.3.1, and allows us to enumerate the summands.

Let $V = Ve_i \oplus V(1-e_i)$ be the vector space decomposition of V with respect to e_i, with corresponding projection π_i onto Ve_i and injection ι_i, fulfilling $\pi_i \cdot \iota_i = e_i$ and $\iota_i \cdot \pi_i = \mathrm{id}$. We have $\mathrm{Fix}_V(H \cap {}^{g_i}K) = \mathrm{Im}_V(e_i) = \mathrm{Im}(\iota_i)$, a basis of which can be found as a standard application of the MeatAxe. The matrix of basis vectors can be interpreted as the matrix of the linear map ι_i. As another standard application of the MeatAxe we obtain the matrix of π_i. We remark that the technique of precomputing the ι_i and π_i has also been used in [14].

19.4.3 Condensation of Group Elements Let $g \in G$; to find the action of $ege \in ek[G]e$ on V^Ge we proceed as follows. Having fixed an index $i \in I$, for each index $j \in I_i$ we compute the corresponding indices i' and j' and $h' \in H$ describing the action of g as defined in 19.3.1. The indices are found using the i-th Schreier chain of K obtained in 19.4.1, this then gives us the element h' explicitly, which is decomposed into $\mathcal{H}$ using the Schreier chain of H, giving a representing matrix η'. The action of h' on $\mathrm{Fix}_V(H \cap {}^{g_i}K)$ is then described by $\iota_i \cdot \eta'$. The action of e, namely projection to the i'-th component of $V^G e$, is given by another multiplication of the latter matrix with $\pi_{i'}$.

19.4.4 Remarks Firstly, the above computations can be simplified, if we have $H^{g_i} \cap K = \{1\}$ for some $i \in I$, i. e. if the i-th orbit is a regular K-orbit. In this case we have $\mathrm{Fix}_V(H \cap {}^{g_i}K) = V$, and hence $\iota_i = \pi_i = \mathrm{id}$. As usually K is small compared to the index $[G : H]$, regular orbits are in fact quite frequent.

Secondly, we make use of the permutation module $(1_H)^G$ because this makes the determination of H-K double cosets in G or the identification, to which H-K double coset in G or $H^{g_i} \cap K$ right coset in K an element belongs, particularly easy. This also could be done in any other representation of G where this information can be extracted in a computationally tractable manner.

19.4.5 Uncondensation Given an $ek[G]e$-submodule $\tilde{M} \leq V^G e$, see 19.2.2, we use the explicitly known injections ι_i from 19.4.2 to find its embedding as a subspace of V^G. Using the description of V^G and the action of G thereon in terms of the Mackey decomposition, see 19.3.1 and 19.4.3, we can use the standard MeatAxe 'spinning' algorithm to find the uncondensed module of $\tilde{M}$. This has been implemented as extensions of the MeatAxe implementations [18] and [7].

19.5 An Application: The 2-Modular Character Table of Co_2

From now on let $G := Co_2$ be the second sporadic simple Conway group, where we follow the notation used in [3]. The determination of its 2-modular Brauer character table has been begun in [20], where 10 of the 13 irreducible 2-modular Brauer characters of Co_2 have been determined, and the remaining three characters were conjectured to have degrees 36 938, 83 948 and 156 538. In [20], too, a condensation technique has been used to find Brauer characters. The main obstacle to give a complete proof was the problem of generation of the algebra $ek[G]e$, see 19.2.3, as even with todays computers it seems to be impossible to do the necessary uncondensation within the permutation module examined in [20]. Now, with our new condensation algorithm at hand, we are able to circumvent the problem of generation by looking at a suitable induced module, which allows a detailed structural analysis, due to its much simpler structure, and where uncondensation can be performed efficiently, due to the much smaller dimensions involved and the powerful parallel computers available today. Finally, the above mentioned conjecture from [20] turns out to be true. Thus this completes the computation of the 2-modular Brauer character table of G.

19.5.1 Let $\varphi_1, \ldots, \varphi_9, \varphi_{13}$ denote the known irreducible Brauer characters as given in the Brauer character table printed at the end of this note. Let ξ_{10}, ξ_{11} and ξ_{12} denote the remaining three irreducible Brauer characters. It has already been shown in [20] that the class functions φ_{10}, φ_{11} and φ_{12} are Brauer atoms and that they can be written as $\xi_i = \varphi_i + \sum_{j=1}^{9} \lambda_{ij}\varphi_j$ for $i \in \{10, 11, 12\}$ and some $\lambda_{ij} \in \mathbb{N}_0$. So our aim is to prove that all $\lambda_{ij} = 0$. For more details on the notion of basic sets and atoms we refer the reader to [8].

We now let $H := U_6(2).2$, which is the largest maximal subgroup of G and has index $[G : H] = 2300$, let $k := \mathbb{F}_2$ and $V \in \text{mod-}k[H]$ be the simple module of dimension 140, see [10] or the remark at the beginning of 19.5.3. We denote the Brauer character of V by ψ. Using the library of character tables and the character theoretic algorithms accessible in GAP, we find that the induced character ψ^G, which is of degree 322 000, decomposes into the set of Brauer atoms as

$$\psi^G = 10 \cdot \varphi_1 + 12 \cdot \varphi_2 + 6 \cdot \varphi_3 + 4 \cdot \varphi_4 + 4 \cdot \varphi_5 + 2 \cdot \varphi_{10} + \varphi_{11} + \varphi_{12} \,.$$

19.5.2 We begin our constructions by accessing the permutation representation P of G on 2 300 points from [24]. It is given there for a pair $\{a, b\}$ of standard generators for G as defined in [23].

As our condensation subgroup we choose $K := 3_+^{1+4} = O_3(N_G(3A)) \lhd N_G(3A) = 3_+^{1+4}{:}2_-^{1+4}.S_5 < G$, the maximal 3-normal subgroup of the normalizer in G of an element in the $3A$ conjugacy class. Hence this defines an unique conjugacy class of subgroups of G of order 243. A representative can be constructed as follows, using the algorithms dealing with permutation groups and soluble groups implemented in GAP. We first find a 3-Sylow subgroup $S < G$, which is of order 3^6, and then it turns out that S contains a unique normal subgroup K of isomorphism type 3_+^{1+4}.

Using GAP again we find the distribution of the elements of K into the conjugacy classes of G. Then we compute the scalar products of the restrictions to K of the φ_i and the trivial K-character. This by 19.2.4 gives us the condensed dimensions of the φ_i. These are given in the forth column of Table 19.1, where the degrees and the behavior under complex conjugation are also indicated. Especially, we conclude that the condensation idempotent $e = e_K$ is faithful.

19.5.3 Let P_H be the restriction of P to H. We find that V is a constituent of the 2-modular reduction of P_H, hence is obtained by a standard application of the MeatAxe. Note that this provides an alternative approach to find ψ: Let V be the unique constituent of the 2-modular reduction of P_H of dimension 140, and compute its Brauer character using the MeatAxe.

We now have prepared all necessary input data to run our condensation program. It turns out that, using the new finite field arithmetic [11] implemented in GAP, the permutation group part (19.4.1) of our algorithm takes $10s$, the precondensation (19.4.2) takes $40s$ and the condensation (19.4.3) of one group element takes $498s$ of CPU time on a SPARC 20 machine. As condensation algebra we take

$$\mathcal{C} := \langle eae, ebe, eabe, ebae \rangle_{k-\text{algebra}} \leq ek[G]e,$$

and we let $M := V^G e|_{\mathcal{C}}$ denote the restriction of the $ek[G]e$-module $V^G e$ to $\mathcal{C}$. The $\mathcal{C}$-module M has dimension 1 180 and the MeatAxe finds the following

Table 19.1: Dimensions of condensed modules

φ	Degree	CC	Dim.
1	1	r	1
2	22	r	4
3	230	r	6
4	748	5	4
5	748	4	4
6	3 520	r	24
7	5 312	r	24
8	8 602	9	16
9	8 602	8	16
10	36 938	r	100
11	83 948	r	288
12	156 538	r	566
13	1 835 008	r	7 424

constituents:

$$1a^{10}, 4a^{12}, 4b^4, 4c^4, 6a^6, 100a^2, 288a^1, 566a^1,$$

where we denote the constituents by their dimensions and lower case letters, and multiplicities by superscripts. The dimensions and multiplicities coincide with the decomposition of ψ^G into Brauer atoms given in 19.5.1 and Table 19.1, so we can already be quite confident that the all the Brauer atoms are in fact Brauer characters. In order to prove this, we are now going to analyze the $\mathcal{C}$-module M in more detail, using the ideas and algorithms developed in [13], which we assume the reader to be familiar with.

19.5.4 Ad φ_{10} Using the techniques developed in [13], we find that M contains exactly two $100a$-local submodules, L^1_{100a} and L^2_{100a}. Furthermore, for each local submodule $L \leq M$, we find that $L^1_{100a} \leq L \leq L^2_{100a}$ holds. Hence we have $\mathrm{Soc}(M) < \mathrm{Rad}(M)$ and $\mathrm{Soc}(M) \cong M/\mathrm{Rad}(M) \cong 100a$ is simple.

Let S denote the condensed $ek[G]e$-module of ξ_{10}. Table 19.1 shows that $100a$ occurs with multiplicity 1 as a constituent of $S|_{\mathcal{C}}$. Hence $S|_{\mathcal{C}}$ is indecomposable, and $100a$ is a socle and a radical quotient constituent of $S|_{\mathcal{C}}$. This means we have $S|_{\mathcal{C}} \cong 100a$, hence $\varphi_{10} = \xi_{10}$.

19.5.5 Ad φ_{11} Let $L_{288a} < M$ denote the unique $288a$-local submodule. It turns out to have dimension 439 and has the following constituents:

$$1a^3, 4a^5, 4b^3, 4c^1, 6a^2, 100a^1, 288a^1.$$

Hence, to prove that $\varphi_{11} = \xi_{11}$ holds, it is sufficient to show that the uncondensed module of L_{288a} in V^G has dimension 124 451 and the uncondensed module of $\mathrm{Rad}(L_{288a})$ has dimension 40 503, since this means that the uncondensed module of L_{288a} has a simple radical quotient module of dimension 83 948.

The latter conditions have been checked using the uncondensation algorithm and the 'spinning' algorithm for induced modules incorporated into the MeatAxe implementation [7]. The computations have taken a total of about 1 000 hours of CPU time using between four to eight nodes on the SP/2 of the computing center of the University of Essen and indeed resulted in two invariant subspaces of the expected dimensions. This proves that $\varphi_{11} = \xi_{11}$ holds.

19.5.6 Ad φ_{12} Let L_{566a} denote the unique $566a$-local submodule of M. It turns out to have dimension 705. Unfortunately, the analysis of the constituents of L_{566a} shows that the uncondensed module of L_{566a} in V^G would have dimension at least 196 249. To avoid the explicit construction of this uncondensed module we instead consider the submodule

$$N := \sum\{L \leq M \text{ local}; L_{566a} \not\leq L\} < M.$$

It turns out that N has dimension 475 and the following constituents:

$$1a^7, 4a^9, 4b^3, 4c^2, 6a^4, 100a^1, 288a^1.$$

Hence N is a $ek[G]e$-submodule of $V^G e$ if and only if its uncondensed module has dimension 125 751. Since we have $L_{288a} \leq N$, the same inclusion must hold for the corresponding uncondensed modules. Again using the parallel MeatAxe implementation [7], we have therefore computed the uncondensed module of N by extending the uncondensed module of L_{288a} found in 19.5.5. Again we have found an invariant subspace of the dimension we have expected, hence N is a $ek[G]e$-submodule of $V^G e$.

Using this information we now proceed as follows. From the list of constituents of N we see that $L_{566a} \not\leq N$ holds; note that this could have been deduced in advance using Lemma 19.1. By the structure theory of modular lattices, see [2, 13], it follows that $(L_{566a} + N)/N = \mathrm{Soc}(M/N) \cong 566a$ holds. Next we note that V is a self-dual $k[H]$-module, hence V^G is a self-dual $k[G]$-module and $V^G e$ is a self-dual $ek[G]e$-module, see 19.2.1. By the above analysis, L_{566a} is the image of N under the duality anti-automorphism of the $ek[G]e$-submodule lattice of $V^G e$. Hence L_{566a} is a $ek[G]e$-submodule of $V^G e$. Finally, we have $\mathrm{Rad}(L_{566a}) = L_{566a} \cap N$. Hence $\mathrm{Rad}(L_{566a})$ also is an $ek[G]e$-submodule of $V^G e$ and thus $\varphi_{12} = \xi_{12}$ holds.

19.5.7 Concluding Remarks The fact $L_{566a} \not\leq N$ used in 19.5.6 could have been deduced in advance from the following statement, we only give without proof.

Lemma 19.1. *Let $M \in \text{mod-}A$, let $L \leq M$ be a local submodule and $M_1, M_2 \leq M$ such that $L \not\leq M_1, M_2$, but $L \leq M_1 + M_2$. Then there are local submodules*

Table 19.2: The character table of Co_2 mod 2

	ind	1A	3A	3B	5A	5B	7A	9A	11A	15A	15B	15C	23A	23B
χ_1	+	1	1	1	1	1	1	1	1	1	1	1	1	1
χ_2	+	22	-5	4	-3	2	1	1	0	-1	0	0	-1	-1
χ_3	+	230	14	5	5	0	-1	-1	-1	0	-1	-1	0	0
χ_4	o	748	-8	1	-2	-2	-1	-2	0	1	b15	$\star\star$	-b23	$\star\star$
χ_5	o	748	-8	1	-2	-2	-1	-2	0	1	$\star\star$	b15	$\star\star$	-b23
χ_6	+	3520	-44	10	-5	0	-1	1	0	0	1	1	1	1
χ_7	+	5312	20	2	12	-3	-1	-1	-1	-3	0	0	-1	-1
χ_8	o	8602	43	-20	2	-3	-1	1	0	0	-b15	$\star\star$	0	0
χ_9	o	8602	43	-20	2	-3	-1	1	0	0	$\star\star$	-b15	0	0
χ_{10}	?	36938	-79	-52	13	-2	-1	2	0	-2	1	1	0	0
χ_{11}	?	83948	-22	-58	-2	-12	-3	-1	-4	-3	-2	-2	-2	-2
χ_{12}	?	156538	100	-80	-12	-2	-3	-2	-3	-5	0	0	0	0
χ_{13}	+	1835008	-128	-128	8	8	0	-2	-1	2	2	2	-1	-1

$L_1 \leq M_1$, $L_2 \leq M_2$ such that $L + L_1 = L + L_2 = L_1 + L_2$ holds, i. e. there is a dotted-line, see [2, 13], through L connecting M_1 and M_2.

Finally, it now remains to determine the 2-modular Frobenius-Schur indicators, see [9, Section 9], for the irreducible Brauer characters now proven to exist. We expect the methods developed here should be sufficient to solve this problem, too; this, however, is work still under progress.

Acknowledgments

The authors gratefully acknowledge discussions with Gerhard Hiss on the topic of this note. Both authors thank the Deutsche Forschungsgemeinschaft (DFG) for financial support in the framework of the joint research project 'Algorithmic number theory and algebra', to which this note is a contribution.

Bibliography

[1] J. Alperin: *Local representation theory*, Cambridge University Press, 1986.

[2] D. Benson, J. Conway: *Diagrams for modular lattices*, J. Pure Appl.Alg. **37**, 1985, 111–116.

[3] J. Conway, R. Curtis, S. Norton, R. Parker, R. Wilson: *Atlas of finite groups*, Clarendon Press, 1985.

[4] G. Cooperman, M. Tselman: *New sequential and parallel algorithms for generating high dimension Hecke algebras using the condensation technique*, in: Proc. of International Symposium on Symbolic and Algebraic Computation (ISSAC '96), ACM Press, 1996, 155–160.

[5] C. Curtis, I. Reiner: *Methods of representation theory I*, Wiley, 1981.

[6] J. Green: *Polynomial representations of* GL_n, Lecture Notes in Mathematics **830**, Springer, 1980.

[7] P. Fleischmann, G. Michler, P. Roelse, J. Rosenboom, R. Staszewski, C. Wagner, M. Weller: *Linear algebra over small finite fields on parallel machines*, Vorlesungen aus dem Fachbereich Mathematik der Universität GH Essen **23**, 1995.

[8] G. Hiss, C. Jansen, K. Lux, R. Parker: *Computational modular character theory*, unpublished manuscript.

[9] C. Jansen, K. Lux, R. Parker, R. Wilson: *An atlas of Brauer characters*, Clarendon Press 1995.

[10] N. N.: *An atlas of Brauer characters part II*, in preparation.

[11] N. Kim: *Implementierung der* MeatAxe *in das Computeralgebra-System* GAP *unter besonderer Berücksichtigung einer schnellen Vektorarithmetik*, Diplomarbeit, RWTH Aachen, 1997.

[12] K. Lux: *Algorithmic methods in modular representation theory*, Habilitationsschrift, RWTH Aachen, 1997.

[13] K. Lux, J. Müller, M. Ringe: *Peakword condensation and submodule lattices: an application of the* MeatAxe, J. Symb. Comp. **17**, 1994, 529–544.

[14] K. Lux, M. Wiegelmann: *Condensing tensor product modules*, in: R. Curtis, R. Wilson (ed.): The Atlas Ten Years On, Birmingham, 1995, Proceedings, London Mathmatical Society, Lecture Note Series 249, Cambridge Univ. Press.

[15] R. Parker: *The computer calculation of modular characters (The* MeatAxe*)*, in: M. Atkinson (ed.): Computational Group Theory 1984, 267–274.

[16] R. Parker, R. Wilson: *Unpublished*,

[17] A. Ryba: *Condensation programs and their application to the decomposition of modular representations*, J. Symb. Comp. **9**, 1990, 591–600.

[18] M. Ringe: *The* C–MeatAxe, Manual RWTH Aachen, 1994.

[19] M. Schönert et. al.: GAP–*Groups, Algorithms and Programming*, Manual, RWTH Aachen, 1995.

[20] I. Suleiman and R. Wilson: *The 2-modular characters of Conway's group* Co_2, Math. Proc. Camb. Philos. Soc. **116(2)**, 1994, 275–283.

[21] J. Thackray: *Modular representations of some finite groups*, PhD thesis, Cambridge University, 1981.

[22] M. Wiegelmann: *Fixpunktkondensation von Tensorproduktmoduln*, Diplomarbeit, RWTH Aachen, 1994.

[23] R. Wilson: *Standard generators for sporadic simple groups*, J. Algebra **184**, 1996, 505–515.

[24] R. Wilson: ATLAS *of finite group representations*, URL: <http://www.mat.bham.ac.uk/atlas>.

Progress in Mathematics, Vol. 173, © 1999 Birkhäuser Verlag Basel/Switzerland

Chapter 20

Bimodule and Matrix Problems

Serge Ovsienko

20.1 Introduction

20.1.1 Justifications This talk is devoted to an approach to the representat on theory over an algebraically closed field $\Bbbk$. I call it *constructive representation theory* and the main technical feature here is that *instead of working with a category of representations we are working with an object, "presenting" (in a sense explained below) this category.* This object often allows combinatorial description, therefore a such approach produces many combinatorial algorithms, which are both of a theoretical value and can be realized in form of computer programs as well.[1]

The notions we will mention below, namely, a bimodule problem, box and their generalizations arose as languages to formalize methods of solutions of so called *matrix problems, shortly* MP: the problem of *descriptions of representatives of equivalence classes of some set of matrices with respect to some set of transformations.* In turn, these sets of matrices form a category over $\Bbbk$ with the direct sums and the equivalence classes, which we describe, coincide with the isomorphisms classes (isoclasses) in this category. To solve MP's there have been developed *matrix reducing techniques* - some combinatorial procedures of generating of indecomposable representations. The idea is *to substitute some of matrices by their canonical forms and to restrict the set of transformations to the subset of transformations, keeping these forms.* These reductions provide us with effective algorithms to obtain the representations and morphisms between them in matrix form. Besides them, as we try to show, *many operations in representation theory can be interpreted as "substitution of some matrices by their special forms".*[2] Another important tool arising here and closely related with reductions is the technique of quadratic and (nonsymmetric) bilinear forms. So these methods allow

1. to transform objects of representation theory into some combinatorial objects;

[1] Some from corresponding programs were developed and successfully used for solution various problems.

[2] In some sense, it can be considered as a generalization of "Harish-Chandra" methodology, sce 20.4.4.

2. to investigate the obtained combinatorial objects using combinatorial methods (especially, computer assisted methods);

3. to re-interpret the results obtained in step 2 in terms of representation theory.

Here we develop a very restricted approach to the investigation of categories of representations. Among others, it is motivated strongly by the interaction between the reduction technique and the theory of quadratic forms. We try to show, that new techniques together with new problems lead to modifications and extensions of the notion of a MP. That work is in progress now and our survey is by no means complete (for example, we do not discuss close to considered here problems theory of tilting transformations and related topics ([36], [33], [44]), the relationship with the theory of covariant and contravariant finite subcategories ([1]) and an important technique of differentiation of vectorspace categories ([29]) etc.).

20.1.2 Categories For a finitedimensional (f.d.) $\Bbbk$-vectorspace V we denote by $|V|_\Bbbk$ or $\dim_\Bbbk V$ its $\Bbbk$-dimension. If $\mathcal{C}$ is a category, then we denote by $\operatorname{Ob}\mathcal{C}$ the set of objects of $\mathcal{C}$. For $X, Y \in \operatorname{Ob}\mathcal{C}$ we denote by $\mathcal{C}(X, Y)$ or $\operatorname{Hom}_\mathcal{C}(X, Y)$ the set of morphisms in $\mathcal{C}$ from X to Y. All categories and functors will be *categories and functors over* $\Bbbk$, that means for $\mathcal{C}$, that all $\mathcal{C}(X, Y)$ are $\Bbbk$-vectorspaces, the composition in $\mathcal{C}$ $\Bbbk$-bilinear and for a functor $F : \mathcal{C}\longrightarrow\mathcal{C}'$ hold $F(f+g) = F(f) + F(g)$, $F(\lambda f) = \lambda F(f)$ for morphisms f, g and $\lambda \in \Bbbk$. We denote by $\operatorname{sum}(\mathcal{C})$ $(\operatorname{Sum}(\mathcal{C}))$ a minimal category, containing $\mathcal{C}$ and closed with respect to finite (countable) direct sums. We use the following convention: for categories $\mathcal{C}$, $\mathcal{C}'$ the notation $F : \mathcal{C}\longrightarrow\mathcal{C}'$ means, we have a functor $F : \mathcal{C}\longrightarrow \operatorname{sum}(\mathcal{C}')$. We call the category $\mathcal{C}$ *completely additive*, provided $\mathcal{C}$ is closed with respect to finite sums and every idempotent in $\mathcal{C}$ splits. For a category $\mathcal{C}$ we denote by $\operatorname{add}(\mathcal{C})$ a minimal completely additive category containing $\mathcal{C}$. We denote by $\operatorname{mod}-\Bbbk$ $(\operatorname{Mod}-\Bbbk)$ the category of f.d. $\Bbbk$-vectorspaces (of $\Bbbk$-vectorspaces), by $\mathbb{D} : \operatorname{mod}-\Bbbk\longrightarrow \operatorname{mod}-\Bbbk$ the standard duality: $V \mapsto \mathbb{D}V = \operatorname{mod}-\Bbbk(V, \Bbbk)$. Consequently, $\operatorname{mod}-\mathcal{C}$ $(\operatorname{Mod}-\mathcal{C})$ we denote the category of locally f.d. left $\mathcal{C}$-modules (left $\mathcal{C}$-modules) or, equivalently, the functors $M : \mathcal{C}\longrightarrow \operatorname{mod}-\Bbbk$ $(M : \mathcal{C}\longrightarrow \operatorname{Mod}-\Bbbk)$. If it does not lead to ambiguity, for $\mathcal{C}$-modules M, N we use the notation $\operatorname{Hom}_\mathcal{C}(M, N)$ instead of $\operatorname{mod}-\mathcal{C}(M, N)$. All modules are supposed to be left unless otherwise stated. The category $\mathcal{C}$ we call *trivial*, if $\mathcal{C}(X, X) = \Bbbk$ and $\mathcal{C}(X, Y) = 0$ for $X \neq Y, X, Y \in \operatorname{Ob}\mathcal{C}$. For a set S we denote by $\mathbb{L}_S$ the trivial category, s.t. $\operatorname{Ob}\mathbb{L}_S = S$ and for a category $\mathcal{C}$ we denote by $\mathbb{L}_\mathcal{C}$ the category $\mathbb{L}_{\operatorname{Ob}\mathcal{C}}$.

20.1.3 Graduations Let $\mathbb{G}$ be an additively written abelian group. A $\mathbb{G}$-*graded module* M over the category k we define as a family of k-modules $M = \{M_i\}_{i\in\mathbb{G}}$. Sometimes it is convenient to consider instead of the family $\{M_i\}_{i\in\mathbb{G}}$ the direct sum $\oplus_{i\in\mathbb{G}} M_i$ or the direct product $\prod_{i\in\mathbb{G}} M_i$. We set for $x \in M_i$ $\deg x = i$ and a *(homogeneous) morphism* f *of degree* $\deg f = d$ between M and $N = \{N_i\}_{i\in\mathbb{G}}$

is a family of k-module homomorphisms $f_i : M_i \longrightarrow N_{i+d}$. If $\deg f = 0$, then f is *a graded morphism.* For $\mathbb{G}$-*graded* k-*bimodules* we write $\otimes$ instead $\otimes_{\mathsf{k}}$. If $\vec{i} = (i_1, \dots, i_k) \in \mathbb{G}^k$, then $|\vec{i}| = i_1 + \cdots + i_k$ and we will use the standard graduation of the tensor product: deg $M^{\otimes \vec{i}} =$ deg $M_{i_1} \otimes \cdots \otimes M_{i_k} = |\vec{i}|$ and use the notation $M_i^{\otimes k} = \oplus_{|\vec{i}|=i} M^{\otimes \vec{i}}$, $M_i^{\hat{\otimes} k} = \prod_{|\vec{i}|=i} M^{\otimes \vec{i}}$. For a $\mathbb{G}$-graded k-bimodule $N = \{N_i\}_{i \in \mathbb{G}}$ we denote by $\mathsf{k}[N]$ the *graded tensor category of the bimodule* N *over* k, provided $\deg N_i = i$ and $\deg \mathsf{k} = 0$, $\mathsf{k}[N] = \oplus_{i=0}^{\infty} N^{\otimes i}$. We denote by $\mathsf{k}[\hat{N}]$ the *completed tensor category* $\mathsf{k}[\hat{N}] = \prod_{i=0}^{\infty} N^{\otimes i}$.

A category K we call $\mathbb{G}$-graded category, provided it is $\mathbb{G}$-graded module $K = \{K_i\}$ over $\mathbb{L}_K$ and $f \cdot g \in K_{i+j}$, when $f \in K_i$, $g \in K_j$. We will consider K as a $\mathbb{k}$-category, provided $K(A,B) = \sum_{i \in \mathbb{G}} K_i(A,B)$, $A, B \in \operatorname{Ob} K$. $\mathbb{G}$-graded modules over k and homogeneous morphisms form a $\mathbb{G}$-graded category $\operatorname{mod}_{\mathbb{G}} \mathsf{k}$, where the graded morphisms form a component of degree 0. A $\mathbb{G}$-graded $\mathbb{L}_K$-module $M = \{M_i\}_{i \in \mathbb{G}}$ is called *graded module over graded category* K, provided $\oplus_{i \in \mathbb{G}} M_i$ Is a module over K and $f \cdot m \in M_{i+j}$, if $f \in K_i, m \in M_j$. If $\mathcal{N} = \{\mathcal{N}_i\}_{i \in \mathbb{G}}$ is a $\mathbb{G}$-graded K-bimodule, then the tensor category of the bimodule $\mathcal{N}$ over K $K[\mathcal{N}] = \oplus_{i=0}^{\infty} \mathcal{N}^{\otimes_K i}$ and the completed tensor category $K[\hat{\mathcal{N}}] = \prod_{i=0}^{\infty} \mathcal{N}^{\otimes_K i}$ have the (inherited from K and $\mathcal{N}$) natural $\mathbb{G}$-graduations. If $\mathcal{C}$ is a $\mathbb{G}$-graded category with the graduation **deg**, then by $\mathcal{C}_{\mathbb{G}}$ we denote the category with the set of objects $X[v]$, $X \in \operatorname{Ob} \mathcal{C}$, $v \in \mathbb{G}$, the morphisms $f[v] : X[v] \longrightarrow Y[v + \mathbf{deg}\, f]$, $X, Y \in \operatorname{Ob} \mathcal{C}$, $f \in \mathcal{C}$, $v \in \mathbb{G}$ and the composition, induced by the composition in $\mathcal{C}$. *There exists the canonical bijection between* $\mathbb{G}$-*graded functors from* $\mathbb{G}$-*graded category* $\mathcal{C}$ *to* $\operatorname{Mod}_{\mathbb{G}} - \mathbb{k}$ *and the functors from* $\mathcal{C}_{\mathbb{G}}$ *to* $\operatorname{Mod} - \mathbb{k}$.

In most cases there $\mathbb{G} = \mathbb{Z}$ (sometimes $\mathbb{G} = \mathbb{Z}^n$). For a $\mathbb{Z}$-graded module M for $x \in M$ we denote $(-1)^{\deg x} x$ by $\hat{x}$. The binary $\mathbb{Z}$-graded multiplication $''\circ''$ on M is called *graded associative*, provided for any x, y, z holds $(x \circ y) \circ z = \hat{x} \circ (y \circ z)$. A graded associative multiplication "$\circ$" can be transformed in the associative "$\cdot$" by setting $x \cdot y = \hat{x} \circ y$.

20.1.4 Graphs A *graded graph*

$$\Gamma = (\Gamma_0, \Gamma_1; \alpha, \beta\, (= \alpha_\Gamma, \beta_\Gamma) : \Gamma_1 \longrightarrow \Gamma_0, \deg\, (= \deg_\Gamma) : \Gamma_1 \longrightarrow \mathbb{Z})$$

consists of the sets of *vertices* Γ_0 and *arrows* Γ_1, where x *leads from* $A = \alpha(x)$ *to* $B = \beta(x)$ (or $x \in \Gamma(A,B)$), and the function $\deg : \Gamma_1 \longrightarrow \mathbb{Z}$, which is called *degree. We assume, that all graphs are graded.* We set $\Gamma^i = \deg^{-1}(i)$, $i \in \mathbb{Z}$. $\check{\Gamma}$ means the graph, s.t. $\check{\Gamma}_0 = \Gamma_0$, $\check{\Gamma}_1 = \Gamma_1 \sqcup \{\omega_A\}_{A \in \Gamma_0}$, where $\alpha_{\check{\Gamma}}, \beta_{\check{\Gamma}}, \deg_{\check{\Gamma}}$ on Γ_1 coincide with $\alpha_\Gamma, \beta_\Gamma, \deg_\Gamma$ correspondingly and

$$\omega_A : A \longrightarrow A, \deg \omega_A = 1 A \in \Gamma_0.$$

If $\deg(\Gamma_1) = \{0\}$ ($\deg(\Gamma_1) = \{0,1\}$), then Γ is called a *quiver* (bigraph). Γ is called *directed*, if it does not contain oriented cycles. We denote by $\mathbb{L}_\Gamma$ the category $\mathbb{L}_{\Gamma_0}$, by $\mathbb{k}^\Gamma$ the free graded $\mathbb{L}_\Gamma$-bimodule, s.t. for every $A, B \in \Gamma_0, i \in \mathbb{G}$ the space $\mathbb{k}_i^\Gamma(A,B)$ has a basis $\Gamma^i(A,B)$, by $\mathbb{k}[\Gamma]$ the *free* $\mathbb{G}$-*graded category over* $\mathbb{k}$, *generated by* Γ, $\mathbb{k}[\Gamma] = \mathbb{L}_\Gamma[\mathbb{k}^\Gamma]$ and by $\mathbb{k}[\hat{\Gamma}]$ the *completed free* $\mathbb{G}$-*graded category over* $\mathbb{k}$. The $\mathbb{G}$-grading $\deg_\Gamma$ extends on $\mathbb{k}[\Gamma]$ and $\mathbb{k}[\hat{\Gamma}]$.

20.1.5 Forms Let S be a set, $|S|$ be its cardinality, $\mathbb{Q}_S$ be the subspace in the $\mathbb{Q}$-vectorspace $\mathbb{Q}^S = \mathrm{Hom}_{Sets}(S, \mathbb{Q})$, consisting of all $\mathbb{Q}$-vectors with finitely many nonzero entries. If $\mathcal{C}$ is a category, then we denote by dim(= $\dim_{\mathcal{C}}$) the function $\dim : \mathrm{mod}-\mathcal{C} \longrightarrow \dim_{\mathcal{C}}$, where $\dim_{\mathcal{C}}$ is the *($\mathbb{Q}$-) vectorspace of dimensions*, $\dim_{\mathcal{C}} = \mathbb{Q}^{\mathrm{Ob}\,\mathcal{C}}$, $\dim(M) = (\dim_{\Bbbk} M(i))_{i \in \mathrm{Ob}\,\mathcal{C}}$.

Let Γ be a graph, s.t. for every $A, B \in \Gamma_0$ holds $|\Gamma(A,B)| < \infty$, $\mathbb{Q}_\Gamma = \mathbb{Q}_{\Gamma_0}$. The *non-symmetric bilinear form* of Γ $<,>_\Gamma$: $\mathbb{Q}_\Gamma \times \mathbb{Q}_\Gamma \longrightarrow \mathbb{Q}$, $\mathbb{Q}_\Gamma = \mathbb{Q}_{\Gamma_0}$, is defined on $x = (x_A), y = (y_A)$ as

$$< x, y >_\Gamma = \sum_{(A,B) \in \Gamma_0^2} \sum_{i \in \mathbb{Z}} (-1)^i |\check{\Gamma}^i(A,B)| x_A y_B,$$

the *symmetrical bilinear form* of Γ as $(x,y)_\Gamma = \frac{1}{2}(< x,y >_\Gamma + < y,x >_\Gamma)$ and the *quadratic form* as $\mathrm{q}_\Gamma(x) = (x,x)_\Gamma =< x,x >_\Gamma$.

A quadratic form $\mathrm{q} : \mathbb{Q}_S \longrightarrow \mathbb{Q}$ is called *weakly positive*, shortly WP (*weakly nonnegative*, shortly WNN) provided for every *positive* $x = (x_A)_{A \in S}$, that means $x_A \geq 0$ for all $A \in S$ and $x \neq 0$, holds $\mathrm{q}(x) > 0$ ($\mathrm{q}(x) \geq 0$).

20.2 The first Example – Bimodule Problems

20.2.1 A historical Remark MP's arose from (sometimes successful) attempts to classify in categories of representations all the indecomposables up to isomorphism. In particular, it is important to decide, whether the classification problem is either of *finite type* (it means, there exists only finitely many isoclasses of indecomposables) or of *infinite type*. In the infinite type one can distinguish a *strictly unbounded type*, that means there exists a dimension (see definition of dimension below), containing infinitely many isoclasses. And, the best known partition of the MP's is into the problems of *tame type* (it means, in any dimension the indecomposables are parametrized by finitely many 1-parameter families plus finitely many indecomposables) or of *wild type* (it means, that the problem of the classifications of representations "contains" the classification of pairs of square matrices with respect to simultaneous conjugation). The definition of a tame and wild problem can be found in [18, 19]. Often these representations were constructed by iterated extensions from some starting set $\mathcal{S} = \{S_i\}$ of representations. Then the representations (not isoclasses) were parametrized by the elements of $Ext^1(\oplus_i S_i^{k_i}, \oplus_j S_j^{l_j})$, that have been written as matrices over $\Bbbk$, see [45] (an example of such Ext^1-matrices give us the quiver representations). In turn, the isoclasses were parametrized by the orbits of a linear action of an algebraic group on these Ext^1-matrices. For fixed k_i and l_j this action can be described as follows. Let Λ be a f.d. associative algebra, $\Lambda^* \subset \Lambda$ be the group of its invertible elements, M be a Λ-bimodule. Then Λ^* acts on M by $(l,b) \mapsto lbl^{-1}$, $l \in \Lambda^*$, $b \in M$. To obtain all representations we should consider in the construction above instead of the pair (Λ, M) all the possible pairs $(\tilde{\Lambda}, \tilde{M})$, where $\tilde{\Lambda}$ is Morita-equivalent to Λ and $\tilde{M}$ is a $\tilde{\Lambda}$-bimodule corresponding to M.

20.2.2 Definition of the Category of Representations of a Bimodule Problem

The situation above can be described using the notion of a bimodule problem, shortly BP (in [16], 1972 is used the term "matrix with coefficients in bimodule"). Let $\mathcal{C}$ be a finite *spectroid* ([29]), that means $\operatorname{Ob}\mathcal{C} = \{A_1, \dots, A_n\}$ is finite, the algebras $\mathcal{C}(A_i, A_i)$ are local, $\mathcal{C}(A_i, A_j)$ are f.d. and $A_i \not\simeq A_j$ for $i \neq j$ (one can interpret it as *a basic f.d. algebra*). The spectroid $\mathcal{C}$ is supposed to be included in *an aggregate* (*Krull-Schmidt category*) $\tilde{\mathcal{C}}$, that means $\tilde{\mathcal{C}}$ has direct sums and every $A \in \operatorname{Ob}\tilde{\mathcal{C}}$ is isomorphic to $A_1^{x_1} \oplus \cdots \oplus A_n^{x_n}$. If $\tilde{\mathcal{M}}$ is a bimodule over $\tilde{\mathcal{C}}$, s.t. $\mathcal{M} = \tilde{\mathcal{M}}|_{\mathcal{C}}$ is f.d., then *a bimodule problem* is a pair $\mathfrak{A} = (\mathcal{C}, \mathcal{M})$ and its *representations category* $\operatorname{rep}\mathfrak{A}$ is defined as follows:

1. a *representation* of $\mathfrak{A}$ is a pair (X, m), where $X \in \operatorname{Ob}\tilde{\mathcal{C}}$, $m \in \tilde{\mathcal{M}}(X, X)$;
2. a *morphism* $f : (X, m) \longrightarrow (Y, n)$ is a $f \in \tilde{\mathcal{C}}(X, Y)$, s.t. $fm = nf$;
3. the *composition* of morphisms in $\operatorname{rep}\mathfrak{A}$ is induced by the composition in the category $\tilde{\mathcal{C}}$.

The category $\operatorname{rep}\mathfrak{A}$ is a Krull-Schmidt category.

Remark 20.1. The same definition of $\operatorname{rep}\mathfrak{A}$ is valid for any pair $\mathfrak{A} = (\mathcal{C}', \mathcal{M}')$, where $\mathcal{C}'$ is a full subcategory in $\tilde{\mathcal{C}}$, s.t. $\tilde{\mathcal{C}}$ is equivalent to an additive closure of $\mathcal{C}'$, $\tilde{\mathcal{M}}|_{\mathcal{C}'} \simeq \mathcal{M}'$.

20.2.3 Examples

Quivers. If $\mathfrak{A} = (\mathcal{C}, \mathcal{M})$ is a BP with trivial $\mathcal{C}$, then $\operatorname{rep}\mathfrak{A}$ is equivalent the category of representations of the quiver $\Gamma_{\mathfrak{A}}$ ([29]). The vertices of $\Gamma_{\mathfrak{A}}$ are $\{A_1, \dots, A_n\}$ and the number of arrows in $\Gamma_{\mathfrak{A}}$, leading from A_i to A_j is equal to $|\mathcal{M}(A_i, A_j)|_{\Bbbk}$.

Bipartite BP's. Let $\mathcal{C}_1$, $\mathcal{C}_2$ be finite spectroids, $\mathcal{C} = \mathcal{C}_1 \oplus \mathcal{C}_2$, $\mathcal{M}$ be a *bipartite* $\mathcal{C}$*-bimodule,* that means $\mathcal{C}_2\mathcal{M} = \mathcal{M}\mathcal{C}_1 = 0$. Then $\mathfrak{A} = (\mathcal{C}, \mathcal{M})$ is called a *bipartite* BP. Such kind of BP's arise often in applications, see below.

Vectorspace categories. Let $\mathcal{K}$ be a spectroid, M a faithful module over $\mathcal{K}$. In this case the image of the functor M in $\operatorname{mod}-\Bbbk$ is isomorphic to $\mathcal{K}$ and $\mathcal{K}$ is called a *vectorspace category.* Let us consider the bipartite BP with $\mathcal{C}_2 = \mathcal{K}$, $\mathcal{M} = M$, $\mathcal{C}_1 = \Bbbk$, so $\mathcal{C}_1$ acts on $\mathcal{M}$ by scalars. Then $\operatorname{rep}\mathfrak{A}$ is called *a subspace* category, [44]. Analogously, if we start from a faithful right $\mathcal{K}$-module M and set $\mathcal{C}_1 = \mathcal{K}$, $\mathcal{M} = M$, $\mathcal{C}_2 = \Bbbk$ we obtain as $\operatorname{rep}\mathfrak{A}$ so called *factorspace category.*

Drozd construction. Let Λ be a finite spectroid. We consider the bipartite BP with $\mathcal{C}_i = \Lambda$, $i = 1, 2$ and $\mathcal{C}_1$ ($\mathcal{C}_2$) acts on $\mathcal{M} = \operatorname{Rad}(\Lambda)$, where $\operatorname{Rad}(\Lambda)$ is the Jacobson radical, from left (from right) in a natural way. The corresponding BP $\mathfrak{A} = (\mathcal{C}, \mathcal{M})$ was introduced in [19]. Every $M \in \operatorname{rep}\mathfrak{A}$ (as a matrix with coefficients in Λ) defines uniquely a map of Λ-projectives $\varphi_M : P_1 \longrightarrow P_2$.

Theorem 20.1. *[19] The map* $M \mapsto \operatorname{Coker}\varphi_M$ *defines a functor* $Cok : \operatorname{rep}\mathfrak{A} \longrightarrow \operatorname{mod}-\Lambda$, *that is surjective on the spaces of morphisms.* Cok *induces a bijection on indecomposables except finitely many* $M \in \operatorname{rep}\mathfrak{A}$, *s.t.* $Cok(M) = 0$.

This construction transforms the statements about associative algebras into the statements about BP's. It is used, for example, in the Drozd proof of tame-wild dichotomy [19], [10].

20.2.4 BP with Differentiation To describe the reduction algorithm for a BP we should consider a more general class of MP's (Drozd, (unpublished), Crawley–Boewey, [11]). The $\Bbbk$-linear map $\delta : \mathcal{C} \longrightarrow \mathcal{M}$, $\delta(\mathcal{C}(X,Y)) \subset \mathcal{M}(X,Y)$ is called *a differentiation*, provided $\delta(xy) = \delta(x)y + x\delta(y)$. The triple $\mathfrak{B} = (\mathcal{C}, \mathcal{M}, \delta)$ is called *a* BP *with differentiation.* The objects of its representations category $\operatorname{rep}\mathfrak{B}$ are pairs (X, m), $X \in \operatorname{Ob}\tilde{\mathcal{C}}$, $m \in \tilde{\mathcal{M}}(X, X)$ and morphisms $f : (X, m) \longrightarrow (Y, n)$ are morphisms $f \in \tilde{\mathcal{C}}(X, Y)$, s.t. $fm - nf + \delta(f) = 0$, the composition in $\operatorname{rep}\mathfrak{B}$ coincides with the composition in $\tilde{\mathcal{C}}$. If $\delta = 0$, this category is the category of representations of BP $\mathfrak{A} = (\mathcal{C}, \mathcal{M})$

20.2.5 Dimension of Representation and Quadratic Form of BP If (X, m) is a representation of a BP or a BP with differentiation $\mathfrak{A}$ and $X = \oplus_{i=1}^{n} A_i^{x_i}$, then the vector $x = (x_1, \dots, x_n) \in \mathbb{Z}^n$ is called *the dimension* of X, that defines a map in the *space of dimensions* $\dim (= \dim(\mathfrak{A})) : \operatorname{rep}\mathfrak{A} \longrightarrow \dim_{\mathfrak{A}} = \mathbb{Q}_{\operatorname{Ob} K} (= \mathbb{Q}^n)$ (analogously to 20.1.2). The dimension is an invariant of an isoclass. In every dimension the isoclasses coincides with the orbits of the action of the group $\tilde{\mathcal{C}}(X, X)^*$ on $\tilde{\mathcal{M}}(X, X)$, analogously to 20.2.1. This action is linear in the case of BP and is affine in the case BP with differentiations. If we choose bases in all spaces $\mathcal{M}(A_i, A_j)$, then every representation X of $\mathfrak{A}$ can be identified with a family of $(x_i \times x_j)$-matrices with coefficients in $\mathcal{M}(A_i, A_j)$, $A_i, A_j \in \operatorname{Ob}\mathcal{C}$ or with the affine space of dimension $\sum_{A,B \in \operatorname{Ob}\mathcal{C}} |\mathcal{M}(A,B)|_{\Bbbk} x_A x_B$ of matrices over $\Bbbk$. The dimension of the acting group $\tilde{\mathcal{C}}(X, X)^*$ as an algebraic variety is equal to $\sum_{A,B \in \operatorname{Ob}\mathcal{C}} |\mathcal{C}(A,B)|_{\Bbbk} x_A x_B$. This leads to the notion of *the quadratic form of the* BP, so called *Tits form*: for $x = (x_A)_{A \in \operatorname{Ob}\mathcal{C}}$ we set

$$\mathrm{q}_{\mathfrak{A}}(x) = \sum_{A,B \in \operatorname{Ob}\mathcal{C}} |\mathcal{C}(A,B)|_{\Bbbk} x_A x_B - \sum_{A,B \in \operatorname{Ob}\mathcal{C}} |\mathcal{M}(A,B)|_{\Bbbk} x_A x_B.$$

The importance of this form is approved by the following statement.

Lemma 20.1. *Let* $\mathfrak{A}$ *be a* BP *with differentiation and* $\mathrm{q}_{\mathfrak{A}}$ *be its quadratic form.*

1. *(see [27]). If for some* $x \in (\mathbb{Z}^+)^n$ *holds* $q_{\mathfrak{A}}(x) \leq 0$, *then in the dimension* x *exist infinitely many isoclasses and* $\mathfrak{A}$ *is of infinite representation type (moreover, strictly unbounded type).*

2. *(see [18]). If for some* $x \in (\mathbb{Z}^+)^n$ *holds* $q_{\mathfrak{A}}(x) < 0$, *then* $\mathfrak{A}$ *is of wild representation type.*

The first statement is proved by simple comparison of the dimensions of the acting group and of the variety of representations, for the proof of second see [18]. The existence of an analogue of this statement is the first test for quadratic forms defined for more general MP's. We associate with $\mathfrak{A}$ a bigraph

$$\Gamma = \Gamma_{\mathfrak{A}}, \Gamma_0 = \operatorname{Ob} K, |\Gamma^0(A,B)| = |\mathcal{M}(A,B)|_{\Bbbk}, |\Gamma^1(A,B)| = |\operatorname{Rad} \mathcal{C}(A,B)|_{\Bbbk}, \quad A, B \in \operatorname{Ob} \mathcal{C}$$

and then holds $q_{\mathfrak{A}} = q_\Gamma$. Corresponding bilinear form has a property (as in [44, 43])

$$< \dim X, \dim Y >_\Gamma = |\operatorname{Hom}_{\operatorname{rep}\mathfrak{A}}(X,Y)|_{\Bbbk} - |\operatorname{Ext}^1_{\operatorname{rep}\mathfrak{A}}(X,Y)|_{\Bbbk}, X, Y \in \operatorname{rep}\mathfrak{A}$$

for appropriate $\operatorname{Ext}^1_{\operatorname{rep}\mathfrak{A}}$ ($\operatorname{Ext}^i_{\operatorname{rep}\mathfrak{A}} = 0$ for $i \geq 2$). The definitions of $\operatorname{Ext}_{\operatorname{rep}\mathfrak{A}}$ are based mostly on some exact structure, but here we use a 'resolutions' approach (see [42]).

20.3 Differential ($\mathbb{Z}$-) graded Category (D($\mathbb{Z}$)GC) and Boxes

20.3.1 Differential ($\mathbb{Z}$-) graded Category (D($\mathbb{Z}$)GC) In [37] was proposed a new formalism for describing MP's. Let $U = \{U_i\}$ be a free $\mathbb{Z}$-graded (shortly D$\mathbb{Z}$GC) or graded, that means, positively $\mathbb{Z}$-graded (shortly DGC) category with a differential $\delta : U \longrightarrow U$, $\delta(U_i) \subset U_{i+1}$, satisfying the *Leibniz rule* for homogeneous elements $\delta(xy) = \delta(x)y + \hat{x}\delta(y)$. According to [37], the $\mathbb{Z}$-graded category U with the differential δ we call *quasi differential $\mathbb{Z}$-graded category (QD$\mathbb{Z}$GC)*, if there exists a family $\mathbf{h} = \{h_A \in U_2(A,A), A \in \operatorname{Ob} U\}$, s.t. $\delta(h_A) = 0$ for any $A \in \operatorname{Ob} U$ and $\delta^2(u) = h_B u - u h_A$ for any u, $u \in U(A,B)$. If there holds $\delta^2 = 0$ (that means, $h_A = 0$ for all $A \in \operatorname{Ob} U$), then U is *differential $\mathbb{Z}$-graded category (D$\mathbb{Z}$GC)*.

From U we construct the new D$\mathbb{Z}$GC T, that is freely generated over U by $\omega_A : A \longrightarrow A$, where A runs through $\operatorname{Ob} U$, with differential d given by

$$d(x) = \omega_A x - \hat{x}\omega_B + \delta(x) \text{ for } x \in U, x : A \longrightarrow B$$

and $d(\omega_A) = \omega_A^2 - h_A$. If U was D$\mathbb{Z}$GC , then T we call *normal*, see 20.3.6.

20.3.2 Representation of DGC Let DGC U be freely generated by a bimodule U_1 over the category U_0, $U_i = U_1^{\otimes i}$, $i \geq 0$, $U_i = 0, i < 0$, K be the category $T_0(= U_0)$. Then T_1 is the K-bimodule $U_1 \oplus F_\Sigma$, where F_Σ is freely generated by all ω_A and we set $V = T_1/(d(T_0))$. The restriction of d on T_1 gives a map

$$d|_{T_1} : T_1 \longrightarrow T_1 \otimes_{T_0} T_1$$

and from $d^2(T_0) = 0$ follows, that $d|_{T_1}$ induces a K-bimodule comultiplication $\mu : V \longrightarrow V \otimes_K V$. From $d^2(T_1) = 0$ and the Leibniz rule we obtain, that μ is coassociative and $\varepsilon : V \longrightarrow K$ given by $\varepsilon(U_1) = 0$ and $\varepsilon(\omega_A) = 1_A$ for any $A \in \operatorname{Ob} T$ is a counit with respect to μ. Then we define the *representations category* $R(T)$ of the DGC T (or *representations category* $R(U)$ of *(Q)DGC* U).

1. The *representations of DGC T* are functors $F : K \longrightarrow \text{mod}-\Bbbk$.
2. The space of *morphisms* $R(T)(F, G)$ is defined as

$$\text{Hom}_{K-K}(V, \text{Hom}_{\Bbbk}(F, G)).$$

3. For $f \in R(T)(F, G)$, $g \in R(T)(G, H)$ the *composition* $g \cdot f$ in $R(T)$ is defined as the composition:

$$V \xrightarrow{\mu} V \otimes_K V \xrightarrow{g \otimes f} \text{Hom}_{\Bbbk}(G, H) \otimes_K \text{Hom}_{\Bbbk}(F, G) \xrightarrow{m_{\Bbbk}} \text{Hom}_{\Bbbk}(F, H). \quad (1)$$

Here $m_{\Bbbk}$ is the composition in $\text{mod}-\Bbbk$. Such category is not always Krull-Schmidt, but in [37], [46] is given the condition *triangularity*, that ensures $R(T)$ to be Krull-Schmidt, see 20.3.6.

20.3.3 Example: Representations of Quiver via DGC We define in terms of DGC the category of representations of a quiver Γ. We set $\deg(\Gamma_1) = \{0\}$, $U = \Bbbk[\Gamma]$ and $\delta = 0$. The differential $d : T \longrightarrow T$ $(T = \Bbbk[\check{\Gamma}])$ is given by

$$d(\omega_{A_i}) = \omega_{A_i}^2, d(a) = \omega_{e(a)}a - a\omega_{s(a)}.$$

If $\mathfrak{A} = (\mathcal{C}, \mathcal{M})$ is a BP with trivial $\mathcal{C}$, then $\text{rep}\,\mathfrak{A}$ is equivalent to $R(T)$ for defined above T with $\Gamma = \Gamma_{\mathfrak{A}}$, 20.2.3.

20.3.4 Box Representations Axiomatizing of the cocategory properties of $V = T_1/(d(T_0))$ led to the notion of *a box* [46]. The box (bocs as in [46]) $\mathfrak{A} = (K, V, \mu, \varepsilon)$ (shortly $\mathfrak{A} = (K, V)$) consists of a category K, a K-bimodule V and K-bimodule morphisms $\mu : V \longrightarrow V \otimes_K V$, $\varepsilon : V \longrightarrow K$, where μ is a coassociative comultiplication and ε is a counit with respect to μ. $\mathfrak{A}$ is a "corepresenting" object for a category $\text{mod}-\mathfrak{A}$, which

1. *objects* are functors $F : K \longrightarrow \text{mod}-\Bbbk$,
2. *morphisms* are defined by $\text{Hom}_{\mathfrak{A}}(X, Y) = \text{Hom}_{K-K}(V, \text{Hom}(X, Y))$;
3. the *composition* is defined by the formula 1 from 20.3.2.

The associativity of the composition follows from the coassociativity of μ and the counit ε produces unit morphisms. A *morphism* $\Phi = (F_0, F_1) : \mathfrak{A} = (K, V, \mu, \varepsilon) \longrightarrow \mathfrak{B} = (L, W, \nu, \epsilon)$ between boxes consists of a functor $F_0 : K \longrightarrow L$ and a K-coalgebra morphism $F_1 : V \longrightarrow W$, where W is endowed with a K-coalgebra structure by the L-coalgebra structure and the functor F_0. The morphism Φ induces an inverse image functor on the representations categories $\Phi^* : \text{mod}-\mathfrak{B} \longrightarrow \text{mod}-\mathfrak{A}$.

In applications arise a modification of the notion of box, namely, *completed box*, (see [24]). If K is a completed category (e.g. in the topology, associated with the powers of an ideal I) and the bimodule V is completed, then the comultiplication can be defined as $\mu : V \longrightarrow V \hat{\otimes}_K V$, where $\hat{\otimes}_K$ is a completed tensor product. After obvious changes in the coassociativity property etc. we can set the objects of the representation category as the continuous functors $F : K \longrightarrow \mathrm{mod}-\Bbbk$ in the discrete topology on $\mathrm{mod}-\Bbbk$ (for I it means $F(I^N) = 0$ for big enough N) and the morphisms as above. Completed boxes arose from noncompleted by the regularizations (see 20.4.4, 2, no. 4).

20.3.5 Examples

1. *The category of the functors in* $\mathrm{mod}-\Bbbk$ *is the category of a box representation.* Let K be a category. The box $\mathfrak{K} = (K, K)$ is called a *principal* box over K. The conjugated associativity proves, that $\mathrm{mod}-K \simeq \mathrm{mod}-\mathfrak{K}$.

2. *Arising of a nonprincipal box (Drozd).* This example shows, how a box can arise directly from associative algebras. Let $f : A \longrightarrow B$ be a homomorphism of algebras, $f^* : \mathrm{mod}-B \longrightarrow \mathrm{mod}-A$ be the induces functor. We will see in 20.4.2, that $\mathrm{Im}\, f^*$ is equivalent to $\mathrm{mod}-\mathfrak{A}^f$ for the box $\mathfrak{A}^f = (B, B \otimes_A B)$. If f is an epi, then $B \otimes_A B \simeq B$ and $\mathfrak{A}^f$ is a principal box $\mathfrak{A} = (B, B)$.

3. *Boxes and* Ext^1 *closed subcategories.* In [9] is shown (with some assumption of finite dimensionality on the box $\mathfrak{A}$), that $\mathrm{mod}-\mathfrak{A}$ can be included in $\mathrm{mod}-\Lambda$ for an algebra Λ as a full subcategory, closed with respect to extensions. The following statement seems to be one more reason showing, that the notion of a box should arise from associative algebras, [41].

   Let Λ *be a f.d algebra,* $\mathbf{M} \subset \mathrm{mod}-\Lambda$ *be a subcategory with* $\mathrm{Ob}\,\mathbf{M} = \{M_1, \dots, M_n\}$, $M = \oplus_{i=1}^n M_i$ *and* $E(\mathbf{M})$ *be a minimal subcategory in* $\mathrm{mod}-\Lambda$, *s.t.* $\mathbf{M} \subset E(\mathbf{M})$ *and* $E(\mathbf{M})$ *is closed with respect to extensions. Then* $E(\mathbf{M}) \simeq \mathrm{mod}-\mathfrak{A}$ *for a (completed) box* $\mathfrak{A} = (K, V, \mu, \varepsilon)$.

   In this case we describe $\mathfrak{A}$ more precise, see 20.4.8.

   (a) K is a completed category, generated over $\mathbb{L}_{\mathbf{M}}$ by the $\mathbb{L}_{\mathbf{M}}$-bimodule $E = \mathbb{D}\,\mathrm{Ext}^1_\Lambda(M, M)$.

   (b) There exists a $\mathbb{L}_{\mathbf{M}}$-bimodule morphism $\sigma : \mathbb{D}\,\mathrm{Ext}^2_\Lambda(M, M) \longrightarrow \hat{\mathbb{L}_{\mathbf{M}}}[E]$, s.t. $K \simeq \hat{\mathbb{L}_{\mathbf{M}}}[E]/(\mathrm{Im}(\sigma))$.

   (c) There exists a K-bimodule epimorphism

$$\pi : K \hat{\otimes}_{\mathbb{L}_{\mathbf{M}}} \mathbb{D}\,\mathrm{Hom}_{\Lambda_{\mathbf{M}}}(M, M) \hat{\otimes}_{\mathbb{L}_{\mathbf{M}}} K \longrightarrow V.$$

If all M_i are simple, then $\mathfrak{A}$ is principal. One more explanation of ambiguity boxes see 20.3.19, no. 7.

20.3.6 Free and Triangular Boxes In the next section we will expose so called reduction technique. It is applied usually in such cases, when the results can be described "almost combinatorically". The usual restriction here is to consider free and triangular boxes or DGC (see [37], [11]). For a box $\mathfrak{A} = (K, V, \mu, \varepsilon)$

the kernel of ε is called *a kernel* of $\mathfrak{A}$ and denoted by $\bar{V}$. $\mathfrak{A}$ is called *a box with a free kernel*, provided $\bar{V}$ is a free K-bimodule. We call any set $\Omega = \{\omega_A \in V(A,A) | A \in \operatorname{Ob} K, \varepsilon(\omega_A) = 1_A\}$ *a section* of $\mathfrak{A}$. If $\mu(\omega_A) = \omega_A \otimes \omega_A$ for each A, then we call this section *normal* and the corresponding box is also called *normal*, 20.3.1. A normal box with a free kernel $\mathfrak{A}$ is called *free*, provided K is a free category, the corresponding DGC also is free. We associate with $\mathfrak{A}$ its *bigraph* $\Gamma = (\Gamma_0, \Gamma_1)$, $\Gamma_1 = \Gamma^0 \sqcup \Gamma^1$, where Γ^0 is a set of free generators of the category K, Γ^1 is a set of free bimodule generators of $\bar{V}$ (the arrows from Γ^0 are called and painted *solid* and from Γ^1 *dotted*). The *Tits form* of $\mathfrak{A}$ for $x = (x_A)_{A \in \Gamma_0}$ defined as $\mathrm{q}_{\mathfrak{A}}(x) = \mathrm{q}_\Gamma(x)$. This form, as in 20.2.5, has an algebraic-geometric sense: if we fix the dimension $x = (x_A), A \in \operatorname{Ob} K$ and the basices in the spaces $F(A) = \Bbbk^{x_A}$, then $p(x) = \sum_{a \in \Gamma^0} x_{\alpha(a)} x_{\beta(a)}$ is the dimension of (affine) variety Π_x of $(x_{\alpha(a)} \times x_{\beta(a)})$ matrices, $a \in \Gamma_0$, representing all isoclasses in the dimension x, $g(x) = \sum_{\varphi \in \Gamma^1} x_{\alpha(\varphi)} x_{\beta(\varphi)}$ is the dimension of variety G_x of $(x_{\alpha(\varphi)} \times x_{\beta(\varphi)})$ matrices, $\varphi \in \Gamma_1$, (some of them) correspond to isomorphisms of the representations in the dimension x. In general G_x cannot be interpreted as an algebraic group and the isoclasses as orbits of an action of a group (see [46]). Anyway, assuming the finiteness of the number of the isoclasses in a dimension x and considering the representations with the isoclass, Zariski dense in Π_x, we conclude $\mathrm{q}_\Gamma(x) > 0$, (see [37]).

The box $\mathfrak{A}$ (and corresponding DGC U and T) is called *triangular*, provided there exists a filtration $\Gamma_1 = \Gamma_{10} \supset \Gamma_{11} \supset \dots \supset \Gamma_{1N-1} \supset \Gamma_{1N} = \emptyset$, s.t. $\delta(\Gamma_{1i})$ belongs to the subcategory in U, generated by Γ_{1i+1}, $i = 1, \dots, N-1$. For a triangular box $\mathfrak{A}$ the representations category mod $-\mathfrak{A}$ is Krull-Schmidt ([37],[46]). The box, correspondinging to a BP (we discuss it in 20.3.20, 2) is triangular ([19]). A free box $\mathfrak{A}$ with directed bigraph Γ is triangular.

Although the notion of box is very handy in applications, at most convenient way to describe a box consists of defining corresponding DGC . Usually the free boxes over an algebraically closed field supposed to be normal, but see [23].

20.3.7 D$\mathbb{Z}$GC as Method to describe nonfree MP's The free triangular boxes (and DGC) are very practicable MP's because of existence of good combinatorial reduction algorithms (see the next section for examples) and the possibility to use the quadratic forms technique. For many reasons it would be useful to have an analogue of such objects for the case of a non-free category. Let $\Gamma = (\Gamma_0, \Gamma_1)$ be a graph, s.t. $\Gamma_1 = \Gamma^{-1} \sqcup \Gamma^0 \sqcup \Gamma^1$. We denote by $U = \{U_i, i \in \mathbb{Z}\}$ the free $\mathbb{Z}$-graded $\Bbbk$-category, generated by Γ and assume there is given a differential $\delta : U \longrightarrow U$ of the degree 1, defined the D$\mathbb{Z}$GC structure on U and d on $T = \Bbbk[\check{\Gamma}]$, 20.3.1. Let I be the differential ideal,

$$I = (x, \delta(x), x \in \Gamma^{-1}).$$

Then $U' = U/I$ has the structure of a DGC , freely generated over the category $K = \Bbbk[\Gamma^0]/(I \cap \Bbbk[\Gamma^0])$ by the free K-bimodule $U_1/(U_1 \cap I)$ (with the generators Γ^1). We define the *category $R(U)$ (or $R(T)$) of representations of the D$\mathbb{Z}$GC U* as the category representations of DGC U' $R(U')$. The quadratic form for

D$\mathbb{Z}$GC is introduced as in 20.3.6: $\mathrm{q}_U = \mathrm{q}_T = \mathrm{q}_\Gamma$ and it is called *Brenner-Tits* quadratic form. In contrast to 20.3.6, for a dimension x the representations of U forms only the subvariety P_x in the matrix variety Π_x and this subvariety is given by $r(x) = \sum_{\varphi\in\Gamma^{-1}} x_{\alpha(\varphi)}x_{\beta(\varphi)}$ equations, that gives an estimate $\dim P_x \geq p(x) - r(x)$. There the analogue of the Lemma 20.1 holds for the form q_U. Remark, that in the Brenner-Tits form coexist the summands, corresponding to the objects of different nature: transformations, transformed matrices and relations and the reflections transformations from [39] "mix up" the relations and transformations. Remark, that for $R(U)$ there exist the reduction algorithms, but it works well only by some restrictions (f.e. in the case of directed Γ).

20.3.8 Example: Representations of Quivers with Relations via D$\mathbb{Z}$GC Let $\Gamma = (\Gamma_0, \Gamma_1)$ be a quiver, $I \subset \Bbbk[\Gamma]$ be a twosided ideal, generated by

$$r_i : A_i \longrightarrow B_i, r_i \in \Gamma, \Bbbk[\Gamma]\Gamma, i = 1, \dots, N, \Lambda = \Bbbk[\Gamma]/I.$$

$\mathrm{mod}-\Lambda$ is called the category of representations of the *quiver* Γ *with the relations* $\{r_i\}$. To describe it in terms of D$\mathbb{Z}$GC we introduce the graph Ω, $\Omega_0 = \Gamma_0$, $\Omega_1 = \Omega^{-1} \sqcup \Omega^0$, where $\Omega^0 = \Gamma_1$, $\Omega^{-1} = \{\varphi_i : A_i \to B_i, i = 1, \dots, N\}$ and introduce the differential δ on $U = \Bbbk[\Omega]$ by $\delta(x) = 0$ for $x \in \Gamma_1$, $\delta(\varphi_i) = r_i, i = 1, \dots, N$. The form q_U is known as the *Brenner form* of Λ, [7].

20.3.9 D$\mathbb{Z}$GC and $A(\infty)$-Structures The definition of a category of representation, based on the $A(\infty)$-structure was given in [40]. At this stage we use techniques from algebraic topology, see [35], [48] (in these papers are considered $A(\infty)$-(co)algebras, but the transition to the categories language is standard). An *$A(\infty)$-category* M over a category k is a $\mathbb{Z}$-graded k-bimodule $M = \{M_i\}_{i\in\mathbb{Z}}$ with a family of k-bimodule morphisms $m_i : M^{\otimes i} \to M$, $i \geq 1$, $\deg\ m_i = i - 2$ s.t. for all $i \geq 1$ holds

$$\sum_{k=0}^{i-1}\sum_{j=1}^{i-k}(-1)^k m_{i-j+1}(\hat{1}^{\otimes k} \otimes m_j \otimes 1^{\otimes(i-j-k)}) = 0.$$

We assume the existence of identities $1_A \in M_0(A, A)$, $A \in \mathrm{Ob}\,\mathsf{k}$, s.t. for every $x : A \longrightarrow B$ hold $m_2(1_B \otimes x) = x, m_2(x \otimes 1_A) = \hat{x}$, so there exists a notion of an isomorphism with respect to the binary operation m_2. We set also $\mathrm{Ob}\,M = \mathrm{Ob}\,\mathsf{k}$. A *morphism or $A(\infty)$-functor* $f : M \to M'$ of $A(\infty)$-categories over k consists of a family $\{f_i\}$, $i \geq 1$ of k-bimodule morphisms $f_i : M^{\otimes i} \to M'$ of degree $i - 1$, s.t. for every $n \geq 1$ holds

$$\sum_{k=0}^{i-1}\sum_{j=1}^{i-k}(-1)^k f_{i-j+1}(\hat{1}^{\otimes k} \otimes m_j \otimes 1^{\otimes(i-j-k)}) = \sum_{t=1}^{n}\sum_{\vec{k}\in S(n,t)} m_t(f_{k_1} \otimes \ldots \otimes f_{k_t}),$$

where $S(n, t)$ is the set of all partitions $\vec{k} = (k_1, \dots, k_t)$ of n in a sum of t positive summands, $|\ \vec{k}\ | = n$. f is an isomorphism, iff f_1 is an k-bimodule isomorphism.

A $\mathbb{Z}$-graded bimodule $N = \{N_i\}_{i\in\mathbb{Z}}$ is called $A(\infty)$-*cocategory over* k, provided there are given k-bimodule morphisms

$$\{n_i\}, i \geq 1, n_i : \longrightarrow N^{\hat{\otimes} i}, n_i(N_k) \subset (N^{\hat{\otimes} i})_{k-i+2},$$

where the n_i satisfy relations, 'dual' to the relations in the definition of an $A(\infty)$-category. To describe these relations we denote by $T = \mathsf{k}[\hat{N}]$ the completed tensor category of the bimodule N over the category k.[3]. We endow T with a grading $\deg_T$, s.t. $\deg_T \mathsf{k} = 0$, $\deg_T N_i = i+1$. Then $\{n_i\}_{i\geq 1}$ define a structure of an $A(\infty)$-*cocategory* on N if and only if the differential $d : T \longrightarrow T$, $d(\mathsf{k}) = 0$, $\deg_T d(N_i) = i+2$, given on N as $d|_N = \sum_{i=1}^{\infty} n_i$ and extended on T by k-linearity, the Leibniz rule and continuity, satisfies the condition $d^2 = 0$. The D$\mathbb{Z}$GC T is called *the co-B-construction* of N. A *morphism or* $A(\infty)$-*functor* between two $A(\infty)$-cocategories over k is defined as a graded k-functor between the corresponding co-B-construction, commuting with the differentials. The condition $d^2 = 0$, gives us among other, that $\{N_i\}$ with respect to $n_1 : N_i \longrightarrow N_{i+1}$, $i \in \mathbb{Z}$ form a complex and if either $n_1 = 0$ or $n_3 = 0$, then n_2 is a graded coassociative comultiplication (and the dual conditions hold in $A(\infty)$-category case). If only finitely many n_i are nonzero and $n_i(N) \subset N^{\otimes i}$, then above we can change $T = \mathsf{k}[\hat{N}]$ to $T = \mathsf{k}[N]$ and say, that N allows non-completed co-B-construction.

20.3.10 $A(\infty)$-Structures over Trivial Category Let us suppose the category k is trivial. Then for an $A(\infty)$-category $M = \{M_i\}, i \in \mathbb{Z}, \{m_i\}, i \geq 1$ the family of k-bimodules $N = \{N_i\}, i \in \mathbb{Z}$, $N_i(A,B) = \mathbb{D}M_i(A,B)$, $A, B \in \operatorname{Ob}\mathsf{k}$ with respect to the family of k-bimodule morphisms $n_i = \mathbb{D}m_i : \mathbb{D}M \longrightarrow \mathbb{D}M^{\otimes i}, i \geq 1$ forms an $A(\infty)$-cocategory, that is called the $A(\infty)$-cocategory *dual to* M. Analogously, for an $A(\infty)$-cocategory over k we construct the *dual* $A(\infty)$-*category*, s.t. in the f.d. case both dualities are mutual inverse.

If all $N_i(A,B)$ are f.d., then to N corresponds its *graded graph* of N $\Gamma = \Gamma_N$, $\Gamma_0 = \operatorname{Ob}\mathsf{k}$ and $|\Gamma^i(A,B)| = |N_{i-1}(A,B)|_{\Bbbk}$. If there are only finitely many nonzero $N_i(A,B)$, then the *Euler form* of N (the same of T) on $x = (x_A)_{A\in\operatorname{Ob}\mathsf{k}}$ is defined (according to 20.1.5)

$$\mathrm{q}_T(x) (= \mathrm{q}_N(x) = \mathrm{q}_\Gamma(x)) = \sum_{A,B\in\operatorname{Ob}\mathsf{k}} \sum_{i\in\mathbb{Z}} (-1)^{i+1} |N_i(A,B)|_{\Bbbk} x_A x_B.$$

The notion of an $A(\infty)$-functor obviously extends for $A(\infty)$-categories M and M' over the trivial categories k and k' correspondingly. Such $f : M \longrightarrow M'$ we call $A(\infty)$-*equivalence*, provided f_1 maps bijectively $M(A,B)$ onto $M'(A,B)$ and every $X \in \operatorname{Ob}\mathsf{k}'$ is isomorphic to some $f(A), A \in \operatorname{Ob} K$.

[3]For us will be convenient to consider categories, that are completed in some topology, mostly by the degrees of the ideal. Technically, the aim is to take sense for some infinite formal sums.

20.3.11 $A(\infty)$**-Structures in Representations Theory**

1. A free $D\mathbb{Z}GC$ $T = \Bbbk[\check{\Gamma}]$ define on $\Bbbk^{\check{\Gamma}}$ the structure of $A(\infty)$-cocategory over $\mathbb{L}_\Gamma$.

2. A BP with differential $\mathcal{A} = (\mathcal{C}, \mathcal{M}, \delta)$ transforms in an $A(\infty)$-category over $\mathbb{L}_\mathcal{C}$ by setting

$$M_0 = \mathcal{C}, M_{-1} = \mathcal{M}, M_i = 0 \text{ for } i \neq 0, -1, m_1 = -\delta, m_2(x \otimes y) = 0,$$

provided $x, y \in M_{-1}$, otherwise for $x, y \in M_{-1}, M_0$ $m_2(x \otimes y) = \hat{x}y$ and $m_i = 0$ for $i \geq 3$.

3. Let $\mathfrak{A} = (K, V, \mu, \varepsilon)$ be a box. Then $\mathfrak{A}$ is an $A(\infty)$-cocategory over K with only nonzero component of the degree 0 and with the unique nonzero operation $n_2 = (\hat{1}_V \otimes 1_V)\mu$.

4. [48] Let $0 \xleftarrow{} (V \xleftarrow{\pi}) \mathcal{P}_0 \xleftarrow{\delta} \mathcal{P}_{-1} \xleftarrow{\delta} \dots$ be a projective K-bimodule resolution of V. Then $\mathcal{P} = \{\mathcal{P}_i\}_{i \in \mathbb{Z}}, \mathcal{P}_i = 0, i > 0$, has a structure of an $A(\infty)$-cocategory over K, $\{\mu_i : \mathcal{P} \longrightarrow \mathcal{P}^{\otimes i}\}, i \geq 1$. It is defined by setting $\mu_1 = \delta$, μ_2 is a lifting of μ to the resolution, and $f : \mathcal{P} \longrightarrow \mathfrak{A}$ s.t. $f_1|_{\mathcal{P}_0} = \pi, f_1|_{\mathcal{P}_i} = 0, i \neq 0, f_i = 0, i \neq 0$, is a morphism of $A(\infty)$-categories.

20.3.12 Derived $A(\infty)$**-Cocategory** This construction is a modification of [40]. We start from an $A(\infty)$-cocategory N over k and set $\boldsymbol{K} = T/(d(T))$. For $x \in T$ the image of x in $\boldsymbol{K}$ we denote also x. The category $\boldsymbol{K}$ inherits the degree $\deg_{\boldsymbol{K}}$ from T. Let $[N]_i$ be the copies of N, parametrized by $i \in \mathbb{Z}$, $\mathcal{N}_i$ be the $\boldsymbol{K}$-bimodule $\boldsymbol{K} \hat{\otimes}_\mathsf{k} [N]_i \hat{\otimes}_\mathsf{k} \boldsymbol{K}$. Then $\mathcal{N}_i$, $i \in \mathbb{Z}$ form the $\mathbb{Z}$-graded $\boldsymbol{K}$-bimodule

$$\mathcal{N}(= \mathcal{N}_N) = \{\mathcal{N}_i\}_{i \in \mathbb{Z}} \deg_{\mathcal{N}}([x]_i) = i.$$

$\mathcal{N}$ has the natural structure of $A(\infty)$-cocategory over $\boldsymbol{K}$ $\nu_i : \mathcal{N} \longrightarrow \mathcal{N}^{\hat{\otimes} i}, i \geq 1$, described by endowing the co-B-construction $\mathcal{T} = \boldsymbol{K}\hat{[}\mathcal{N}]$ with a differential $\boldsymbol{\partial}$. To introduce $\boldsymbol{\partial}$, we define first for all $j \in \mathbb{Z}$ a k-bimodule map $\Delta_j = \Delta_j(N)$, $\Delta_j : T \to \mathcal{T}$, $j \in \mathbb{Z}$, s.t. $\Delta_j \mid_\mathsf{k} = 1_\mathsf{k}$, for $x \in N$ hold $\Delta_j(x) = x + \sum_{i \in \mathbb{Z}} (-1)^{(j+1)i} [x]_i$ and $\Delta_j(ab) = \Delta_j(a)\Delta_{j + \deg_T a}(b), a, b \in T$. The last equality allows to extend Δ_n from the elements $[x]_i$, $x \in N$, $i \in \mathbb{Z}$ on the monomials from T and by 'continuity' on all T. It follows easily

Lemma 20.2. *There exists unique* $\boldsymbol{K}$*-linear continuous differential* $\boldsymbol{\partial} : \mathcal{T} \to \mathcal{T}, \deg \boldsymbol{\partial} = 1$, *satisfying Leibniz rule,* $\boldsymbol{\partial}^2 = 0$ *and s.t. for any* $n \in \mathbb{Z}$ *holds* $\Delta_{n+1} d = (-1)^n \boldsymbol{\partial} \Delta_n$.

For $x \in N$ the equality $\Delta_{n+1} d = (-1)^n \boldsymbol{\partial} \Delta_n$ allows to define $\boldsymbol{\partial}([x]_i)$, then we extend $\boldsymbol{\partial}$ on $\mathcal{T}$ by Leibniz formula. $\Delta_{n+1} d = (-1)^n \boldsymbol{\partial} \Delta_n$ is proved for a monomial $a \in T$ by induction by t, s.t. $a \in \mathcal{N}^{\otimes t}$. Then

$$\boldsymbol{\partial}^2([x]_i) = (\boldsymbol{\partial}^2 \Delta_1(x))_{(i+2)} \text{ and } \boldsymbol{\partial}^2 \Delta_1(x) = -\boldsymbol{\partial} \Delta_2 d(x) = -\Delta_1 d^2(x) = 0.$$

20.3.13 Remark and Notation Starting from (k, N, T, d) we obtain a *derived $A(\infty)$-structure* $(\boldsymbol{K}, \mathcal{N}, \mathcal{T}, \boldsymbol{\partial})$. But we can start from $(\boldsymbol{K}, \mathcal{N}, \mathcal{T}, \boldsymbol{\partial})$ as well, moreover, iterating this construction, we obtain the sequence $(\mathsf{k}^{(i)}, N^{(i)}, T^{(i)}, d^{(i)})$, s.t. $(\mathsf{k}^{(0)}, N^{(0)}, T^{(0)}, d^{(0)}) = (\mathsf{k}, N, T, d)$, $(\mathsf{k}^{(1)}, N^{(1)}, T^{(1)}, d^{(1)}) = (\boldsymbol{K}, \mathcal{N}, \mathcal{T}, \boldsymbol{\partial})$ etc. To construct the representations we start from a trivial category k, 20.3.14, but the reduction procedure "shifts back" this sequence, see 20.4.7.

We endow $\{\mathcal{N}[i]\}_{i\in\mathbb{Z}}$ with an extra graduation $\mathrm{dg}_{\mathcal{N}}$ by setting $\mathrm{dg}_{\mathcal{N}}[x]_i = \deg_N x + i \,(= \deg_T x + i - 1)$, $x \in N$, $i \in \mathbb{Z}$, assuming $\mathrm{dg}_{\mathcal{N}} f = \deg_{\boldsymbol{K}} f$, $f \in \boldsymbol{K}$ and we endow $\mathcal{T}$ by $\mathbb{Z}^2$-graduation $\mathrm{dg}_{\mathcal{T}}$, namely

$$\mathrm{dg}_{\mathcal{T}}\, f = (\deg_{\boldsymbol{K}} f, 0), \mathrm{dg}_{\mathcal{T}}[x]_i = (deg_{\boldsymbol{K}} x - i, i).$$

20.3.14 $A(\infty)$-Category of Representations $\mathsf{A}(T)$ Let us suppose k is trivial. We define *$A(\infty)$-category of representations* $\mathsf{R}(T)$, s.t its binary operations in $\mathsf{R}(T)$ is graded associative, so (20.1.3) we can consider $\mathsf{R}(T)$ as a category. Firstly we define $A(\infty)$-category $\mathsf{A}(T), \{\mu_i\}_{i\geq 1}$. Then

1. *Objects of the $A(\infty)$-category* $\mathsf{A}(T)$ are $\mathbb{Z}$-graded functors
$$F : \boldsymbol{K} \longrightarrow \mathrm{mod}_{\mathbb{Z}} - \mathbb{k}, {}^4$$
so $\mathsf{A}(T)$ is $A(\infty)$-category *over the trivial category* $\mathbb{L}_{\mathsf{A}(T)}$ with the of objects $\mathrm{mod}\,_{\mathbb{Z}}\boldsymbol{K}$.

2. For $F, G \in \mathrm{Ob}\,\mathbb{L}_{\mathsf{A}(T)}$ *the i-th component* $\mathsf{A}(T)(F,G)_i$ *is*
$$\mathrm{Hom}_{\boldsymbol{K}-\boldsymbol{K}}(\mathcal{N}[i], \mathrm{Hom}_{\mathbb{k}}(F,G)),$$
where $\mathrm{Hom}_{\boldsymbol{K}-\boldsymbol{K}}(\mathcal{N}[i], \mathrm{Hom}_{\mathbb{k}}(F,G))$ is the space of graded $\boldsymbol{K}$-bimodule morphisms, provided $\mathcal{N}[i]$ has the degree $\mathrm{dg}_{\mathcal{N}}$ and for $A, B \in \mathrm{Ob}\,\boldsymbol{K}$ the degree of $\mathrm{Hom}_{\mathbb{k}}(F(A)_{k_1}, G(B)_{k_2})$ is $k_2 - k_1$.

3. The *operation* $\{\mu_i\}, i \geq 1$ for
$$\begin{aligned}
&f_j \in \mathsf{A}(T)(F_j, F_{j+1})_{n_j},\\
&f_j : \mathcal{N}[n_j] \longrightarrow \mathrm{Hom}_{\mathbb{k}}(F_j, F_{j+1}), j = 1, \dots, i, n = n_1 + \dots + n_i + (i-2),\\
&\mu_i(f_i \otimes \dots \otimes f_1)\ (\mu_i : \otimes_{j=1}^{i} \mathsf{A}(T)(F_j, F_{j+1})_{n_j} \longrightarrow \mathsf{A}(T)(F_1, F_{i+1})_n)
\end{aligned}$$
is defined (compare with the box case 20.3.2) as the composition
$$\begin{aligned}
&\mathcal{N}[n] \xrightarrow{\nu_i} \mathcal{N}[n_i] \otimes_{\boldsymbol{K}} \dots \otimes_{\boldsymbol{K}} \mathcal{N}[n_1] \xrightarrow{f_i \otimes \dots \otimes f_1}\\
&(F_i, F_{i+1})_{n_i} \otimes_{\mathbb{L}_{\mathsf{A}(T)}} \dots \otimes_{\mathbb{L}_{\mathsf{A}(T)}} (F_1, F_2)_{n_1} \xrightarrow{\mu_v} (F_1, F_{i+1})_n.
\end{aligned}$$
We denote by $(F_j, F_{j+1})_{n_j}$ $\mathrm{Hom}_{\mathbb{k}}(F_j, F_{j+1})_{n_j}$, $m_{\mathbb{k}}$ is the composition in $\mathrm{mod} - \mathbb{k}$. The axioms of $A(\infty)$-category for $\mathsf{A}(T)$ follows (as in 20.3.4) from $A(\infty)$-cocategory axioms for $\mathcal{N}$.

[4] Sometimes are useful other subcategories of $\mathrm{mod}\,_{\mathbb{Z}} - \mathbb{k}$, f.e. bounded, bounded from above, from below, etc.

20.3.15 Passage to Homologies The $A(\infty)$-category $\mathsf{A}(T)$ is easy to calculate, but the operation μ_2 is not graded associative, so we cannot speak about underlying category structure. Now we apply the following statement, [35], [48].

Theorem 20.2. *Let $M = \{M_i\}, i \in \mathbb{Z}$, $m'_j : M^{\otimes j} \longrightarrow M, j \geq 1$ be an $A(\infty)$-category over k, s.t. for all $i \in \mathbb{Z}$ the homologies with respect to m_1 $H_i(M)$ are free k-bimodules. Then*

1. *On the $\mathbb{Z}$-graded k-bimodule $H(M) = \{H_i(M)\}, i \in \mathbb{Z}$ exists the structure of an $A(\infty)$-category $\{m_i, i \geq 1\}$, s.t. $m_1 = 0$, m_2 is induced by m'_2 in the homologies and there exists a morphism of $A(\infty)$-categories $f : H(M) \longrightarrow M$, s.t. f_1 induces a homology isomorphism. Especially, m_2 is graded associative.*

2. *$A(\infty)$-structure on $H(M) = \{H_i(M)\}, i \in \mathbb{Z}$ satisfying 1, is unique up to an isomorphism.*

Remark 20.2.

1. If k is trivial, then the conditions of the theorem hold.

2. Analogous operation (a 'partial' transition to cohomologies) in the dual case of $A(\infty)$-cocategories in terms of DGC was introduced in [37] and called *regularization*, see 20.4.4, 2, no. 4. The *full regularization* in the case of trivial k was considered in [24]: if N is a $A(\infty)$-cocategory over a trivial category k with f.d. $N_i(A, B)$, then by N_{reg} we denote the $A(\infty)$-cocategory of the cohomology of N with respect to n_1 and by T_{reg} the corresponding co-B-construction.

3. It is remarked in [35], that if in M holds $m'_i = 0$ for all $i \geq 3$, there can exist an $j \geq 3$, s.t. $m_j \neq 0$. In the case of box it means that regularizations transforms linear boxes (DGC) in nonlinear, see 20.4.4, 2, no. 4.

20.3.16 An Example: Ext in the Box Case This example has an intermediate position between DGC representations category $R(T)$ and $A(\infty)$-category $\mathsf{R}(T)$. Let $\mathcal{P} = \{\mathcal{P}_i\}_{i \in \mathbb{Z}}, \mathcal{P}_i = 0, i > 0$, $\{\mu_i\}, i \geq 1$ be the $A(\infty)$-cocategory over K, constructed in 20.3.11 for a box $\mathfrak{A} = (K, V, \mu, \varepsilon)$. For a subcategory $\mathbf{M} = \{M_1, \ldots, M_k\} \subset \text{mod}-\mathfrak{A}$ we consider an $A(\infty)$-category $(\mathcal{E}, \{\mu_i\}) (= \mathcal{E}(\mathbf{M}))$ over $\mathsf{k} = \mathbb{L}_{\mathbf{M}}$, s.t. $\mathcal{E}_i(M_{k_1}, M_{k_2}) = \text{Hom}_{K-K}(\mathcal{P}_i, \text{Hom}_{\Bbbk}(M_{k_1}, M_{k_2}))$, $i \in \mathbb{Z}, k_1, k_2 = 1, \ldots, k$ and $\{\mu_i\}, i \in \mathbb{Z}$ are defined analogously to 20.3.14. The $A(\infty)$-category of homologies $\mathsf{E}(\mathbf{M}) = H(\mathcal{E}(\mathbf{M}))$ we call the $A(\infty)$-*category of* Ext*'s* of $\mathfrak{A}$ and, as in [35, 48], $\mathsf{E}(\mathbf{M})$ is defined uniquely up to an $A(\infty)$-isomorphism. This definition corresponds to the definition from [42] of Ext's in the category mod$-\mathfrak{A}$, namely $\mathsf{E}_{-i}(\mathbf{M})(M_{k_1}, M_{k_2}) = \text{Ext}^i_{\mathfrak{A}}(M_{k_1}, M_{k_2})$. If $\mathfrak{A}$ is a principal box over the spectroid K and $\mathbf{M}$ consists of representatives of isoclasses of all simples K-modules, then we denote $\mathsf{E}(\mathbf{M})$ by $\mathsf{E}(K)$ and call it $A(\infty)$-*category of the simple modules* of K.

20.3.17 $A(\infty)$**-Category** $\mathsf{R}(T)$ The Theorem 20.2, 20.3.15 allows us to define an $A(\infty)$-category with an (graded) associative binary operations. *We set* $\mathsf{R}(T) = H(\mathsf{A}(T))$, *where* $H(\mathsf{A}(T))$ *is the homology* $A(\infty)$*-category*, chosen following theorem above.

The only additional condition is, that corresponding $f_1 : \mathsf{R}(T) \longrightarrow \mathsf{A}(T)$ 'commutes' with natural shift of the graduation. The i-th *slice* of $\mathsf{R}(T)$ $\mathsf{R}(T)[i]$ we define as full $A(\infty)$-subcategory, formed by the objects $F : \boldsymbol{K}_{\mathbb{Z}} \longrightarrow \mathrm{mod} - \Bbbk$, s.t. $F(X[j]) = 0$ for all $X \in \mathrm{Ob}\,\boldsymbol{K}, j \neq i$. The representations from $\mathsf{R}(T)[0]$ we call *observable representations.* Although $\mathsf{R}(T)$ transforms to a category, an important advantage of the $A(\infty)$-category $\mathsf{A}(T)$ is the existence there of so called *cone* transformation. Let X, Y be two representations from $\mathsf{A}(T)$, $F \in \mathsf{A}(T)(X,Y)_0$ be a cocycle with respect to μ_1. Then the *cone* of F $Z = c_T(F) \in \mathrm{Ob}\,\mathsf{A}(T)$ is defined on the objects as $Z(A[k]) = X(A[k]) \oplus Y(A[k-1])$ and for $a : A \longrightarrow B$ $Z(a[k]) : X(A[k]) \oplus Y(A[k-1]) \longrightarrow X(A[k + \deg_T a]) \oplus Y(A[k - 1 + \deg_T a])$ is defined by a matrix

$$Z(a[k]) = \begin{pmatrix} X(a[k]) & 0 \\ F([\hat{a}]_0[k]) & Y(\hat{a})[k-1] \end{pmatrix}.$$

It is a part of more general construction. Let us consider $\boldsymbol{\partial} : \mathcal{T} \longrightarrow \mathcal{T}$ and $\mathcal{K} = \mathcal{T}/\boldsymbol{\partial}(\mathcal{T})$. Consider $c_T : T_{\mathbb{Z}} \longrightarrow \mathrm{Sum}(\mathcal{T}_{\mathbb{Z}^2})$, defined on the objects $c_T(A[k]) = \oplus_{p+q=k} A[p,q]$ and for $a \in N, a : A \longrightarrow B$ $c_T(a[k])$ is written as matrix with entries, indexed by $(\vec{v}, \vec{w})$, $|\vec{v}| = k$, $\vec{i} = (-i-1, i+1)$, $\vec{v} = (p,q)$. Then

$$c_T(a[k])_{\vec{v}, \vec{v} + \mathrm{dg}_T a} = (-1)^{q \deg_T a}(a + [a]_{-1})[p,q]$$

and for $i \neq -1$

$$c_T(a[k])_{\vec{v}, \vec{v} + \mathrm{dg}_T a + \vec{i}} = (-1)^{(q+i+1)\deg_T a}[a]_i[p,q].$$

Then $(\vec{v}, \vec{w})$-component-wise there holds $c_T d_{\mathbb{Z}}(a[k]) = (-1)^{q+i \deg_T a}(\mathcal{D}[a]_i)$, where $d_{\mathbb{Z}} : T_{\mathbb{Z}} \longrightarrow T_{\mathbb{Z}}$ is induced by d. It means especially, that c_T induced the functor $c_{\boldsymbol{K}} : \boldsymbol{K}_{\mathbb{Z}} \longrightarrow \mathrm{Sum}(\mathcal{K}_{\mathbb{Z}^2})$. On the category of $\mathcal{K}_{\mathbb{Z}^2}$ modules $c_{\boldsymbol{K}}$ induced "condensation" ([38]) and, especially, cone. It is checked easily, that isoclass of cone in $\mathsf{R}(T)$ is well defined. The cone can be used for introducing a structure of triangulated category on $\mathsf{R}(T)$, [49].

20.3.18 Remarks The category $\mathsf{R}(T)$, called in [40] *derived category of the representations category*, has many nice features. It looks like "MP" (BP, DGC, box). The algorithms, developed in this language are as effective as algorithms in the case of free triangular boxes. The quadratic form for MP's, especially, Tits, Brenner and homological forms ([44]) can be interpreted and used in $\mathsf{R}(T)$. It gives us "a corepresenting object" for a "derived category of a representation category" in a way, in which a box is "corepresenting object" for its representations category. This technique includes the cases of an associative algebra and free box in an uniform way (see examples below) and there exists a natural interpretation of the operations, connecting boxes and algebras (reflections ([41]), small reductions etc.), namely these operation are "a substitution of some matrix by its special forms". Any arrow from Γ can correspond to 'transformed' (in K), to 'transforming' (in $\mathcal{N}$) matrices and to "relations" (in sense 20.3.7). We try to show, how it is reached using described language. Quadratic forms allows to check, how good works our definitions (see Section 20.5).

20.3.19 Examples

1. *Triangular $D\mathbb{Z}GC$.* As in case of box, there is technically very convenient the assumption of *normality and triangularity* of the corresponding $D\mathbb{Z}GC$ T, both definitions are the same as in 20.3.6. Under these assumptions in the category $\mathsf{R}(T)$ for a representations X the unit morphism of X is defined by cycle 1_X, $1_X([\omega_A]_0) = 1_{X(A)}$, 1_X is zero on other generators, all idempotents in $\mathsf{R}(T)$ splits and, if for a cycle $F \in \mathsf{A}(T)(X,Y)_0$ all linear maps $F([\omega_A]_0)$ are invertible, then corresponding class in $\mathsf{R}(T)$ is an invertible morphism. These properties make the category $\mathsf{R}(T)$ similar to the category of triangular box or DGC representation in sense [37, 46, 19]. But in order to make this similarity stronger, we should consider so called normalized representations. $D\mathbb{Z}GC$ we will consider will be usually normal and triangular.

2. *Dimension.* In $\mathsf{R}(T)$ the dimension of representation (as the functor) *is not an invariant of an isoclass.* The right analogue of the notion dimension in the category $\mathsf{R}(T)$ is the function $\mathrm{DIM}(=\mathrm{DIM}_{\mathsf{R}(T)}) : \mathsf{R}(T) \longrightarrow \mathbb{Q}^{\mathrm{Ob}\,\boldsymbol{K}_{\mathbb{Z}}}$, where $(\mathrm{DIM}\,X)_{A[i]}$ is equal the dimension of $|H^i(X(A),\omega_A)|_{\Bbbk}$, that means the dimension of the cohomology of the complex $(X(A[j]), j \in \mathbb{Z}$ with respect to $X(\omega_A)$. If T is triangular, then $\mathrm{DIM}\,X$ is an invariant of isoclasses. The *observable dimension* $\dim X \in \mathbb{Q}^{\Gamma_0}$ is defined as $(\dim X)_A = \sum_{i\in\mathbb{Z}}(-1)^i(\mathrm{DIM}\,X)_{A[i]}$, provided $\mathrm{DIM}\,X$ has for every A finitely many nonzero $(\mathrm{DIM}\,X)_{A[i]}$. Obviously holds

$$< \dim X, \dim Y >_\Gamma = \sum_{i\in\mathbb{Z}}(-1)^i|\mathsf{R}_i(X,Y)|_{\Bbbk} = \sum_{i\in\mathbb{Z}}(-1)^i|\mathsf{A}_i(X,Y)|_{\Bbbk},$$

   if the sums contain finitely many nonzero summand (compare [44, 43]).

   There exist in $\mathsf{R}(T)$ the analogues of main projectives and injectives (the objects, "reflecting dimension"). For a normal T for every $A \in \mathrm{Ob}\,T, j \in \mathbb{Z}$ we define $P_A[j] \in \mathsf{R}(T)$, such that $\mathsf{R}(T)(P_A[j],V)_0 = H^j(V(A),\omega_A)$. $P_A[j]$ is defined as follows: $P_A[j](B[i]) = U_{i-j}(A,B)$, the morphisms from U acting on $P_A[j]$ in a usual way and for $x \in U(A,B)$ we set $\omega_B x = -\delta(x)$. The projectives (for simplicity in the layer 0) forms $A(\infty)$-category over $\mathbb{L}_\Gamma$. Up to homology it could be described in the terms of U: on the opposite category U° is set

$$\begin{aligned} U_i^\circ = U_{-i}, i \in \mathbb{Z}, m_1(u) =& \delta(u)^\circ, \\ m_2(u_1^\circ \otimes u_2^\circ) =& (-1)^{\deg_T u_1 \deg_T u_2}(u_2 u_1)^\circ, m_i = 0 \end{aligned}$$

for $i \geq 3$. Using obvious duality one can introduce analogues of main injectives $Q_A[j]$. It leads to the notion of translations Auslander-Reiten, etc. in style of [33].

3. *Normalized representation.* We call a representation $X \in \mathsf{R}(T)$ *normalized* if $X(\omega_A) = 0$ for all $A \in \operatorname{Ob} T$. We denote by $\mathsf{R}_n(T)$ the $A(\infty)$-subcategory of the normalized representations. In the triangular case *for every* $X \in \mathsf{R}(T)$ *there exists* $Y \in \mathsf{R}_n(T)$, *s.t.* $X \simeq Y$. Of course, on $\mathsf{R}_n(T)$ the dimension in the usual sense and $\mathrm{DIM}_{\mathsf{R}(T)}$ coincide. From the point of view of calculations, this category is a graded variant of a category representations of weak box (see below). The representation $X \in \mathsf{R}(T)$ we call *pure*, provided there exists $Y \in \mathsf{R}_n(T)$, s.t. $X \simeq Y$ and for every $A \in \operatorname{Ob} T$ there exists at most one $i \in \mathbb{Z}$, such that $Y(A)[i] \neq 0$.

4. *Free box.* If $\mathfrak{A} = (K, V, \mu, \varepsilon)$ is a free triangular box and T is its DGC , then as the category $\mathsf{R}(T)$ for the corresponding *DGC T* can be described in the terms of $R(T)$. Let $\mathrm{Com}_{R(T)}$ be the category of complexes over $R(T)$, $\mathrm{Hot}_{R(T)}$ be its homotopic category, $\mathcal{P}^{\cdot} = \ldots \longrightarrow 0 \longrightarrow \mathcal{P}_{-1} \xrightarrow{\partial} \mathcal{P}_0 \longrightarrow 0 \longrightarrow \ldots$ be a K-bimodule free resolution of V (the construction of one of them in the terms of bigraph and differential see [41]). The comultiplication $\mu : V \longrightarrow V \otimes_K V$ induces unique up to homotopy morphism of the complexes $\mu_{\mathcal{P}} : \mathcal{P}^{\cdot} \longrightarrow \mathcal{P}^{\cdot} \otimes_K \mathcal{P}^{\cdot}$. Then $\operatorname{Ob} \mathsf{R}(T)$ canonically identifies with $\operatorname{Ob} \mathrm{Com}_{R(T)}$. For $X : \boldsymbol{K}_{\mathbb{Z}} \longrightarrow \mathrm{mod}-\mathbb{k}$ the spaces $X[i] = X|_{A[i], A \in \operatorname{Ob} \boldsymbol{K}}$ together with $X(f[i]), f \in K$ forms a representation $X_i \in R(T)$ and the set of $\mathbb{k}$-linear maps $X(\varphi) : X_A[i] \longrightarrow X_B[i+1], \varphi \in \check{\Gamma}^1(A, B)$ define a K-bimodule morphism $x[i] : \mathcal{P}_0 \longrightarrow \mathrm{Hom}_{\mathbb{k}}(X_i, X_{i+1})$. The conditions $X(d(a)) = 0$ for all $a \in \Gamma^0$ imply, that $x[i]$ defines a morphism from $R(T)$ $x[i] : X_i \longrightarrow X_{i+1}$ and the conditions $X(d(\varphi)) = 0$ for all $\varphi \in \check{\Gamma}^1$ imply $x[i+1]x[i] = 0$. If $X^{\cdot}$ is the complex, corresponding to $X \in \mathsf{R}(T)$, then we show analogously, that $\mathsf{R}(T)(X, Y) = \mathrm{Hot}_{R(T)}(\mathcal{P}^{\cdot} \otimes_K X^{\cdot}, Y^{\cdot})$.

   If $f \in \mathsf{R}(T)(X, Y)$, $g \in \mathsf{R}(T)(Y, Z)$ and $\bar{f}, \bar{g}$ are their representatives in $\mathrm{Com}_{R(T)}(\mathcal{P}^{\cdot} \otimes_K X^{\cdot}, Y^{\cdot})$ and $\mathrm{Com}_{R(T)}(\mathcal{P}^{\cdot} \otimes_K Y^{\cdot}, Z^{\cdot})$ correspondingly, then gf is defined as the class of the composition

$$\mathcal{P}^{\cdot} \otimes_K X^{\cdot} \xrightarrow{\mu_{\mathcal{P}} \otimes 1_{X^{\cdot}}} \mathcal{P}^{\cdot} \otimes_K \mathcal{P}^{\cdot} \otimes_K X^{\cdot} \xrightarrow{1_{\mathcal{P}^{\cdot}} \otimes \bar{f}} \mathcal{P}^{\cdot} \otimes_K Y^{\cdot} \xrightarrow{\bar{g}} Z^{\cdot}.$$

   We will consider $R(T)$ included in $\mathsf{R}(T)$ on the zero slice.

5. *Spectroid.* If Λ is a spectroid (20.2.2), then it is an $A(\infty)$-category over $\mathbb{L}_{\Lambda}$. We consider $A(\infty)$-cocategory $\mathbb{D}\Lambda$, the corresponding co-B-construction T and $\mathsf{R}(T)$. The category $\mathsf{R}(T)$ is equivalent to the category of projective complexes over Λ with respect to homotopy. In fact, it is a modification of Drozd construction (20.2.3), especially the representations from $\mathsf{R}_n(T)$ correspond to such complexes $\mathcal{P} = \{P_i, \partial_1 : P_i \longrightarrow P_{i+1}, i \in \mathbb{Z}\}$, that $\partial_i(P_i) \subset \mathrm{Rad}(P_{i+1})$.

6. *$A(\infty)$-category of simple modules* $\mathsf{E}(K)$. For the defined in 20.3.16 above $A(\infty)$-category $\mathsf{E}(K)$ we set T the co-B-construction over $\mathbb{D}\mathsf{E}(K)$. Then

$\mathsf{R}(T)[k]$ is equivalent to mod $-K$. Let us suppose K has a finite homological dimension and consider only bounded representations. Then in this and previous cases $\mathsf{R}(T)$ is equivalent to corresponding derived category of bounded complexes. In our case there exists obvious functor $\imath : \mathrm{Com}_{\mathrm{mod}\,-K} \longrightarrow \mathsf{R}(T)$ with the image, dense in $\mathsf{R}(T)$. From this point of view an object $X \in \mathsf{R}_n(T)$ we can consider as homology of a complex C_X over K, s.t. $\imath(C_X) \simeq X$ with an extra structure, defining the isoclass of C_X in the derived category. These extra structure is a graded variant of the weak box (see below).

7. *Comeback to (weak) box.* In the case of free box we can return from the category $\mathsf{R}(T)$ to the category $R(T)$ by considering the representations, concentrated in one slice. In general case it leads us to a generalization of the notion of box. Let K be a $\Bbbk$-category, V be a K-bimodule, P be a free K-bimodule and are given the K-bimodule morphisms:

$$\begin{aligned}
&\mu : V \longrightarrow V \otimes_K V;\\
&h : V \longrightarrow P,\ g_r : P \longrightarrow V \otimes_K P;\\
&g_l : P \longrightarrow P \otimes_K V,\ g : P \longrightarrow V \otimes_K V \otimes_K V;\\
&\varepsilon : V \longrightarrow K,
\end{aligned}$$

$$\begin{aligned}
\text{s.t.: } 1)\quad & (h \otimes 1_V)\mu - g_l h = 0,\ (1_V \otimes h)\mu - g_r h = 0;\\
2)\quad & (\mu \otimes 1_V)\mu - (1_V \otimes \mu)\mu + gh = 0;\\
3)\quad & (\varepsilon \otimes 1_V)\mu = (1_V \otimes \varepsilon)\mu = 1_V.
\end{aligned}$$

We call $\mathfrak{A} = (K, V, P, \mu, h, g_l, g_r, g, \varepsilon)$ a *weak box*. Let us define the representations category $R(\mathfrak{A})$.

(a) *Representations* of $\mathfrak{A}$ are functors $F : K \longrightarrow \mathrm{mod}\,-\Bbbk$.

(b) *Morphisms* in $R(\mathfrak{A})$ for $F, G \in R(\mathfrak{A})$ we define as $R(\mathfrak{A})(F, G) = \mathrm{Coker}\ h^\star(F, G)$, where a map

$$h^\star(F,G) : \mathrm{Hom}_{K-K}(P, \mathrm{Hom}_\Bbbk(F,G)) \longrightarrow \mathrm{Hom}_{K-K}(V, \mathrm{Hom}_\Bbbk(F,G))$$

is induced by $h : V \longrightarrow P$.

(c) Consider $f \in R(\mathfrak{A})(F, G)$, $g \in R(\mathfrak{A})(G, H)$ and their representatives $\bar{f} \in \mathrm{Hom}_{K-K}(V, (F, G)_\Bbbk)$, $\bar{g} \in \mathrm{Hom}_{K-K}(V, (G, H)_\Bbbk)$. We define the *composition* $gf \in R(\mathfrak{A})(F, H)$ as the class of

$$\bar{g} \circ \bar{f} \in \mathrm{Hom}_{K-K}(V, \mathrm{Hom}_\Bbbk(F, H))$$

in $R(\mathfrak{A})(F, H)$, where $\bar{g} \circ \bar{f}$ is the composition

$$V \xrightarrow{\mu} V \otimes_K V \xrightarrow{\bar{g} \otimes \bar{f}} \mathrm{Hom}_\Bbbk(G, H) \otimes_K \mathrm{Hom}_\Bbbk(F, G) \xrightarrow{m_\Bbbk} \mathrm{Hom}_\Bbbk(F, H)$$

and $m_\Bbbk$ is the composition in mod $-\Bbbk$.

The conditions a) show, that the definition of the composition is correct. The condition b) shows, that the composition is associative and c) gives us for $F \in R(\mathfrak{A})$ the unit morphism $1_F \in R(\mathfrak{A})(F, F)$ as the class of $\Delta\varepsilon$, where $\Delta : K \longrightarrow \mathrm{Hom}_{\Bbbk}(F, F)$ is given by $\Delta(1_A) = 1_{V(A)}$, $A \in \mathrm{Ob}\ K$. In the case $P = 0$ the notion of the weak box turns into the box. *The category* $\mathsf{R}(T)[k]$, *especially the category of observables, is equivalent (as a category) to the category of representations of a weak box* $\mathfrak{A} = (K, V, P, \mu, h, g_l, g_r, g, \varepsilon)$. Namely, $K \subset \boldsymbol{K}$ is the subcategory, generated by N_{-1}, $V \subset \mathcal{N}_0/(\nu_1(\mathcal{N}_{-1}))$ is a K-bimodule, generated by $[N_0]_0$, $P \subset \mathcal{N}_1$ is a K-bimodule, generated by $[N_1]_1$ and the K-bimodule maps μ, h, g_l, g_r, g are induced by $\mathcal{D} : \mathcal{T} \longrightarrow \mathcal{T}$ (or ν_1, ν_2, ν_3). If T does not contain the generators of degree ≥ 2 (it is true for $A(\infty)$-cocategories, corresponding to free boxes and algebras), *then the category of observables is equivalent to the (completed, 20.3.4) box representations category.*

20.3.20 Transitions between Definitions of MP's

1. *Transition: "$A(\infty)$-category" $\Longrightarrow$ "$A(\infty)$-cocategory" $\Longrightarrow$ $D\mathbb{Z}GC$* . The first transition is described in 20.3.10.

   Let us suppose N is an $A(\infty)$-cocategory over a trivial category k. If only finitely many $i \geq$ are s.t. $n_i : N_i \longrightarrow N^{\otimes i}$ are nonzero, then the co-B–construction gives us a structure of D$\mathbb{Z}$GC on the free DGC on $T = \mathsf{k}[N]$, otherwise the structure of an completed free D$\mathbb{Z}$GC on $T = \mathsf{k}[\hat{N}]$. We obtain in the first case a DGC (box), if $N_i = 0$ for $i \neq -2, -1, 0$ and free DGC (free box) if $N_i = 0$ for $i \neq -1, 0$. Using the analogy between the notions $A(\infty)$-category and the category, we can motivate some notions, arising in the box and DGC representations theory ([37],[46]). A normality of a free box (and corresponding DGC) corresponds to the existence of an analogue for an $A(\infty)$-category over k M a "Levy decomposition" $M = \mathsf{k} \oplus \bar{M}$, $M_0 = \mathsf{k} \oplus \bar{M}_0$, $M_i = \bar{M}_i$ and the triangularity corresponds to the "nilpotency" of all m_i together on $\bar{M}$. We do not discuss here the opposite transition, that leads, analogously to [16], to the notion of "$A(\infty)$-category of matrices with coefficients in an $A(\infty)$-category".

2. *Transition:* "BP *(with differentiation)"* $\Longrightarrow$ *"DGC" ([19])*. This transition is a partial case of the (noncompleted) co-B–construction above, but it arises often in applications. Let $\mathfrak{A} = (\mathcal{C}, \mathcal{M}, \delta)$ be a BP. We define a DGC T, s.t. $R(T)$ is equivalent to mod $-\mathfrak{A}$. T is freely generated over $\mathbb{L}_{\mathcal{C}}$ by $\mathbb{L}_{\mathcal{C}}$-bimodules $\mathbb{D}(\mathcal{M})$ in the grade 0 and $\mathbb{D}(\mathcal{C})$ in the grade 1. If $L_r : \mathcal{C} \otimes_{\mathbb{L}_{\mathcal{C}}} \mathcal{M} \longrightarrow \mathcal{M}$, $R_r : \mathcal{M} \otimes_{\mathbb{L}_{\mathcal{C}}} \mathcal{C} \longrightarrow \mathcal{M}$ are the right and left actions and $m : \mathcal{C} \otimes_{\mathbb{L}_{\mathcal{C}}} \mathcal{C} \longrightarrow \mathcal{C}$ is the category multiplication, then the differential $d : T \longrightarrow T$ is given by $d|_{\mathbb{L}_{\mathcal{C}}} = 0, d|_{\mathbb{D}(\mathcal{M})} = \mathbb{D}\delta + \mathbb{D}R_r - \mathbb{D}L_r, d|_{\mathbb{D}(\mathcal{C})} = \mathbb{D}m$.[5]

---

[5] $\Gamma^0(A_i, A_j)$ in a quiver Γ forms a basis $\mathbb{D}\,\mathrm{Ext}^1_{\Bbbk\Gamma}(U_i, U_j)$ for the corresponding to A_i and A_j simples, 20.2.3.

To check, that T is triangular, we choose the filtration $\{S_i\}$, s.t. for the $\Bbbk$-vectorspaces $\Bbbk(S_{1i+1} \setminus S_{1i})$, generated by $S_{1i+1} \setminus S_{1i}$ holds:

$$\begin{aligned}\Bbbk(S_{01} \setminus S_{11}) =& \mathbb{D}(\mathcal{M}/\mathcal{M}_1),\\ \Bbbk(S_{i1} \setminus S_{i+11}) =& \mathbb{D}(\mathcal{M}_i/\mathcal{M}_{i+1}) \oplus \mathbb{D}((\operatorname{Rad}\mathcal{C})^i/(\operatorname{Rad}\mathcal{C})^{i+1}),\\ i =& 1, \dots, N,\end{aligned}$$

where $\mathcal{M}_i = \sum_{k=0}^{i} (\operatorname{Rad}\mathcal{C})^k \mathcal{M} (\operatorname{Rad}\mathcal{C})^{i-k}$ for big enough N.

3. *Transition: "DGC" $\Longrightarrow$ "box"*. It is described in 20.3.2.

4. *Transition: "box" $\Longrightarrow$ "DGC"*. Let $\mathfrak{A} = (K, V, \mu, \varepsilon)$ be a box. For any section Ω, we define a map $\delta(= \delta_\Omega) K \longrightarrow \bar{V}$, given for $f \in K(A_1, A_2)$ as $\delta(f) = f\omega_{A_1} - \omega_{A_2} f$ and for $v \in V(A_1, A_2)$ as $\delta(v) = \mu(v) - v\omega_{A_1} - \omega_{A_2} v$. Obviously $\delta(f) \in \bar{V}$, $\delta(v) \in \bar{V} \otimes_K \bar{V}$. Then (in notations of 20.3) we set $U_0 = K$, $U_1 = \bar{V}$, extend δ by Leibniz formula to $\delta : U \longrightarrow U$.

   Then for $x \in U(A_1, A_2)$ holds

$$\delta^2(x) = x(\mu(\omega_{A_1}) - \omega_{A_1} \otimes_K \omega_{A_1}) - (\mu(\omega_{A_2}) - \omega_{A_2} \otimes_K \omega_{A_2})x.$$

   So we obtain a $QDGC$ U and if Ω is normal, we obtain DGC, 20.3.1.

20.4 Reductions

20.4.1 Reductions in Language of BP's Let $\mathfrak{A} = (\mathcal{C}, \mathcal{M}, \delta)$ be a BP, $\mathcal{N} \subset \mathcal{M}$ be a subbimodule, $\pi : \mathcal{M} \longrightarrow \mathcal{M}/\mathcal{N}$ be the canonical projection, $\pi^{-1} : \mathcal{M}/\mathcal{N} \longrightarrow \mathcal{M}$ be a $\Bbbk$-linear section. Then $\bar{\delta} = \pi\delta$ is a differential, $\bar{\mathfrak{A}} = (\mathcal{C}, \mathcal{M}/\mathcal{N}, \bar{\delta})$ is a BP with a differential and we denote $\mathcal{L} = \operatorname{rep}\bar{\mathfrak{A}}$. Then the canonical inclusion $i : \mathcal{L} \longrightarrow \tilde{\mathcal{C}}$ endows $\tilde{\mathcal{N}}$ with a structure of $\mathcal{L}$-bimodule. We define a differentiation $d : \mathcal{L} \longrightarrow \tilde{\mathcal{N}}$ as follows: if $l \in \mathcal{L}((X, m), (Y, n))$, then $d(l) = \delta(l) + \pi^{-1}(n)l - l\pi^{-1}(m)$. It defines a BP with a differentiation $\mathfrak{B} = (\mathcal{L}, \tilde{\mathcal{N}}|_{\mathcal{L}}, d)$. Then there holds ([10])

Lemma 20.3. *The functor $\imath : \operatorname{rep}\mathfrak{B} \longrightarrow \operatorname{rep}\mathfrak{A}$ given on the objects by*

$$F : ((X, m), n) \mapsto (X, \pi^{-1}(m) + n)$$

and induced on the morphisms by the functor $\tilde{i} : \tilde{\mathcal{L}} \longrightarrow \tilde{\mathcal{C}}$ is an equivalence.

20.4.2 Reduction as a Base Change At most fruitful seems to be a description of *reductions as a base change* ([19]). Let $\mathfrak{A} = (K, V, \mu, \varepsilon)$ be a box, $F : K \longrightarrow L$ be a functor, then we define a new box $\mathfrak{A}^F = (L, V^F, \mu^F, \varepsilon^F)$, where $V^F = L \otimes_K V \otimes_K L$, μ^F and ε^F are induced by μ and ε. If $F_1 : V \longrightarrow V^F$ the K-bimodule morphism $v \overset{F_1}{\mapsto} 1 \otimes v \otimes 1$, then the pair $(F_0 = F, F_1)$ induces a morphism of boxes $\Phi : \mathfrak{A} \longrightarrow \mathfrak{A}^F$. The following theorem is from [19].

Theorem 20.3. *The induced by* Φ *functor*

$$\Phi^*(=F^*): \text{mod}-\mathfrak{A}^F \longrightarrow \text{mod}-\mathfrak{A}$$

is an equivalence on the full subcategory of mod $-\mathfrak{A}$ *of all functors*

$$M : K \longrightarrow \text{mod}-\mathbb{k},$$

which factorize through F.

The proof is an application of conjugated associativity in this special case.

20.4.3 Reduction via Amalgam The most important partial case is *a reduction of a subproblem,* see [19]. For the pair of functors $i : K' \longrightarrow K$ and $\sigma : K' \longrightarrow L'$ their amalgamation ([30]) $L = K \sqcup_{K'} L'$ is the category L with a pair of functors $F : K \longrightarrow L$, $j : L' \longrightarrow L$, s.t. the diagram

$$\begin{array}{ccc} K' & \xrightarrow{\sigma} & L' \\ {\scriptstyle i}\downarrow & & \downarrow{\scriptstyle j} \\ K & \xrightarrow{F} & L \end{array}$$

commutes and if $F_1 : K \longrightarrow L_1$, $j_1 : L' \longrightarrow L_1$ are s.t. $j_1\sigma = F_1 i$, then there exists a unique $\pi_1 : L \longrightarrow L_1$, s.t. $j_1 = \pi_1 j$ and $F_1 = \pi_1 F$. A construction of an amalgamation is described in [19]. It is used mostly in the case, when i is an inclusion of a subcategory K'. If K is free and L' is trivial, then in many important cases $L = K \sqcup_{K'} L'$ is easy to calculate.

Corollary 20.1. *For* F *as above the image of* $F^* : \text{mod}-\mathfrak{A}^F \longrightarrow \text{mod}-\mathfrak{A}$ *is the full subcategory* $\text{mod}_\sigma -\mathfrak{A}$ *of* mod $-\mathfrak{A}$ *of all* $M : K \longrightarrow \text{mod}-\mathbb{k}$, *whose restriction to* K' *factors through* σ.

It is a tool to use the information about the restriction of representations on a subcategory.

20.4.4 Examples of Amalgamations Here we bring some examples of box reductions.

1. *Substitution of canonical forms.* Let $K' \subset K$, $\mathbf{M} = \{M_1, \ldots, M_k\} \subset \text{mod}-K'$ be a family of (not necessary indecomposable) representations, $L' = \mathbb{L}_{\mathbf{M}}$ and $\sigma = M_1 \oplus \cdots \oplus M_k$. Then obtained $\mathfrak{A}^F$ has 'Harish-Chandra property': the category $\text{mod}-\mathfrak{A}^F$ is equivalent to the full subcategory $\text{mod}-\mathfrak{A}_{\mathbf{M}} \subset \text{mod}-\mathfrak{A}$, s.t. $M \in \text{mod}-\mathfrak{A}_{\mathbf{M}}$ iff $M|_{K'} \simeq M_1^{i_1} \oplus M_2^{i_2} \oplus \cdots \oplus M_k^{i_k}$ for some $i_1, \ldots, i_k$.

2. *Reduction of an arrow.* Let K' be a free category with objects A_1, A_2, $A_1 \neq A_2$, $a : A_1 \longrightarrow A_2$, $\delta(a) = 0$, $L' = \mathbb{L}_{A_1,A_2,A_{12}}$ and σ be the functor, which maps:

$$A_1 \mapsto A_1 \oplus A_{12}, A_2 \mapsto A_{12} \oplus A_2, \ a \mapsto \begin{pmatrix} 0 & 1 \\ 0 & 0 \end{pmatrix}.$$

Then we said the box $\mathfrak{A}^F$ is obtained from $\mathfrak{A}$ *by reduction of the arrow a or the edge a.* Since $\mathrm{add}(M_\sigma)$ is equivalent to $\mathrm{mod}-K'$, the functor F^* is an equivalence. This operation was described firstly in [37]for free DGC combinatorially in terms of bigraph and differential. In [46] the category L' is described by generators $a : A_1 \longrightarrow A_2$, $a^* : A_2 \longrightarrow A_1$ and the relations $aa^*a = a$, $a^*aa^* = a^*$. In fact, this construction can be applied without assumption $\delta(a) = 0$, the box $\mathfrak{A}^F$ (or DGC) obtained in that case can be neither normal nor triangular (even for the free box).

3. *Small reduction and tightening.* Let us consider in the previous example the canonical projections

$$\pi_i : L' \longrightarrow L/(1_{A_i}), i = 1, 2, \pi_{12} : L \longrightarrow L/(1_{A_1}, 1_{A_2}).$$

If we change σ above to one of the functor $\pi_1\sigma$, $\pi_2\sigma$, $\pi_{12}\sigma$, then is said that the box $\mathfrak{A}^F$ is obtained from $\mathfrak{A}$ *by small reduction of a from A_2 to A_1, by small reduction of a from A_1 to A_2, by tightening of the arrow a* correspondingly. In these cases F^* *is not* an equivalence. For free boxes (more general, free D$\mathbb{Z}$GC) both small reductions induce derived equivalences, that means, they extend to the $A(\infty)$-equivalence of $\mathsf{R}(T)$ and $\mathsf{R}(T^F)$ for corresponding to $\mathfrak{A}$ and $\mathfrak{A}^F$ DGC T and T^F, 20.4.10. The reason of it is, that they can be interpreted as reductions in $A(\infty)$-case (see below). On the quadratic form they induce so called *Gabrielov transformations*, see [44]. These operations were used in [32, 4] for investigation so called shurian matrix problems, see 20.5.2. In [4] these reductions were described in the style of [46] by the adding to K left, right and twosided inverse a^* correspondingly. An interesting application is given in [21] by construction of the representations of general position of some Lie groups.

4. *Regularization.* Let $\mathfrak{A}$ be a free box, K' is a free category, $a : A_1 \longrightarrow A_2$ and there exists $\varphi \in \Gamma^1$, $\varphi : A_1 \longrightarrow A_2$ s.t. $\delta(a) = \varphi$, or slightly more general, $\delta(a) = \sum_{i=1}^k \lambda_i\varphi_i, \lambda \in \Bbbk \setminus \{0\}$ and $\varphi_i \in \Gamma^1$ are pairwise different (the second case transforms in the first by a linear change of the basis). Let $L' = K'/(a)$ and σ be the projection. In this case F^* also is an equivalence. Such operation is called *a regularization of the (nonregular) arrow a*, (see [37], [46], [19]). In the free case the bigraph changes trivially (a and φ vanish) and the Tits form do not changed.

   A more general notion of regularization was given in [37] (for a completed DGC see also [24]). Let $T = \mathsf{k}[N]$ (or $T = \mathsf{k}[\hat{N}]$) as in 20.3.10 and $n_1 : N \longrightarrow N$ is nonzero (if $n_1 = 0$ then N and T we call *regular* and *nonregular* otherwise). For simplicity we suppose, there exists $x \in \Gamma_1$, s.t.

$$d(x) = \lambda\varphi + r,\ \lambda \in \Bbbk \setminus \{0\}, \varphi \in \Gamma_1, r \in N \cdot T \cdot N.$$

   Then the regularization of D$\mathbb{Z}$GC T in x in sense [37] is the D$\mathbb{Z}$GC $T' = T/(x, d(x))$, provided T' is free. It is not always possible, f.e.

$$\Gamma_0 = \{A\}, \Gamma^0 = \{x, y\}, \Gamma^1 = \{\varphi\}, \delta(y) = 0, \delta(\varphi) = 0, \delta(x) = \varphi + y\varphi.$$

In opposite, if $T = \mathsf{k}[\hat{N}]$ and I is the *completed* ideal, generated by x and $d(x)$, then $T/I = \mathsf{k}[\hat{N'}]$ for some $A(\infty)$-cocategory N' ([24]). Analogously we can construct a maximal regular factor of T T_{reg} together with the projection $T_\pi : T \longrightarrow T_{reg}$. It is a dual version of the operation, contained in the Theorem 20.2, 20.3.15.

5. *Reduction of a loop.* Let K' be generated by the only solid *loop* $a : A \longrightarrow A$ with $\delta(a) = 0$. Following [19] we consider as L' the trivial category containing $n+1$ vertices

$$\{A_1, \dots, A_n\}, \sigma(A) = \oplus_{k=1}^n kA_k \text{ and } \sigma(a) = \oplus_{k=1}^n J_k,$$

where J_k is the Jordan cell of size k with eigenvalue λ, where $\lambda \in \mathbb{k}$. Then $\operatorname{Im} F^*$ consists of all M s.t. $M(a-\lambda)^n = 0$. Analogously for any set S of Jordan cells we can construct such F_S, that $\operatorname{Im} F_S^*$ consists of M, s.t. the Jordan form of $M(a)$ contains only cells from S. For the reduction of loop see also [11, 42].

6. *Morita equivalence.* Let $\operatorname{Ob} K' = \operatorname{Ob} L' = \{A_1, \dots, A_n\}$, K' be a trivial category and $\mathbf{k} = (k_1, \dots, k_n)$ be an integral nonnegative vector. We define $\sigma(A_i) = \oplus_{j=1}^{k_i} A_i$. Then we say $\mathfrak{A}^F$ is Morita-equivalent to $\mathfrak{A}$ (with respect to vector $\mathbf{k}$). F^* is an equivalence. If $\mathfrak{A}$ is a principal box, then K and K^F are Morita-equivalent in the usual sense.

7. *Swelling of objects.* Let $K' = \mathbb{L}_K$, $\operatorname{Ob} K = \{A_1, \dots, A_n\}$ and $\mathbf{k} = (k_1, \dots, k_n)$ be an integral nonnegative vector, $L' = \mathbb{L}_A^{\mathbf{k}}$ is a trivial category with the objects $\{A_{ij}, 1 \le i \le n, 1 \le j \le k_i\}$ and define $\sigma(A_i) = \oplus_{j=1}^{k_i} A_{ij}$. We say then $\mathfrak{A}^F$ is obtained from $\mathfrak{A}$ *by swelling of objects* (with respect to vector $\mathbf{k}$). Obviously, F^* is an equivalence. The word 'swelling' originated in the quadratic forms theory, [15]. It can be considered as a sort of 'dual Morita equivalence'.

The examples above show, that the notion of a box together with the reduction technique is a very flexible (maybe, too flexible) tool. At least, it allows to present the same (up to equivalence) category as representations category of many boxes. This situation could be called a generalized Morita-equivalence, but the variety of possible boxes with the same representations category is much more diverse. It forces us to be close to the safe area of free and triangular boxes.

20.4.5 The practical importance of the operations above follows from following lemma.

Lemma 20.4.

1. *In the triangular case in $\mathfrak{A}$ there always exists at least one solid edge a, satisfying one from the condition 20.4.4 no. 2, no. 4, no. 5 (the third case is impossible in the box of finite type).*

2. *If $\mathfrak{A}$ was a free triangular box, then in the cases above $\mathfrak{A}^F$ is also free and triangular.*

3. *Let $\mathfrak{A} = (K, V)$ be a free triangular box of finite representation type. Then there exists a chain of triangular free boxes $\mathfrak{A} = \mathfrak{A}_1, \mathfrak{A}_2, \ldots, \mathfrak{A}_n = (K_n, V_n)$, s.t. $\mathfrak{A}_{i+1}$ is obtained from $\mathfrak{A}_i$ by one from the operation 20.4.4 no. 2, no. 4, $i = 1, 2, \ldots, n-1$ and s.t. the category K_n is trivial.*

Since there holds $R(\mathfrak{A}) \simeq R(\mathfrak{A}_1) \simeq R(\mathfrak{A}_2) \simeq \cdots \simeq R(\mathfrak{A}_n)$, we can identify the vertices M_i in the bigraph of $\mathfrak{A}_n$ with the indecomposables in $R(\mathfrak{A})$, $R(\mathfrak{A})(M_i, M_j) = \mathbb{D}V_n(M_i, M_j)$ and the comultiplication in V_n is the dual to the multiplication in $R(\mathfrak{A})$. In the case of a free box of infinite representation type there exists some substitution of this lemma using the notion of box, defined over an affine algebra, see [11, 42], but in this case we can obtain only "the category of representations till some fixed dimension".

We give here one more application of the reduction technique, that gives some generalization of the classical technique of the calculations classical Harish-Chandra modules.

1. *Harish-Chandra (HC) modules.* We suppose $\Bbbk$ to be algebraically closed of characteristic 0. The *cofinite spectrum* of an algebra Λ, $\mathrm{cfs}(\Lambda)$ is the set of maximal ideals of finite codimension in Λ, that means $\Lambda/\mathbf{m}$ is a f.d. simple $\Bbbk$-algebra. Let $S_{\mathbf{m}}$ be the simple left $\Lambda/\mathbf{m}$-module. We call Λ *quasi-commutative* provided $\mathrm{Ext}^1_\Lambda(S_{\mathbf{m}}, S_{\mathbf{n}}) = 0$ for all $\mathbf{m}, \mathbf{n} \in \mathrm{cfs}(\Lambda), \mathbf{m} \neq \mathbf{n}$. *Examples of quasi-commutative algebras:* commutative, semi-simple algebras, universal enveloping algebras of a f.d. reductive or solvable Lie algebras $\mathcal{G}$. Let Λ be a subalgebra of an algebra A. We call Λ *quasi-central* (in A) if for any $a \in A$ the Λ-bimodule $\Lambda a \Lambda$ is finitely generated as left and as right Λ-module. *Examples of quasi-central subalgebras:* Λ is contained in the center of A, $A = U(\mathcal{G})$ and $\Lambda = U(\mathcal{H})$, where $\mathcal{G}$ is a f.d. Lie algebra and $\mathcal{H}$ its Lie subalgebra. The subalgebra $\Lambda \subseteq A$ is called a *HC subalgebra* provided it is quasi-central and quasi-commutative. *Examples of HC subalgebras:* any central subalgebra is a HC one, $A = U(\mathcal{G})$ for a f.d. Lie algebra $\mathcal{G}$ and $\Lambda = U(\mathcal{H})$ where $\mathcal{H}$ is either reductive or solvable Lie subalgebra of $\mathcal{G}$. Another example of HC subalgebra in $U(gl_n)$, so called Gelfand-Zetlin subalgebra, was considered in [25].

   Let Λ be a HC subalgebra of A, M be an A-module, $\mathbf{m} \in \mathrm{cfs}(\Lambda)$ and $M(\mathbf{m}) = \{ x \in M \mid \exists k(\mathbf{m}^k x = 0) \}$. M is called a *HC module (with respect to Λ)* if $M = \oplus_{\mathbf{m} \in \mathrm{cfs}(\Lambda)} M(\mathbf{m})$ (equivalently M is a sum of f.d. Λ-submodules). Any submodule, factor-module of a HC module, or extension of HC modules is also HC. If $A = U(\mathcal{G})$ and $\Lambda = U(\mathcal{H})$, where $\mathcal{G}$ is a f.d. Lie algebra and $\mathcal{H}$ its semi-simple Lie subalgebra, then the given there notion of HC modules coincides with the usual definition of HC $\mathcal{G}$-modules with respect to $\mathcal{H}$ (see [12]). The category of HC modules can be described as $\mathrm{mod}\,{-\mathcal{A}}$ for the amalgamation $\mathcal{A}$ of the canonical inclusion $\imath : \Lambda \hookrightarrow A$ and the diagonal map $c : \Lambda \longrightarrow \prod_{\mathbf{m} \in \mathrm{cfs}(\Lambda)} \Lambda_{\mathbf{m}}$ with the separability conditions, where $\Lambda_{\mathbf{m}}$ is the completion of Λ in $\mathbf{m}$-adic topology. It was done in [25]. The category $\mathcal{A} = \mathcal{A}_{A,\Lambda}$ can be calculated as follows:

$$\begin{aligned} \mathrm{Ob}\mathcal{A} &= \mathrm{cfs}(\Lambda), \\ \text{for } \mathbf{m}, \mathbf{n} \in \mathrm{Ob}\,\mathcal{A}\mathcal{A}(\mathbf{m}, \mathbf{n}) &= \lim_{\leftarrow n,m} A/(\mathbf{n}^n A + A\mathbf{m}^m) \\ &(= \lim_{\leftarrow n,m} \Lambda/\mathbf{n}^n \otimes_\Lambda A \otimes_\Lambda \Lambda/\mathbf{m}^m). \end{aligned}$$

The definition of the composition of the morphisms see [25]. For any $\mathcal{A}$-module N the corresponding HC module from mod $-A$ is defined as $\oplus_{\mathbf{m}} N(\mathbf{m})$.

Theorem 20.4. *([25]) The category* $\mathbf{H}(A, \Lambda)$ *of HC modules is equivalent to the category* $\mathrm{mod}_d -\mathcal{A}$ *of discrete* $\mathcal{A}$*-modules.*

This technique was applied to the investigation of the categories of weight modules over $U(gl_n)$ and over generalized Weyl algebras, [25], [26], see also [20].

20.4.6 Regularization and $\mathsf{R}(T)$ Let N be an $A(\infty)$-cocategory over a trivial category k, N_{reg} be the its cohomology $A(\infty)$-cocategory together with the morphism of $A(\infty)$-cocategory $\pi : N \longrightarrow N_{reg}$ (see Theorem 20.2, 20.3.15 or 20.4.4, 2, no. 4). This π induces the morphisms

$$T_\pi : T \longrightarrow T_{reg}, \boldsymbol{K}_\pi : \boldsymbol{K} \longrightarrow \boldsymbol{K}_{reg}, \mathcal{N}_\pi : \mathcal{N} \longrightarrow \mathcal{N}_{reg}$$

etc., especially, $A(\infty)$-functor $\mathsf{R}_\pi : \mathsf{R}(T_{reg}) \longrightarrow \mathsf{R}(T)$. There holds the following statement (compare with the regularization of box, 20.4.4, 2, no. 4).

Lemma 20.5. *The induced by* π $A(\infty)$*-functor* $\mathsf{R}_\pi : \mathsf{R}(T_{reg}) \longrightarrow \mathsf{R}(T)$ *is an* $A(\infty)$*-equivalence.*

In the statement we consider *completed co-B-construction*, even if N allows noncompleted co-B-construction.

20.4.7 Reduction in the Categories $\mathsf{A}(T)$ and $\mathsf{R}(T)$ By the reduction in the category of representations of a box, associated with a subcategory

$$\mathbf{M} = \{M_1, \ldots, M_k\} \subset \mathrm{mod} -K'$$

(see 20.4.4, no. 1 above) the category $\mathbf{M}$ is kept after the reduction, but

$$\mathrm{Ext}^1(M_i, M_j), i, j = 1, \ldots, k$$

vanishing. Analogous applications of amalgamations in the case of the $A(\infty)$-categories of representations gives us the transformations of category, keeping the morphisms in all grades. For the simplicity we assume, we consider the representations in the category of bounded $\mathbb{Z}$-graded vectorspaces.

Let $\mathbf{M} = \{M_1, \dots, M_k\}$ be a family of representations of $D\mathbb{Z}GC\ T$, that obtained from f.d. $A(\infty)$-cocategory N. By $C^{\mathsf{A}}_{\mathbf{M}}$ ($C^{\mathsf{R}}_{\mathbf{M}}$) we denote the minimal $A(\infty)$-subcategory in $\mathsf{A}(T)$ (the minimal $A(\infty)$-subcategory in $\mathsf{R}(T)$) containing $\mathbf{M}$ and closed with respect to shift and cone. On other hand let us denote by $\mathsf{A}_{\mathbf{M}}$ and $\mathsf{R}_{\mathbf{M}}$ the restriction of the $A(\infty)$-categories $\mathsf{A}(T)$ and $\mathsf{R}(T)$ on $\mathbf{M}$, by $S_{\mathbf{M}}$ ($T_{\mathbf{M}}$) we denote the co-B-construction of $\mathbb{D}\mathsf{A}_M$ over $\mathbb{L}_{\mathbf{M}}$ (of $\mathbb{D}\mathsf{R}_M$ over $\mathbb{L}_{\mathbf{M}}$).

Lemma 20.6. *There exist $A(\infty)$-equivalences* $\mathsf{A}(S_{\mathbf{M}}) \simeq C^{\mathsf{A}}_{\mathbf{M}}$, $\mathsf{R}(T_{\mathbf{M}}) \simeq C^{\mathsf{R}}_{\mathbf{M}}$.

The first equivalence is, as in 20.4.2, a corollary of the conjugated associativity. The free DℤGC corresponding to the category $C^{\mathsf{A}}_{\mathbf{M}}$ can be obtained as the amalgamation

$$\begin{array}{ccc} \oplus_{j\in\mathbb{Z}} \boldsymbol{K}_{\mathbb{Z}}[,j] & \xrightarrow{\sigma} & \mathrm{mod}_{\mathbb{Z}} - \mathbb{k} \\ \imath \big\downarrow & & j \big\downarrow \\ \mathcal{T}_{\mathbb{Z}^2} & \xrightarrow{F} & L \end{array}$$

where $\boldsymbol{K}_{\mathbb{Z}}[,j]$ by σ maps into j times shifted $\oplus_{i=1}^{n} M_i$ and $\imath(a[i][,j]) = a[i,j]$. The second equivalence is from 20.4.6 above: $A(\infty)$-categories $\mathsf{R}(S_{\mathbf{M}})$ and $\mathsf{R}((S_{\mathbf{M}})_{reg})$ are $A(\infty)$-equivalent.

20.4.8 Example: Application to the Associative Algebras As an example, we use this technique to obtain the statement from 20.3.5, example 3. Let us consider the example 6 from 20.3.19. Without loss of generality assume Λ is a spectroid. We apply the construction above to the family of Λ-modules $\mathbf{M} = \{M_1, \dots, M_n\}$, considered as representations of T, where $T = \mathbb{L}_{\Lambda}[\mathbb{D}\hat{\mathsf{E}}(\Lambda)]$ as in 6, 20.3.19. The closed with respect to extensions category $E(\mathbf{M})$ is equivalent to $\mathsf{R}(T_{\mathbf{M}})[0]$. On other hand, the $A(\infty)$-category of Ext's $\mathsf{E}(\mathbf{M})$ (20.3.16) coincides with $\mathsf{R}_{\mathbf{M}}$ as above. Then we should apply 20.3.19, example 7.

20.4.9 Some Pathologies We should be carefully choosing the objects $\mathbf{M} = \{M_i\}$. For example, if in $\mathbf{M}$ a decomposable M_i, then in $\mathsf{R}(T_{\mathbf{M}})$ can exist non-split idempotents. But if all M_i are indecomposable, there is possible the strange behavior of $\mathsf{R}(T_{\mathbf{M}})$. Let Γ be a quiver, $\Gamma_0 = \{A_1, A_2\}$, $\Gamma_1 = \{a : A_1 \longrightarrow A_2\}$. The indecomposables here defines uniquely by its dimension, so we set $\mathbf{M} = \{M_1, M_2, M_3\}$, $\dim M_1 = (1,0)$, $\dim M_2 = (0,1)$, $\dim M_3 = (1,1)$. The graph $\Gamma_{\mathbf{M}}$ of the corresponding free DℤGC $T_{\mathbf{M}}$ has 3 vertices M_1, M_2, M_3, 3 arrows $a_i : M_i \longrightarrow M_{i+1}$, $i = 1,2,3 \pmod 3$, $\deg a_1 = 0$, $\deg a_2 = \deg a_3 = 1$, $d(a_i) = \omega_{M_{i+1}} a_i - (-1)^{\deg_T a_i} a_i \omega_{M_i}$, $d(\omega_i) = \omega_i^2 + a_{i+2} a_{i+1} a_i$, $i = 1,2,3 \pmod 3$. Then in $\mathsf{R}(T_{\mathbf{M}})$ the indecomposables (it means, the representations with local endomorphisms rings) in the dimensions $(1,1,0)$ and $(0,0,1)$ are isomorph. The dimension is not an invariant of an isoclass! In $T_{\mathbf{M}}$ the arrow a_1 generates an differential ideal and the free DGC $T/(a)$ corresponds to the operations of edge reductions ([37]). In this case there exists an equivalence of the categories of the observables.

Let Γ be a quiver, $\Gamma_0 = \{A_0, A_1, A_2\}$, $\Gamma_1 = \{a_i : A_i \longrightarrow A_0, i = 1, 2\}$. If $\mathbf{M} = \{M_1, M_2\}$ consists of indecomposables, s.t. $\dim M_1 = (1,0,0), \dim M_2 = (1,1,1)$, then $T_{\mathbf{M}}$ is normal but not triangular: the graph $\Gamma_{\mathbf{M}}$ is given by

$$\Gamma_{\mathbf{M}\,0} = \{M_1, M_2\}, \Gamma^1_{\mathbf{M}\,1}(M_1, M_2) = \{\varphi\}, \Gamma^0_{\mathbf{M}\,1}(M_2, M_1) = \{a\}$$

and the differential δ is given by $\delta(\varphi) = 0$, $\delta(a) = a\varphi a$. In the endomorphisms ring of the (observable) representation $X \in \mathsf{R}(T_{\mathbf{M}})$ $\dim X = (1,1)$, $X(a) = 1_{\Bbbk}$ exist nonsplit idempotents, f.e.

$$E : X \longrightarrow X, X(\omega_{M_1}) = 0, X(\omega_{M_2}) = 1_{\Bbbk}, X(\varphi) = 1_{\Bbbk}.$$

20.4.10 Small Reductions The examples above shows, that not for any $\mathbf{M}$ the obtained in 20.4.7 $A(\infty)$-category $\mathsf{R}_{\mathbf{M}}$ (or D$\mathbb{Z}$GC $T_{\mathbf{M}}$) is normal and triangular. One good checked case is to use (as in the box case) minimal edges. Let T be a regular D$\mathbb{Z}$GC with directed Γ, $A \neq B \in \Gamma_0$ and $R = \{a_1, \ldots, a_n : A \longrightarrow B\}$ be a set of distinct arrows. Denote the representation $X (= X(R)) \in \mathsf{R}(T)$ as follows:

$$\begin{aligned} X_A[0] =& \Bbbk \cdot 1_A, \\ X_B[j] =& \oplus_{a_i, \deg_T a_i = j} \Bbbk \cdot a_i, \\ X(a_i)[0](1_A) =& a_i \in X_B[\deg_T a_i], i = 1, \ldots, n \ , \end{aligned}$$

all other spaces and maps in X are 0. Let us suppose, all $\delta(a_i) = 0$ and set $\mathbf{U}$ the set of all trivial representations $L_A, A \in \operatorname{Ob} T$ in the layer 0, that means $\operatorname{DIM} L_A$ is equal 0 except $(\operatorname{DIM} L_A)_{A[0]} = 1$, $\mathbf{M} = (\mathbf{U} \setminus \{L_A\}) \cup \{X\}$. Then $C^{\mathsf{A}}_{\mathbf{U}}$ and $C^{\mathsf{A}}_{\mathbf{M}}$ are equivalent and the $A(\infty)$-categories $\mathsf{A}(S_{\mathbf{M}})$ and $\mathsf{R}(T_{\mathbf{M}})$ are easy to calculate. The same is true, if we set $Y (= Y_R) \in \mathsf{R}(T)$ as follows:

$$\begin{aligned} Y_B[0] &= \Bbbk \cdot 1^\circ_B, & Y_A[j] &= \oplus_{a_i, \deg_T a_i = -j} \Bbbk \cdot a^\circ_i, \\ Y(a_i)[-\deg_T a_i](a^\circ_i) &= 1^\circ_A \in Y_A[0], & i &= 1, \ldots, n \end{aligned}$$

and $\mathbf{M} = (\mathbf{U} \setminus \{L_B\}) \cup \{Y\}$. If $R = \Gamma^1(A, B)$, then the constructed equivalences induce on the quadratic forms Gabrielov transformations, [44], also 20.4.4, no. 3. If $n = 1, \deg_T a_1 = 0$ the D$\mathbb{Z}$GC T_{a_1}, s.t. $\mathsf{R}(T_{a_1}) = \mathsf{R}(T_{\mathbf{M}})$ is obtained by small reduction of a_1, as in 20.4.4, 3. Also in general case there exists an analogous construction.

20.4.11 Reflections The following known example, where D$\mathbb{Z}$GC $T_{\mathbf{M}}$ is easy to calculate, form so called *reflections*. Let T be a regular D$\mathbb{Z}$GC with directed Γ, given by $\delta : U \longrightarrow U$, see 20.3. For an $x \in \Gamma_1(A, B)$ we call an oriented way (it means, the composition the arrows) $l : A \longrightarrow B$ a *detour* of x, provided $\deg_T l = \deg_T x + 1$, by L_x we denote the set of all detours of x. There exists the function $\lambda_x : L_x \longrightarrow \Bbbk$, such that $\delta(x) = \sum_{l \in L_x} \lambda_x(l) l$ and $l \in L_x$ is called *active* (with respect to x), provided $\lambda_x(l) \neq 0$, [32]. The set $\{\lambda_x\}, x \in \Gamma_1$ defines U and T. Let $A \in \Gamma_0$ be a $(+)-$ *admitted* vertex ([3]), that means $\beta^{-1}_\Gamma(A) = \emptyset$.

Then by ${}^{A}T$ we denote so called *((+)−) reflected* in A DZGC . Its graph ${}^{A}\Gamma$ defined as follows:

$$ {}^{A}\Gamma_0 = \Gamma_0, \quad {}^{A}\Gamma_1(B,C) = \Gamma_1(B,C) $$

for $C \neq A$, and on the subgraph

$$ \Delta = {}^{A}\Gamma|_{\Gamma_0 \setminus \{A\}} = \Gamma|_{\Gamma_0 \setminus \{A\}} \deg, L_x, \lambda_x $$

in T and ${}^{A}T$ coincide. Define

$$ {}^{A}\Gamma_1(A,C) = \overline{\Gamma_1(C,A)} \text{ and } \deg_{{}^{A}T} \overline{x} = -\deg_T x, {}^{A}\Gamma_1(C,A) = \emptyset \ . $$

For any B, $x \in \Gamma_1(B,A)$ and $l \in L_x$ we can write $l = yl'$ for some $y \in \Gamma_1(C,A)$. Then we $l'\overline{x} \in L_{\overline{y}}$ and $\lambda_{\overline{y}}(l'\overline{x}) = -\lambda_x(l)$ and that defines all nonzero values of λ_x. It corresponds to the choice as $\mathbf{M}$ the set $\{X(\Gamma(B,A)), B \in \Gamma_0\}$, where $X(\Gamma(A,A))$ means the representation L_A, shifted in one degree back, thereafter $\mathsf{R}({}^{A}T) \simeq \mathsf{R}(T)$. It corresponds to the reflections in [3], [39], [22]. Analogous construction exists in a (−)-admitted vertex, [3]. These operations can be presented as a superposition of small reductions and "shift of the dimension in the object A". The transformation, induced on the dimensions, coincides with usual reflection, [44]. Technically at most complicated part of applications small reductions and reflections is to prove, that a pure representations transforms into pure one.

20.5 Quadratic Forms

20.5.1 General Remarks Let K and L be categories, s.t. $\operatorname{Ob} K = \operatorname{mod} - \mathcal{C}$, $\operatorname{Ob} L = \operatorname{mod} - \mathcal{D}$ and $F : K \longrightarrow L$ be a functor, s.t. F induces a linear map between the spaces of dimensions $\dim_F : \dim_{\mathcal{C}} \longrightarrow \dim_{\mathcal{D}}$, $\dim_F(\dim(X)) = \dim(F(X))$. Such a property have most of the functors, induced by the reductions. Moreover, in this case on the spaces $\dim_{\mathcal{C}}$ and $\dim_{\mathcal{D}}$ exist quadratic and nonsymmetric bilinear forms. We try to answer the questions: *To what extend the properties of the properties of the forms defines the properties of the* MP*'s?*

We start from the class of quadratic forms, that we will consider, namely the class of *integral unit quadratic forms*

$$ \chi(x) = \sum_{i=1}^{n} f_{ii} x_i^2 + \sum_{i<j} f_{ij} x_i x_j, $$

where $f_{ij} \in \mathbb{Z}$, $f_{ii} = 1$ and $f_{ij} = f_{ji}$, $i,j = 1,\ldots,n$. The *bigraph* Γ_χ of such a form has *n points or vertices* and if $i \neq j$, then i is connected with j by $|f_{ij}|$ *edges.* These edges supposed to be *dotted*, provided $f_{ij} > 0$ and *solid* otherwise. We identify the form itself (up to permutations of the variables) and its bigraph. The sources of information about such forms concerning the representations theory is [44],[29]. An integer positive vector x is called *(positive) root* of the form χ, provided $\chi(x) = 1$, the set of the roots of χ we denote by Σ_χ. If χ is WNN , then x is called an *imaginary root,* provided $\chi(x) = 0$. A vector x is called *sincere in* i, provided $x_i \neq 0$ and *sincere,* if it sincere in all i. A form χ is called *sincere,* provided it has at least one sincere root.

20.5.2 Interaction Representations Theory and Quadratic Forms Theory What we claim from the quadratic form for representations theory? The statement, converse to the Lemma 20.1, obviously fails in general. Thereafter 'quadratic forms oriented approach' has been divided in same steps. More precise, let us suppose we consider some class of the MP's and try to investigate it: to find all indecomposables, to describe the morphism between them, etc.

1. We define the class of MP, with which we are working.
2. We distinguish 'good MP's', that means 'they are controlled by corresponding quadratic form'.
3. We should be able to transform any MP from our class into the good one.
4. Using quadratic forms technique we should to describe all (at least sincere) 'good MP's' in the form of some criterion or by giving the complete list.
5. At last find a way to produce all MP's from the considered class using the 'good MP's'.

We use now the notations from 20.3.9. As a class of the matrix problems to consider is proposed the class of *free triangular* $D\mathbb{Z}GC$, s.t. the subcategory in K, generated by Γ^0 is f.d. The main problem is to describe all of them, that possessing finitely many isoclasses of observable indecomposables, among them at least one sincere. In fact, such a problem is solved for some classes of MP's (f.e. associative algebras) in the form of some procedures, the lists of critical MP's etc. ([6]) and it is the best result we try to obtain. Analogous problem can be posed to find the 'observable' tame MP's. In the case of the MP's of finite type the standard of a MP'controlled by corresponding quadratic form' was given by the Theorem of Gabriel, [27].

Theorem 20.5. *Let* $\mathsf{Q} = (\mathsf{Q}_0, \mathsf{Q}_1)$ *be a quiver* $\mathsf{Q}_0 = \{1, 2, \dots, n\}$, q_{Q} *be its Tits form,*

$$\mathrm{q}_{\mathsf{Q}} = \sum_{i=1}^{n} x_i^2 - \sum_{\substack{a:i\to j\\ a\in \mathsf{Q}_1}} x_i x_j .$$

Then Q *is of of finite representation type iff the form* q_{Q} *is* WP, *or equivalently, positive definite, the underlying graph of* Q *is a Dynkin graph of type* A_n, D_n, E_6, E_7, *or* E_8. *In this case there exits a bijection between the set of the isoclasses of indecomposable representations of* Q *and the set* $\Sigma_{\mathrm{q}_{\mathsf{Q}}}$, *namely* $\dim : \operatorname{ind}\mathsf{Q}/\sim \longrightarrow \Sigma_{\mathrm{q}_{\mathsf{Q}}}$.

For a sincere MP T of finite representation type the existence of the bijection $\dim : \operatorname{ind} T/\sim \longrightarrow \Sigma_{\mathrm{q}_T}$ is considered as a feature to be 'a good MP' ([4, 5, 17, 32, 39]), where for a free triangular D$\mathbb{Z}$GC by $\operatorname{ind} T$ is denoted the set of indecomposable observable indecomposable representations in $\mathsf{R}(T)$. It is a reason, that the roots of q_T play an important role in the representations theory. An explanation of the bijection $\dim : \operatorname{ind} T/\sim \longrightarrow \Sigma_{\mathrm{q}_T}$ was given in [28] and it is appropriate for free triangular DGC [37]: if for every $X \in \operatorname{ind} T$ the endomorphism ring is trivial, $\operatorname{End}_{R(T)}(X) = \Bbbk$, then there exists the bijection $\dim : \operatorname{ind} T/\sim \longrightarrow \Sigma_{q_T}$. After this (more strong) form of the Schur Lemma in $R(T)$ the corresponding DGC is called in [37] *schurian.* In [4] was proved a criterion for DGC (and, automatically, for the free triangular boxes) to be shurian. For a box $\mathfrak{A} = (K, V)$ it is formulated as follows: the Tits form $q_{\mathfrak{A}}$ should be WP and for every $S \subset \operatorname{Ob} K$, s.t. the box $\mathfrak{A}|_S = (K|_S, V|_S)$ is sincere, the $K|_S$-bimodule $V|_S$ does not contain free direct summand. After reformulation in terms of the graph Γ and the $A(\infty)$-cocategory structure this criterion of bijectivity of $\dim : \operatorname{ind} T/\sim \longrightarrow \Sigma_{q_T}$ can be carried over any $D\mathbb{Z}GC$. It is strong belief, that such criterion is true in general (some motivation, connected with coverings theory in [31], some partial cases in [32]).

Thereafter, between the chosen MP's as *good* one we consider such T, that there exists the bijection $\dim : \operatorname{ind} T/\sim \longrightarrow \Sigma_{q_T}$. For a given graph Γ the structure of a good $D\mathbb{Z}GC$ (if exists) is conjecturally dense in the variety of the $D\mathbb{Z}GC$ structures on Γ. It means, that this structure is defined by the graph Γ. In the known cases the coefficient in the corresponding $A(\infty)$-structure on Γ can be chosen from $\{-1, 0, 1\}$ (some analogue of multiplicative basis Theorem, [2, 47]). Analogously, there exists a conjecture, that in this case $\deg_T(\Gamma_1) \subset \{-1, 0, 1\}$ and $\mathsf{R}(T) \simeq \mathsf{R}(\Bbbk[\check{\mathsf{Q}}])$ for some quiver Q, (20.3.3). The first statement has analogue for the associative algebras and the free boxes. In the case of algebra it means, that sincere simply connected f.d. algebra of finite type has the homological dimension at most 2. In the second case it means for a free regular shurian box $\mathfrak{A} = (K, V)$, that for any free $K-K$-bimodule $\mathcal{F}$ holds $\operatorname{Hom}_{K-K}(V, \mathcal{F}) = 0$ (if you want, $\operatorname{Ext}_{\mathfrak{A}}^{-1} = 0$). The equivalence $\mathsf{R}(T) \simeq \mathsf{R}(\Bbbk[\check{\mathsf{Q}}])$ allows to introduce a kind of orbit graph (in sense [44]) for T. In fact, this equivalence can be obtained as a superposition of a chain of Gabrielov transformations and reflections, so it performs purely combinatorically. In fact this equivalence together with a results of Kac for quivers [34] proves the bijectivity of $\dim : \operatorname{ind} T/\sim \longrightarrow \Sigma_{q_T}$. On other hand, it is very interesting to characterize all arising here quivers Q (conjecturally, they are described in [5, 13]). The category of the form $\mathsf{R}(\Bbbk[\check{\mathsf{Q}}])$ for such Q we call *tilting universe.* Remark, that for analogous problem in the tame case the quivers are not sufficient: in describing of derived equivalence of the tame MP's arose derived categories of canonical algebras, that corresponds to so called swelled quiver (see [15]). One more good feature of good problems is an analogue of the statement from [43]. If $X, Y \in R(T)$ are two indecomposable, then in $\mathsf{R}(T)$ the $\operatorname{Hom}^i_{\mathsf{R}(T)}(X, Y)$ can be nonzero only for the unique $i \in \mathbb{Z}$ and in this case

$$< \dim X, \dim Y >_\Gamma = (-1)^i |\operatorname{Hom}^i_{\mathsf{R}(T)}(X, Y)|_\Bbbk .$$

Since a quadratic form is more basic and more simple object, as MP, one may demand from the quadratic forms theory, especially from the computer part of it *to give the lists of sincere quadratic forms* before a classifications of the representations will be obtained. In fact, such strategy is used often, at least as an heuristic technique.

To transform any MP from our class to the good one we use the *covering technique*. The obstacle to using such a technique is the necessity to construct a 'good' basis in the $A(\infty)$-cocategory N (or 'good' inclusion of the generating graph Γ in T). Sometimes this problem is very complicated ([2]), sometimes not ([31, 47]). In the worth case one assumes, that such a basis exists. Let (T, d) be a free D$\mathbb{Z}$GC , generated by a graph Γ, G be a group, $p : G \longrightarrow \mathrm{Aut}_d(T)$ a monomorphism of G in the automorphisms group of D$\mathbb{Z}$GC T, s.t. G acts freely on Γ_0 and for every $g \in G$ holds $p(G)(\Gamma_1) \subset \Gamma_1$. In this case we define a factor DGC (T_G, d_G), s.t. $T_G = T/G = \Bbbk[\check{\Gamma}/G]$, where $\Gamma/G = (\Gamma_0/G, \Gamma_1/G)$ is the factorgraph and d induces the differential $d_G : T_G \longrightarrow T_G$, s.t. for the canonical projection $\pi_G : T \longrightarrow T_G$ holds $d_G\pi = \pi d$. Then a DGC -functor $\pi : T_1 \longrightarrow T_2$ isomorphic to $\pi_G : T \longrightarrow T_G$ for some T and G we call a *Galois covering* of T_1 with the Galois group G. In opposite, if (T, d) is given by its graph Γ, then there exists an *universal Galois covering* $\pi_\Gamma : \tilde{T}_\Gamma \longrightarrow T$ with the *Galois group* G_T, that we call the fundamental group of (T, d, Γ), s.t.for any other Galois covering $\pi_1 : T_1 \longrightarrow T$ holds $\pi_\Gamma = \tau_1\pi_1$ for some τ_1. Then for MP's, defined through D$\mathbb{Z}$GC T, can be formulated the conjectural sequence of the steps intended for classification of MP's of finite type.

1. Either there exists a choice of generators of T, $T = \Bbbk[\check{\Gamma}]$, s.t. for the universal covering $\pi_\Gamma : \tilde{T} = \tilde{T}_\Gamma \longrightarrow T$ one from the following holds:
   - the quadratic form $\mathrm{q}_{\tilde{T}}$ is not WP , then both $\tilde{T}$ and T are of the strictly unbounded type;
   - the form $\mathrm{q}_{\tilde{T}}$ is WP and in this case the map $\dim : \operatorname{ind} T/\sim \longrightarrow \Sigma^+_{\mathrm{q}_{\tilde{T}}}$ is a bijection and
     (a) either the numbers $|x|$ for $x \in \Sigma^+_{\mathrm{q}_{\tilde{T}}}$ are bounded by some $C > 0$, then T is of finite representations type and π_Γ induces a Galois covering of the representations categories,[29];
     (b) or the numbers $|x|$ are unbounded, then T is of strictly unbounded type.
2. Or T contains as a 'a subproblem' one from the "obstacles" the construction of a 'good' basis Γ. Such subproblems are of strictly unbounded type and then the same holds for T.

This program was performed in some special cases: associative algebras, vectorspace categories and so called tree D$\mathbb{Z}$GC (that means in the graph $\tilde{\Gamma}$ the subgraph $\tilde{\Gamma}^0$ is a tree). Anyway, we hope, that the obtained by computer lists of the quadratic forms allows to produce sincere and critical "good" MP.

Bibliography

[1] M. Auslander, S. Smalø: *Preprojective modules over artin algebras*, J. of Alg. **66**, (1980), 61–122.

[2] R. Bautista, P. Gabriel, A. V. Roiter, L. Salmeron: *Representation-finite algebras and multiplicative bases*, Invent.Math. **81**, (1985), 217–285.

[3] I. N. Bernstein, I. M. Gelfand, V. A. Ponomarev: *Coxeter functors and Gabriel's Theorem*, Uspechi Mat. Nauk **28**, 2 (1973), 19–34.

[4] V. M. Bondarenko, N. S. Golovachtchuk, S. A. Ovsienko, A. V. Roiter: *Schurian matrix problems*, in: 'Schurian Matrix Problems and Quadratic Forms', Preprint 78.25, IM AN UkSSR, Kiev, (1978), 18–48.

[5] K. Bongartz: *Treue einfach zusammenhängende Algebraen I*, Comment Math, Helv. **57**, (1982), 282–330.

[6] K. Bongartz: *A criterion of finite representation type*, Math. Ann. **269**, (1984), 1–12.

[7] S. Brenner: *Quivers with commutativity conditions and some phenomenology of forms*, in: V. Dlab, P. Gabriel (ed.) "Representations of Algebras", Springer Lecture Notes in Math., **488** (1975), 29–53.

[8] S. Brenner, M. C. R. Butler: *Generalization of the Bernstein-Gelfand-Ponomarev reflection functors*, Springer Lecture Notes in Math., **832** (1980), 103–169.

[9] W. L. Burt, M. C. R. Butler: *Almost split sequences for Bocses*, preprint 1990.

[10] W. Crawley-Boevey: *On tame algebras and BOCS's*, Proc. London Math. Soc. **56**, (1988), 451–483.

[11] W. Crawley Boevey: *Matrix problems and Drozd's Theorem*, Banach Center publ. **26**, p. 1, (1990), 199–222.

[12] J. Dixmier: *Algébres enveloppantes*, Gauthier-Villarrd, Paris, 1974.

[13] P. Dräxler: *Aufrichtige gerichtete Ausnahmealgebraen*, Bayreuther Math. Schriften **29**, (1989).

[14] P. Dräxler: *Sur les algébres exceptionalles de Bongartz*, C.R.Acad.Sci.Paris, t.311 Sér. I, (1990), 495–498.

[15] P. Dräxler, Yu. A. Drozd, N. S. Golovachtchuk, S. A. Ovsienko and M. V. Zeldich: *Towards the classification of sincere weakly positive unit forms*, Europ. J. Combinatorics **16**, (1995), 1–16.

[16] Yu. A. Drozd: *Matrix problems and categories of matrices*, Zapiski Nauch. Semin. LOMI **28**, (1972), 144–152.

[17] Yu. A. Drozd: *Coxeter transformations and representations of partially ordered sets*, Funk. Anal. Prilozh. **8**, No. 4 (1974), 34–42.

[18] Yu. A. Drozd: *On tame and wild matrix problems*, in: "Matrix Problems", Kiev, (1977), 104–114.

[19] Yu. A. Drozd: *Tame and wild matrix problems*, in: "Representations and Quadratic Forms", Kiev, (1979), 39–74 (AMS Translations **128**,).

[20] Yu. A. Drozd: *On representation of Lie algebra* $sl(2)$, Visnik Kiev. Univ. Ser. Mat. Mekh. **25**,. (1983), 70–77.

[21] Yu. A. Drozd: *Matrix problems, small reductions and representations of a class of mixed Lie groups*, in: "Representations of algebras and related topics", Cambridge Univ. Press, (1992), 225–249.

[22] Yu. A. Drozd: *Representations of bisected posets and Reflection Functors*, Preprint **96 062**, SFB 343, Bielefeld, (1996).

[23] Yu. A. Drozd, S. A. Ovsienko: *Coverings of tame boxes*, to appear.

[24] Yu. A. Drozd, S. A. Ovsienko, B. Yu. Furchin: *Categorical constructions in representations theory*, in: 'Algebraic Structures and Their Applications', Kiev State Univ., Kiev, (1984).

[25] Yu. A. Drozd, V. M. Futorny, S. A. Ovsienko: *Harish–Chandra subalgebras and Gelfand-Zetlin modules*, in: 'Finitedimensional algebras and related topics', NATO ASI Ser. C:, Math. and Phys. Sci. **424**, (1994), 79–93.

[26] Yu. A. Drozd, B. L. Guzner, S. A. Ovsienko: *Weight modules over generalized Weyl algebras*, J. of Alg. **184**, (1994), 491–504.

[27] P. Gabriel: *Unzerlegbare Darstellungen*, Manuscr. Math. **6**, (1972), 71–103.

[28] P. Gabriel: *Indecomposable representations II*, Symp. Math. **11**, (1973), 81–104.

[29] P. Gabriel and A. V. Roiter: *Representations of finite-dimensional algebras*, Algebra VIII, Encyclopaedia of Math.Sc., Vol.73., Springer (1992).

[30] P. Gabriel and M. Zisman: *Calculus of fractions and homotopy theory*, in: "Ergebnisse der Math. und ihrer Grenzgebiete" **35**, Springer (1967).

[31] N. S. Golovachtchuk, S. A. Ovsienko: *Covering of bimodule problems*, to appear.

[32] N. S. Golovachtchuk, S. A. Ovsienko, A. V. Roiter: *On the shurian DGC*, in: "Matrix problems", IM AN UkrSSR, Kiev, (1977), 162–165.

[33] D. Happel: *On derived category of a finite-dimensional algebra*, Comment. Math. Helv. **62**, (1987), 339–389.

[34] V. Kac: *Root systems, representations of quivers and invariant theory,* in: "Invariant Theory", Springer Lecture Notes in Math. **996**, (1983), 74–108.

[35] T. V. Kadeishvili: *On the homology theory of fibre spaces,* Uspechi Math. Nauk **35**, 6, (1980), 183-188.

[36] B. Keller, D. Vossieck: *Sous les catégories dérivées,* Comp.-Rend. Acad. Sc. Paris, Sér. I **305**, (1987), 225–228.

[37] M. M. Kleiner, A. V. Roiter: *Representations of differential graded categories*, Springer Lecture Notes in Math. **488**, (1975), 316–339.

[38] S. MacLane: *Homology,* Springer, 1963.

[39] S. A. Ovsienko: *Representations of quivers with relations* , in: "Matrix Problems", Kiev, (1977), 88–103.

[40] S. A. Ovsienko: *On derived categories of representations categories*, in: "XVIII Allunion algebraic conference", Proceedings, part II, Kishinev, (1985), 71.

[41] S. A. Ovsienko: *On connection between the representations categories,* Proc. of Novosibirsk Int. Alg. Conf., II, (1989).

[42] S. A. Ovsienko: *Generic representations of free bocses,* Preprint 93 010, SFB 343, Bielefeld, (1993).

[43] S. A. Ovsienko, A. V. Roiter: *Bilinear forms and categories of representations,* in: "Matrix Problems", Kiev, (1977), 71–80.

[44] C. M. Ringel: *Tame Algebras and Integral Quadratic Forms*, Springer Lecture Notes in Math. **1099**, (1984).

[45] A. V. Roiter: *Matrix problems and representations of bisystems,* Zapiski Nauch. Semin. LOMI **28**, (1972).

[46] A. V. Roiter: *Matrix problems and representations of BOCS's,* In: "Representation theory I", Proc. Conf. Ottawa 1979, V.Dlab and P.Gabriel (eds.), Springer Lecture Notes in Math. **831**, (1980), 288–324.

[47] A. V. Roiter, V. V. Sergejchuk: *Existence of a multiplicative basis for a finitely spaced module over an aggregate,* Ukr. Math. J. **46**, 6, (1994), 567–579.

[48] W. A. Smirnov: *Homology of fibre spaces,* Uspechi Math.Nauk **35**, 6, (1980), 277–280.

[49] J. L. Verdier: *Catégories dérivées,* état 0, Springer Lecture Notes in Math. **569**, (1977), 262–311.